Abstract Algebra
Edition 2.71

Abstract Algebra
by
Justin R. Smith

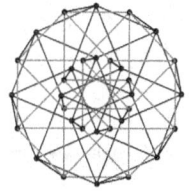

Five Dimensions Press

This is dedicated to the memory of my wonderful wife, Brigitte.

©2020. Justin R. Smith. All rights reserved.

ISBN: 9798579698353

Also published by Five Dimensions Press
▷ *Introduction to Algebraic Geometry* (paperback and hardcover), Justin Smith.
▷ *Eye of a Fly* (Kindle edition and paperback), Justin Smith.
▷ *The God Virus*, (Kindle edition and paperback) by Justin Smith.
▷ *Ohana*, (Kindle edition and paperback) by Justin Smith.
▷ *The Accidental Empress*, (Kindle edition and paperback) by Justin Smith.
▷ *Die zufällige Kaiserin*, German translation of *The Accidental Empress* (Kindle edition and paperback) by Justin Smith.

Five Dimensions Press page:
HTTP://www.five-dimensions.org
Email:jsmith@drexel.edu

Foreword

> Algebra is the offer made by the devil to the mathematician. The devil says "I will give you this powerful machine, and it will answer any question you like. All you need to do is give me your soul; give up geometry and you will have this marvelous machine."
>
> M. F. Atiyah (2001)

This book arose out of courses in Abstract Algebra, Galois Theory, Algebraic Geometry, and Manifold Theory the author taught at Drexel University.

It is useful for self-study or as a course textbook.

The first four chapters are suitable for the first of a two semester undergraduate course in Abstract Algebra and chapters five through seven are suitable for the second semester.

Chapter 2 on page 5 covers a few preliminaries on basic set theory (*axiomatic* set theory is discussed in appendix 14 on page 463).

Chapter 3 on page 13 discusses basic number theory, which is a useful introduction to the theory of groups. It also presents an application of number theory to cryptography.

Chapter 4 on page 33 discusses the theory of groups and homomorphisms of groups.

Chapter 5 on page 107 covers ring-theory with material on computations in polynomial rings and Gröbner bases. It also discusses some applications of algebra to motion-planning and robotics.

Chapter 6 on page 163 covers basic linear algebra, determinants, and discusses modules as generalizations of vector-spaces. We also discuss resultants of polynomials and eigenvalues.

Chapter 7 on page 261 discusses basic field-theory, including algebraic and transcendental field-extensions, finite fields, and the unsolvability of some classic problems in geometry. This, section 4.9 on page 84, and chapter 8 on page 297 might be suitable for a topics course in Galois theory.

Chapter 15 on page 467 discusses some more advanced areas of the theory of rings, like Artinian rings and integral extensions.

Chapter 8 on page 297 covers basic Galois Theory and should be read in conjunction with chapter 7 on page 261 and section 4.9 on page 84.

Chapter 9 on page 323 discusses division-algebras over the real numbers and their applications. In particular, it discusses applications of quaternions to computer graphics and proves the Frobenius Theorem classifying associative division algebras over the reals. It also develops octonions and discusses their properties.

Chapter 10 on page 339 gives an introduction to Category Theory and applies it to concepts like direct and inverse limits and multilinear algebra (tensor products an exterior algebras).

Chapter 11 on page 387 gives a brief introduction to group representation theory.

Chapter 12 on page 415 gives a brief introduction to algebraic geometry and proves Hilbert's Nullstellensatz.

Chapter 13 on page 433 discusses some 20th century mathematics: homology and cohomology. It covers chain complexes and chain-homotopy classes of maps and culminates in a little group-cohomology, including the classification of abelian extensions of groups.

 Sections marked in this manner are more advanced or specialized and may be skipped on a first reading.

 Sections marked in this manner are even *more* advanced or specialized and may be skipped on a first reading (or skipped entirely).

I am grateful to Matthias Ettrich and the many other developers of the software, LyX — a free front end to LaTeX that has the ease of use of a word processor, with spell-checking, an excellent equation editor, and a thesaurus. I have used this software for years and the current version is more polished and bug-free than most commercial software.

I am grateful to Darij Grinberg for his extremely careful reading of the manuscript. He identified several significant errors. I would also like to thank Xintong Li for pointing out errors in some of my definitions.

LyX is available from $\texttt{HTTP://www.lyx.org}$.

Edition 2.7: I have added material on the classification of modules over a principal ideal domain and the Jordan Canonical Form, as well as an expanded section on quaternions, octonions and sedenions.

Edition 2.71: I have added material on group representation theory.

Contents

Foreword	vii
List of Figures	xiii
Chapter 1. Introduction	1
Chapter 2. Preliminaries	5
2.1. Numbers	5
2.2. Set theory	9
2.3. Operations on sets	10
2.4. The Power Set	11
Chapter 3. A glimpse of number theory	13
3.1. Prime numbers and unique factorization	13
3.2. Modular arithmetic	18
3.3. The Euler ϕ-function	22
3.4. Applications to cryptography	25
3.5. Quadratic Residues	27
Chapter 4. Group Theory	33
4.1. Introduction	33
4.2. Homomorphisms	39
4.3. Cyclic groups	40
4.4. Subgroups and cosets	43
4.5. Symmetric Groups	49
4.6. Abelian Groups	58
4.7. Group-actions	73
4.8. The Sylow Theorems	80
4.9. Subnormal series	84
4.10. Free Groups	87
4.11. Groups of small order	104
Chapter 5. The Theory of Rings	107
5.1. Basic concepts	107
5.2. Homomorphisms and ideals	111
5.3. Integral domains and Euclidean Rings	117
5.4. Noetherian rings	121
5.5. Polynomial rings	125
5.6. Unique factorization domains	145
5.7. The Jacobson radical and Jacobson rings	154

5.8. Artinian rings — 157

Chapter 6. Modules and Vector Spaces — 163
  6.1. Introduction — 163
  6.2. Vector spaces — 165
  6.3. Modules — 222
  6.4. Rings and modules of fractions — 248
  6.5. Integral extensions of rings — 251

Chapter 7. Fields — 261
  7.1. Definitions — 261
  7.2. Algebraic extensions of fields — 265
  7.3. Computing Minimal polynomials — 272
  7.4. Primitive roots of unity — 276
  7.5. Algebraically closed fields — 283
  7.6. Finite fields — 287
  7.7. Transcendental extensions — 291

Chapter 8. Galois Theory — 297
  8.1. Before Galois — 297
  8.2. Galois — 300
  8.3. Isomorphisms of fields — 300
  8.4. Roots of Unity — 303
  8.5. Group characters — 304
  8.6. Galois Extensions — 308
  8.7. Solvability by radicals — 314
  8.8. Galois's Great Theorem — 317
  8.9. The fundamental theorem of algebra — 321

Chapter 9. Division Algebras over $\mathbb{R}$ — 323
  9.1. The Cayley-Dickson Construction — 323
  9.2. Quaternions — 325
  9.3. Octonions and beyond — 332

Chapter 10. A taste of category theory — 339
  10.1. Introduction — 339
  10.2. Functors — 345
  10.3. Adjoint functors — 349
  10.4. Limits — 351
  10.5. Abelian categories — 361
  10.6. Direct sums and Tensor products — 365
  10.7. Tensor Algebras and variants — 378

Chapter 11. Group Representations, a Drive-by — 387
  11.1. Introduction — 387
  11.2. Finite Groups — 393
  11.3. Characters — 398
  11.4. Examples — 409
  11.5. Burnside's Theorem — 411

Chapter 12. A little algebraic geometry		415
12.1.	Introduction	415
12.2.	Hilbert's Nullstellensatz	419
12.3.	The coordinate ring	427
Chapter 13. Cohomology		433
13.1.	Chain complexes and cohomology	433
13.2.	Rings and modules	447
13.3.	Cohomology of groups	454
Chapter 14. Axiomatic Set Theory		463
14.1.	Introduction	463
14.2.	Zermelo-Fraenkel Axioms	463
Chapter 15. Further topics in ring theory		467
15.1.	Discrete valuation rings	467
15.2.	Metric rings and completions	470
15.3.	Graded rings and modules	471

Solutions to Selected Exercises	475
Glossary	521
Index	523
Bibliography	531

List of Figures

2.1.1	The real line, $\mathbb{R}$	6
2.1.2	The complex plane	9
4.1.1	The unit square	36
4.1.2	Symmetries of the unit square	36
4.5.1	Regular pentagon	58
4.7.1	A simple graph	74
5.4.1	Relations between classes of rings	123
5.5.1	Reaching a point	138
5.5.2	Points reachable by the second robot arm	140
6.2.1	Image of a unit square under a matrix	191
6.2.2	A simple robot arm	204
6.2.3	A more complicated robot arm	205
6.2.4	Projection of a vector onto another	210
6.2.5	The plane perpendicular to **u**	213
6.2.6	Rotation of a vector in $\mathbb{R}^3$	214
9.3.1	Fano Diagram	333
12.1.1	An elliptic curve	416
12.1.2	Closure in the Zariski topology	418
12.2.1	An intersection of multiplicity 2	424

Abstract Algebra
Edition 2.71

CHAPTER 1

Introduction

> "L'algèbre n'est qu'une géométrie écrite; la géométrie n'est qu'une algèbre figurée." (Algebra is merely geometry in words; geometry is merely algebra in pictures)
> — Sophie Germain, [44]

The history of mathematics is as old as that of human civilization itself. Ancient Babylon (circa 2300 BCE) used a number-system that was surprisingly modern except that it was based on 60 rather than 10 (and still lacked the number 0). This is responsible the fact that a circle has 360° and for our time-units, where 60 seconds form a minute and 60 minutes are an hour. The ancient Babylonians performed arithmetic using pre-calculated tables of squares and used the fact that

$$ab = \frac{(a+b)^2 - a^2 - b^2}{2}$$

to calculate products. Although they had no algebraic notation, they knew about completing the square to solve quadratic equations and used tables of squares in reverse to calculate square roots.

In ancient times, mathematicians almost always studied algebra in its guise as geometry — apropos of Sophie Germain's quote. Ancient Greek mathematicians solved quadratic equations geometrically, and the great Persian poet-mathematician, Omar Khayyam[1], solved *cubic* equations this way.

Geometry in the West originated in Egypt, where it began as a kind of folk-mathematics farmers used to survey their land after the Nile's annual flooding (the word geometry means "earth measurement" in Greek). The more advanced geometry used in building pyramids remained the secret of the Egyptian priesthood.

The Greek merchant[2] and amateur mathematician, Thales, traveled to Egypt and paid priests to learn about geometry.

Thales gave the first proof of a what was a well-known theorem in geometry. In general, Greece's great contribution to mathematics was in the concept of *proving* that statements are true. This arose from many of the early Greek mathematicians being lawyers.

In ancient Greek geometry, there were a number of famous problems the ancient Greeks couldn't solve: squaring the circle (finding a square

---

[1]*The* Omar Khayyam who wrote the famous poem, *The Rubaiyat* — see [39].
[2]He is credited with the first recorded use of financial arbitrage.

whose area was the same as a given circle), doubling a cube (given a cube, construct one with double its volume), and trisecting an angle.

It turned out that these problems have no solutions, although proving that required modern algebra and number theory (see chapter 7 on page 261).

Ancient India had a sophisticated system of mathematics that remains largely unknown since it was not written down[3]. Its most visible modern manifestation is the universally used decimal number system and especially, the number zero. Otherwise, Indian mathematics is known for isolated, deep results given without proof, like the infinite series

$$\frac{\pi}{4} = 1 - \frac{1}{3} + \frac{1}{5} - \frac{1}{7} + \cdots$$

Arabic mathematicians transmitted Indian numerals to Europe and originated what we think of as algebra today — i.e. the use of non-geometric abstract symbols. One of the first Arabic texts describes completing the square in a quadratic equation in *verbal* terms[4].

There are hints of developments in Chinese mathematics from 1000 B.C.E. — including decimal, binary, and negative numbers. Most of this work was destroyed by the order of First Emperor of the Qin dynasty, Qin Shi Huangdi, in his Great Book Burning.

Isolated results like the Chinese Remainder Theorem (see 3.3.5 on page 24) suggest a rich mathematical tradition in ancient China. The *Jiuzhang suanshu* or, the *Nine Chapters on the Mathematical Art*, from 200 B.C.E., solves systems of three linear equations in three unknowns. This is the beginning of linear algebra (see chapter 6 on page 163)

The ancient Greeks regarded algebraic functions in purely geometric terms: a square of a number was a physical square and a numbers' cube was a physical cube. Other exponents had no meaning for them.

The early European view of all things as *machines* changed that: numbers became *idea-machines*, whose properties did not necessarily have any physical meaning. Nicole Oresme had no hesitation in dealing with powers other than 2 and 3 or even fractional exponents. This led to entirely new approaches to algebra.

> Nicole Oresme, (1320 – 1382) was a French philosopher, astrologer, and mathematician. His main mathematical work, *Tractatus de configurationibus qualitatum et motuum*, was a study on heat flow. He also proved the divergence of the harmonic series.

After Oresme, Renaissance Europe saw the rise of the concept of *virtù* — not to be confused with "virtue". The hero with virtù was competent in all things — and notable figures conducted public displays of skills like swordsmanship, poetry, art, chess, or ... mathematics. Tartaglia's solution of the general cubic equation was originally written in a poem.

---

[3] It was passed from teacher to student in an oral tradition.
[4] This makes one appreciate mathematical notation!

Bologna University, in particular, was famed for its intense public mathematics competitions.

People often placed bets on the outcome, rewarding winners with financial prizes. These monetary rewards motivated a great deal of mathematical research. Like magicians, mathematicians often kept their research secret — as part of their "bag of tricks[5]."

Renaissance Italy also saw the solution of the general quartic (i.e., fourth degree) equation — see section 8.1 on page 297. Attempts to push this further — to polynomials of degree five and higher — failed. In the early 1800's Abel and Galois showed that it is impossible — see chapter 8 on page 297.

The nineteenth and twentieth centuries saw many developments in algebra, often motivated by algebraic geometry and topology. See chapters 12 on page 415 and 13 on page 433.

---

[5]Tartaglia told Cardano his solution to the general cubic equation and swore him to strict secrecy. When Cardano published this solution, it led to a decade-long rift between the two men.

CHAPTER 2

Preliminaries

"The number system is like human life. First you have the natural numbers. The ones that are whole and positive. Like the numbers of a small child. But human consciousness expands. The child discovers longing. Do you know the mathematical expression for longing? The negative numbers. The formalization of the feeling that you're missing something. Then the child discovers the in-between spaces, between stones, between people, between numbers and that produces fractions, but it's like a kind of madness, because it does not even stop there, it never stops... Mathematics is a vast open landscape. You head towards the horizon and it's always receding..."

— Smilla Qaavigaaq Jaspersen, in the novel *Smilla's Sense of Snow,* by Peter Høeg (see [57]).

2.1. Numbers

The *natural numbers*, denoted $\mathbb{N}$, date back to prehistoric times. They are numbers

$$1, 2, 3, \ldots$$

Now we go one step further. The integers, denoted $\mathbb{Z}$, include the natural numbers, zero, and negative numbers. Negative numbers first appeared in China in about 200 BCE.

Zero as a digit (i.e., a place-holder in a string of digits representing a number) appears to date back to the Babylonian culture, which denoted it by a space or dot. Zero as a numerical value seems to have originated in India around 620 AD in a work of Brahmagupta that called positive numbers *fortunes* and negative numbers *debts*. This work also correctly stated the rules for multiplying and dividing positive and negative numbers, except that it stated $0/0 = 0$.

So we have the integers:

$$\mathbb{Z} = \{\ldots -3, -2, -1, 0, 1, 2, 3, \ldots\}$$

The next step involves fractions and the *rational* number system. Fractions (positive ones, at least) have appeared as early as ancient Egypt with numerators that were always 1. For instance, the fraction

$$\frac{8}{3}$$

would be represented in ancient Egypt as

$$2 + \frac{1}{2} + \frac{1}{6}$$

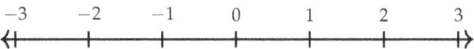

FIGURE 2.1.1.   The real line, $\mathbb{R}$

So the rational numbers, denoted by $\mathbb{Q}$, are equivalence classes of symbols
$$\frac{p}{q}$$
where $p \in \mathbb{Z}$ and $q \in \mathbb{N}$, and
$$\frac{p_1}{q_1}$$
is equivalent to
$$\frac{p_2}{q_2}$$
if $p_1 q_2 = p_2 q_1$.

Next in order of complexity and historical order is the *real numbers*, denoted $\mathbb{R}$. Irrational numbers were known in antiquity but the idea of incorporating them into the regular number system seems to date from the Italian Renaissance, where they were regarded as points on an infinite line — see figure 2.1.1. The term *real* number originated with Rene Descartes, who drew a distinction between these numbers and *imaginary* numbers like $i = \sqrt{-1}$.

The first rigorous definitions of the real numbers (in terms of rational numbers) came about in the 19th century with the work of Dedekind.

> Julius Wilhelm Richard Dedekind (1831 – 1916) was a German mathematician who worked in abstract algebra, algebraic number theory and analysis (he gave one of the first rigorous definitions of the real numbers). The concept of an ideal originated in Dedekind's research on Fermat's last theorem — see [29].

The Italian Renaissance also saw the invention of complex numbers in an effort to be able to write down solutions to algebraic equations. Prior to this era, an equation like
$$x^2 + 1 = 0$$
was thought to have *no* solutions. With complex numbers, we could say that the solutions to this equation are $\pm i$.

As with the real numbers, the rigorous construction of the complex number system had to wait until the 19th century.

DEFINITION 2.1.1. Let $\mathbb{C}$ denote the set of all expressions of the form $a + bi$ where $a, b \in \mathbb{R}$ and we define $i^2 = -1$. Addition is as follows:
$$(a + bi) + (c + di) = (a + b) + i(c + d)$$

The identity $i^2 = -1$ and the distributive law imply that it has the product
$$(a + bi)(c + di) = ac - bd + i(ad + bc)$$

where $a+bi, c+di \in \mathbb{R}^2 = \mathbb{C}$. It is not hard to see that 1 is the *identity element* of $\mathbb{C}$. Given $z = a+bi \in \mathbb{C}$, define $\Re(z) = a$ and $\Im(z) = b$, the *real* and *imaginary* parts, respectively.

If $x = a+bi$, the *complex conjugate* of x, denoted $\bar{x}$, is $a - bi$. Then

$$x \cdot \bar{x} = a^2 + b^2 \in \mathbb{R}$$

and

$$(a+bi)\left(\frac{a}{a^2+b^2} - \frac{bi}{a^2+b^2}\right) = 1$$

so

(2.1.1) $$(a+bi)^{-1} = \frac{a}{a^2+b^2} - \frac{bi}{a^2+b^2}$$

If $x = a+bi$, the quantity $|x| = \sqrt{x \cdot \bar{x}} = \sqrt{a^2+b^2}$ is called the *absolute value* of x. Note that $|x| = 0$ implies that $x = 0$.

REMARK. Complex numbers can represent points in the plane and one can add, subtract, multiply, and divide them (see section 9.1 on page 323). Note that complex multiplication is *associative* and *commutative* (left to the reader to prove!):

$$z_1(z_2 z_3) = (z_1 z_2) z_3$$

$$z_1 z_2 = z_2 z_1$$

for $z_i \in \mathbb{C}$.

Leonhard Euler (1707 – 1783) was, perhaps, the greatest mathematician all time. Although he was born in Switzerland, he spent most of his life in St. Petersburg, Russia and Berlin, Germany. He originated the notation $f(x)$ for a function and made contributions to mechanics, fluid dynamics, optics, astronomy, and music theory. His final work, "Treatise on the Construction and Steering of Ships," is a classic whose ideas on shipbuilding are still used to this day.

To do justice to Euler's life would require a book considerably longer than the current one — see the article [42]. His collected works fill more than 70 volumes and, after his death, he left enough manuscripts behind to provide publications to the Journal of the Imperial Academy of Sciences (of Russia) for 47 years.

REMARK. One of his many accomplishments is:

THEOREM 2.1.2 (Euler's Formula). *If $x \in \mathbb{R}$, then*

(2.1.2) $$e^{ix} = \cos(x) + i\sin(x)$$

PROOF. Just plug ix into the power series for exponentials

(2.1.3) $$e^y = 1 + y + \frac{y^2}{2!} + \frac{y^3}{3!} + \cdots$$

to get

$$\begin{aligned}
e^{ix} &= 1 + ix - \frac{x^2}{2!} - i\frac{x^3}{3!} + \cdots \\
&= \left(1 - \frac{x^2}{2!} + \frac{x^4}{4!} - \cdots\right) \\
&\quad + i\left(x - \frac{x^3}{3!} + \frac{x^5}{5!} - \cdots\right) \\
&= \cos x + i \sin x
\end{aligned}$$

$\square$

Since $x = \cos(\theta)$, $y = \sin(\theta)$ is the parametric equation of a circle, it follows that *all* points on the unit circle in $\mathbb{C}$ are of the form $e^{i\theta}$ for a suitable θ. If we draw a line from the origin to a point, u, on the unit circle, θ is the angle it makes with the real axis.

If $z \in \mathbb{C}$ is an arbitrary nonzero element, $u = z/|z|$ is on the unit circle, so $u = e^{i\theta}$ and

$$z = |z| \cdot e^{i\theta}$$

— see figure 2.1.2 on the facing page.

For instance, if $z = 2 + i$, we have $|z| = \sqrt{5}$ and

$$u = \frac{1}{\sqrt{5}}(2 + i) = e^{i\theta}$$

where

$$\theta = \arctan(1/2)$$

If we multiply z by $e^{i\phi}$, we get

$$ze^{i\phi} = |z| \cdot e^{i(\theta+\phi)}$$

which now makes an angle of $\theta + \phi$ with the real axis. It follows that multiplication has a *geometric significance*:

> Multiplying by $e^{i\phi}$ rotates the entire complex plane in a counterclockwise direction by ϕ.

The complex numbers transformed entire fields of mathematics, including function theory, number theory, and early topology. Doing these developments justice would fill several books larger than the current volume.

EXERCISES.

1. If $x, y \in \mathbb{C}$, show that $\bar{x} \cdot \bar{y} = \overline{x \cdot y}$.
2. If $x, y \in \mathbb{C}$, show that $|x| \cdot |y| = |x \cdot y|$.

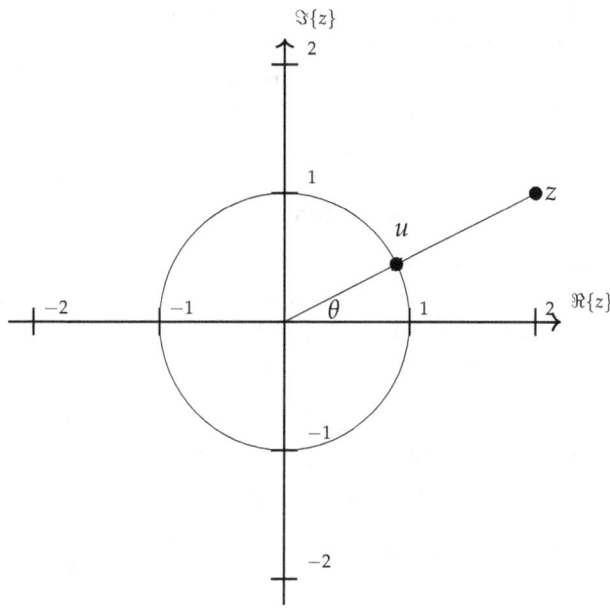

FIGURE 2.1.2.   The complex plane

2.2. Set theory

In mathematics (and life, for that matter) it is often necessary to discuss *collections* of objects, and we formalize this with the concept of a *set*. For instance there is the set $\mathbb{N}$ of natural numbers — integers > 0, or the set, $\mathbb{Q}$, of rational numbers. Giving a *rigorous* definition of sets is more complicated, as the Russell Paradox in section 14.1 on page 463 shows.

The objects in a set are called its *elements* or *members*.

If S is a set with an element x, we write

$$x \in S$$

to represent the fact that x is an *element* of S. The negation of this is represented by the symbol $\notin$ — so $y \notin S$ states that y is *not* an element of the set S.

We can define a finite set by listing its elements in curly brackets. For instance

$$A = \{1, 2, 3\}$$

If a set S is finite, the number of elements of S is denoted $|S|$.

If all the elements of a set, S are also elements of another set T, we say that S is a *subset* of T and write it as

$$S \subset T$$

to express the fact that every element of S is *also* an element of T. This definition could also be written in terms of first order logic as

(2.2.1) $$S \subset T \Leftrightarrow \forall_x (x \in S) \implies (x \in T)$$

where the symbol $\forall_x$ means "for all possible values of x".

Note that *every* set is a subset of *itself*, so $S \subset S$.

The *empty set* is denoted $\emptyset$ — it contains *no* elements. Equation 2.2.1 on the preceding page implies that for *any* set, S,

$$\emptyset \subset S$$

2.3. Operations on sets

We can perform various operations on sets

Union: If A and B are sets, their *union* — written $A \cup B$ — consists of all elements of A and all elements of B.
Example: If $A = \{1,2,3\}$ and $B = \{1,6,7\}$, then
$$A \cup B = \{1,2,3,6,7\}$$
We could give a *first-order logic* definition of union via
$$\forall_x x \in A \cup B \Leftrightarrow (x \in A) \vee (x \in B)$$
In some sense, the *union* operation is a set-version of the logical *or* operation.

Intersection: If A and B are sets, their *intersection* — written $A \cap B$ — consists of all elements contained in *both* A and B.
Example: If $A = \{1,2,3\}$ and $B = \{1,6,7\}$, then
$$A \cap B = \{1\}$$
We could give a *first-order logic* definition of union via
$$\forall_x x \in A \cap B \Leftrightarrow (x \in A) \wedge (x \in B)$$
In some sense, the *intersection* operation is a set-version of the logical *and* operation.

Difference: If A and B are sets, their *difference* — written $A \setminus B$ — consists of all elements of A *not* also in B.
Example: If $A = \{1,2,3\}$ and $B = \{1,6,7\}$, then
$$A \setminus B = \{2,3\}$$
We could give a *first-order logic* definition of union via
$$\forall_x x \in A \setminus B \Leftrightarrow (x \in A) \wedge (x \notin B)$$

Complement: If A is a set A^c or $\overline{A}$ denotes its *complement* — all elements *not* in A. For this to have any definite meaning, we must define the *universe* where A lives. For instance, if $A = \{1\}$, then $\overline{A}$ depends on whether A is considered a set of *integers, real numbers, objects in the physical universe*, etc.

Product: If A and B are sets, their *product*, $A \times B$ is the set of all possible *ordered pairs* (a,b) with $a \in A$ and $b \in B$. For instance, if $A = \{1,2,3\}$ and $B = \{1,4,5\}$, then
$$\begin{aligned}A \times B = \{&(1,2),(1,4),(1,5),\\&(2,1),(2,4),(2,5),\\&(3,1),(3,4),(3,5)\}\end{aligned}$$

If A and B are *finite*, then $|A \times B| = |A| \times |B|$.

2.4. The Power Set

If S is a set, the *powerset* of S — denoted $\mathcal{P}(S)$ or 2^S — is the set of all subsets of S. For instance, if
$$S = \{1, 2, 3\}$$
then
$$2^S = \{\emptyset, S, \{1\}, \{2\}, \{3\}, \{1,2\}, \{1,3\}, \{2,3\}\}$$
If S is *finite*, then $|2^S| = 2^{|S|}$.

PROOF. We do induction on the number of elements in S. To do the Ground Step, let $S = \emptyset$. The $2^S = \{\emptyset\}$ and the result is true.

For the Induction Step, assume the conclusion is true for all sets of size k. Given a set, S, of size $k+1$, let $x \in S$ and let $S' = S \setminus \{x\}$. Every subset of S either has x in it or it does not. The subsets of S that *do not* have x in them is precisely the set of subsets of S' and we know that there are 2^k of them. Every subset, T, of S that *has* x in it can be written as $T = T' \cup \{x\}$ where $T' \subset S'$. We know that there are 2^k of these subsets. It follows that S has
$$2^k + 2^k = 2^{k+1}$$
subsets. This completes the induction. □

PROPOSITION 2.4.1. *If S is a set, $\mathcal{P}(S)$ is strictly larger than S in the sense that there exists no surjection*

(2.4.1) $$f: S \to \mathcal{P}(S)$$

REMARK. This result is due to Gregor Cantor, and proves that the set $\mathcal{P}(\mathbb{Z})$ is uncountable — i.e. it is impossible to form a "list" of all the elements of $\mathcal{P}(\mathbb{Z})$. Although $\mathbb{Z}$ and $\mathcal{P}(\mathbb{Z})$ are both infinite, the infinity represented by $\mathcal{P}(\mathbb{Z})$ is "larger" than that of $\mathbb{Z}$.

PROOF. Suppose f is some function as in equation 2.4.1 and let $T \subset S$ be the set of all elements $s \in S$ such that $s \notin f(s)$. If $T = f(t)$ for some $t \in S$, then the definition of T implies that $t \notin T$, but this means that $t \in T$ — a contradiction. It follows that T cannot be in the image of f. □

CHAPTER 3

A glimpse of number theory

"Mathematics is the queen of the sciences and number theory is the queen of mathematics. She often condescends to render service to astronomy and other natural sciences, but in all relations she is entitled to the first rank."
— Carl Friedrich Gauss, in *Gauss zum Gedächtniss* (1856) by Wolfgang Sartorius von Waltershausen.

3.1. Prime numbers and unique factorization

We begin by studying the algebraic properties of what would appear to be the simplest and most basic objects possible: the integers. Study of the integers is called *number theory* and some of the deepest and most difficult problems in mathematics belong to this field.

Most people learned the following result in grade school — *long division* with a quotient and remainder:

PROPOSITION 3.1.1. *Let n and d be positive integers. Then it is possible to write*
$$n = q \cdot d + r$$
where $0 \le r < d$. If $r = 0$, we say that $d \mid n$ — stated "d divides n".

The division algorithm mentioned above gives rise to the concept of greatest common divisor.

DEFINITION 3.1.2. Let n and m be positive integers. The *greatest common divisor* of n and m, denoted $\gcd(n,m)$, is the largest integer d such that $d \mid n$ and $d \mid m$. The *least common multiple* of n and m, denoted $\mathrm{lcm}(n,m)$, is the smallest positive integer k such that $n \mid k$ and $m \mid k$.

Since 0 is divisible by any integer, $\gcd(n,0) = \gcd(0,n) = n$.

There is a very fast algorithm for computing the greatest common divisor due to Euclid — see [36, 37].

We need a lemma first:

LEMMA 3.1.3. *If $n > m > 0$ are two integers and*
$$n = q \cdot m + r$$
with $0 < r < m$, then
$$\gcd(n,m) = \gcd(m,r)$$

PROOF. If $d \mid n$ and $d \mid m$, then $d \mid r$ because
$$r = n - q \cdot m$$

On the other hand, if $d \mid m$ and $d \mid r$, then
$$n = q \cdot m + r$$
implies that $d \mid n$. It follows that the common divisors of n and m are the same as the common divisors of m and r. The same is true of the *greatest common divisors*. □

ALGORITHM 3.1.4. *Given positive integers n and m with $n > m$, use the division algorithm to set*
$$\begin{aligned} n &= q_0 \cdot m + r_0 \\ m &= q_1 \cdot r_0 + r_1 \\ r_0 &= q_2 \cdot r_1 + r_2 \\ &\vdots \\ r_{k-2} &= q_k \cdot r_{k-1} + r_k \end{aligned}$$
with $m > r_0 > r_1 > \cdots > r_k$. At some point $r_N = 0$ and we claim that $r_{N-1} = \gcd(n,m)$.

REMARK. Euclid's original formulation was geometric, involving line-segments. Given two line-segments of lengths r_1 and r_2, it found a real number r such that
$$\frac{r_1}{r}, \frac{r_2}{r} \in \mathbb{Z}$$
An ancient proof of the irrationality of $\sqrt{2}$ showed that this process never terminates if one of the line-segments is of unit length and the other is the diagonal of a unit square.

PROOF. This follows from lemma 3.1.3 on the preceding page, applied inductively. □

As trivial as proposition 3.1.1 on the previous page appears to be, it allows us to prove Bézout's Identity:

LEMMA 3.1.5. *Let n and m be positive integers. Then there exist integers u and v such that*
$$\gcd(n,m) = u \cdot n + v \cdot m$$

REMARK. Bézout proved this identity for polynomials — see [13]. However, this statement for integers can be found in the earlier work of Claude Gaspard Bachet de Méziriac (1581–1638) — see [103].

Étienne Bézout (1730–1783) was a French algebraist and geometer credited with the invention of the determinant (in [14]).

PROOF. Let z be the smallest positive value taken on by the expression
(3.1.1) $$z = u \cdot n + v \cdot m$$
as u and v run over all possible integers. Clearly, $\gcd(n,m) \mid z$ since it divides any possible linear combination of m and n. It follows that $\gcd(n,m) \leq z$.

We claim that $z\,|\,n$. If not, then proposition 3.1.1 on page 13 implies that $n = q\cdot z + r$, where $0 < r < z$, or $r = n - q\cdot z$. Plugging that into equation 3.1.1 on the preceding page gives

$$\begin{aligned} r &= n - q\cdot(u\cdot n + v\cdot m)\\ &= (1 - q\cdot u)\cdot n - q\cdot v\cdot m \end{aligned}$$

which is a linear combination of n and m *smaller* than z — a contradiction. Similar reasoning shows that $z\,|\,m$ so z is a common divisor of m and n $\geq \gcd(m,n)$ so it must *equal* $\gcd(m,n)$. □

DEFINITION 3.1.6. A *prime number* is an integer that is not divisible by any integer other than 1 or ($\pm$)itself.

Bézout's Identity immediately implies:

PROPOSITION 3.1.7. *Let p be a prime number and let n and m be integers. Then*

$$p\,|\,m\cdot n \implies p\,|\,m \text{ or } p\,|\,n$$

PROOF. Suppose $p \nmid m$. We will show that $p\,|\,n$. Since p is prime and $p \nmid m$, we have $\gcd(p,m) = 1$. Lemma 3.1.5 on the facing page implies that there exist integers u and v such that

$$1 = u\cdot m + v\cdot p$$

Now multiply this by n to get

$$n = u\cdot mn + v\cdot n\cdot p$$

Since p divides each of the terms on the right, we get $p\,|\,n$. A similar argument show that $p \nmid n \implies p\,|\,m$. □

A simple induction shows that:

COROLLARY 3.1.8. *If p is a prime number, $k_i \in \mathbb{Z}$ for $i = 1,\ldots, n$ and*

$$p\,\Big|\,\prod_{i=1}^{n} k_i$$

then $p|k_j$ for at least one value of $1 \leq j \leq n$. If p and q are both primes and

$$q\,|\,p$$

for some integer $i \geq 1$, then $p = q$.

PROOF. We do induction on n. Proposition 3.1.7 proves the result for $n = 2$.

Suppose the result is known for $n - 1$ factors, and we have n factors. Write

$$\prod_{i=1}^{n} k_i = k_1 \cdot \left(\prod_{i=2}^{n} k_i\right)$$

Since

$$p\,\Big|\,k_i \cdot \left(\prod_{i=2}^{n} k_i\right)$$

we either have $p|k$ or
$$p \mid \prod_{i=2}^{n} k_i$$
The inductive hypothesis proves the result. If the k_j are all copies of a prime, p, we must have $q \mid p$, which only happens if $q = p$. □

This immediately implies the well-known result:

LEMMA 3.1.9. *Let n be a positive integer and let*

$$\begin{aligned} n &= p_1^{\alpha_1} \cdots p_k^{\alpha_k} \\ &= q_1^{\beta_1} \cdots q_\ell^{\beta_\ell} \end{aligned}$$
(3.1.2)

be factorizations into powers of distinct primes. Then $k = \ell$ and there is a reordering of indices $f\colon \{1,\dots,k\} \to \{1,\dots,k\}$ such that $q_i = p_{f(i)}$ and $\beta_i = \alpha_{f(i)}$ for all i from 1 to k.

PROOF. First of all, it is easy to see that a number can be factored into a product of primes. We do induction on k. If $k = 1$ we have
$$p_1^{\alpha_1} = q_1^{\beta_1} \cdots q_\ell^{\beta_\ell}$$
Since $q_1 | p_1^{\alpha_1}$, corollary 3.1.8 on the preceding page implies that $q_1 = p_1$, $\beta_1 = \alpha_1$ and that the primes $q_i \neq p_1$ cannot exist in the product. So $\ell = 1$ and the conclusion follows.

Assume the result for numbers with $k-1$ distinct prime factors. Equation 3.1.2 implies that
$$q_1 \mid p_1^{\alpha_1} \cdots p_k^{\alpha_k}$$
and corollary 3.1.8 on the preceding page implies that $q_1 \mid \mid p$ for some value of j. It also implies that $p_j = q_1$ and $\alpha_j = \beta_1$. We define $f(1) = j$ and take the quotient of n by $q_1^{\beta_1} = p_j^{\alpha_j}$ to get a number with $k-1$ distinct prime factors. The inductive hypothesis implies the conclusion. □

This allows us to prove the classic result:

COROLLARY 3.1.10 (Euclid). *The number of prime numbers is infinite.*

PROOF. Suppose there are only a finite number of prime numbers
$$S = \{p_1, \dots, p_n\}$$
and form the number
$$K = 1 + \prod_{i=1}^{n} p_i$$
Lemma 3.1.9 implies that K can be uniquely factored into a product of primes
$$K = q_1 \cdots q_\ell$$
Since $p_i \nmid K$ for all $i = 1, \dots, n$, we conclude that $q_j \notin S$ for all j, so the original set of primes, S, is incomplete. □

Unique factorization also leads to many other results:

PROPOSITION 3.1.11. *Let n and m be positive integers with factorizations*

$$n = p_1^{\alpha_1} \cdots p_k^{\alpha_k}$$
$$m = p_1^{\beta_1} \cdots p_k^{\beta_k}$$

Then $n|m$ if and only if $\alpha_i \leq \beta_i$ for $i = 1, \ldots, k$ and

$$\gcd(n,m) = p_1^{\min(\alpha_1,\beta_1)} \cdots p_k^{\min(\alpha_k,\beta_k)}$$
$$\operatorname{lcm}(n,m) = p_1^{\max(\alpha_1,\beta_1)} \cdots p_k^{\max(\alpha_k,\beta_k)}$$

Consequently

(3.1.3) $$\operatorname{lcm}(n,m) = \frac{nm}{\gcd(n,m)}$$

PROOF. If $n \mid m$, then $p_i^{\alpha_i} \mid p_i^{\beta_i}$ for all i, by corollary 3.1.8 on page 15. If $k = \gcd(n,m)$ with unique factorization

$$k = p_1^{\gamma_1} \cdots p_k^{\gamma_k}$$

then $\gamma_i \leq \alpha_i$ and $\gamma_i \leq \beta_i$ for all i. In addition, the γ_i must be as large as possible and still satisfy these inequalities, which implies that $\gamma_i = \min(\alpha_i, \beta_i)$ for all i. Similar reasoning implies statement involving $\operatorname{lcm}(n,m)$. Equation 3.1.3 follows from the fact that

$$\max(\alpha_i, \beta_i) = \alpha_i + \beta_i - \min(\alpha_i, \beta_i)$$

for all i. □

The Extended Euclid algorithm explicitly calculates the factors that appear in the Bézout Identity:

ALGORITHM 3.1.12. *Suppose n, m are positive integers with $n > n$ and we use Euclid's algorithm (3.1.4 on page 14) to compute $\gcd(n,m)$. Let q_i, r_i for $0 < i \leq N$ (in the notation of 3.1.4 on page 14) denote the quotients and remainders used. Now define*

(3.1.4)
$$\begin{aligned} x_0 &= 0 \\ y_0 &= 1 \\ x_1 &= 1 \\ y_1 &= -q_1 \end{aligned}$$

and recursively define

(3.1.5)
$$\begin{aligned} x_k &= x_{k-2} - q_k x_{k-1} \\ y_k &= y_{k-2} - q_k y_{k-1} \end{aligned}$$

for all $2 \leq k \leq N$. Then

$$r_i = x_i \cdot n + y_i \cdot m$$

so that, in particular,

$$\gcd(n,m) = x_{N-1} \cdot n + y_{N-1} \cdot m$$

PROOF. If $r_i = x_i \cdot n + y_i \cdot m$ then
$$\begin{aligned} r_k &= r_{k-2} - q_k r_{k-1} \\ &= x_{k-2} \cdot n + y_{k-2} \cdot m - q_k(x_{k-1} \cdot n + y_{k-1} m) \\ &= (x_{k-2} - q_k x_{k-1}) \cdot n + (y_{k-2} - q_k y_{k-1}) \cdot m \end{aligned}$$

This implies the inductive formula 3.1.5 on the previous page, and to get the correct values for r_1 and r_2:
$$\begin{aligned} r_1 &= n - m \cdot q_1 \\ r_2 &= m - r_1 \cdot q_2 \\ &= m - q_2 \cdot (n - m \cdot q_1) \\ &= -q_2 \cdot n + (1 + q_1 q_2) \cdot m \end{aligned}$$

we must set x_0, x_1, y_0, y_1 to the values in equation 3.1.4 on the preceding page. $\square$

EXERCISES.

1. Find the elements of $\mathbb{Z}_m$ that have a multiplicative inverse, where $m > 1$ is some integer.

2. Find the greatest common divisor of 123 and 27 and find integers a and b such that
$$\gcd(123, 27) = a \cdot 123 + b \cdot 27$$

3. If $x > 0$ is an integer and y is a rational number that is *not* an integer, show that x^y is either an integer or irrational.

3.2. Modular arithmetic

We begin with an equivalence relation on integers

DEFINITION 3.2.1. If $n > 0$ is an integer, two integers r and s are *congruent modulo n*, written
$$r \equiv s \pmod{n}$$
if
$$n \mid (r - s)$$

REMARK. It is also common to say that r and s are *equal* modulo n. The first systematic study of these type of equations was made by Gauss in his *Disquistiones Arithmeticae* ([41]). Gauss wanted to find solutions to equations like
$$a_n x^n + \cdots + a_1 x + a_0 \equiv 0 \pmod{p}$$

PROPOSITION 3.2.2. *Equality modulo n respects addition and multiplication, i.e. if $r, s, u, v \in \mathbb{Z}$ and $n \in \mathbb{Z}$ with $n > 0$, and*

(3.2.1)
$$r \equiv s \pmod{n}$$
$$u \equiv v \pmod{n}$$

then

(3.2.2)
$$r + u \equiv s + v \pmod{n}$$
$$r \cdot u \equiv s \cdot v \pmod{n}$$

PROOF. The hypotheses imply the existence of integers k and ℓ such that
$$r - s = k \cdot n$$
$$u - v = \ell \cdot n$$
If we simply add these equations, we get
$$(r + u) - (s + v) = (k + \ell) \cdot n$$
which proves the first statement. To prove the second, note that
$$\begin{aligned} ru - sv &= ru - rv + rv - sv \\ &= r(u - v) + (r - s)v \\ &= r\ell n + kvn \\ &= n(r\ell + kv) \end{aligned}$$

□

This elementary result has some immediate implications:

EXAMPLE. Show that $5 | 7^k - 2^k$ for all $k \geq 1$. First note, that $7 \equiv 2 \pmod{5}$. Equation 3.2.2, applied inductively, implies that $7^k \equiv 2^k \pmod 5$ for all $k > 1$.

DEFINITION 3.2.3. *If n is a positive integer, the set of equivalence classes of integers modulo n is denoted $\mathbb{Z}_n$.*

REMARK. It is not hard to see that the size of $\mathbb{Z}_n$ is n and the equivalence classes are represented by integers
$$\{0, 1, 2, \ldots, n-1\}$$
Proposition 3.2.2 implies that addition and multiplication is well-defined in $\mathbb{Z}_n$. For instance, we could have an addition table for $\mathbb{Z}_4$ in table 3.2.1 and a multiplication table in table 3.2.2 on the following page

+	0	1	2	3
0	0	1	2	3
1	1	2	3	0
2	2	3	0	1
3	3	0	1	2

TABLE 3.2.1. Addition table for $\mathbb{Z}_4$

+	0	1	2	3
0	0	0	0	0
1	0	1	2	3
2	0	2	0	2
3	0	3	2	1

TABLE 3.2.2.   Multiplication table for $\mathbb{Z}_4$

Note from table 3.2.2 that $2 \cdot 2 \equiv 0 \pmod 4$ and $3 \cdot 3 \equiv 1 \pmod 4$. It is interesting to speculate on when a number has a multiplicative inverse modulo n.

We can use lemma 3.1.5 on page 14 for this:

PROPOSITION 3.2.4. *If $n > 1$ is an integer and $x \in \mathbb{Z}_n$, then there exists $y \in \mathbb{Z}_n$ with*
$$x \cdot y \equiv 1 \pmod n$$
if and only if $\gcd(x, n) = 1$.

PROOF. If $\gcd(x, n) = 1$, then lemma 3.1.5 on page 14 implies that there exist $a, b \in \mathbb{Z}$ such that
$$ax + bn = 1$$
and reduction modulo n gives
$$ax \equiv 1 \pmod n$$
On the other hand, suppose there exists $y \in \mathbb{Z}_n$ such that $xy = 1 \in \mathbb{Z}_n$. Then we have
$$xy = 1 + k \cdot n$$
or
$$(3.2.3) \qquad xy - kn = 1$$
and the conclusion follows from the proof of lemma 3.1.5 on page 14 which shows that $\gcd(x, n)$ is the smallest *positive* value taken on by an expression like equation 3.2.3. □

DEFINITION 3.2.5. If $n > 1$ is an integer, $\mathbb{Z}_n^\times$ — the *multiplicative group of* $\mathbb{Z}_n$ — is defined to be the set of elements $x \in \mathbb{Z}_n$ with $x \neq 0$ and $\gcd(x, n) = 1$. The operation we perform on these elements is multiplication.

EXAMPLE 3.2.6. Crossing out numbers in $\mathbb{Z}_{10}$ that have factors in common with 10 gives
$$\{1, \cancel{2}, 3, \cancel{4}, \cancel{5}, \cancel{6}, 7, \cancel{8}, 9\}$$
so
$$\mathbb{Z}_{10}^\times = \{1, 3, 7, 9\}$$
The multiplication-table is

×	1	3	7	9
1	1	3	7	9
3	3	9	1	7
7	7	1	9	3
9	9	7	3	1

Suppose we consider $\mathbb{Z}_p^\times$ for p a prime number:

EXAMPLE 3.2.7. Let $p = 5$. Then
$$\mathbb{Z}_5^\times = \{1, 2, 3, 4\}$$

— with *no* numbers crossed out because 5 is a prime number. In this case, the multiplication-table is

×	1	2	3	4
1	1	2	3	4
2	2	4	1	3
3	3	1	4	2
4	4	3	2	1

and all of the elements are powers of 2:
$$2^1 = 2$$
$$2^2 = 4$$
$$2^3 = 3$$
$$2^4 = 1$$

EXERCISES.

1. Show that $5 \mid (7^k - 2^k)$ for all $k \geq 1$.

2. Show that $7 \mid (93^k - 86^k)$ for all $k \geq 1$.

3. Suppose n is a positive integer and $0 < d < n$ is an integer such that $d \mid n$. Show that all solutions of the equation
$$d \cdot x \equiv 0 \pmod{n}$$
are multiples of
$$\frac{n}{d}$$
Hint: If there is a number $x \in \mathbb{Z}$ such that $dx \equiv 0 \pmod{n}$ and x is *not* a multiple of n/d use proposition 3.1.1 on page 13 to get a contradiction.

4. Let p be an odd prime and let $z = \frac{p-1}{2}$. Show that
$$(z!)^2 \equiv (-1)^z (p-1)! \pmod{p}$$

3.3. The Euler ϕ-function

DEFINITION 3.3.1. If n is a positive integer then
$$\phi(n)$$
is the number of generators of $\mathbb{Z}_n$ — or
- ▷ If $n > 1$ it is the number of integers, d, with $1 \leq d < n$ with $\gcd(d, n) = 1$.
- ▷ If $n = 1$, it is equal to 1.

This is called the *Euler phi-function*. Euler also called it the *totient*.

REMARK. Since an element x, of $\mathbb{Z}_n$ has a multiplicative inverse if and only if $\gcd(x, n) = 1$ (see lemma 3.1.5 on page 14), it follows that the multiplicative group $\mathbb{Z}_n^\times$ has $\phi(n)$ elements, if $n > 1$.

This ϕ-function has some interesting applications

PROPOSITION 3.3.2. *If n and m are integers > 1 with $\gcd(n, m) = 1$, then*

(3.3.1) $$m^{\phi(n)} \equiv 1 \pmod{n}$$

It follows that, for any integers a and b

(3.3.2) $$m^a \equiv m^b \pmod{n}$$

whenever
$$a \equiv b \pmod{\phi(n)}$$

REMARK. Fermat proved this for n a prime number — in that case, it is called Fermat's Little Theorem[1].

Proposition 3.3.2 is a special case of a much more general result — corollary 4.4.3 on page 43.

PROOF. Let the elements of $\mathbb{Z}_n^\times$ be
$$S = \{n_1, \ldots, n_{\phi(n)}\}$$
where $1 \leq n_i < n$ for $i = 1, \ldots, \phi(n)$. Since $\gcd(n, m) = 1$, m is one of them. Now multiply all of these integers by m and reduce modulo n (so the results are between 1 and $n - 1$). We get
$$T = \{mn_1, \ldots, mn_{\phi(n)}\}$$
These products are all *distinct* because m has a multiplicative inverse, so
$$mn_i \equiv mn_j \pmod{n}$$
implies
$$m^{-1}mn_i \equiv m^{-1}mn_j \pmod{n}$$
$$n_i \equiv n_j \pmod{n}$$
The Pigeonhole Principle implies that *as sets*
$$\mathbb{Z}_n^\times = \{n_1, \ldots, n_{\phi(n)}\} = \{mn_1, \ldots, mn_{\phi(n)}\}$$

---

[1] Fermat's *Big Theorem* is the statement that $a^n + b^n = c^n$ has no positive integer solutions for n an integer > 2. This was only proved in 1993 by Andrew Wiles.

— in other words, the list T is merely a *permutation* of S. Now we multiply everything in S and T together

$$n_1 \cdots n_{\phi(n)} \equiv b \pmod{n}$$
$$mn_1 \cdots mn_{\phi(n)} \equiv m^{\phi(n)} b \pmod{n}$$
$$\equiv b \pmod{n} \text{ since } T \text{ is a permutation of } S$$

Multiplication by b^{-1} proves equation 3.3.1 on the facing page. □

EXAMPLE. What is the low-order digit of 7^{1000}? This is clearly 7^{1000} modulo 10. Example 3.2.6 on page 20 shows that $\phi(10) = 4$. Since $4|1000$, equation 3.3.2 on the facing page implies that

$$7^{1000} \equiv 7^0 \equiv 1 \pmod{10}$$

so the answer to the question is 1.

If p is a prime number, every integer between 1 and $p-1$ is relatively prime to p, so $\phi(p) = p - 1$ (see example 3.2.7 on page 21). In fact, it is not hard to see that

PROPOSITION 3.3.3. *If p is a prime number and $k \geq 1$ is an integer, then*
$$\phi(p^k) = p^k - p^{k-1}$$

PROOF. The only integers $0 \leq x < p - 1$ that have the property that $\gcd(x, p^k) \neq 1$ are *multiples* of p — and there are $p^{k-1} = p^k/p$ of them. □

It turns out to be fairly easy to compute $\phi(n)$ for all n. To do this, we need the Chinese Remainder Theorem:

LEMMA 3.3.4. *If n and m are positive integers with $\gcd(n, m) = 1$ and*

(3.3.3)
$$x \equiv a \pmod{n}$$
$$x \equiv b \pmod{m}$$

are two congruences, then they have a unique solution modulo nm.

PROOF. We explicitly construct x. Since $\gcd(n, m) = 1$, there exist integers u and v such that

(3.3.4) $$u \cdot n + v \cdot m = 1$$

Now define

(3.3.5) $$x = b \cdot u \cdot n + a \cdot v \cdot m \pmod{nm}$$

Equation 3.3.4 implies that

▷ $vm \equiv 1 \pmod{n}$ so $x \equiv a \pmod{n}$
▷ $un \equiv 1 \pmod{m}$ so $x \equiv b \pmod{m}$.

Suppose x' is *another* value modulo nm satisfying equation 3.3.3. Then

$$x' - x \equiv 0 \pmod{n}$$
(3.3.6) $$x' - x \equiv 0 \pmod{m}$$

which implies that $x' - x$ is a multiple of both n and m. The conclusion follows from equation 3.1.3 on page 17, which shows that $x' - x$ is a multiple of nm, so $x' \equiv x \pmod{nm}$. □

Using this, we can derive the full theorem

THEOREM 3.3.5 (Chinese Remainder Theorem). *If $n_1, \ldots, n_k$ are a set of positive integers with $\gcd(n_i, n_j) = 1$ for all $1 \leq i < j \leq k$, then the equations*

$$x \equiv a_1 \pmod{n_1}$$
$$\vdots$$
$$x \equiv a_k \pmod{n_k}$$

have a unique solution modulo $\prod_{i=1}^{k} n_i$.

REMARK. The Chinese Remainder Theorem was first published sometime between the 3rd and 5th centuries by the Chinese mathematician Sun Tzu (not to be confused with the author of "The Art of Warfare").

PROOF. We do induction on k. Lemma 3.3.4 proves it for $k = 2$. We assume it is true for $k = j - 1$ — which means we have equations

$$x \equiv b \pmod{n_1 \cdots n_{j-1}}$$
$$x \equiv a_j \pmod{n_j}$$

where b is whatever value the theorem provided for the first $j - 1$ equations.

Note that the hypotheses imply that the sets of primes occurring in the factorizations of the n_i are all *disjoint*. It follows that the primes in the factorization of $n_1 \cdots n_{j-1}$ will be disjoint from the primes in the factorization of n_j. It follows that

$$\gcd(n_1 \cdots n_{j-1}, n_j) = 1$$

and the conclusion follows from an additional application of lemma 3.3.4. □

COROLLARY 3.3.6. *If n and m are integers > 1 such that $\gcd(n, m) = 1$, then*

$$\phi(nm) = \phi(n)\phi(m)$$

PROOF. The correspondence in the Chinese Remainder Theorem (lemma 3.3.4 on the previous page) actually respects products: If

(3.3.7)
$$x \equiv a \pmod{n}$$
$$x \equiv b \pmod{m}$$

and

(3.3.8)
$$y \equiv a^{-1} \pmod{n}$$
$$y \equiv b^{-1} \pmod{m}$$

then

$$xy \equiv 1 \pmod{nm}$$

since the value is *unique* modulo nm and reduces to 1 modulo n and m. It follows that there is a 1-1 correspondence between *pairs*, (a, b) with $a \in \mathbb{Z}_n^\times$, $b \in \mathbb{Z}_m^\times$ and *elements* of $\mathbb{Z}_{nm}^\times$ — i.e., as sets,

$$\mathbb{Z}_{nm}^\times = \mathbb{Z}_n^\times \times \mathbb{Z}_m^\times$$

proving the conclusion. □

At this point, computing $\phi(n)$ for any n becomes fairly straightforward. If
$$n = p_1^{k_1} \cdots p_t^{k_t}$$
then

(3.3.9) $\quad \phi(n) = \left(p_1^{k_1} - p_1^{k_1-1}\right) \cdots \left(p_t^{k_t} - p_t^{k_t-1}\right)$
$\qquad\qquad = p_1^{k_1-1}(p_1 - 1) \cdots p_t^{k_t-1}(p_t - 1)$

EXERCISES.

1. Compute $\phi(52)$.

2. Compute the low-order *two* digits of 7^{1000}.

3.4. Applications to cryptography

In this section, we will describe a cryptographic system that everyone reading this book has used — probably without being aware of it.

A regular cryptographic system is like a locked box with a key — and one cannot open the box without the key. The cryptographic system we discuss here is like a *magic* box with *two keys* — if *one* key is used to *lock* the box *only* the *other* key can *open* it. It is called a public-key system and was first publicly described by Ron Rivest, Adi Shamir and Leonard Adleman in 1977.

One application of this system is to make one of the keys *public* — so anyone who wants to communicate with you can use it to encrypt the message. Your *evil* enemies (!) cannot read the message because the *other* key (the one you keep private) is the *only* one that can decrypt it.

Another application involves digital signatures:

> How do you *sign* a document (like an email) in the digital age?

Typing your name at the bottom is clearly inadequate — *anyone* can type your name. Even an ink signature on a paper document is highly flawed since robotic signing machines can mimic a person's handwriting perfectly.

The answer: encrypt the document with your *private key*. In this case, the goal is not to *hide* the message. If your public key can *decrypt* it, the message *must* have come from you.

So here is the RSA system, the first public key system ever devised (and the one most widely used to this day):

We start with two large primes (large=50 digits or so), p and q, and integers m and n that satisfy

$$\gcd(n, \phi(pq)) = 1$$
$$n \cdot m \equiv 1 \pmod{\phi(pq)}$$

Proposition 3.3.2 implies that

$$x^{nm} \equiv x \pmod{pq}$$

whenever $\gcd(x, pq) = 1$.

Our *public key* is the pair (pq, n) and the *encryption* of a number k involves

$$k \mapsto e = k^n \pmod{pq}$$

where e is the encrypted message. The *private key* is m and *decryption* involves

$$e = k^n \mapsto e^m = k^{nm} \equiv k \pmod{pq}$$

One may wonder how secure this system is. We know that

$$\phi(pq) = (p-1)(q-1)$$

(from proposition 3.3.3 on page 23 and lemma 3.3.4 on page 23), so if your *evil* enemies know $(p-1)(q-1)$, they can *easily*[2] compute m, given n. The problem boils down to computing $(p-1)(q-1)$ when one only knows pq.

Oddly enough, this can be very hard to do. The only known way of finding $(p-1)(q-1)$ from pq involves factoring pq into p and q.

Recall how we factor numbers: try primes like $2, 3, 5, \ldots$ until we find one that divides the number. If the *smallest prime* that divides pq is 50 digits long, even a *computer* will have major problems factoring it.

Now we return to the statement made at the beginning of the section: *every reader has used this system*. That is because secure web transactions use it: A web server sends its public key to your web browser which encrypts your credit-card number (for instance) and sends it back to the server.

EXERCISES.

1. Suppose $p = 11$ and $q = 13$. Compute public and private keys for an RSA cryptosystem using these.

---

[2]OK, maybe not easily, but there are well-known methods for this, using Algorithm 3.1.12 on page 17.

3.5. Quadratic Residues

If p is a prime, consider the multiplicative group $\mathbb{Z}_p^\times$ — see definition 3.2.5 on page 20. If we write $X^2 = 1$ or $X^2 - 1 = 0$, we can factor it as

$$(X-1)(X+1) = 0$$

which implies that the only elements of $\mathbb{Z}_p^\times$ that are their own inverses are ± 1. It follows that in the list

$$2, 3, \ldots, p-3, p-2$$

every element will be paired with its multiplicative inverse. This implies that

$$(p-2)! \equiv 1 \pmod{p}$$

and

$$(p-1)! \equiv p-1 \equiv -1 \pmod{p}$$

or

$$(p-1)! + 1 \equiv 0 \pmod{p}$$

This leads into

THEOREM 3.5.1 (Wilson's Theorem). *If $n > 1$ is an integer, then n is a prime if and only if*

(3.5.1) $$(n-1)! + 1 \equiv 0 \pmod{n}$$

REMARK. In principle, this could be used as a test for primality. In practice, computing $(n-1)!$ is too computationally expensive.

This theorem was stated by Ibn al-Haytham (c. 1000 AD), and, in the 18th century, by John Wilson. In [105], Edward Waring announced the theorem in 1770, although neither he nor his student Wilson could prove it. Lagrange gave the first proof in 1771 (see [67]). There is evidence that Leibniz was also aware of the result a century earlier, but he never published it.

PROOF. We have already proved it in the case where n is a prime. If $n = k\ell$ with $k, \ell > 1$ then

$$(n-1)! + 1 \equiv 0 \pmod{n}$$

implies that

$$(n-1)! + 1 \equiv 0 \pmod{k}$$

which is a contradiction since k occurs as a factor in $(n-1)!$ so that

$$(n-1)! \equiv 0 \pmod{k}$$

□

The reasoning in Wilson's Theorem allows us to explore the question of *quadratic residues* — *squares* of elements modulo a prime.

DEFINITION 3.5.2. Let p be a prime number and let $n > 0$ be an integer. Define the *Legendre symbol*

$$\left(\frac{n}{p}\right) = \begin{cases} 1 & \text{if } p \nmid n \text{ and } \exists x \text{ with } n \equiv x^2 \pmod{p} \\ -1 & \text{if } p \nmid n \text{ and } \nexists x \text{ with } n \equiv x^2 \pmod{p} \\ 0 & \text{if } p \mid n \end{cases}$$

REMARK. Legendre symbols are used to solve quadratic equations in $\mathbb{Z}_p$.

We can get a useful computational formulation of the Legendre symbol

THEOREM 3.5.3 (Euler's Criterion). *If p is an odd prime number and $n > 0$ is an integer, then*

$$\left(\frac{n}{p}\right) \equiv n^{\frac{p-1}{2}} \pmod{p}$$

PROOF. Clearly true if $p \mid n$. If $n \equiv x^2 \pmod{p}$ then

$$n^{\frac{p-1}{2}} \equiv x^{p-1} \equiv 1 \pmod{p}$$

by proposition 3.3.2 on page 22. If n is *not* the square of any element of $\mathbb{Z}_p^\times$, define two elements $x, y \in \{1, \ldots, p-1\}$ to be *associates* if

$$x \cdot y \equiv n \pmod{p}$$

Every element of $\{1, \ldots, p-1\}$ has a *unique* associate distinct from itself. If $m = (p-1)/2$, the list

$$\{x_1, y_1, x_2, y_2, \ldots, x_m, y_m\}$$

where x_i is associated to y_i, is just a *permutation* of the list

$$\{1, \ldots, p-1\}$$

so that

$$x_1 \cdot y_1 \cdot x_2 \cdot y_2 \cdots x_m \cdot y_m \equiv n^m \pmod{p}$$
$$\equiv 1 \cdot 2 \cdot 3 \cdots (p-1) \equiv -1 \pmod{p}$$

by theorem 3.5.1 on the previous page. The conclusion follows. □

This immediately implies

COROLLARY 3.5.4. *If $p > 1$ is a prime and $n, m > 0$ are integers, then*

$$\left(\frac{n}{p}\right)\left(\frac{m}{p}\right) = \left(\frac{nm}{p}\right)$$

REMARK. This facilitates computation of Legendre symbols: if

$$n = p_1^{\alpha_1} \cdots p_k^{\alpha_k}$$

is the prime factorization of n and $i_1, \ldots, i_t$ are the subscripts for which the corresponding exponents α_{i_j} are *odd*, then

$$\left(\frac{n}{p}\right) = \left(\frac{p_{i_1}}{p}\right) \cdots \left(\frac{p_{i_t}}{p}\right)$$

so the computation of $\left(\frac{n}{p}\right)$ is reduced to computing $\left(\frac{q}{p}\right)$ for *prime* values of q.

One of the most important tools for computing these is

THEOREM 3.5.5 (Law of Quadratic Reciprocity). *If p and q are distinct odd primes, then*

$$\left(\frac{p}{q}\right)\left(\frac{q}{p}\right) = (-1)^{\frac{p-1}{2} \cdot \frac{q-1}{2}}$$

REMARK. This beautiful theorem was conjectured by Euler and Legendre and first proved by Gauss in 1796 in [41]. He called it his "golden theorem," and published six proofs of it in his lifetime. Two more were found in his posthumous papers.

There are now over 240 published proofs of this result.

PROOF. The proof given here is due to George Rousseau — see [95].

The Chinese Remainder Theorem (see lemma 3.3.4 on page 23) implies that the map
$$\tau: \mathbb{Z}_{pq}^\times \to \mathbb{Z}_p^\times \times \mathbb{Z}_q^\times$$
$$n \mapsto (n \bmod p, n \bmod q)$$
is a bijection that preserves products, where multiplication in $\mathbb{Z}_p^\times \times \mathbb{Z}_q^\times$ is defined by $(a,b) \cdot (c,d) = (a \cdot c, b \cdot d)$. Define
$$L = \left\{ k \in \mathbb{Z}_{pq}^\times \mid 1 \le k < \frac{pq}{2} \right\}$$
and
$$R = \left\{ (a,b) \in \mathbb{Z}_p^\times \times \mathbb{Z}_q^\times \mid 1 \le b < \frac{q}{2} \right\}$$
They are chosen so for all $x \in \mathbb{Z}_{pq}^\times$ either x or $-x \in L$ but *not both*, and for any $(a,b) \in \mathbb{Z}_p^\times \times \mathbb{Z}_q^\times$ either $(a,b) \in R$ or $-(a,b) \in R$ but *not both*. For any $(a,b) \in R$, there exists a *unique* $k \in \mathbb{Z}_{pq}^\times$ such that $\tau(k) = (a,b)$, and either k or $-k$ is in L. We have

(3.5.2)
$$\prod_{(a,b) \in R} (a,b) = \epsilon \prod_{k \in L} (k \bmod p, k \bmod q)$$

where $\epsilon = \pm 1$.

Set $P = (p-1)/2$ and $Q = (q-1)/2$. Then we can evaluate the left side of equation 3.5.2 via
$$\prod_{(a,b) \in R} (a,b) = \prod_{\substack{a < p \\ b < q/2}} (a,b) = \left((p-1)!^Q, Q!^{2P} \right)$$
$$= \left((-1)^Q, \left((q-1)!(-1)^Q \right)^P \right)$$
$$= \left((-1)^Q, (-1)^P (-1)^{PQ} \right)$$

The middle step follows from exercise 4 on page 21.

Now we analyze the *right* side of equation 3.5.2. We start with the left factor:
$$\prod_{k \in L} k \bmod p = \prod_{\substack{k < pq/2 \\ \gcd(k,pq)=1}} k \bmod p$$
$$= \left(\prod_{\substack{k < pq/2 \\ p \nmid k}} k \bmod p \right) \left(\prod_{\substack{k < pq/2 \\ q \mid k}} k \bmod p \right)^{-1}$$
$$= \left(\prod_{j=0}^{Q-1} \left(\prod_{jp < k < (j+1)p} k \bmod p \right) \right) \left(\prod_{Qp < k < pq/2} k \bmod p \right)$$
$$\cdot \left(\prod_{\substack{k < pq/2 \\ q \mid k}} k \bmod p \right)^{-1}$$
$$= \frac{(p-1)!^Q P!}{(q)(2q) \cdots (Pq)} = \frac{(-1)^Q}{q^P}$$
$$= (-1)^Q \left(\frac{q}{p} \right) \quad \text{(see Euler's Criterion, theorem 3.5.3)}$$

By symmetry, the *right* factor must satisfy
$$\prod_{k\in L} k \bmod q = (-1)^P \left(\frac{p}{q}\right)$$
so that equation 3.5.2 on the preceding page becomes
$$\left((-1)^Q, (-1)^P(-1)^{PQ}\right) = \epsilon \left((-1)^Q \left(\frac{q}{p}\right), (-1)^P \left(\frac{p}{q}\right)\right)$$
Equating the left factors shows that
$$\epsilon = \left(\frac{q}{p}\right)$$
and equating the right factors implies that
$$(-1)^{PQ} = \left(\frac{q}{p}\right)\left(\frac{p}{q}\right)$$
which is the theorem's statement. □

Now we can compute!

EXAMPLE 3.5.6. Is 11 a square modulo 53? We must compute $\left(\frac{11}{53}\right)$. Quadratic Reciprocity show that
$$\left(\frac{11}{53}\right)\left(\frac{53}{11}\right) = (-1)^{5\cdot 26} = 1$$
but
$$\left(\frac{53}{11}\right) = \left(\frac{53 \bmod 11}{11}\right) = \left(\frac{9}{11}\right) = \left(\frac{3}{11}\right)^2 = 1$$
so we conclude that 11 *is* a square modulo 53.

One possible difficulty might involve computing
$$\left(\frac{2}{p}\right)$$
In this case, we use the fact that Legendre symbols only depend on the equivalence class modulo p so that
$$\left(\frac{2}{p}\right) = \left(\frac{2-p}{p}\right) = \left(\frac{-1}{p}\right)\left(\frac{p-2}{p}\right) = (-1)^{\frac{p-1}{2}} \left(\frac{p-2}{p}\right)$$

EXAMPLE 3.5.7. Compute
$$\left(\frac{102}{113}\right)$$
Since $102 = 2 \cdot 3 \cdot 17$, we have
$$\left(\frac{102}{113}\right) = \left(\frac{2}{113}\right) \cdot \left(\frac{3}{113}\right) \cdot \left(\frac{17}{113}\right)$$
Now $\left(\frac{2}{113}\right) = (-1)^{56} \left(\frac{111}{113}\right) = \left(\frac{3}{113}\right)\left(\frac{37}{113}\right)$, so $\left(\frac{102}{113}\right) = \left(\frac{3}{113}\right)^2 \cdot \left(\frac{17}{113}\right) \cdot \left(\frac{37}{113}\right) = \left(\frac{17}{113}\right) \cdot \left(\frac{37}{113}\right)$. Now we use Quadratic Reciprocity to conclude that
$$\left(\frac{17}{113}\right)\left(\frac{113}{17}\right) = (-1)^{8\cdot 56} = 1$$
and $\left(\frac{113}{17}\right) = \left(\frac{11}{17}\right)$ and Quadratic Reciprocity implies that
$$\left(\frac{11}{17}\right)\left(\frac{17}{11}\right) = (-1)^{5\cdot 8} = 1$$

But $\left(\frac{17}{11}\right) = \left(\frac{6}{11}\right) = \left(\frac{2}{11}\right) \cdot \left(\frac{3}{11}\right)$. We conclude that $\left(\frac{2}{11}\right) = -\left(\frac{9}{11}\right) = -1$. The factor
$$\left(\frac{3}{11}\right)\left(\frac{11}{3}\right) = (-1)^{1 \cdot 5} = -1$$
We get $\left(\frac{11}{3}\right) = \left(\frac{2}{3}\right) = -1$ (by direct computation), so $\left(\frac{3}{11}\right) = \left(\frac{17}{11}\right) = \left(\frac{11}{17}\right) = \left(\frac{113}{17}\right) = \left(\frac{17}{113}\right) = -1$.

The factor $\left(\frac{37}{113}\right)$ satisfies
$$\left(\frac{37}{113}\right)\left(\frac{113}{37}\right) = (-1)^{18 \cdot 56} = 1$$
and $\left(\frac{113}{37}\right) = \left(\frac{2}{37}\right) = (-1)^{18}\left(\frac{35}{37}\right) = \left(\frac{5}{37}\right)\left(\frac{7}{37}\right)$. For the first factor
$$\left(\frac{5}{37}\right)\left(\frac{37}{5}\right) = (-1)^{2 \cdot 18} = 1$$
and $\left(\frac{37}{5}\right) = \left(\frac{2}{5}\right) = -1$ (by direct computation). For the second factor
$$\left(\frac{7}{37}\right)\left(\frac{37}{7}\right) = (-1)^{3 \cdot 18} = 1$$
and $\left(\frac{37}{7}\right) = \left(\frac{2}{7}\right) = 1$, since $3^2 \equiv 2 \pmod{7}$.

We finally conclude that $\left(\frac{113}{37}\right) = \left(\frac{37}{113}\right) = -1$ and
$$\left(\frac{102}{113}\right) = (-1)(-1) = 1$$
so that 102 *is* a square modulo 113.

CHAPTER 4

Group Theory

"The introduction of the digit 0 or the group concept was general nonsense too, and mathematics was more or less stagnating for thousands of years because nobody was around to take such childish steps..."
— Alexander Grothendieck, writing to Ronald Brown.

4.1. Introduction

One of the simplest abstract algebraic structures is that of the *group*. In historical terms its development is relatively recent, dating from the early 1800's. The official definition of a group is due to Évariste Galois, used in developing *Galois Theory* (see chapter 8 on page 297).

Initially, Galois and others studied *permutations* of objects. If the set has a finite number of elements — 5, for instance — we can regard S as the set of natural numbers from 1 to 5 and write permutations as little arrays, where

$$a = \begin{pmatrix} 1 & 2 & 3 & 4 & 5 \\ 2 & 1 & 5 & 3 & 4 \end{pmatrix}$$

represents the permutation

$$1 \to 2$$
$$2 \to 1$$
$$3 \to 5$$
$$4 \to 3$$
$$5 \to 4$$

The set of all such permutations has several properties that are important:

(1) One can *compose* (i.e. multiply) permutations to get another permutation. Here, we regard them as functions, so the second operation is written to the left of the first. If

$$b = \begin{pmatrix} 1 & 2 & 3 & 4 & 5 \\ 4 & 5 & 1 & 2 & 3 \end{pmatrix}$$

than $a \circ b = ab$ means "perform b first and follow it with a" to get
$$\begin{aligned} 1 &\to 4 \to 3 \\ 2 &\to 5 \to 4 \\ 3 &\to 1 \to 2 \\ 4 &\to 2 \to 1 \\ 5 &\to 3 \to 5 \end{aligned}$$
or
$$ab = \begin{pmatrix} 1 & 2 & 3 & 4 & 5 \\ 3 & 4 & 2 & 1 & 5 \end{pmatrix}$$
of strings
  Note that
$$ba = \begin{pmatrix} 1 & 2 & 3 & 4 & 5 \\ 5 & 4 & 3 & 1 & 2 \end{pmatrix}$$
so that $ab \neq ba$, in general.
(2) Since multiplication of permutations is composition of functions, we have $a(bc) = (ab)c$.
(3) There exists an *identity permutation* that does *nothing at all*
$$1 = \begin{pmatrix} 1 & 2 & 3 & 4 & 5 \\ 1 & 2 & 3 & 4 & 5 \end{pmatrix}$$
(4) Every permutation has an *inverse* gotten by flipping the rows of the array defining it. For instance
$$a^{-1} = \begin{pmatrix} 2 & 1 & 5 & 3 & 4 \\ 1 & 2 & 3 & 4 & 5 \end{pmatrix}$$
or, if we sort the upper row into ascending order, we get
$$a^{-1} = \begin{pmatrix} 1 & 2 & 3 & 4 & 5 \\ 2 & 1 & 4 & 5 & 3 \end{pmatrix}$$
and it is easy to see that $a^{-1}a = aa^{-1} = 1$.

We are in a position to define groups of permutations:

DEFINITION 4.1.1. If $n > 0$ is an integer, the group S_n is the set of all permutations of the set
$$\{1, \ldots, n\}$$
and is called the *symmetric group of degree n*.

Note that this will have $n!$ elements.

The properties of the symmetric group motivate us to define groups in the *abstract*:

DEFINITION 4.1.2. A *group* is a set, G, equipped with a two maps
$$\mu \colon G \times G \to G$$
$$\iota \colon G \to G$$
called, respectively, the multiplication and inversion maps. We write $\mu(g_1, g_2)$ as $g_1 g_2$ and $\iota(g)$ as g^{-1} for all $g_1, g_2, g \in G$, and these operations satisfy

(1) there exists an element $1 \in G$ such that $1g = g1 = g$ for all $g \in G$
(2) for all $g \in G$, $gg^{-1} = g^{-1}g = 1$
(3) for all $g_1, g_2, g_3 \in G$, $g_1(g_2 g_3) = (g_1 g_2) g_3$

If the group G is finite, the number of elements of G is denoted $|G|$ and called the *order* of G. If for any elements $a, b \in G$, $ab = ba$, we say that G is *abelian*.

REMARK. Rather confusingly, the group-operation for an abelian group is often written as *addition* rather than multiplication.

In the beginning all groups were groups of permutations, many of which were used for geometric purposes.

We have already seen a few examples of groups:

EXAMPLE 4.1.3. for any positive integer, $\mathbb{Z}_n$ is a group under the operation of addition. We can indicate this by writing it as $(\mathbb{Z}_n, +)$.

We can take a similar set of numbers and give it a group-structure in a different way:

EXAMPLE 4.1.4. $\mathbb{Z}_n^\times$ is a group under the operation of integer-multiplication[1], or $(\mathbb{Z}_n^\times, \times)$ — although multiplication is generally *implied* by the $\times$-superscript.

Here are some others:

EXAMPLE 4.1.5. The set of real numbers, $\mathbb{R}$, forms a group under addition: $(\mathbb{R}, +)$.

We can do to $\mathbb{R}$ something similar to what we did with $\mathbb{Z}_n$ in example 4.1.4:

EXAMPLE 4.1.6. The set of *nonzero* reals, $\mathbb{R}^\times = \mathbb{R} \setminus \{0\}$, forms a group under *multiplication,* Again, the group-operation is implied by the $\times$-superscript.

We can roll the real numbers up into a circle and make *that* a group:

EXAMPLE 4.1.7. $S^1 \subset \mathbb{C}$ — the *complex unit circle,* where the group-operation is multiplication of complex numbers.

Let us consider a group defined *geometrically*. Consider the unit square in the plane in figure 4.1.1 on the next page. If we consider all possible rigid motions of the plane that leave it fixed, we can represent these by the induced permutations of the vertices. For instance the 90° counterclockwise rotation is represented by $R = \begin{pmatrix} 1 & 2 & 3 & 4 \\ 2 & 3 & 4 & 1 \end{pmatrix}$. Composing this with itself gives

$$R^2 = \begin{pmatrix} 1 & 2 & 3 & 4 \\ 3 & 4 & 1 & 2 \end{pmatrix}$$

and

$$R^3 = \begin{pmatrix} 1 & 2 & 3 & 4 \\ 4 & 1 & 2 & 3 \end{pmatrix}$$

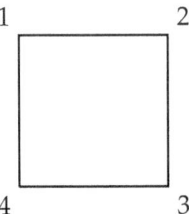

FIGURE 4.1.1.   The unit square

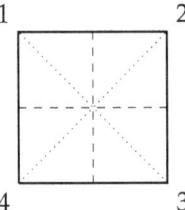

FIGURE 4.1.2.   Symmetries of the unit square

We have two additional symmetries of the unit square, namely reflection through diagonals (dotted lines) through the square and reflection though axes going though the center (dashed lines) — see figure 4.1.2.

This gives additional symmetries

$$d_1 = \begin{pmatrix} 1 & 2 & 3 & 4 \\ 3 & 2 & 1 & 4 \end{pmatrix}$$

$$d_2 = \begin{pmatrix} 1 & 2 & 3 & 4 \\ 1 & 4 & 3 & 2 \end{pmatrix}$$

$$c_1 = \begin{pmatrix} 1 & 2 & 3 & 4 \\ 4 & 3 & 2 & 1 \end{pmatrix}$$

$$c_2 = \begin{pmatrix} 1 & 2 & 3 & 4 \\ 2 & 1 & 4 & 3 \end{pmatrix}$$

So, now we have 7 elements (8 if we include the identity element) in our set of motions. It we compose them, we get

$$c_2 c_1 = \begin{pmatrix} 1 & 2 & 3 & 4 \\ 3 & 4 & 1 & 2 \end{pmatrix} = R^2$$

$$c_1 c_2 = \begin{pmatrix} 1 & 2 & 3 & 4 \\ 3 & 4 & 1 & 2 \end{pmatrix} = R^2$$

$$c_1 d_1 = \begin{pmatrix} 1 & 2 & 3 & 4 \\ 2 & 3 & 4 & 1 \end{pmatrix} = R$$

$$c_2 d_1 = \begin{pmatrix} 1 & 2 & 3 & 4 \\ 4 & 1 & 2 & 3 \end{pmatrix} = R^3$$

---

[1]In spite of the remark above!

*	1	R	R^2	R^3	c_1	c_2	d_1	d_2
1	1	R	R^2	R^3	c_1	c_2	d_1	d_2
R	R	R^2	R^3	1	d_2	d_1	c_1	c_2
R^2	R^2	R^3	1	R	c_2	c_1	d_2	d_1
R^3	R^3	1	R	R^2	d_1	d_2	c_2	c_1
c_1	c_1	d_1	c_2	d_2	1	R^2	R^3	R
c_2	c_2	d_2	c_1	d_1	R^2	1	R	R^3
d_1	d_1	c_2	d_2	c_1	R^3	R	1	R^2
d_2	d_2	c_1	d_1	c_2	R	R^3	R^2	1

TABLE 4.1.1. Multiplication table for D_4

and further analysis shows that we cannot generate any additional elements by composition. We get a multiplication table, as in figure 4.1.1. This is, incidentally, called the *dihedral group*, D_8

Since this defines a *subset* of S_4 (S_4 has $4! = 24$ elements) that is *also* a group, we had better augment our abstract definition of a group with:

DEFINITION 4.1.8. If G is a group and $H \subset G$ is a subset of its elements, H is called a *subgroup* of G if

(1) $x \in H \implies x^{-1} \in H$
(2) $x, y \in H \implies xy \in H$

REMARK. In other words, H is a subset of G that forms a group in its own right.

Note that these conditions imply that $1 \in H$.

Notice that, in D_8 the *powers* of the element R form a subgroup

$$\mathbb{Z}_4 = \{1, R, R^2, R^3\} \subset D_4$$

Since $R^i \cdot R^j = R^{i+j} = R^j \cdot R^i$, $\mathbb{Z}_4$ is a group where multiplication is commutative — i.e., $\mathbb{Z}_4$ is *abelian*.

We say that $\mathbb{Z}_4$ is *generated* by R and this prompts us to define:

DEFINITION 4.1.9. If G is a group and $\{g_1, \ldots, g_n\} \subset G$ is a set of elements, define

$$\langle g_1, \ldots, g_n \rangle \subset G$$

to be the set containing 1 and all possible products of strings $g_{i_1}^{\alpha_1} \cdots g_{i_k}^{\alpha_{i_k}}$. where the α_i are integers. Since the products of any two elements of $\langle g_1, \ldots, g_n \rangle$ is also in $\langle g_1, \ldots, g_n \rangle$, this is a subgroup of G, called the *subgroup generated* by $\{g_1, \ldots, g_n\}$. If $G = \langle g_1, \ldots, g_n \rangle$, we say that the group G is generated by $g_1, \ldots, g_n$. A group that is generated by a *single element* is called *cyclic*.

We can define an interesting subgroup of the group in example 4.1.7 on page 35:

EXAMPLE 4.1.10. If p is a prime, define $\mathbb{Z}/p^\infty$ to be the multiplicative subgroup of the unit circle in $\mathbb{C}$ generated by elements of the form $e^{2\pi i/p^k}$ for all integers $k > 0$. This is called a *Prüfer group*.

Every element of a group generates a cyclic subgroup:

DEFINITION 4.1.11. If G is a group and $g \in G$ is an element, the set of all *powers* of g forms a subgroup of G denoted $\langle g \rangle \subset G$. If G is finite, the *order* of this subgroup is called the *order of g*, denoted $\text{ord}(g)$.

When we have two groups, G and H, we can build a new group from them:

DEFINITION 4.1.12. If G and H are groups, their *direct* sum, $G \oplus H$, is the set of all possible pairs (g, h) with $g \in G, h \in H$ and group-operation

$$(g_1, h_1)(g_2, h_2) = (g_1 g_2, h_1 h_2)$$

The group

$$K = \mathbb{Z}_2 \oplus \mathbb{Z}_2$$

is called the *Klein 4-group*.

EXERCISES.

1. If G is a group and 1_1 and 1_2 are *two* identity elements, show that

$$1_1 = 1_2$$

so that a group's identity element is *unique*.

2. If G is a group and $a, b, c \in G$ show that $ab = ac$ implies that $b = c$.

3. Find elements $a, b \in D_4$ such that $(ab)^{-1} \neq a^{-1}b^{-1}$.

4. If G is a group and $a, b \in G$ have the property that $ab = 1$, show that $ba = 1$.

5. If G is a group and $a, b \in G$, show that $(ab)^{-1} = b^{-1}a^{-1}$ — so we must *reverse* the order of elements in a product when taking an inverse.

6. List all of the generators of the cyclic group, $\mathbb{Z}_{10}$.

7. Show that the set

$$\{\pm 3^k | k \in \mathbb{Z}\} \subset \mathbb{R}^\times$$

is a subgroup.

8. Show that the set

$$\left\{ \begin{pmatrix} 1 & 2 & 3 \\ 1 & 2 & 3 \end{pmatrix}, \begin{pmatrix} 1 & 2 & 3 \\ 1 & 3 & 2 \end{pmatrix} \right\}$$

is a subgroup of S_3.

9. Show that $D_4 = \langle R, d_1 \rangle$.

10. If $G = \mathbb{Z}_p^\times$ for a prime number p, define
$$G_2 = \left\{ x^2 \,\middle|\, x \in G \right\}$$
Show that $G_2 \subset G$ is a subgroup of order $(p-1)/2$ if p is odd, and order 1 if $p = 2$.

11. If $g \in G$ is an element of a finite group, show that there exists an integer $n > 0$ such that $g^n = 1$.

12. Prove that the operation $x * y = x + y + xy$ on the set $x \in \mathbb{R}$, $x \neq -1$, defines an abelian group.

13. List all of the subgroups of a Klein 4-group.

4.2. Homomorphisms

Now we consider functions from one group to another, and the question of when two groups are mathematically equivalent (even when they are defined in very different ways).

We start with a pointlessly abstract definition of a function:

DEFINITION 4.2.1. If S and T are sets, a function
$$f: S \to T$$
is a set of pairs
$$f \subset S \times T$$
– i.e., $\{(s_1, t_1), \ldots, (s_j, t_j)\}$ with the property that
   (1) every element of S occurs as the first member of some pair
   (2) for any two pairs (s_1, t_1) and (s_2, t_2), $s_1 = s_2$ implies that $t_1 = t_2$.

If $(s, t) \in f \subset S \times T$, we write $f(s) = t$. The set S is called the *domain* of f and T is called the *range*. The set of $t \in T$ such $f(s) = t$ for some $s \in S$ is called the *image* or *codomain* of f.

REMARK. In other words, f just associates a unique element of T to every element of S.

For instance, $f(x) = x^2$ defines a function whose domain and range is $\mathbb{R}$. The equation $f(x) = \sqrt{x}$ defines a function whose domain and range are $\mathbb{R}^+$ — real numbers ≥ 0.

Having defined functions, we also distinguish various types of functions:

DEFINITION 4.2.2. A function, $f: S \to T$, is *injective* if $f(s_1) = f(s_2)$ implies $s_1 = s_2$. It is *surjective* if for every $t \in T$, there exists an $s \in S$ such that $f(s) = t$ — so the image of f is all of T. It is *bijective* if it is injective and surjective.

The reader may wonder what all of this has to do with groups.

DEFINITION 4.2.3. If G and H are groups and
$$f: G \to H$$
is a function, f is called a *homomorphism* if it preserves group-operations, i.e. for all $g_1, g_2 \in G$
$$f(g_1 g_2) = f(g_1) f(g_2) \in H$$
The set of all elements $g \in G$ such that $f(g) = 1 \in H$ is called the *kernel* of f, denoted $\ker f$. If f is bijective, it is called an *isomorphism* and the groups G and H are said to be *isomorphic*. An isomorphism from a group to itself is called an *automorphism*.

A homomorphism $f: G \to H$ has certain elementary properties that we leave as exercises to the reader:

EXERCISES.

1. Show that $f(1) = 1 \in H$

2. Show that $f(g^{-1}) = f(g)^{-1}$

3. Show that $\ker f \subset G$ is a subgroup

4. Show that $\operatorname{im} f \subset H$ is a subgroup

5. If S^1 is the complex unit circle (see example 4.1.7 on page 35), show that the function
$$f: \mathbb{R} \to S^1$$
mapping $x \in \mathbb{R}$ to $e^{ix} \in S^1$ is a homomorphism. What is its kernel?

6. Show that the map
$$\begin{aligned} f: \mathbb{Z} &\to \mathbb{Z}_n \\ i &\mapsto i \pmod{n} \end{aligned}$$
is a homomorphism.

7. If m and n are positive integers with $\gcd(m, n) = 1$, show that
$$\mathbb{Z}_m \oplus \mathbb{Z}_n \cong \mathbb{Z}_{mn}$$
$$\mathbb{Z}_m^\times \oplus \mathbb{Z}_n^\times \cong \mathbb{Z}_{mn}^\times$$

4.3. Cyclic groups

Cyclic groups are particularly easy to understand since they only have a single generator. In fact, we have already studied such groups because:

PROPOSITION 4.3.1. *Let G be a cyclic group*
(1) *If $|G| = n$, then G is isomorphic to $\mathbb{Z}_n$.*

(2) If $|G| = \infty$, then G is isomorphic to $\mathbb{Z}$.

REMARK. When $|G| \cong \mathbb{Z}$, it is called an *infinite cyclic* group.

PROOF. Since G is cyclic it consists of powers of a single element G
$$\{1, g, \ldots, g^k, \ldots\}$$
and the isomorphism maps g^k to k in $\mathbb{Z}_n$ or $\mathbb{Z}$, respectively. □

PROPOSITION 4.3.2. *If G is a cyclic group and $H \subset G$ is a subgroup, then H is cyclic.*

PROOF. Suppose G is generated by an element g and $H = \{1, g^{n_1}, g^{n_2}, \ldots\}$. If $\alpha = \min\{|n_1|, \ldots\}$, we claim that g^α generates H. If not, there exists a $g^n \in H$ such that $\alpha \nmid n$. In this case
$$n = \alpha \cdot q + r$$
with $0 < r < \alpha$ and $g^r = g^n \cdot (g^\alpha)^{-q} \in H$ with $r < \alpha$. This is a contradiction. □

EXERCISE 4.3.3. If n and d are a positive integers such that $d \mid n$, show that there exists a *unique* subgroup, $S \subset \mathbb{Z}_n$, with d elements Hint: proposition 4.3.2 implies that S is *cyclic*, so every element is a multiple of a generator $g \in S$ with $d \cdot g \equiv 0 \pmod{d}$ If $x = k \cdot g \in S$, this means that $d \cdot x \equiv 0 \pmod{n}$. Now look at exercise 3 on page 21.

If $G = \mathbb{Z}_n$, and $d \mid n$, then the set

(4.3.1) $$\left\{0, \frac{n}{d}, 2\frac{n}{d}, \ldots, (d-1)\frac{n}{d}\right\}$$

forms this unique subgroup isomorphic to $\mathbb{Z}_d$.

REMARK. In $\mathbb{Z}_n$, the group-operation is written *additively*, so the order of $m \in \mathbb{Z}_n$ (see definition 4.1.11 on page 38) 4.1.11 on page 38is the smallest $k > 0$ such that
$$k \cdot m \equiv 0 \pmod{n}$$

PROPOSITION 4.3.4. *If $m \in \mathbb{Z}_n$ is a nonzero element, then*
$$\mathrm{ord}(m) = \frac{n}{\gcd(n, m)}$$
It follows that $m \neq 0$ is a generator of $\mathbb{Z}_n$ if and only if $\gcd(n, m) = 1$.

REMARK. It follows that $\mathbb{Z}_n$ has precisely $\phi(n)$ distinct generators.

PROOF. The order of m is the smallest k such that $k \cdot m = 0 \in \mathbb{Z}_n$, i.e.,
$$k \cdot m = \ell \cdot n$$
for some integer ℓ. It follows that $k \cdot m$ is the least common multiple of m and n. Since
$$\mathrm{lcm}(n, m) = \frac{nm}{\gcd(n, m)}$$
— see proposition 3.1.11 on page 17, we get
$$\frac{nm}{\gcd(n, m)} = k \cdot m$$

and
$$k = \frac{n}{\gcd(n,m)}$$
The number m is a generator of $\mathbb{Z}_n$ if and only if $\operatorname{ord}(m) = n$ — because it has n *distinct* multiples in this case — which happens if and only if $\gcd(n,m) = 1$. □

As basic as this result is, it implies something interesting about the Euler ϕ-function:

LEMMA 4.3.5. *If n is a positive integer, then*
(4.3.2) $$n = \sum_{d\mid n} \phi(d)$$
where the sum is taken over all positive divisors, d, of n.

PROOF. If $d\mid n$, let $\Phi_d \subset \mathbb{Z}_n$ be the set of generators of the unique cyclic subgroup of order d (generated by n/d). Since every element of $\mathbb{Z}_n$ generates *one* of the $\mathbb{Z}_d$, it follows that $\mathbb{Z}_n$ is the disjoint union of all of the Φ_d for all divisors $d\mid n$. This implies that
$$|\mathbb{Z}_n| = n = \sum_{d\mid n} |\Phi_d| = \sum_{d\mid n} \phi(d)$$
□

For instance
$$\phi(20) = \phi(4) \cdot \phi(5)$$
$$= (2^2 - 2)(5 - 1)$$
$$= 8$$
$$\phi(10) = \phi(5)\phi(2)$$
$$= 4$$
and
$$20 = \phi(20) + \phi(10) + \phi(5) + \phi(4) + \phi(2) + \phi(1)$$
$$= 8 + 4 + 4 + 2 + 1 + 1$$

EXERCISES.

1. If G is a group of order n, show that it is cyclic if and only if it has an element of order n.

2. If G is an abelian group of order nm with $\gcd(n,m) = 1$ and it has an element of order n and one of order m, show that G is cyclic.

3. If G is a cyclic group of order n and k is a positive integer with $\gcd(n,k) = 1$, show that the function
$$f\colon G \to G$$
$$g \mapsto g^k$$

is an isomorphism from G to itself.

4.4. Subgroups and cosets

We being by considering the relationship between a group and its subgroups.

DEFINITION 4.4.1. If G is a group with a subgroup, H, we can define an equivalence relation on elements of G.

$$x \equiv y \pmod{H}$$

if there exists an element $h \in H$ such that $x = y \cdot h$. The equivalence classes in G of elements under this relation are called *left-cosets* of H. The number of left-cosets of H in G is denoted $[G:H]$ and is called the *index* of H in G.

REMARK. It is not hard to see that the left cosets are sets of the form $g \cdot H$ for $g \in G$. Since these are equivalence classes, it follows that

$$g_1 \cdot H \cap g_2 \cdot H \neq \emptyset$$

if and only if $g_1 \cdot H = g_2 \cdot H$.

Since each of these cosets has a size of $|H|$ and are *disjoint*, we conclude that

THEOREM 4.4.2 (Lagrange's Theorem). *If $H \subset G$ is a subgroup of a finite group, then $|G| = |H| \cdot [G:H]$.*

> Joseph-Louis Lagrange (born Giuseppe Lodovico Lagrangia) 1736 – 1813 was an Italian mathematician and astronomer. He made significant contributions to the fields of analysis, number theory, and celestial mechanics. His treatise, [68], laid some of the foundations of group theory — including a limited form of his theorem listed above.

Lagrange's theorem immediately implies that

COROLLARY 4.4.3. *If G is a finite group and $g \in G$ is any element, then*

$$g^{|G|} = 1$$

PROOF. The element g generates a cyclic subgroup of G of order $\text{ord}(g)$. In particular

$$g^{\text{ord}(g)} = 1$$

and the conclusion follows from theorem 4.4.2, which implies that $\text{ord}(g) \mid |G|$. □

Sometimes, we can deduce properties of a group just by the number of elements in it:

PROPOSITION 4.4.4. *If the group G has p elements, where p is a prime number, then G is cyclic generated by any $x \in G$ such that $x \neq 1$.*

PROOF. In $x \in G$, then $\mathrm{ord}(x) = 1$ or p. If $x \neq 1$, $\mathrm{ord}(x) = p$ and the distinct powers of x are *all* of the elements of G. □

The equivalence relation in definition 4.4.1 on the preceding page looks very similar to that of definition 3.2.1 on page 18 so proposition 3.2.2 on page 19 leads to the natural question

Does equivalence modulo a subgroup respect multiplication (i.e., the group-operation)?

Consider what happens when we multiply *sets* of group-elements together. If $H \subset G$ is a *subgroup*, then

$$H \cdot H = H$$

— multiplying every element of H by every other element just gives us H back. This follows from the fact that $1 \in H$ and H is closed under the group-operation. If we multiply two *cosets* together

$$g_1 \cdot H \cdot g_2 \cdot H$$

we get a set of group-elements that may or may not be a coset. Note that

$$g_1 \cdot H \cdot g_2 \cdot H = g_1 g_2 \cdot g_2^{-1} H g_2 H$$

If $g_2^{-1} H g_2 = H$ as a *set*, then

(4.4.1) $$g_1 H \cdot g_2 H = g_1 g_2 H$$

This suggests making a few definitions

DEFINITION 4.4.5. If G is a group with a subgroup $H \subset G$ and $g \in G$, then the *conjugate of H by g* is defined to be $g \cdot H \cdot g^{-1}$, and denoted H^g.

A subgroup $H \subset G$ of a group is said to be *normal* if $H = H^g$ for all $g \in G$. This fact is represented by the notation $H \triangleleft G$.

REMARK. For H to be a normal subgroup of G, we do *not* require $ghg^{-1} = h$ for $h \in H$ and every $g \in G$ — we *only* require $ghg^{-1} \in H$ whenever $h \in H$.

If G is abelian, *all* of its subgroups are normal because $ghg^{-1} = gg^{-1}h = h$.

Here's an example of a *non*-normal subgroup:
Let

$$S = \left\{1, a = \begin{pmatrix} 1 & 2 & 3 \\ 2 & 1 & 3 \end{pmatrix}\right\} \subset S_3$$

and let

$$g = \begin{pmatrix} 1 & 2 & 3 \\ 1 & 3 & 2 \end{pmatrix}$$

so that $g^2 = 1$ which means $g^{-1} = g$. When we conjugate a by this, we get

$$\begin{pmatrix} 1 & 2 & 3 \\ 1 & 3 & 2 \end{pmatrix} \begin{pmatrix} 1 & 2 & 3 \\ 2 & 1 & 3 \end{pmatrix} \begin{pmatrix} 1 & 2 & 3 \\ 1 & 3 & 2 \end{pmatrix} = \begin{pmatrix} 1 & 2 & 3 \\ 3 & 2 & 1 \end{pmatrix} \notin S$$

PROPOSITION 4.4.6. *If $H \triangleleft G$ is a normal subgroup of a group, then the set of left cosets forms a group.*

PROOF. Equation 4.4.1 on the preceding page implies that
$$g_1 H \cdot g_2 H = g_1 g_2 H$$
so the group identities for *cosets* follow from the identities in G:
- ▷ the *identity element* is
$$1 \cdot H$$
  the inverse of $g \cdot H$ is
$$g^{-1} \cdot H$$
- ▷ and
$$\begin{aligned} g_1 H(g_2 H g_3) &= g_1(g_2 g_3)H \\ &= (g_1 g_2)g_3 H \\ &= (g_1 H g_2 H)g_3 H \end{aligned}$$

□

This group of cosets has a name:

DEFINITION 4.4.7. If $H \triangleleft G$ is a normal subgroup, the *quotient group*, G/H, is well-defined and equal to the group of left cosets of H in G. The map
$$p: G \to G/H$$
that sends an element $g \in G$ to its coset is a homomorphism — the *projection to the quotient*.

If $G = \mathbb{Z}$ then G is abelian and all of its subgroups are normal. If H is the subgroup $n \cdot \mathbb{Z}$, for some integer n, we get
$$\frac{\mathbb{Z}}{n \cdot \mathbb{Z}} \cong \mathbb{Z}_n$$
In fact, a common notation for $\mathbb{Z}_n$ is $\mathbb{Z}/n\mathbb{Z}$.

If $G = \mathbb{Z}_n$ and $d \mid n$, we know that G has a subgroup isomorphic to $\mathbb{Z}_d$ (see exercise 4.3.3 on page 41) and the quotient
$$\frac{\mathbb{Z}_n}{\mathbb{Z}_d} \cong \mathbb{Z}_{n/d}$$

Quotient groups arise naturally whenever we have a homomorphism:

THEOREM 4.4.8 (First Isomorphism Theorem). *If $f: G \to H$ is a homomorphism of groups with kernel K, then*
1. *$K \subset G$ is a normal subgroup, and*
2. *there exists an isomorphism $i: G/K \to f(G)$ that makes the diagram*

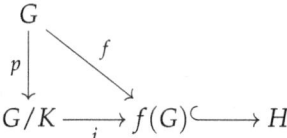

commute, where $p: G \to G/K$ is projection to the quotient.

REMARK. The phrase "diagram commutes" just means that, as *maps*, $f = i \circ p$.

If f is surjective, then H is *isomorphic* to G/K.

PROOF. We prove the statements in order. If $k \in K$ and $g \in G$,
$$f(g \cdot k \cdot g^{-1}) = f(g) \cdot f(k) \cdot f(g^{-1})$$
$$= f(g) \cdot 1 \cdot f(g)^{-1}$$
$$= 1$$

so $g \cdot k \cdot g^{-1} \in K$ and K is normal.

Now we define i to send a coset $K \cdot g \subset G$ to $f(g) \in H$. First, we have to show that this is even *well-defined*. If $g' \in g \cdot K$, we must show that $f(g') = f(g)$. the fact that $g' \in K \cdot g$ implies that $g' = g \cdot k$, for some $k \in K$ so
$$f(g') = f(k) \cdot f(g)$$
$$= 1 \cdot f(g) = f(g)$$

Since $(g_1 \cdot K)(g_2 \cdot K) = g_1 g_2 \cdot K$, it is not hard to see that i is a *homomorphism*. It is also clearly *surjective* onto $f(G)$. If $i(K \cdot g) = 1 \in H$, it follows that $g \in K$ and $g \cdot K = 1 \cdot K$ so that i is also *injective*. □

Now we consider some *special* subgroups of a group:

DEFINITION 4.4.9. If G is a group, define $Z(G)$ — the *center* of G — to be the set of all elements $x \in G$ such that $xg = gx$ for all $g \in G$.

REMARK. Note that $|Z(G)| \geq 1$ since 1 commutes with everything in G.

Since the elements of $Z(G)$ commute with *all* of the elements of G, $Z(G)$ is *always* a normal subgroup of G.

If $H \subset G$ is a subgroup of a group, H is frequently not a normal subgroup. The *normalizer* of H is the largest subgroup of G containing H in which H is normal:

DEFINITION 4.4.10. If $H \subset G$ is a subgroup of a group, the *normalizer* $N_G(H)$ is defined by
$$N_G(H) = \left\{ g \in G \,|\, \forall h \in H, ghg^{-1} \in H \right\}$$

REMARK. In other words, $gHg^{-1} = H$, as sets.

If $g \in G$ is actually an element of H, we always have $gHg^{-1} = H$. This means that $H \subset N_G(H) \subset G$.

We can also define the normal closure or conjugate closure of a subgroup.

DEFINITION 4.4.11. If $S \subset G$ is a subgroup of a group, its *normal closure* (or *conjugate closure*), S^G, is defined to be the smallest normal subgroup of G containing S. It is given by
$$S^G = \langle s^g \text{ for all } s \in S \text{ and } g \in G \rangle$$

REMARK. The normal closure is the *smallest* normal subgroup of G that contains S.

We also have

DEFINITION 4.4.12. If $g_1, g_2 \in G$ are elements of a group, their *commutator* $[g_1, g_2]$ is defined by
$$[g_1, g_2] = g_1 g_2 g_1^{-1} g_2^{-1}$$
and the *commutator subgroup*, $[G, G]$, is defined by
$$[G, G] = \langle [g_1, g_2] \text{ for all } g_1, g_2 \in G \rangle$$

Some subgroups of a group are "more than normal:"

DEFINITION 4.4.13. If G is a group with subgroup, H, then H is said to be a *characteristic subgroup* if, for *any* automorphism
$$f\colon G \to G$$
$f(H) = H$.

REMARK. Characteristic subgroups are always normal, but not all normal subgroups are characteristic. Usually characteristic subgroups are defined by some structural property that is preserved by all automorphisms. For instance, the *center* of a group (see definition 4.4.9 on the preceding page) is characteristic.

We also have the interesting concept of the *socle* of a group:

DEFINITION 4.4.14. If G is a group, its *socle*, denoted $\text{Soc}(G)$ is the subgroup generated by all minimal subgroups of G. If a group has no minimal subgroups, its socle is the identity, $\{1\}$.

REMARK. If $G = \mathbb{Z}_{12}$, generated by 1, then it has two minimal subgroups, namely the one generated by 6 and isomorphic to $\mathbb{Z}_2$ and the one generated by 4, isomorphic to $\mathbb{Z}_3$. The socle is the subgroup generated by $\{4, 6\}$, which is the one generated by 2.

It is not hard to see that minimal subgroups and the socle are characteristic.

There are more isomorphism theorems that are proved in the exercises.

EXERCISES.

1. If $H, K \subset G$ are subgroups of a group, HK stands for all products of the form $h \cdot k$ with $h \in H$ and $k \in K$. This is not usually a subgroup of G. If K is a *normal* subgroup of G, show that HK is a subgroup of G.

2. If $H, K \subset G$ are subgroups of a group, where K is a normal subgroup, show that K is a normal subgroup of HK.

3. If $H, K \subset G$ are subgroups of a group, where K is a normal subgroup, show that $H \cap K$ is a normal subgroup of H.

4. If $H, K \subset G$ are subgroups of a group, where K is a normal subgroup, show that
$$\frac{HK}{K} \cong \frac{H}{H \cap K}$$
This is called The Second Isomorphism Theorem.

5. If $K \triangleleft G$ is a normal subgroup and
$$p: G \to G/K$$
is the projection to the quotient, show that there is a 1-1 correspondence between subgroups of G/K and subgroups of G that contain K, with normal subgroups of G/K corresponding to normal subgroups of G. This is called the *Correspondence Theorem* for groups.

6. If G is a group with normal subgroups H and K with $K \subset H$, then $H/K \triangleleft G/K$ and
$$\frac{G}{H} \cong \frac{G/K}{H/K}$$
This is called the Third Isomorphism Theorem.

7. Show that the set
$$S = \left\{ \begin{pmatrix} 1 & 2 & 3 \\ 1 & 2 & 3 \end{pmatrix}, \begin{pmatrix} 1 & 2 & 3 \\ 2 & 3 & 1 \end{pmatrix}, \begin{pmatrix} 1 & 2 & 3 \\ 3 & 1 & 2 \end{pmatrix} \right\}$$
is a *normal* subgroup of S_3. What is the quotient, S_3/S?

8. Show that $Z(G) \subset G$ is a subgroup.

9. If $Z(G)$ is the center of a group G, show that $Z(G)$ is abelian.

10. Suppose G is a group with normal subgroups H, K such that $H \cap K = 0$ and $G = HK$. Show that
$$G \cong H \oplus K$$

11. If G is a group with 4 elements, show that G is abelian.

12. If $H \subset G$ is a subgroup of a group and $g \in G$ is any element, show that H^g is a subgroup of G.

13. If G is a group, show that conjugation by any $g \in G$ defines an automorphism of G. Such automorphisms (given by conjugation by elements of the group) are called *inner automorphisms*.

14. Show that the automorphisms of a group, G, form a group themselves, denoted $\mathrm{Aut}(G)$

15. If m is an integer, show that $\mathrm{Aut}(\mathbb{Z}_m) = \mathbb{Z}_m^\times$.

16. If G is a group show that the inner automorphisms of G form a group, $\mathrm{Inn}(G)$ — a subgroup of all automorphisms.

17. If G is a group show that $\mathrm{Inn}(G) \triangleleft \mathrm{Aut}(G)$, i.e., the subgroup of inner automorphisms is a *normal* subgroup of the group of *all* automorphisms. The quotient is called the group of *outer automorphisms*.
$$\mathrm{Out}(G) = \frac{\mathrm{Aut}(G)}{\mathrm{Inn}(G)}$$

18. If G is a group, show that the commutator subgroup, $[G, G]$, is a normal subgroup.

19. If G is a group, show that the quotient
$$\frac{G}{[G, G]}$$
is abelian. This is often called the *abelianization* of G.

20. If G is a group and
$$p \colon G \to A$$
is a homomorphism with A abelian and kernel K, then
$$[G, G] \subset K$$
so $[G, G]$ is the *smallest* normal subgroup of G giving an abelian quotient.

4.5. Symmetric Groups

In the beginning of group theory, the word "group" meant a group of permutations of some set — i.e., a *subgroup* of a symmetric group. In some sense, this is accurate because:

THEOREM 4.5.1 (Cayley). *If G is a group, there exists a symmetric group S_G and an injective homomorphism*
$$f \colon G \to S_G$$
where S_G is the group of all possible permutations of the elements of G.

> Arthur Cayley (1821 – 1895) was a British mathematician noted for his contributions to group theory, linear algebra and algebraic geometry. His influence on group theory was such that group multiplication tables (like that in 4.1.1 on page 37) were originally called *Cayley tables*.

PROOF. First, list all of the elements of G
$$\{1, g_1, g_2, \dots\}$$
If $x \in G$, we define the function f by
$$f(x) = \begin{pmatrix} 1 & g_1 & g_2 & \cdots \\ x & x \cdot g_1 & x \cdot g_2 & \cdots \end{pmatrix}$$
We claim that this is a permutation in S_G: If $x \cdot g_i = x \cdot g_j$, then we can multiply on the left by x^{-1} to conclude that $g_i = g_j$. Furthermore, every element of G occurs in the bottom row: if $g \in G$ is an arbitrary element of G, then it occurs under the entry $x^{-1} \cdot g$ in the top row. This function is *injective* because $f(x) = 1$ implies that
$$\begin{pmatrix} 1 & g_1 & g_2 & \cdots \\ x & x \cdot g_1 & x \cdot g_2 & \cdots \end{pmatrix} = 1 = \begin{pmatrix} 1 & g_1 & g_2 & \cdots \\ 1 & g_1 & g_2 & \cdots \end{pmatrix}$$

which implies that $x = 1$.

We must still verify that f is a *homomorphism*, i.e. that $f(xy) = f(x)f(y)$. We do this by direct computation:

$$f(y) = \begin{pmatrix} 1 & g_1 & g_2 & \cdots \\ y & y \cdot g_1 & y \cdot g_2 & \cdots \end{pmatrix}$$

$$f(x)f(y) = \begin{pmatrix} 1 & g_1 & g_2 & \cdots \\ \downarrow & \downarrow & \downarrow & \\ y & y \cdot g_1 & y \cdot g_2 & \cdots \\ \downarrow & \downarrow & \downarrow & \\ xy & xy \cdot g_1 & xy \cdot g_2 & \cdots \end{pmatrix}$$

where we perform the permutation defined by $f(y)$ and then (in the middle row) apply the permutation defined by $f(x)$. The *composite* — from top to bottom — is a permutation equal to

$$f(xy) = \begin{pmatrix} 1 & g_1 & g_2 & \cdots \\ xy & xy \cdot g_1 & xy \cdot g_2 & \cdots \end{pmatrix}$$

which proves the conclusion. □

Since symmetric groups are so important, it makes sense to look for a more compact notation for permutations:

DEFINITION 4.5.2. A *cycle*, denoted by a symbol $(i_1, \ldots, i_k)$ with $i_j \neq i_\ell$ for $j \neq \ell$ represents the permutation that maps i_j to i_{j+1} for all $1 \leq j < k$ and maps i_k to i_1. If a cycle has only two indices (so $k = 2$) it is called a *transposition*. We usually follow the convention that in the cycle $(i_1, \ldots, i_k)$, the index $i_1 < i_j$ for $j > 1$. For instance the cycle $(i_1, \ldots, i_5)$ represents the permutation:

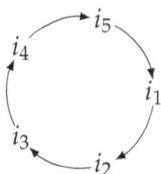

Any index *not* listed in the cycle is mapped to *itself*.

REMARK. It follows that the *identity* permutation is the *empty cycle*, $()$, and transpositions are their *own inverses* — i.e., $(i,j)(i,j) = 1$. Most permutations are *products* of cycles, like

$$(1,3)(2,6,7)$$

representing the permutation

(4.5.1) $$\begin{pmatrix} 1 & 2 & 3 & 4 & 5 & 6 & 7 \\ 3 & 6 & 1 & 4 & 5 & 7 & 2 \end{pmatrix}$$

It is fairly straightforward to convert a permutation in matrix-form into a product of cycles:

$$\begin{pmatrix} 1 & 2 & 3 & 4 & 5 & 6 & 7 \\ 3 & 4 & 7 & 5 & 2 & 1 & 6 \end{pmatrix}$$

(1) Start with 1 and note that it maps to 3, so we get $(1,3,\ldots)$. Now, note that 3 maps to 7, giving $(1,3,7,\ldots)$. Since 7 maps to 6, we get $(1,3,7,6,\ldots)$. Since 6 maps to 1, and 1 is already in the cycle we are constructing, our first cycle is complete:
$$(1,3,7,6)$$

(2) Now *delete* those columns from the matrix representation in 4.5.1 on the preceding page to give
$$\begin{pmatrix} 2 & 4 & 5 \\ 4 & 5 & 2 \end{pmatrix}$$

(3) And *repeat* the previous steps until all columns have been eliminated. We get
$$\begin{pmatrix} 1 & 2 & 3 & 4 & 5 & 6 & 7 \\ 3 & 4 & 7 & 5 & 2 & 1 & 6 \end{pmatrix} = (1,3,7,6)(2,4,5)$$
which is much more compact than the matrix-representation.

Step 2 above guarantees that the cycles we generate will have the property

DEFINITION 4.5.3. Two cycles $a = (i_1,\ldots,i_s)$ and $b = (j_1,\ldots,j_t)$ are said to be *disjoint* if, as sets, $\{i_1,\ldots,i_s\} \cap \{j_1,\ldots,j_t\} = \emptyset$.

REMARK. Since a does not affect *any* of the elements that b permutes and vice-versa, we have $ab = ba$.

THEOREM 4.5.4. *The symmetric group is generated by transpositions.*

PROOF. A little computation shows that we can write every *cycle* as a product of *transpositions*:
$$(4.5.2) \qquad (i_1,\ldots,i_s) = (i_{s-1},i_s)\cdots(i_2,i_s)(i_1,i_s)$$

Recall exercise 5 on page 38, in which the reader showed that $(ab)^{-1} = b^{-1}a^{-1}$. A simple induction shows that
$$(a_1\ldots a_n)^{-1} = a_n^{-1}\cdots a_1^{-1}$$
in any group. This fact, coupled with the fact that transpositions are their own inverses means
$$(4.5.3) \qquad (i_1,\ldots,i_s) = (a_1,b_1)\cdots(a_{s-1},b_{s-1})(a_s,b_s)$$
implies that
$$(4.5.4) \qquad (i_1,\ldots,i_s)^{-1} = (a_s,b_s)(a_{s-1},b_{s-1})\cdots(a_1,b_1)$$

Since every permutation is a product of disjoint cycles, and every cycle can be expressed as a product of transpositions, it follows that *every permutation* can be written as a product of transpositions. This representation is far from unique — for instance a bit of computation shows that

$$(4.5.5) \qquad \begin{aligned}(1,2,3,4,5) &= (1,5)(1,4)(1,3)(1,2) \\ &= (4,5)(2,5)(1,5)(1,4)(2,3)(1,4)\end{aligned}$$

□

We can carry this result further

COROLLARY 4.5.5. *The symmetric group, S_n, is generated by the transpositions*
$$(1,2), (1,2), \ldots, (1,n)$$

PROOF. By theorem 4.5.4 on the previous page, it suffices to show that every transposition can be expressed in terms of the given set. We have:
$$(i,j) = (1,i)(1,j)(1,i)$$
□

It is possible to generate S_n with fewer than n generators:

COROLLARY 4.5.6. *The symmetric group, S_n, is generated by the $n-1$ adjacent transpositions*
$$(1,2), (2,3), \ldots, (n-1,n)$$

PROOF. As before, we must show that *any* transposition, (i,j), can be expressed in terms of these. We assume $i < j$ and do induction on $j - i$. If $j - 1 = 1$ the transposition *is* adjacent and the result is true.

Now note that
$$(i,j) = (i,i+1)(i+1,j)(i,i+1)$$
and $j - (i+1) < j - i$ so the induction hypothesis implies that $(i+1,j)$ can be written as a product of adjacent transpositions. □

Notice that the longer representation in equation 4.5.2 on the preceding page has precisely *two* more transpositions in it than the shorter one. This no accident — representing a more general phenomena. To study that, we need the following

LEMMA 4.5.7. *If*
$$(4.5.6) \qquad 1 = (a_1, b_1) \cdots (a_k, b_k) \in S_n$$
then k must be an even number.

PROOF. We will show that there exists another representation of 1 with $k - 2$ transpositions. If k is *odd*, we can *repeatedly* shorten an equation like 4.5.6 by two and eventually arrive at an equation
$$1 = (\alpha, \beta)$$
which is a contradiction.

Let i be any number appearing in equation 4.5.6 and suppose the *rightmost* transposition in which i appears is $t_j = (i, b_j)$ — so $i = a_j$ and $i \neq a_\ell, b_\ell$ for all $\ell > j$. We distinguish four cases:

(1) $(a_{j-1}, b_{j-1}) = (a_j, b_j) = (i, b_j)$. In this case $(a_{j-1}, b_{j-1})(a_j, b_j) = 1$ and we can simply *delete* these two permutations from equation 4.5.6. We are done.

(2) $(a_{j-1}, b_{j-1}) = (i, b_{j-1})$ where $b_{j-1} \neq b_j, i$. In this case, a little computation shows that
$$(a_{j-1}, b_{j-1})(i, b_j) = (i, b_j)(b_{j-1}, b_j)$$
so that the index i now occurs one position to the *left* of where it occurred before.

(3) $(a_{j-1}, b_{j-1}) = (a_{j-1}, b_j)$, where $a_{j-1} \neq a_j, b_j$. In this case
$$(a_{j-1}, b_{j-1})(i, b_j) = (i, b_{j-1})(a_{j-1}, b_j)$$
and, again, the index i is moved one position to the left.

(4) The transpositions (a_{j-1}, b_{j-1}) and $(i, b_j) = (a_j, b_j)$ are disjoint, as per definition 4.5.3 on page 51 in which case they *commute* with each other so that
$$(a_{j-1}, b_{j-1})(i, b_j) = (i, b_j)(a_{j-1}, b_{j-1})$$

We use cases 2-4 to move the rightmost occurrence of i left until we encounter case 1. This *must happen* at some point — otherwise we could move i all the way to the left until there is only one occurrence of it in all of equation 4.5.6 on the preceding page. If this leftmost occurrence of i is in a transposition (i, i'), it would imply that the *identity permutation* maps i to whatever i' maps to — which *cannot* be i since i doesn't appear to the right of this transposition. This is a contradiction. □

We are ready to prove an important result

PROPOSITION 4.5.8. *If $\sigma \in S_n$ is a permutation and*
$$\sigma = (a_1, b_1) \cdots (a_s, b_s)$$
$$= (c_1, d_1) \cdots (c_t, d_t)$$
are two ways of writing σ as a product of transpositions then
$$s \equiv t \pmod{2}$$

PROOF. We will show that $s + t$ is an even number. Equations 4.5.3 and 4.5.4 on page 51 show that
$$\sigma^{-1} = (c_t, d_t) \cdots (c_1, d_1)$$
so that
$$1 = \sigma \cdot \sigma^{-1} = (a_1, b_1) \cdots (a_s, b_s)(c_t, d_t) \cdots (c_1, d_1)$$
and the conclusion follows from lemma 4.5.7 on the preceding page. □

So, although the number of transpositions needed to define a permutation is not unique, its *parity* (odd or even) is:

DEFINITION 4.5.9. If $n > 1$ and $\sigma \in S_n$ is a permutation, then the *parity* of σ is defined to be *even* if σ is equal to the product of an even number of transpositions, and *odd* otherwise. We define the *parity-function*
$$\wp(\sigma) = \begin{cases} 0 & \text{if } \sigma \text{ is even} \\ 1 & \text{if } \sigma \text{ is odd} \end{cases}$$

REMARK. Equation 4.5.2 on page 51 shows that cycles of *even* length (i.e., number of indices in the cycle) are *odd* permutations and cycles of *odd* length are *even*.

If $\sigma \in S_n$ is a permutation, equation 4.5.4 on page 51 implies $\wp(\sigma^{-1}) = \wp(\sigma)$. By counting transpositions, it is not hard to see that:

(1) The product of two *even* permutations is *even*,
(2) The product of an *even* and *odd* permutation is *odd*,
(3) The product of two *odd* permutations is *even*.

This implies that the even permutations form a *subgroup* of the symmetric group:

DEFINITION 4.5.10. If $n > 1$ the subgroup of even permutations of S_n is called the degree-n *alternating group* A_n.

REMARK. The parity-function defines a homomorphism

$$\wp \colon S_n \to \mathbb{Z}_2$$

whose kernel is A_n.

It is fairly straightforward to compute *conjugates* of cycles:

PROPOSITION 4.5.11. *If $\sigma \in S_n$ is a permutation and $(i_1, \ldots, i_k) \in S_n$ is a cycle, then*

$$(i_1, \ldots, i_k)^\sigma = (\sigma(i_1), \ldots, \sigma(i_k))$$

PROOF. Recall that

$$(i_1, \ldots, i_k)^\sigma = \sigma \circ (i_1, \ldots, i_k) \circ \sigma^{-1}$$

If $x \notin \{\sigma(i_1), \ldots, \sigma(i_k)\}$, then $\sigma^{-1}(x) \notin \{i_1, \ldots, i_k\}$ so $(i_1, \ldots, i_k)\sigma^{-1}(x) = \sigma^{-1}(x)$ and

$$\sigma \circ (i_1, \ldots, i_k) \circ \sigma^{-1}(x) = x$$

On the other hand, if $x = \sigma(i_j)$, then $\sigma^{-1}(x) = i_j$ and $(i_1, \ldots, i_k) \circ \sigma^{-1}(x) = i_{j+1}$ (unless $j = k$, in which case we wrap around to i_1) and

$$\sigma \circ (i_1, \ldots, i_k) \circ \sigma^{-1}(\sigma(i_j)) = \sigma(i_{j+1})$$

□

COROLLARY 4.5.12. *The cycles $(i_1, \ldots, i_s), (j_1, \ldots, j_t) \in S_n$ are conjugate if and only if $s = t$.*

PROOF. Proposition 4.5.11 implies that $s = t$ if they are conjugate. If $s = t$, proposition 4.5.11 shows that

$$(j_1, \ldots, j_s) = (i_1, \ldots, i_s)^{(i_1, j_1) \cdots (i_s, j_s)}$$

Here, we follow the convention that $(i_\alpha, j_\alpha) = ()$ if $i_\alpha = j_\alpha$. □

Just as S_n is generated by transpositions, we can find standard generators for alternating groups:

PROPOSITION 4.5.13. *The alternating group, A_n, is generated by cycles of the form $(1, 2, k)$ for $k = 3, \ldots, n$.*

PROOF. The group, A_n, is generated by pairs of transpositions: $(a,b)(c,d)$ or $(a,b)(b,c)$ but

$$(a,b)(c,d) = (a,c,b)(a,c,d)$$
$$(a,b)(b,c) = (a,c,b)$$

so we know that A_n is generated by cycles of length 3. Now, note that $(1,a,2) = (1,2,a)^{-1}$ and $(1,a,b) = (1,2,b)(1,2,a)^{-1}$. In addition, $(2,a,b) = (1,2,b)^{-1}(1,2,a)$.

With these basic cases out of the way, we are in a position to handle the *general* case. If $a,b,c \neq 1,2$, a slightly tedious calculation shows that

$$(a,b,c) = (1,2,a)^{-1}(1,2,c)(1,2,b)^{-1}(1,2,a)$$

which proves the result. □

DEFINITION 4.5.14. A group, G, is defined to be *simple* if it has no normal subgroups other than $\{1\}$ and G itself.

REMARK. It is not hard to find examples of simple groups: if p is a prime number, $\mathbb{Z}_p$ is simple since it doesn't have *any* subgroups other than $\{0\}$ and $\mathbb{Z}_p$. One reason that simple groups are interesting is that one can construct *all* finite groups out of them.

For many years, the classification of all finite simple groups was one of the most famous unsolved problems in group theory. It was solved in a series of several hundred papers published between 1955 and 2004 — involving thousands of pages of logical reasoning.

THEOREM 4.5.15. *If $n \neq 4$, the group A_n is simple.*

PROOF. If $n = 3$, $A_3 = \mathbb{Z}_3$ so it is simple. If $n = 4$, A_4 has the subgroup

$$\{1, (1,2)(3,4), (1,3)(2,4), (1,4)(2,3)\}$$

which can be verified to be normal, so A_4 is *not* simple. Now suppose $n \geq 5$.

Suppose $H \subset A_n$ is a normal subgroup. If $(1,2,3) \in H$, then proposition 4.5.11 on the facing page implies that $(1,2,3)^{(3,k)} = (1,2,k)$ and this conjugate *must* be contained in H because H is *normal*. Since these cycles *generate* A_n (see proposition 4.5.13 on the preceding page) it follows that $H = A_n$. If H has *any* 3-cycle, corollary 4.5.12 on the facing page shows that this 3-cycle is conjugate to $(1,2,3)$ so that $(1,2,3) \in H$ and $H = A_n$.

Now we show that, if $n > 4$, $H \subset A_n$ must have a 3-cycle so that $H = A_n$. We consider the elements of H — products of disjoint cycles:

$$x = c_1 \cdots c_r$$

Case 1: At least one of the c_i has length ≥ 4 — i.e. $c_i = (i_1, \ldots, i_k)\tau \in H$, where $k \geq 4$ and τ is disjoint from $(i_1, \ldots, i_k)$. The normality of H and corollary 4.5.12 on the preceding page means that it *also* contains $\sigma = (1,2,\ldots,k)\tau'$ and

$$\sigma^{-1}\sigma^{(1,2,3)} = (1,2,k)$$

which implies that $H = A_n$.

Case 2: x has at least *two* cycles of length 3 — i.e., $x = (i_1, i_2, i_3)(j_1, j_2, j_3)\tau$ where τ is disjoint from the cycles. Proposition 4.5.11 on page 54 and the fact that H is normal shows that, without loss of generality, we can assume $x = (1,2,3)(4,5,6)\tau$. Let $\sigma = (1,2,4)$. Then σ commutes with τ so

$$x^\sigma = [(1,2,3)(4,5,6)]^\sigma \tau$$
$$= (1,2,3)^\sigma (4,5,6)^\sigma \tau$$
$$= (2,4,3)(1,5,6)\tau$$

Then

$$x^\sigma x^{-1} = (2,4,3)(1,5,6)\tau\tau^{-1}(4,5,6)^{-1}(1,2,3)^{-1}$$
$$= (2,4,3)(1,5,6)(4,6,5)(1,3,2)$$
$$= (1,2,5,3,4)$$

which shows that H has a cycle of length ≥ 4 — and this takes us back to case 1.

Case 3: x has *one* cycle of length 3 and all of the others have length 2. Without loss of generality, we can assume $x = (1,2,3)\tau$. Since cycles of length 2 are their own inverses we have $\tau^2 = 1$ and $x^2 = (1,3,2)$ — which implies that $H = A_n$.

Case 4: x consists *entirely* of disjoint transpositions — so, for instance, $x^2 = 1$. Without loss of generality, we can assume

$$x = (1,2)(3,4)\tau$$

If $\sigma = (1,2,3)$, we have

$$x^\sigma = \sigma x \sigma^{-1}$$
$$= (1,2)^\sigma (3,4)^\sigma \tau$$
$$= (2,3)(1,4)\tau$$

and

$$x^\sigma x^{-1} = (2,3)(1,4)\tau\tau^{-1}(3,4)(1,2)$$
$$= (2,3)(1,4)(3,4)(1,2)$$
$$= (1,3)(2,4)$$

so we conclude that $\alpha = (1,3)(2,4) \in H$ (i.e., *without* any τ-factor). Proposition 4.5.11 on page 54 shows that

$$\alpha^{(1,3,5)} = (3,5)(2,4)$$

and we have

(4.5.7)
$$\alpha\alpha^{(1,3,5)} = (1,3)(2,4)(3,5)(2,4)$$
$$= (1,3,5)$$

so $(1,3,5) \in H$ and $H = A_n$. Note that the $(2,4)$-transpositions in equation 4.5.7 are disjoint from the others and just cancel each other out.

□

We can use symmetric groups and permutations to analyze other groups.

Recall the group, D_8, defined in table 4.1.1 on page 37 — the group of symmetries of the square.

EXAMPLE 4.5.16. The *general* dihedral group D_{2n} is the group of symmetries of a regular n-sided polygon.

These symmetries are:

▷ rotations by $2\pi/n$ — giving rise to an element $r \in D_{2n}$ with $r^n = 1$.
▷ when n is even, we have reflections through axes through two corners (the *dotted* lines in figure 4.1.2 on page 36), giving rise to $f_i \in D_n$, $i = 1, \ldots, n$ with $f_i^2 = 1$. In this case, $f_1 = (2,n)(3,n-1)\cdots(k,k+2)$ as a permutation of vertices.
▷ In the *odd* case $n = 2k+1$, f_i is a reflection through vertex i and the midpoint of the face $(k+1,k+2)$, and $f_1 = (2,n)(3,n-1)\cdots(k+1,k+2)$, as a permutation of vertices.
▷ if $n = 2k$, we also have reflections through axes passing through midpoints of two opposite faces (the *dashed* lines in figure 4.1.2 on page 36). These give rise to elements $g_i \in D_n$, $i = 1, \ldots, n$ with $g_i^2 = 1$. It appears that we have $2n$ reflections in all but the symmetry of the even case implies that $\{f_1, \ldots, f_k\} = \{f_{k+1}, \ldots, f_n\}$ and $\{g_1, \ldots, g_k\} = \{g_{k+1}, \ldots, g_n\}$ as *sets*, so that, again, we have n reflections in all.

In *all* cases, proposition 4.5.11 on page 54 shows that

$$r^{f_1} = (1, n, n-1, \ldots, 2) = r^{-1} = r^{n-1}$$

PROPOSITION 4.5.17. *If $n > 2$, D_{2n} is generated by r and f_1 and the elements of D_{2n} are precisely*

$$\left\{1, r, \ldots, r^{n-1}, f_1, rf_1, \ldots, r^{n-1}f_1\right\}$$

so it is of order $2n$ and defined by the equations $r^n = 1$, $f_1^2 = 1$, and $r^{f_1} = r^{-1}$.

REMARK. In a rather confusing convention, the dihedral group D_{2n} is sometimes written as D_n (especially where its geometric properties are being studied).

PROOF. We have already seen $2n$ potential elements of D_{2n}: the rotations and reflections. The relations $f_1^2 = 1$ and $r^{f_1} = r^{n-1}$ imply that $f_1 r f_1 = r^{n-1}$ so that $f_1 r^k f_1 = r^{k(n-1)}$, which implies that

$$f_1 r^k = r^{k(n-1)} f_1$$

and this exhausts possible products of r and f_1. It is not hard to see that these are all distinct. An *arbitrary* product

$$f_1 r^{n_1} \cdots f_1 r^{n_k} f_1^i$$

can be rewritten as $r^t f_1^s$. □

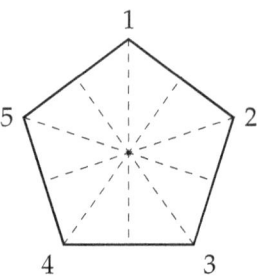

FIGURE 4.5.1.   Regular pentagon

EXERCISES.

1. Convert
$$\begin{pmatrix} 1 & 2 & 3 & 4 & 5 & 6 & 7 & 8 & 9 \\ 3 & 1 & 6 & 9 & 4 & 7 & 2 & 5 & 6 \end{pmatrix}$$
to cycle notation

2. Is the permutation
$$\begin{pmatrix} 1 & 2 & 3 & 4 & 5 & 6 & 7 & 8 & 9 \\ 9 & 1 & 6 & 3 & 4 & 7 & 6 & 5 & 2 \end{pmatrix}$$
odd or even?

3. If $g \in S_n$ is an element we arrange the disjoint cycles of g in order from longest to shortest (and list elements that are fixed by g as cycles of length 1):
$$g = (a_1, \ldots, a_{k_1})(b_1, \cdots, b_{k_2}) \cdots (z_1, \ldots, z_{k_t})$$
and call the sequence of numbers $k_1 \geq k_2 \geq \cdots \geq k_t$, the *cycle structure* of g. Show that two elements of S_n are *conjugate* if and only if they have the *same* cycle structure.

4. What is the center of S_3?

5. Represent the dihedral group, D_8 — see 3.2.2 on page 20 — as a subgroup of S_4.

6. Let $G \subset S_5$ be a subgroup containing a 5-cycle and a transposition. Show that this implies that $G = S_5$.

4.6. Abelian Groups

There are a "islands of success" in mathematics — areas that are "completely solved:" Every reasonable question one has can be answered. The theory of finitely generated abelian groups is one of these areas.

We begin with *free* abelian groups:

DEFINITION 4.6.1. A group, A, is a *free* abelian group if it has a generating set
$$X = \{x_1, \dots\}$$
that is *linearly independent* — i.e. any equation

(4.6.1) $$\sum_{i=1}^{n} a_{k_i} x_{k_i} = 0$$

where $a_i \in \mathbb{Z}$, implies that $a_{k_1} = \cdots = a_{k_n} = 0$. If the set X is finite, its order is called the *rank* of A. This generating set is called a *free basis* of A.

REMARK. The word "free" is used because equation 4.6.1 implies that the basis do not satisfy *any* equations other than those dictated by simple logic (and the fact that the groups are abelian). If B is *any* abelian group, this freeness property means that *any* map of the basis elements
$$f(x_i) = b_i \in B$$
gives rise to a *unique* homomorphism
$$f: A \to B$$

(4.6.2) $$f\left(\sum_{i=1}^{n} n_i x_i\right) = \sum_{i=1}^{n} n_i b_i$$

At this point, we come up against the distinction between a direct *sum* (as in definition 4.1.12 on page 38) and direct *product*.

DEFINITION 4.6.2. Given an infinite set of abelian groups $\{A_i\}$
(1) their *direct product*
$$\prod_{i=1}^{\infty} A_i$$
is the set of all infinite sequences
$$(a_1, a_2, \dots)$$
with $a_i \in A_i$.
(2) their *direct sum*
$$\bigoplus_{i=1}^{\infty} A_i$$
is the set of infinite sequences
$$(a_1, a_2, \dots)$$
in which only a *finite* number of the a_i are *nonzero*.

REMARK. When a *finite* number of groups are involved, direct products are identical to direct sums. When an *infinite* number of groups are involved, they become very different and
$$\bigoplus_{i=1}^{\infty} A_i \subset \prod_{i=1}^{\infty} A_i$$

In a *category-theoretic* sense (see chapter 10 on page 339) direct products are *products* (see definition 10.1.1 on page 339) in the category of abelian groups and direct sums are *coproducts* (see definition 10.1.4 on page 342).

Using direct sums, we can *characterize* free abelian groups:

PROPOSITION 4.6.3. *If X is a set and A is a free abelian group with X as a basis, then there exists an isomorphism*

$$f: \bigoplus_{x \in X} \mathbb{Z}_x \to A$$

REMARK. This is true whether X is finite or infinite. Note that $\mathbb{Z}_x$ represents a copy of $\mathbb{Z}$ indexed by $x \in X$.

PROOF. Just define

$$f(n_1, \ldots) = n_1 x_1 + \cdots$$

The fact that the x_i are linearly independent in A implies that this map is injective. The fact that every element of A is a *finite* linear combination of the $x_i \in X$ implies that it is surjective. □

It follows that the infinite direct sum

$$\bigoplus_{i=1}^{\infty} \mathbb{Z}$$

is a free abelian group.

We can also define

DEFINITION 4.6.4. The infinite direct product

$$B = \prod_{i=1}^{\infty} \mathbb{Z}$$

is called the *Baer-Specker group*.

REMARK. In exercises 3 through 6 on page 73, the reader will get a chance to prove that this group is *not* free abelian. This group first appeared in Baer's paper [9] in which he proved it was not free abelian[2]. Specker later published the paper [100] which explored its rather bizarre properties.

COROLLARY 4.6.5. *If A is a free abelian group of rank n, B is a free abelian group of rank m and if*

$$A \cong B$$

is an isomorphism, then $n = m$.

REMARK. In other words the rank of a finitely generated free abelian group is well-defined.

PROOF. The isomorphism $A \cong B$ induces an isomorphism

$$\frac{A}{2A} = \mathbb{Z}_2^n \cong \frac{B}{2B} = \mathbb{Z}_2^m$$

forcing $n = m$. □

The property of freeness of an abelian group is hereditary in the sense that all of its subgroups inherit it:

---

[2] The proof in the exercises is simpler.

4.6. ABELIAN GROUPS

PROPOSITION 4.6.6. *If $S \subset A$ is a subgroup of a free abelian group, then S is free abelian. If* $\operatorname{rank}(A) < \infty$*, then*
$$\operatorname{rank}(S) \leq \operatorname{rank}(A)$$

REMARK. Compare this result with example 4.10.24 on page 103. The essential difference is that A is free *abelian*.

PROOF. Let X be a free basis for A. By the Well Ordering Axiom of set theory (see axiom 14.2.9 on page 465), we can *order* the elements of X
$$x_1 \succ x_2 \succ \cdots$$
so that every subset of X has a *minimal* element. Note that this ordering is completely *arbitrary* — we only use the fact that such an ordering *exists*[3].

If $s \in S$, we have
$$s = \sum n_i x_i$$
— a finite sum. Define the *leading term*, $\lambda(s) = n_\alpha x_\alpha$ — the *highest* ordered term in the sum. Let $Y \subset X$ be the set of basis elements that occur in these leading terms of elements of S. For each $y \in Y$, there is at least one $s \in S$ whose leading term is $n \cdot y$. Let

(4.6.3) $\qquad I(y) = \{0\} \cup \{n \in \mathbb{Z} | \lambda(s) = n \cdot y \text{ for some } s \in S\}$

It is not hard to see that
$$I(y) \subset \mathbb{Z}$$
is a cyclic subgroup (see proposition 4.3.2 on page 41). If n_y is the *positive generator*[4] of $I(y)$, there exists a set of elements $S_y \subset S$ whose leading terms are $n_y y$. The Axiom of Choice (see theorem 14.2.11 on page 465) implies that we can select *one* such element for each $y \in Y$ — i.e. we can define a function
$$f: Y \to S$$
where $f(y)$ is some (arbitrarily selected) element of S_y. We claim that the set
$$B = \{f(y) | y \in Y\}$$
is a basis for S. We must demonstrate two things:

(1) *the set B is linearly independent*. This follows immediately from the fact that its leading terms are multiples of *distinct* basis elements of A. If we have a linear combination
$$\sum_{i=1}^{n} n_i y_i + H = 0$$
where H is the *non*-leading terms. Of those leading terms, one is ordered the *highest* — say $n_\alpha y_\alpha$ and it can *never* be canceled out by any other term. It follows that $n_\alpha = 0$ and a similar argument applied to the remaining terms eventually leads to the conclusion that *all* of the coefficients are 0.

---

[3]If X is finite, this is completely *trivial*; if $X = \mathbb{R}$, not so much. The hard part is that every subset must have a *minimal* element. See remark 14.2.10 on page 465.

[4]Both n_y and $-n_y$ are generators of the same cyclic subgroup of $\mathbb{Z}$.

(2) *the set B generates S*. Suppose *not* and suppose $g \in S \setminus \langle B \rangle$ has the property that $\lambda(g) = my$ for $m \in \mathbb{Z}$ and $y \in Y$ is minimal (in the ordering on X — axiom 14.2.9 on page 465). Then $m \in I(g)$ (see equation 4.6.3 on the preceding page) so that $m = k \cdot n_g$ and $k \cdot f(y) \in \langle B \rangle$ also has leading term my. It follows that
$$g - k \cdot f(y) \in S \setminus \langle B \rangle$$
has a leading term *strictly lower* than that of g — a contradiction.

□

It is interesting that *every* abelian group is a quotient of a free abelian group:

PROPOSITION 4.6.7. *If A is any abelian group, there exists a surjective homomorphism*
$$f: F \to A$$
where F is free abelian. If A is finitely generated then F can be of finite rank.

REMARK. Proposition 4.6.6 on the previous page implies that the kernel of this map is also free so proposition 4.4.8 on page 45 implies that every abelian group is a quotient of a free abelian group by a free abelian subgroup.

PROOF. Just define F to be the free abelian group on the basis $A \setminus \{0\}$. The map f sends the *basis element* $a \in A \setminus \{0\}$ to $a \in A$, and we extend this map of basis elements to all of F via an analogue to equation 4.6.2 on page 59.

If A has a finite generating set $S = \{a_1, \ldots, a_n\}$, we can define F to be a free abelian group whose *basis* is precisely S. The same argument as that used above constructs the homomorphism f.

□

Our structure-theorem for finitely generated abelian groups is:

THEOREM 4.6.8. *If A is a finitely generated abelian group, then*
$$(4.6.4) \qquad A \cong F \oplus \mathbb{Z}_{n_1} \oplus \cdots \oplus \mathbb{Z}_{n_k}$$
where F is free-abelian and $n_1 \mid n_2 \mid \cdots \mid n_k$.

REMARK. Finite generation is essential to this result. When an abelian group is not finitely generated the result fails — see example 4.6.14 on page 67.

PROOF. Let $\{a_1, \ldots, a_n\}$ be a generating set for A. Define G to be the free-abelian group on basis $\{x_1, \ldots, x_n\}$ and define the surjective homomorphism
$$f: G \to A$$
$$x_i \mapsto a_i$$

Proposition 4.6.6 on the previous page implies that the *kernel*, K is also free abelian on some basis $\{y_1, \ldots, y_m\}$, where $m \leq n$ and proposition 4.4.8 on page 45 implies that
$$A \cong \frac{G}{K}$$

If
$$y_i = \sum_{j=1}^{n} a_{i,j} x_j$$
the $m \times n$ matrix $M = [a_{i,j}]$ expresses how K fits in G and the quotient. We consider ways of "simplifying" this matrix in such a way that the quotient G/K remains unchanged:

(1) Rearranging the rows of M corresponds to re-labeling the y_i, so it leaves G/K unchanged.
(2) Rearranging the columns of M corresponds to re-labeling the x_j, leaving G/K unchanged.
(3) Multiplying a row by -1 defines an automorphism of K (see definition 4.2.3 on page 40), leaving G/K unchanged.
(4) Multiplying a column by -1 defines an automorphism of G, leaving G/K unchanged.
(5) Adding a multiple of the ith row to the jth row (with $i \neq j$) defines an automorphism of K — we replace the basis $\{y_1, \ldots, y_m\}$ by $\{y_1, \ldots, y'_j, \ldots, y_m\}$ where $y'_j = y_j + n \cdot y_i$, where n is the multiplying factor. This is invertible because $y_j = y'_j - n \cdot y_i$ so it defines an isomorphism that doesn't change the subgroup $H \subset G$ or the quotient.
(6) For the same reason, adding a multiple of the ith column to the jth column (with $i \neq j$) defines an automorphism of G that leaves the quotient, G/K, unchanged.

It follows that we can perform *all* of the operations listed above to M without changing the quotient in any way.

We'll call this part of our algorithm, Diagonalization:

This consists of two phases: *row-reduction* and *column-reduction*.

We will call the $(1,1)$ position in the matrix, the *pivot*.

By performing operations 1, 2, and 3 if necessary, we can assume that $a_{1,1} > 0$. Now we scan down column 1.

(1) If $a_{1,1} | a_{k,1}$ with $k > 1$, we subtract $(a_{1,1}/a_{k,1}) \times$ row 1 from row k, replacing $a_{k,1}$ by 0.
(2) If $a_{1,1} \nmid a_{k,1}$, perform *two* steps:
  (a) write $a_{k,1} = q \cdot a_{1,1} + r$ with $0 < r < a_{1,1}$. Now subtract $q \times$ row 1 from row k, replacing $a_{k,1}$ by r.
  (b) Swap row 1 with row k so that $a_{1,1}$ has been replaced by r.

We perform step 2 until the matrix-entry in position $(k, 1)$ is 0. Since the pivot position decreases every time we do step 2, we can only do it a finite number of times. The astute reader will recognize repetition of step 2 above as the Euclidean algorithm (see algorithm 3.1.4 on page 14) for computing the greatest common divisor.

The effect is[5]:
$$a_{1,1} \to \gcd(a_{1,1}, a_{k,1})$$
$$a_{k,1} \to 0$$

---

[5]Where the symbol '$\to$' means "is replaced by."

Repeating these operations for all lower rows ultimately clears all elements of column 1 below $a_{1,1}$ to 0. This completes row-reduction for column 1.

Now we do column-reduction on row 1, ultimately clearing out all elements to the *right* of $a_{1,1}$ to 0.

After these steps, our matrix looks like

$$\begin{bmatrix} \bar{a}_{1,1} & 0 \\ 0 & \bar{M} \end{bmatrix}$$

where $\bar{M}$ is an $(m-1) \times (n-1)$ matrix.

After recursively carrying out our algorithm above on this and the smaller sub-matrices that result, we get the diagonal form:

$$\begin{bmatrix} \bar{a}_{1,1} & 0 & 0 & \cdots & 0 \\ 0 & \bar{a}_{2,2} & 0 & \cdots & 0 \\ \vdots & \vdots & \ddots & \cdots & \vdots \\ 0 & 0 & 0 & \cdots & 0 \end{bmatrix}$$

where *all* off-diagonal entries are 0 and we have $n - m$ columns of 0's on the right. This matrix means that the new bases for K and F satisfy

$$y_i = \bar{a}_{i,i} \cdot x_i$$

for $1 \le i \le m$. This completes the phase of the algorithm called diagonalization.

Now we may run into the problem $\bar{a}_{i,i} \nmid \bar{a}_{i+1,i+1}$. We can resolve this as follows:

The subgroup of A determined by these two rows of the matrix has $\bar{a}_{i,i} \cdot \bar{a}_{i+1,i+1}$ elements. Now add column $i+1$ to column i to give

$$\begin{bmatrix} \bar{a}_{i,i} & 0 \\ \bar{a}_{i+1,i+1} & \bar{a}_{i+1,i+1} \end{bmatrix}$$

If we row-reduce this, a tedious calculation shows that we get

$$\begin{bmatrix} \delta & u \\ 0 & v \end{bmatrix}$$

where $\delta = \gcd(\bar{a}_{i,i}, \bar{a}_{i+1,i+1})$ and u and v are linear combinations of $\bar{a}_{i,i}$ and $\bar{a}_{i+1,i+1}$. At this point, we subtract $(u/\delta) \times$ column i from column $i+1$ to get

$$\begin{bmatrix} \delta & 0 \\ 0 & v \end{bmatrix}$$

Since the subgroup of A determined by these two rows is unchanged, it still has $\bar{a}_{i,i} \cdot \bar{a}_{i+1,i+1}$ elements. This means $\delta \cdot v = \bar{a}_{i,i} \cdot \bar{a}_{i+1,i+1}$ so

$$v = \frac{\bar{a}_{i,i} \cdot \bar{a}_{i+1,i+1}}{\delta} = \mathrm{lcm}(\bar{a}_{i,i}, \bar{a}_{i+1,i+1})$$

On the other hand, the number of distinct prime factors in δ must be strictly less than those in $\bar{a}_{i,i}$ and $\bar{a}_{i+1,i+1}$ (equality only occurs if $\bar{a}_{i,i}$ and $\bar{a}_{i+1,i+1}$ have the *same* prime factors). So this operation shifts prime factors to the right and down.

After a finite number of steps, $\bar{a}_{1,1}$ must *stabilize* (i.e. remain constant over iterations of this process) since it has a finite number of prime factors. One can say the same of $\bar{a}_{2,2}$ as well. Eventually the entire list of elements
$$\{\bar{a}_{1,1}, \ldots, \bar{a}_{m,m}\}$$
must stabilize, finally satisfying the condition
$$\bar{a}_{1,1} \mid \cdots \mid \bar{a}_{m,m}$$
When we form the quotient
$$A \cong \frac{G}{K} = \frac{\mathbb{Z} \oplus \mathbb{Z} \oplus \cdots \oplus \mathbb{Z}}{\bar{a}_{1,1} \cdot \mathbb{Z} \oplus \bar{a}_{2,2} \cdot \mathbb{Z} \oplus \cdots \oplus \mathbb{Z}}$$
$$= \frac{\mathbb{Z}}{\bar{a}_{1,1} \cdot \mathbb{Z}} \oplus \frac{\mathbb{Z}}{\bar{a}_{2,2} \cdot \mathbb{Z}} \oplus \cdots \oplus \mathbb{Z}$$

▷ each *nonzero* entry $a_{i,i}$ results in a direct summand $\mathbb{Z}_{a_{i,i}} = \mathbb{Z}/\bar{a}_{i,i} \cdot \mathbb{Z}$, if $\bar{a}_{i,i} \neq \pm 1$ and 0 otherwise.
▷ each column of *zeros* contributes a direct summand of $\mathbb{Z}$ to F in equation 4.6.4 on page 62.

□

If an abelian group is *finite*, the F-factor is zero and, in light of exercise 7 on page 40, we conclude that

COROLLARY 4.6.9. *If A is a finite abelian group, then*

(4.6.5) $$A \cong \mathbb{Z}_{p_1^{k_1}} \oplus \cdots \oplus \mathbb{Z}_{p_n^{k_n}}$$

where the p_i are (not necessarily distinct) primes.

PROOF. Simply factor each n_i in equation 4.6.4 on page 62 into powers of primes. □

We summarize and extend these results with:

THEOREM 4.6.10. *If A is a finite abelian group, then*
$$A \cong \prod_i \underbrace{\mathbb{Z}_{p_i} \oplus \cdots \oplus \mathbb{Z}_{p_i}}_{\alpha_{i,1} \text{ factors}} \oplus \underbrace{\mathbb{Z}_{p_i^2} \oplus \cdots \oplus \mathbb{Z}_{p_i^2}}_{\alpha_{i,2} \text{ factors}} \oplus \cdots \oplus \underbrace{\mathbb{Z}_{p_i^{t_i}} \oplus \cdots \oplus \mathbb{Z}_{p_i^{t_i}}}_{\alpha_{i,t_i} \text{ factors}}$$
where the (finite number of) primes, p_i, and the integers $\alpha_{i,j}$ are uniquely determined by A.

REMARK. Of course many of the $\alpha_{i,j}$ may be zero.

PROOF. We get the equation above from equation 4.6.5 by arranging the factors in order by primes and their powers. The only additional things to be proved are the statements regarding the primes p_i and the integers $\alpha_{i,j}$.

If p and q are distinct primes, then for any integers $i, k > 0$
$$q^k \times : \mathbb{Z}_{p^i} \to \mathbb{Z}_{p^i}$$

is an *isomorphism* (see exercise 3 on page 42), so its kernel is 0. In addition, the kernel of
$$p \times : \mathbb{Z}_{p^k} \to \mathbb{Z}_{p^k}$$
is isomorphic to $\mathbb{Z}_p$ (generated by $p^{k-1} \in \mathbb{Z}_{p^k}$).

Consider the map
$$p_i^j \times : A \to A$$
and call its kernel $A_{p_i^j}$ and its image $p_i^j A$. Factors corresponding to primes *other* than p_i will *not* appear in the kernels of these multiplication-maps because they are mapped *isomorphically*. Then any isomorphism
$$f: A \to A'$$
induces isomorphisms
$$f|A_{p_i^j}: A_{p_i^j} \to A'_{p_i^j}$$
$$\bar{f}: p_i^j A \to p_i^j A'$$

Now note that
$$A_{p_i} \cong \underbrace{\mathbb{Z}_{p_i} \oplus \cdots \oplus \mathbb{Z}_{p_i}}_{\alpha_{i,1} + \alpha_{i,2} + \cdots \text{ factors}}$$
because every factor of $\mathbb{Z}_{p_i^j}$ (regardless of j) gives rise to a copy of $\mathbb{Z}_{p_i}$ in the kernel. Multiplying by p_i kills off all copies of $\mathbb{Z}_{p_i^1}$ and we get
$$(p_i A)_{p_i} \cong \underbrace{\mathbb{Z}_{p_i} \oplus \cdots \oplus \mathbb{Z}_{p_i}}_{\alpha_{i,2} + \alpha_{i,3} + \cdots \text{ factors}}$$

In general
$$|(p_i^{j-1} A)_{p_i}| = p_i^{\alpha_{i,j} + \alpha_{i,j+1} + \cdots}$$
so we can compute $\alpha_{i,j}$ via
$$p_i^{\alpha_{i,j}} = \frac{|(p_i^{j-1} A)_{p_i}|}{|(p_i^j A)_{p_i}|}$$

Any isomorphism of A will map all of these groups derived from it isomorphically, and preserve *all* of the $\alpha_{i,j}$. □

Given any positive integer, n, we could list all of the isomorphism classes of abelian groups of order n. For instance, if $n = 8$ we get
$$\mathbb{Z}_2 \oplus \mathbb{Z}_2 \oplus \mathbb{Z}_2$$
$$\mathbb{Z}_4 \oplus \mathbb{Z}_2$$
$$\mathbb{Z}_8$$

DEFINITION 4.6.11. A group, G, is *torsion free* if it has no elements of finite order.

Theorem 4.6.8 on page 62 immediately implies that

PROPOSITION 4.6.12. *A finitely generated torsion free abelian group is free.*

REMARK. This *requires* finite generation. For instance $\mathbb{Q}$ under *addition* is torsion free but definitely *not* free abelian: If $\{q_1, q_2, \dots\}$ is *any* generating set with $q_i \in \mathbb{Q}$, let $q_1 = a_1/b_1$ and $q_2 = a_2/b_2$ where $a_1, a_2, b_1, b_2 \in \mathbb{Z}$. Then
$$(b_1 a_2) \cdot q_1 - (b_2 a_1) \cdot q_2 = 0$$
which implies that this arbitrary generating set is *not* linearly independent.

We need the following result to prove several other things in the future:

LEMMA 4.6.13. *Let G be an abelian group with subgroups A and B. Then we have an isomorphism*
$$\frac{A \oplus B}{((1 \oplus -1) \circ \Delta)(A \cap B)} \cong A + B \subset G$$
where
$$\Delta \colon A \cap B \to A \oplus B$$
is the diagonal embedding defined by $\Delta(x) = x \oplus x$ for all $x \in A \cap B$, and
$$(1 \oplus -1) \colon A \oplus B \to A \oplus B$$
sends (a, b) to $(a, -b)$ for all $a \in A$ and $b \in B$.

REMARK. In particular, whenever $G = A + B$ and $A \cap B = 0$, $G \cong A \oplus B$.

In the nonabelian case, this result *fails* — see definition 4.7.15 on page 79.

PROOF. Consider the map
$$A \oplus B \to A + B$$
defined by
$$(a, b) \mapsto a + b$$
The kernel of this map is
$$(c, -c)$$
where $c \in A \cap B$, so Proposition 4.4.8 on page 45 implies
$$A + B = \frac{A \oplus B}{((1 \oplus -1) \circ \Delta)(A \cap B)}$$
$\square$

When groups are *not* finitely generated, their behavior becomes more complicated. For instance we have a counterexample to theorem 4.6.8 on page 62:

EXAMPLE 4.6.14. Let p be a prime number and define
$$G = \prod_{k=1}^{\infty} \mathbb{Z}_{p^{2k}}$$
This group is not torsion because it has elements of infinite order like
$$(1, 1, 1, \dots)$$
We claim that its torsion subgroup, tG, is *not* a direct summand. If so, there would be a subgroup of G isomorphic to G/tG (as per lemma 4.6.13).

Note that all elements of G are divisible by, at most, a *finite* power of p, i.e.
$$x = (n_1 p^{k_1}, n_2 p^{k_2}, \ldots)$$
where all of the n_i are relatively prime to p, is a multiple of p^n with $n = \min(k_1, \ldots)$ and no higher power of p. On the other hand, we claim there are elements of G/tG divisible by *arbitrarily* high powers of p:

Let $x = (p, p^2, \ldots, p^k, \ldots)$, let $j > 0$ be an integer, and let $y = (y_1, \ldots, y_k, \ldots)$
$$y_k = \begin{cases} p^{k-j} & \text{if } k \geq j \\ 0 & \text{otherwise} \end{cases}$$
Then
$$x - p^j y = (p, p^2, \ldots, p^{2j-2}, 0, 0, \ldots)$$
has order p^{2j-2} and is, therefore, torsion. It follows that the element of G/tG represented by x cannot be any element of G.

DEFINITION 4.6.15. If G is a torsion abelian group and p is a prime, its *p-primary component*, G_p is the subgroup of elements $x \in G$ such that $p^k \cdot x = 0$ for some integer $k > 0$.

We can prove a limited version of theorem 4.6.10 on page 65 for groups that may not be finitely generated:

THEOREM 4.6.16. *If G is a torsion abelian group, then G is a direct sum of its p-primary components for all primes p:*
$$G \cong \bigoplus_p G_p$$

PROOF. Suppose x has order
$$n = p_1^{k_1} \cdots p_t^{k_t}$$
The integers
$$n_i = \underbrace{p_1^{k_1} \cdots p_t^{k_t}}_{\text{omit } i^{\text{th}} \text{ factor}}$$
have a greatest common divisor of 1 so that there exist integers m_i such that
$$1 = \sum m_i n_i$$
We conclude that
$$x = \sum m_i n_i \cdot x$$
and $n_i \cdot x$ has order $p_i^{k_i}$. It follows that
$$G = \sum_p G_p$$
The conclusion follows from lemma 4.6.13 on the previous page and the fact that $G_p \cap G_{p'} = 0$ whenever p, p' are *distinct* primes. □

4.6.1. Divisible groups.
This is a class of groups that is used in category theory and cohomology. All groups will be assumed to be abelian.

DEFINITION 4.6.17. A group G is *divisible* if every element is an arbitrary multiple of some other element, i.e., if $x \in G$ and $n > 0$ is an integer, then
$$x = ny$$
for some $y \in G$.

REMARK. For example, $\mathbb{Q}$, $\mathbb{R}$, and $\mathbb{Q}/\mathbb{Z}$ are clearly divisible.

PROPOSITION 4.6.18. *Divisible groups have the following properties*
 (1) sums and products of divisible groups are divisible, and
 (2) quotients of divisible groups are divisible

PROOF. If
$$D = \prod D_i$$
is a product of divisible groups and $x \in D$ is and element and $n > 0$ is an integer, then
$$x = (d_1, \dots)$$
and each component, d_i, of x can be 'divided' by n, so the same is true of x. If all but a finite number of the d_i are 0, the same reasoning applies (where the result of 'dividing' 0 by n is 0).

If D is divisible with a subgroup, S, and
$$G = \frac{D}{S}$$
we can 'divide' $x \in G$ by lifting it to D, dividing it there, and projecting back to G. □

The following general result will be useful

LEMMA 4.6.19 (Pasting lemma). *Let C be an abelian group with subgroups A and B and*
$$f: A \to C$$
$$g: B \to C$$
are homomorphisms of groups with the property that
$$f|A \cap B = g|A \cap B$$
then there exists a unique extension $f + g$ to , $A + B \subset C$, defined by
$$(f + g)(a + b) = f(a) + g(b)$$
for $a \in A$ and $b \in B$.

REMARK. A corresponding result holds for nonabelian groups but is *much* more complicated.

PROOF. Lemma 4.6.13 on page 67 implies that
$$f \oplus g: A \oplus B \to A + B$$
has kernel
$$((1 \oplus -1) \circ \Delta)(A \cap B)$$
The hypotheses imply that
$$f \oplus g|((1 \oplus -1) \circ \Delta)(A \cap B) = 0$$
so it factors through the quotient. □

The following result gives one of the most significant properties of divisible groups:

THEOREM 4.6.20. *An abelian group D is divisible if and only if it satisfies the "injectivity condition:"*

Any homomorphism $f: A \to D$ from any abelian group, A, extends to any group containing A. In other words, if $A \subset B$ and

$$f: A \to D$$

is a homomorphism, there exists a homomorphism

$$F: B \to D$$

such that

$$F|A = f$$

REMARK. See propositions 10.5.5 on page 363 and 10.5.6 on page 363 for a generalization of this argument.

PROOF. Suppose D satisfies this condition. If $n > 0$ is an integer and $x \in D$, we can define a mapping

$$\begin{aligned} f_x : \mathbb{Z} &\to D \\ 1 &\mapsto x \end{aligned}$$

Since $\mathbb{Z}$ can be regarded as the subgroup $A = n \cdot \mathbb{Z} \subset \mathbb{Z} = B$, the injectivity condition implies that f_x extends to a map

$$F: \mathbb{Z} \to D$$

and $F(1) \in D$ is an element such that $x = n \cdot F(1)$, so D is divisible.

Conversely, suppose D is divisible. Let P be the partially ordered set of elements (C, g) where

$$A \subset C$$

and $g|A = f$. This contains, at least, (A, f). Any tower of elements $\{(C_\alpha, g_\alpha)\}$ has an upper bound

$$\left(\bigcup C_\alpha, \cup g_\alpha \right)$$

Zorn's lemma (14.2.12 on page 465) implies that it has a maximal element $(\bar{C}, \bar{g})$. We claim that $\bar{C} = B$.

If not, there exists $x \in B \setminus \bar{C}$. Consider its image, $\hat{x}$, in the quotient, $B/\bar{C}$. If the order of $\hat{x}$ is *infinite*, then

$$\langle x \rangle \cap \bar{C} = 0$$

so that lemma 4.6.13 on page 67 implies that

$$\bar{C} + \langle x \rangle \cong \bar{C} \oplus \langle x \rangle$$

and we can extend $\bar{g}$ to $\bar{g} \oplus 0: \cong \bar{C} \oplus \langle x \rangle \to D$, contradicting the maximality of $(\bar{C}, \bar{g})$.

On the other hand, suppose $\hat{x}$ has order n in $B/\bar{C}$. Then $nx \in \bar{C}$. Since D is divisible, there exists a $y \in D$ such that $n \cdot y = \bar{g}(n \cdot x)$.

Lemma 4.6.19 on the previous page implies that $\bar{g}$ can be extended to

$$\begin{aligned} g': \bar{C} + \langle x \rangle &\to D \\ x &\mapsto y \end{aligned}$$

which also contradicts the maximality of $(\bar{C}, \bar{g})$. □

This has a number of interesting implications:

COROLLARY 4.6.21. *If D is a divisible group and $D \subset G$, where G is abelian, then*
$$G \cong D \oplus (G/D)$$
In other words, a divisible subgroup of an abelian group is a direct summand.

PROOF. Since D is a subgroup of G, there exists an injective map
$$1 \colon D \to D$$
and theorem 4.6.20 on the facing page implies the existence of a map
$$F \colon G \to D$$
such that $F|D = 1$ — which implies that $F^2 = F$ (since the image of F lies in D).

If $x \in G$, then $F(x - F(x)) = F(x) - F^2(x) = 0$ so $x - F(x) \in \ker F$. It follows that
$$G = D + \ker F$$
Since $D \cap \ker F = 0$, lemma 4.6.13 on page 67 implies the conclusion. □

This will allow us to classify all divisible groups

COROLLARY 4.6.22. *Let D be a divisible abelian group and let tD be its torsion subgroup. Then tD is divisible and*

(4.6.6) $$D \cong tD \oplus (D/tD)$$

where D/tD is a direct sum of copies of $\mathbb{Q}$.

PROOF. Suppose $x \in tD$. Then $m \cdot x = 0$ for some integer m. Since D is divisible, for any integer $n > 0$, there exists a $y \in D$ such that $x = n \cdot y$ — and $mn \cdot y = 0$. It follows that tD is divisible and corollary 4.6.21 implies equation 4.6.6.

Proposition 4.6.18 on page 69 implies that D/tD is *also* divisible.

We claim that elements of D/tD are uniquely divisible in the sense that, if $x = ny_1 = ny_2$ then $y_1 = y_2$. This true because $n(y_1 - y_2) = 0$, so $y_1 - y_2$ must be torsion — but D/tD is torsion free. We can define an action of $\mathbb{Q}$ on D/tD via
$$\frac{n}{m} \cdot x = n \cdot y$$
where $m \cdot y = x$. It follows that D/tD is a $\mathbb{Q}$-*vector space* (see definition 6.2.1 on page 165) and a direct sum of copies of $\mathbb{Q}$. □

Now we consider divisible groups that are *torsion*.

DEFINITION 4.6.23. If G is an abelian group and n is an integer, let $G[n]$ denote the kernel of
$$f = n\times \colon G \to G$$

The Prüfer groups, $\mathbb{Z}/p^\infty$ (see example 4.1.10 on page 38), play a vital part in the classification of divisible groups:

LEMMA 4.6.24. *If p is a prime and D is a divisible p-group (i.e., one where every element is of order p^k for some $k > 0$) with*
$$D[p] = \bigoplus_{i=1}^{n} \mathbb{Z}_p$$
then
$$D \cong \bigoplus_{i=1}^{n} \mathbb{Z}/p^\infty$$

PROOF. It is not hard to see that $(\mathbb{Z}_{p^\infty})[p] = \mathbb{Z}_p$ so
$$\left(\bigoplus_{i=1}^n \mathbb{Z}/p^\infty\right)[p] = \bigoplus_{i=1}^n \mathbb{Z}_p$$
Since $\mathbb{Z}_{p^\infty}$ is divisible, the isomorphism
$$D[p] \to \bigoplus_{i=1}^n \mathbb{Z}_p \subset \bigoplus_{i=1}^n \mathbb{Z}/p^\infty$$
extends to a map
$$F: D \to \bigoplus_{i=1}^n \mathbb{Z}/p^\infty$$
If the kernel is K, we get $K[p] = K \cap D[p] = 0$ since the original map is an isomorphism. We claim that this forces K to be 0. If $x \neq 0 \in K$ then $p^t \cdot x = 0$ for some integer t — assume t is *minimal* in this respect. Then $p^{t-1} \cdot x \in K[p] = 0$, which contradicts the minimality of t.

The image $F(D) \subset \bigoplus_{i=1}^n \mathbb{Z}/p^\infty$ is divisible, so corollary 4.6.21 on the previous page implies that
$$\bigoplus_{i=1}^n \mathbb{Z}/p^\infty = F(D) \oplus W$$
where W is some other (divisible) subgroup of $\bigoplus_{i=1}^n \mathbb{Z}/p^\infty$. If $x \neq 0 \in H$, reasoning like that used above implies that $x = 0$. □

This immediately leads to

THEOREM 4.6.25. *If D is a divisible group, then D is a direct sum of copies of $\mathbb{Q}$ and copies of $\mathbb{Z}/p^\infty$ for primes p.*

PROOF. Corollary 4.6.22 on the preceding page implies the immediate split of D into torsion free and torsion components, and theorem 4.6.16 on page 68 implies that the torsion subgroup splits into a direct sum of its p-primary components for all primes p. Lemma 4.6.24 on the preceding page classifies the p-primary divisible subgroups as direct sums of copies of $\mathbb{Z}/p^\infty$. □

DEFINITION 4.6.26. If D is a divisible group let
(1) $\dim_\mathbb{Q} D$ be the dimension of the torsion-free summand (see corollary 4.6.22 on the previous page).
(2) for a prime, p, $\dim_p D = \dim D[p]$ (see definition 4.6.23 on the preceding page and lemma 4.6.24 on the previous page).

We can characterize divisible groups now:

THEOREM 4.6.27. *Two divisible groups, D_1 and D_2 are isomorphic if and only if*
(1) $\dim_\mathbb{Q} D_1 = \dim_\mathbb{Q} D_2$, *and*
(2) $\dim_p D_1 = \dim_p D_2$ *for all primes p.*

EXERCISES.

1. List all isomorphism classes of abelian groups of order 60.

2. Show that the group
$$Q^+ = \{q \in \mathbb{Q} | q > 0\}$$
where the operation is multiplication, is free abelian (of infinite rank).

3. If a free abelian group, A, is uncountable, show that $A/2A$ is *also* uncountable.

4. Let $S \subset B$ be the subgroup of the Baer-Specker group whose entries are eventually divisible by arbitrarily high powers of 2 — i.e. S consists of infinite sequences of integers
$$(n_1, \dots)$$
such that, for any integer k, there exists an m such that $2^k \mid n_i$ for all $i > m$. Show that S is uncountable.

5. If S is the subgroup of the Baer-Specker group defined in exercise 4, compute $S/2S$.

6. Prove that the Baer-Specker group is *not* free abelian.

7. Show that the group $\mathbb{Z}/p^\infty$ defined in example 4.1.10 on page 38 is divisible (see definition 4.6.17 on page 69).

8. Show that
$$\frac{\mathbb{Q}}{\mathbb{Z}} \cong \bigoplus_p \mathbb{Z}/p^\infty$$
where the direct sum is taken over all primes, p.

4.7. Group-actions

Group theory can be used to count complex symmetries and combinatorial structures.

We begin with

DEFINITION 4.7.1. If S is a set and G is a group, an *action* of G on S is a map
$$f: G \times S \to S$$
written as $f(g, s) = g \cdot s$ with the property that
$$g_1 \cdot (g_2 \cdot s) = (g_1 g_2) \cdot s$$
for all $s \in S$ and all $g_1, g_2 \in G$.

A group-action will be called *transitive* if, given any $s_1, s_2 \in S$, there exists some $g \in G$ such that $g \cdot s_1 = s_2$.

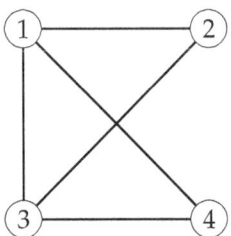

FIGURE 4.7.1. A simple graph

REMARK. The simplest example of this is G acting on itself by multiplication.

A more subtle example of a group action is G acting on itself via *conjugation*:
$$g \cdot x = x^g$$
for $x, g \in G$.

EXAMPLE 4.7.2. The symmetric group S_n acts on the set $\{1, \ldots, n\}$ by permuting them. This action is transitive since, given any two elements $a, b \in \{1, \ldots, n\}$, there is a permutation (many, in fact), $\sigma \in S_n$ such that $\sigma(a) = b$.

In order to give another good example of group-actions, we need the following combinatorial concept:

DEFINITION 4.7.3. A *graph*, (V, E), consists of a set of *vertices*, V, and a set of pairs of vertices $E = \{(v_1, w_1), \ldots\}$, called *edges*. If $e = (v, w) \in E$, the vertices v and w are called the *ends* of e.

Two graphs (V_1, E_1) and (V_2, E_2) are *isomorphic* if there is bijective function
$$f: V_1 \to V_2$$
such that $(f(v), f(w)) \in E_2$ if and only if $(v, w) \in E_1$.

REMARK. Graphs are used to represent networks in general: computer networks, networks of roads, pipes carrying oil, etc. Isomorphic graphs are essentially identical.

Graphs can easily be represented by diagrams — see figure 4.7.1. If a graph. $G = (V, E)$, has n vertices, the symmetric group, S_n, acts on it by permuting the vertices and edges. The result of this group-action is another graph, G', that is *isomorphic* to G.

DEFINITION 4.7.4. If G is a finite group acting on a set S, the *orbit of an element* $s \in S$ is defined to be
$$\text{Orbit}(s) = \{g \cdot s | \forall g \in G\}$$
and the *stabilizer* of $s \in S$ is defined by
$$G_s = \{g \in G | g \cdot s = s\}$$
Given an element $g \in G$, we also define
$$S^g = \{s \in S | g \cdot s = s\}$$

— the *invariant set* of g.

The *set of orbits* is written as S/G.

REMARK. If T is the set of all possible graphs with n vertices, the symmetric group S_n, acts on this by permuting vertices. The *orbit* of a graph within T is just the set of graphs *isomorphic* to it (even though the vertex-numbers that appear in its list of edges may vary).

The stabilizer of a graph consists of all permutations that leave all of the two-element subsets in E unchanged. If there are isolated vertices (i.e., vertices that are not endpoints of any edges), they can be permuted arbitrarily without changing the graph.

EXAMPLE 4.7.5. If $H \subset G$ is a subgroup of a group, H acts on G by right-multiplication. The orbit of an element $g \in G$ is just its coset gH. The stabilizer of $g \in G$ is the set of elements $h \in H$ such that $gh = g$, i.e. just $1 \in H$.

PROPOSITION 4.7.6. *If $f: G \times S \to S$ is a group-action on a set and $s_1, s_2 \in S$ are two elements such that*

$$\mathrm{Orbit}(s_1) \cap \mathrm{Orbit}(s_2) \neq \emptyset$$

then

$$\mathrm{Orbit}(s_1) = \mathrm{Orbit}(s_2)$$

It follows that the group-action partitions S into a disjoint union of orbits

$$S = \bigsqcup_{i=1}^{|S/G|} \mathrm{Orbit}(s_i)$$

PROOF. If $x \in \mathrm{Orbit}(s_1) \cap \mathrm{Orbit}(s_2)$, then

$$x = g_1 \cdot s_1 = g_2 \cdot s_2$$

for $g_1, g_2 \in G$. It follows that $g_2^{-1} g_1 \cdot s_1 = s_2$, so $s_2 \in \mathrm{Orbit}(s_1)$ and $\mathrm{Orbit}(s_2) \subset \mathrm{Orbit}(s_1)$. A similar argument shows that $s_1 \in \mathrm{Orbit}(s_2)$. □

It is not hard to show that

PROPOSITION 4.7.7. *If G acts on a set S and $s \in S$, then*

$$|\mathrm{Orbit}(s)| = [G: G_s] = \frac{|G|}{|G_s|}$$

or

$$|\mathrm{Orbit}(s)| \cdot |G_s| = |G|$$

PROOF. We will show that there is a 1-1 correspondence between the cosets of G_s and elements of the orbit. If $g_1 \cdot s = g_2 \cdot s$ then $g_1^{-1} g_2 \cdot s = s$ so that $g_1^{-1} g_2 \in G_s$ and $g_1 G_s = g_2 G_s$. On the other hand, if $g_1 G_s \neq g_2 G_s$, it is not hard to see that $g_1 \cdot s \neq g_2 \cdot s$. □

In many cases, we are most interested in the *number* of orbits of a group-action:

LEMMA 4.7.8 (Burnside's Lemma). *If G is a finite group and S is a finite set with group-action $f\colon G \times S \to S$ then*

$$|S/G| = \frac{1}{|G|} \sum_{g \in G} |S^g|$$

REMARK. Although this result is called Burnside's Lemma, it was not due to him — he mentioned it in his 1897 book [23] and attributed it to Frobenius. It was also known to Cauchy, many years earlier. This is often cited as an example of Stigler's law of eponymy: "No scientific discovery is named after its original discoverer."

Sometimes it's called "The Lemma *not* due to Burnside."

> William Burnside (1852 – 1927) was an English mathematician and researcher in the theory of finite groups. Burnside's main area of research was in group representations. For many years Burnside's classic, [23], was the standard text on group theory. One of Burnside's best known contributions to group theory is his $p^a q^b$ theorem, which proves that every finite group whose order is divisible by fewer than three distinct primes is solvable (see chapter 8 on page 297 for this concept). He also posed *Burnside's Problem* (see 4.10.1 on page 87) in combinatorial group theory.

> Baron Augustin-Louis Cauchy (1789 – 1857) was a French mathematician, engineer and physicist who made pioneering contributions to several branches of mathematics, including: mathematical analysis and continuum mechanics. He almost singlehandedly founded complex analysis and the study of permutation groups in abstract algebra.

PROOF. We begin by expressing a sum over elements of G as a sum over elements of S:

$$\sum_{g \in G} |S^g| = |\{(g,s) \in G \times S | g \cdot s = s\}| = \sum_{s \in S} |G_s|$$

Now we use proposition 4.7.7 on the preceding page to conclude that $|G_s| = |G|/|\operatorname{Orbit}(s)|$ so we get

$$\sum_{g \in G} |S^g| = \sum_{s \in S} \frac{|G|}{|\operatorname{Orbit}(s)|} = |G| \cdot \sum_{s \in S} \frac{1}{|\operatorname{Orbit}(s)|}$$

Since the group-action partitions S into *disjoint orbits* (see proposition 4.7.6 on the previous page), we can rewrite this as

(4.7.1) $$\sum_{g \in G} |S^g| = |G| \cdot \sum_{z \in G/S} \left(\sum_{s \in z} \frac{1}{|z|} \right)$$

where z runs over all of the *orbits* in S. Now, simply note that, if z is an orbit of the G-action then

$$\sum_{s \in z} \frac{1}{|z|} = 1$$

so equation 4.7.1 on the facing page becomes

$$\sum_{g\in G}|S^g|=|G|\cdot\sum_{z\in G/S}1=|G|\cdot|G/S|$$

which proves the result. □

EXAMPLE 4.7.9. Now we answer the question: how many *distinct* ways of coloring a square with five colors exist?

The colors are Red, Green, Yellow, Blue, and Orange and $c(i)$ is the color of vertex i, where $i = 1,\ldots,4$. We regard two squares to be colored the same if one can be rotated or flipped so its colors coincide with those of the other square, i.e. the sets of colors are mapped into each other by the action of D_4 — see figure 4.1.1 on page 37.

(1) If $g = 1 \in D_4$, then the invariant set consists of all possible colors on the 4 vertices, so $|S^1| = 5^4$.
(2) If $g = R \in D_4$, then a coloring is invariant under g if and only if all vertices are colored the same. It follows that $|S^R| = 5$. The same is true for $|S^{R^3}| = 5$.
(3) If $g = R^2$, then vertex 1 must be the same color as vertex 3 and vertex 2 must be the same color as vertex 4. It follows that $|S^{R^2}| = 5^2$.
(4) If $g = d_1 \in D_4$, then vertex 1 must be the same color as vertex 3 and no other conditions need be satisfied. It follows that $|S^{d_1}| = 5^3$. A similar argument shows that $|S^{d_2}| = 5^3$.
(5) If $g = c_1 \in D_4$, then vertex 1 must be the same color as vertex 4 and vertex 2 must be the same color as vertex 3. It follows that $|S^{c_1}| = 5^2$. A similar argument shows that $|S^{c_2}| = 5^2$.

Combining all of this together, we get that the number of *distinct* colorings of the square are

$$\frac{1}{8}\left(5^4 + 2\cdot 5^3 + 3\cdot 5^2 + 2\cdot 5\right) = 120$$

EXAMPLE 4.7.10. Now we will apply Burnside's Lemma to determining the number of *isomorphism classes* of graphs with a bounded number of vertices. Suppose we limit the number of vertices to 4. In this case, S is the number possible graphs on four vertices and the group acting on it is S_4.

▷ $S^1 = S$ and we can determine the size of this set by noting possible edges involve pairs of vertices. There are $\binom{4}{2} = 6$ and each such pair can either *have* an edge or *not* have one. This gives rise to $2^6 = 64$ possible graphs.
▷ S_4 has $\binom{4}{2} = 6$ two-cycles. If a graph is invariant under the two-cycle (a,b) (with $a < b$ taken from the set $\{1,2,3,4\}$) then vertices a and c are connected if and only if vertices b and c are connected. The possibilities are
  • a is connected to b or not: ×2
  • c is connected to d or not: ×2
  • a and b are both connected to c or not: ×2

- a and b are both connected to d or not: $\times 2$

This gives rise to $2^4 = 16$ possibilities.
▷ S_4 has 3 disjoint two-cycles $(a,b)(c,d)$. The possibilities are:
- a is connected to b or not: $\times 2$
- c is connected to d or not: $\times 2$
- a is connected to d *and* b is connected to c or not: $\times 2$
- a is connected to c *and* b is connected to d or not: $\times 2$

Again, we get $2^4 = 16$ possibilities.
▷ S_4 has 8 three-cycles that look like (a,b,c). The possibilities are:
- vertices a,b,c are *all* connected to each other or not: $\times 2$
- vertices a,b,c are *all* connected to vertex d or not: $\times 2$

We get 4 possibilities.
▷ S_4 has 6 cycles of length four (a,b,c,d) — the first entry is 1 and the other three entries are permutations of $\{2,3,4\}$ of which there are 3! possibilities. The possibilities are:
- a is connected to b which is connected to c which is connected to d or not: $\times 2$
- each vertex is connected to the other 3 or not: $\times 2$

We get 4 possibilities in this case.

The number of isomorphism classes of graphs with 4 vertices is, therefore,

$$\frac{1}{|S_4|} \sum_{\sigma \in S_4} |S^\sigma| = \frac{1}{24}(1 \cdot 64 + 6 \cdot 16 + 3 \cdot 16 + 8 \cdot 4 + 6 \cdot 4) = 11$$

We conclude this section with development of the *class equation*.

PROPOSITION 4.7.11. *If G is a group with element x, y the relation*

$$x \sim y$$

— *called conjugation — if $x = y^g$ for some $g \in G$ is an equivalence relation.*

PROOF. Certainly $x \sim x$ because $x = x^1$. If $x \sim y$, $x = y^g$ it is not hard to see that $y = x^{g^{-1}}$, so $y \sim x$. If $x \sim y$ and $y \sim z$ then $x = y^{g_1}$ and $y = z^{g_2}$ so $x = y^{g_1 g_2}$. □

It follows that conjugation partitions G into a union of *disjoint* equivalence classes called *conjugacy* classes:

(4.7.2) $$G = \bigsqcup C_i$$

Since the elements of $Z(G)$ commute with *everything* in G, each of them is its *own* conjugacy class. We get an equation

(4.7.3) $$|G| = |Z(G)| + |C_1| + \cdots + |C_k|$$

called the *class equation* of a group. Here $C_1, \ldots, C_k$ are the various conjugacy classes that contain more than one element.

Consider the action of G on itself via *conjugation*. The *orbit* of any element $g \in G$ is just its *conjugacy* class, say C_i as in equation 4.7.2. In this case, define

DEFINITION 4.7.12. If $g \in G$ is an element of a group, we define the *centralizer* of g, denoted $Z_G(g)$ — via
$$Z_G(g) = \{x \in G | xg = gx\}$$

REMARK. The centralizer of g is just the *stabilizer* of g in the group-action via conjugation — i.e. $Z_G(g) = G_g$.

COROLLARY 4.7.13. *If $g \in G$ is an element of a group G, the size of the conjugacy-class of g,*
$$|C_g(G)| = [G : Z_G(g)] = |G|/|Z_G(g)|$$

PROOF. This follows immediately from proposition 4.7.7 on page 75. □

This implies that we can rewrite the class equation, 4.7.3 on the facing page, as

(4.7.4) $\quad |G| = |Z(G)| + [G : Z_G(g_1)] + \cdots + [G : Z_G(g_k)]$

where $g_1, \ldots, g_k \in G$ are representatives of the different conjugacy classes. This new version of the class equation has some interesting implications:

THEOREM 4.7.14 (Burnside's Theorem for *p*-groups). *If p is a prime number and G is a group with $|G| = p^n$ for some integer $n > 0$ then $|Z(G)| > 1$ and is a multiple of p.*

PROOF. Each of the groups $Z_G(g_i)$ is a subgroup of G so $|Z_G(g_i)| \mid |G| = p^n$. If $|Z_G(g_i)| = |G|$, then $Z(G) = Z_G(g_i) = G$ and the result follows.

If $|Z_G(g_i)| < |G|$, then $|Z_G(g_i)| = p^{j_i}$ for some $j_i > 0$ and the class equation (4.7.4) becomes
$$p^n = |Z(G)| + p^{j_1} + \cdots + p^{j_k}$$
from which it follows that $p \mid |Z(G)|$. □

DEFINITION 4.7.15. If G is a group with normal subgroup, N, and another subgroup, H, and a homomorphism
$$\varphi \colon H \to \text{Aut}(N)$$
we define the *semidirect product*
$$G = N \rtimes_\varphi H$$
to have a set of elements $N \times H$ with group-operation defined by
$$(n_1, h_1) \cdot (n_2, h_2) = (n_1 \varphi(h_1)(n_2), h_1 h_2)$$

REMARK. Note that G contains both N and H as subgroups and
$$N \cap H = 1 \in G$$
This shows how lemma 4.6.13 on page 67 fails when the groups are not abelian.

In general, the semidirect product depends on the map
$$\varphi \colon H \to \text{Aut}(N)$$
and different such maps give rise to non-isomorphic semidirect products.

If $\varphi\colon H \to \mathrm{Aut}(N)$ maps all of H to $1 \in \mathrm{Aut}(N)$, then
$$N \rtimes_\varphi H = N \oplus H$$

EXAMPLE 4.7.16. The dihedral group, D_{2n}, satisfies
$$D_{2n} \cong \mathbb{Z}_n \rtimes_\varphi \mathbb{Z}_2$$
where $1 \in \mathbb{Z}_2$ maps to the element of $\mathrm{Aut}(\mathbb{Z}_n)$ that sends each element to its additive inverse, giving rise to the familiar presentation
$$D_{2n} = \left\langle x, y \mid x^2 = 1, y^n = 1, xyx^{-1} = y^{-1} \right\rangle$$

EXERCISES.

1. If G is a group of order p^2 for some prime p, show that G is abelian.

2. Verify that the semidirect product in definition 4.7.15 on the previous page really defines a group.

3. What is the inverse of $(n, h) \in N \rtimes_\varphi H$?

4. In the semidirect product $N \rtimes_\varphi H$, show that conjugation of N by an element $h \in H$, induces the automorphism $\varphi(h)$.

5. If G is a group with subgroups N and H with
    a. $N \triangleleft G$ and abelian
    b. $N \cap H = \{1\}$
    c. $G = NH$.

Show that
$$G = N \rtimes_\varphi H$$
where $\varphi\colon H \to \mathrm{Aut}(N)$ defined by conjugating N by elements of H.

4.8. The Sylow Theorems

4.8.1. Statement. These are deep theorems that are indispensable for classifying finite groups — they form a partial converse to Lagrange's Theorem (4.4.2 on page 43). For small values of n, in many cases, these three theorems allow one to give an exhaustive list of *all possible* groups of order n.

> Peter Ludwig Mejdell Sylow (1832 – 1918) was a Norwegian mathematician who was a high school teacher from 1858 to 1898 and a substitute lecturer at Christiania University in 1862 — where he covered Galois theory. During this time he posed a question that led to his celebrated Sylow Theorems, which he published in 1872.

Sylow's first theorem is:

THEOREM 4.8.1 (First Sylow Theorem). *If G is a group with $|G| = p^k \cdot m$, where p is a prime number and $\gcd(p,m) = 1$, then there exists a subgroup $H \subset G$ with $|H| = p^k$.*

This subgroup, H, is called a *Sylow p-subgroup* of G.

THEOREM 4.8.2 (Second Sylow Theorem). *Given G as above, all Sylow p-subgroups of G are conjugates of each other.*

Sylow's third theorem is

THEOREM 4.8.3 (Third Sylow Theorem). *Given G as above, if n_p is the number of distinct Sylow p-subgroups, then $n_p \mid m$ and $n_p \equiv 1 \pmod{p}$.*

EXAMPLE 4.8.4. In A_4 we have Sylow subgroups of order $2^2 = 4$ and 3. Consider subgroups of order 3: If n_3 is the number of them, the third Sylow theorem states that $n_3 \mid 4$ and $n_3 \equiv 1 \pmod 3$. This means that $n_3 = 1$ or 4. It turns out that A_4 has four cyclic subgroups

$$\langle (1,2,3) \rangle, \langle (1,2,4) \rangle, \langle (1,3,4) \rangle, \langle (2,3,4) \rangle$$

If we consider 2-Sylow subgroups, they must have order $2^2 = 4$ and the number of them, n_2 must satisfy $n_2 \mid 3$ and $n_2 \equiv 1 \pmod 2$ — conditions that could be satisfied by $n_2 = 1$ or 3. It turns out (and the Sylow theorems cannot tell us this) that $n_1 = 1$ and the 2-Sylow subgroup is

$$\langle (1,2)(3,4), (1,4)(2,3) \rangle$$

4.8.2. Preliminary lemmas. In order to prove the Sylow Theorems, we need several lemmas.

LEMMA 4.8.5. *If H and K are subgroups of a finite group G, HK is generally just a set of elements. The size of this set is given by*

(4.8.1) $$|HK| = \frac{|H| \cdot |K|}{|H \cap K|}$$

REMARK. If H or K is a *normal* subgroup of G, then HK is a subgroup. This result works even if neither H nor K are normal.

PROOF. Note that HK is a union of cosets

$$HK = \bigcup_{h \in H} hK$$

Since each coset has $|K|$ elements, it suffices to compute how many distinct cosets there are. Since $h_1 K = h_2 K$ if and only if $h_1^{-1} h_2 \in K$, it follows that $h_1^{-1} h_2 \in H \cap K$ (since it is also in H). Lagrange's theorem implies that the number of distinct such cosets is

$$\frac{|H|}{|H \cap K|}$$

and since the size of these cosets is $|K|$, we get equation 4.8.1. □

We will also need this somewhat odd lemma:

LEMMA 4.8.6. *If G is a finite group with p-Sylow subgroup $P \subset G$ and $Q \subset G$ is any other subgroup of order a power of p, then*

$$Q \cap N_G(P) = Q \cap P$$

REMARK. Recall that $N_G(P)$ is the *normalizer* of P — see definition 4.4.10 on page 46.

PROOF. Set $H = Q \cap N_G(P)$. Since $P \triangleleft N_G(P)$ it follows that $Q \cap P \triangleleft H$. Since $H \subset Q$, the statement of the lemma is equivalent to $H \subset P$. Since $H \subset N_G(P)$ and $P \triangleleft N_G(P)$, it follows that HP is a subgroup of $N_G(P)$ — and also a subgroup of G. Lemma 4.8.5 on the previous page implies that

$$|HP| = \frac{|H| \cdot |P|}{|H \cap P|}$$

Since all of the numbers in this quotient are powers of p, it follows that HP is a p-group containing P. Since P is the *maximal* such p-group, it follows that $HP = P$, which implies that $H \subset P$. □

4.8.3. Proof of the Sylow theorems. We begin with theorem 4.8.1 on the preceding page — the statement that Sylow subgroups *exist*. It is trivially true if $|G| = 1$ so we will do induction on $|G|$. If $|G| = p^k \cdot m$ with $\gcd(p, m) = 1$, we consider two cases:

Case 1: $p \mid |Z(G)|$ In this case, we know that $Z(G)$ has a subgroup, H, of order p (see theorem 4.6.10 on page 65 and proposition 4.3.4 on page 41) which — since it is in the *center* — is *normal*. Form the quotient

$$\frac{G}{H} = G'$$

By the inductive hypothesis, G' has a Sylow p subgroup, S, of order p^{k-1}. The Correspondence Theorem (see exercise 5 on page 48) implies that G has a subgroup of order p^k.

Case 2: $p \nmid |Z(G)|$. In this case, we write out the class equation (see equation 4.7.4 on page 79):

$$|G| = |Z(G)| + [G : Z_G(g_1)] + \cdots + [G : Z_G(g_n)]$$

If $p \mid [G : C(g_i)]$ for all i, then $|Z(G)|$ is a linear combination of multiples of p so $p \mid |Z(G)|$. Since we are assuming that this is not true, we conclude that $p \nmid [G : Z_G(g_i)]$ for at least one value of i. If $|Z_G(g_i)| = p^j \cdot \ell$, with $\gcd(p, \ell) = 1$ we get

$$[G : Z_G(g_i)] = \frac{|G|}{|Z_G(g_i)|} = p^{k-j} \cdot \frac{m}{\ell}$$

and since $p \nmid [G : Z_G(g_i)]$, we have $k = j$. Since $|Z_G(g_i)| < |G|$ induction implies that $Z_G(g_i)$ has a subgroup of order p^k — which is *also* a Sylow p-subgroup of G.

Now we prove theorem 4.8.3 on the previous page and 4.8.2 on the preceding page. Let S be a Sylow p-subgroup of G (which we now know exists) and let

$$Z = \{S_1, \ldots, S_r\}$$

be the set of all conjugates of S. In addition, let T be a p-subgroup of G and let
$$Z = \text{Orbit}_1 \cup \cdots \cup \text{Orbit}_s$$
where Orbit_i are the orbits (see definition 4.7.4 on page 74) of the action of T on the S_j by conjugation — so that
$$r = |\text{Orbit}_1| + \cdots + |\text{Orbit}_s|$$
and renumber the elements of Z so that the first s terms are representatives of distinct orbits of T. It follows from lemma 4.7.7 on page 75 that
$$|\text{Orbit}_i| = [T: N_T(S_i)]$$
By definition, $N_T(S_i) = N_G(S_i) \cap T$, so lemma 4.8.6 on the preceding page implies that $N_T(S_i) = S_i \cap T$ and we get
$$|\text{Orbit}_i| = [T: T \cap S_i]$$
Since T was an *arbitrary* p-subgroup of G, we can set $T = S_1$ and
$$|\text{Orbit}_1| = 1$$
Since $S_1 \neq S_i$ for $i > 1$, we get $|S_1 \cap S_i| < |S_1|$ if $i > 1$, which implies that
$$|\text{Orbit}_i| = [S_1: S_1 \cap S_i] > 1$$
for $i > 1$. Since S_1 and S_i are p-groups, we conclude that $p \,|\, [S_1: S_1 \cap S_i]$ and $|\text{Orbit}_i|$ for $i > 1$.

We conclude that
$$r = |\text{Orbit}_1| + (|\text{Orbit}_2| + \cdots + |\text{Orbit}_s|) \equiv 1 \pmod{p}$$

Now we are in a position to prove that all Sylow subgroups are conjugates of each other. Suppose T is a p-subgroup of G that is not contained in any of the S_i. Then the argument used above implies that
$$|\text{Orbit}_i| = [T: T \cap S_i] > 1$$
for all i — and p divides it. This implies that $p \,|\, r$, which contradicts the fact that $r \equiv 1 \pmod{p}$.

It follows that *all p-subgroups* are contained in conjugates of this *one* Sylow subgroup. Since they have the same number of elements, all Sylow subgroups must be on this list of conjugates of one Sylow subgroup.

The *final statement* to be proved is that $r \,|\, m$:

Since r is the size of the orbit of S under the action of G, lemma 4.7.7 on page 75 implies that
$$r = [G: G_S]$$
where G_S is the *stabilizer* of S — see definition 4.7.4 on page 74 — but this is identical to the definition of the *normalizer* of S, so
$$r = [G: N_G(S)] = \frac{|G|}{|N_G(S)|}$$
Since $S \subset N_G(S)$, we have $|N_G(S)| = p^k \cdot \ell$ and
$$r = \frac{p^k \cdot m}{p^k \cdot \ell} = \frac{m}{\ell}$$
or $r \cdot \ell = m$ and $r \,|\, m$.

EXERCISES.

1. Show that every group of order 35 is cyclic.

2. Show that no group of order 70 can be simple.

4.9. Subnormal series

We can break groups into smaller factors in various ways.

DEFINITION 4.9.1. If G is a group, a *subnormal series*
$$\{1\} \subset G_0 \subset G_1 \subset \cdots \subset G$$
is a sequence of subgroups of G such that $G_i \triangleleft G_{i+1}$ for all i.

REMARK. Note that $G_i \triangleleft G_{i+1}$ but this does *not* mean $G_i \triangleleft G$.
Since each term is a normal subgroup of the next, the quotients
$$\frac{G_{i+1}}{G_i}$$
are well-defined for all i.

Now we will discuss certain standard classes of subnormal series.

DEFINITION 4.9.2. If G is a group, its *derived series*
$$\cdots \subset G_n \subset \cdots \subset G_1 \subset G_0 = G$$
is defined by
$$G_{n+1} = [G_n, G_n]$$

REMARK. Note that all of the quotients G_n/G_{n+1} are abelian.

Derived series do not necessarily terminate in a finite number of steps: Since A_n is simple for $n \geq 5$ (see theorem 4.5.15) and $[A_n, A_n] \triangleleft A_n$ it follows that all of the terms of the derived series for A_n are *equal* to A_n.

The next class of subnormal series is particularly important in the classification of finite groups.

DEFINITION 4.9.3. If G is a finite group, its *composition series* is a subnormal series
$$\{1\} \subset G_0 \triangleleft G_1 \triangleleft \cdots \triangleleft G_k \triangleleft G$$
with the property that each quotient
$$\frac{G_{i+1}}{G_i} = F_i$$
is a *simple* group. The quotients, F_i, are called *composition factors*.

REMARK. In a manner of speaking, we are fracturing G into a series of smaller groups — the composition factors. Each of them fits into what is called a *group-extension*

$$1 \to G_i \to G_{i+1} \to F_i \to 1$$

With suitable information, it is possible to *reconstruct* G_{i+1} from G_i and F_i — see chapter 13 on page 433 and section 13.3.2 on page 458.

This is related to the Isomorphism Problem for finite groups:

List all isomorphism classes of finite groups.

Since it is possible to construct groups from normal subgroups and quotients, this follows (to some extent) from the corresponding problem for *simple groups*:

List all isomorphism classes of finite *simple* groups.

Since this problem has been solved, in principle, so has the former. In 1983, Daniel Gorenstein announced that finite simple groups had been classified, but his classification turned out to be incomplete.

The first complete classification was published in 2004 by Aschbacher and Smith. At 1221 pages, their proof is beyond the scope of the present text. See the survey [4].

LEMMA 4.9.4. *If G is a finite group, a composition series for G exists.*

PROOF. Let

$$\{1\} \subset G_0 \triangleleft G_1 \triangleleft \cdots \triangleleft G_k \triangleleft G$$

be a subnormal series of *maximal length* (i.e., maximal value of k). We claim that

$$\frac{G_{i+1}}{G_i} = F_i$$

is *simple* for all i. If not, let $A \triangleleft F_i$ be a proper normal subgroup. If

$$p_i \colon G_{i+1} \to F_i$$

is the standard projection, then $p_i^{-1}(A) \triangleleft G_{i+1}$ (see exercise 5 on page 48) and $G_i \triangleleft p_i^{-1}(A)$ so we get a subnormal series

$$\{1\} \subset G_0 \triangleleft G_1 \triangleleft \cdots \triangleleft G_i \triangleleft p_i^{-1}(A) \triangleleft G_{i+1} \triangleleft \cdots \triangleleft G_k \triangleleft G$$

of length $k+1$, a contradiction. □

In an analogy to the unique factorization of *integers* into *primes*, a group *uniquely determines* its composition factors.

THEOREM 4.9.5 (Jordan-Hölder). *If G is a finite group and*

$$\{1\} \subset G_0 \triangleleft G_1 \triangleleft \cdots \triangleleft G_k \triangleleft G$$

and

$$\{1\} \subset H_0 \triangleleft H_1 \triangleleft \cdots \triangleleft H_\ell \triangleleft G$$

are two composition series for G, then $k = \ell$ and both series have the same composition factors — possibly occurring in a different order.

PROOF. The statement that $k = \ell$ follows quickly from lemma 4.9.4 on the previous page: its proof implies that a subnormal series is a composition series *if and only if* it is of maximal length.

We prove the rest of the theorem by induction on $|G|$. It's trivial if $|G| = 1$.

Suppose $|G| = n$ and the theorem is true for all groups of order $< n$. If $G_k = H_k$, we apply the inductive hypothesis to conclude that the theorem is true for G_k which implies it is true for G.

Consequently, assume $G_k \neq H_k$ and let $K = G_k \cap H_k$.

Note that $K \triangleleft H_k$ (see exercise 3 on page 47) and

$$\text{(4.9.1)} \qquad \frac{H_k}{K} \cong \frac{G_k \cdot H_k}{G_k}$$

(see exercise 4 on page 48) where $G_k \cdot H_k \subset G$ is a normal subgroup (see exercise 1 on page 47). Note that

$$G_k \cdot H_k / G_k \triangleleft G/G_k$$

Since the latter group is *simple*, we must have $G_k \cdot H_k / G_k = G/G_k$ so that:
(1) $G = G_k \cdot H_k$
(2) $G_k \cdot H_k / G_k$ and H_k/K are *also* simple (by equation 4.9.1).

It follows that we can get a composition series for H_k via

$$\{1\} \triangleleft K_0 \triangleleft \cdots \triangleleft K_t \triangleleft K \triangleleft H_k$$

where

$$\{1\} \triangleleft K_0 \triangleleft \cdots \triangleleft K_t \triangleleft K$$

is a composition series for K, and $t + 1 = k - 1$, and induction implies that

$$\{K_1/K_0, \ldots, K_t/K_{t-1}, H_k/K\} = \{H_1/H_0, \ldots, H_k/H_{k-1}\}$$

as *sets* of *isomorphism classes* of groups. Furthermore,

$$\{1\} \triangleleft K_0 \triangleleft \cdots \triangleleft K_t \triangleleft K \triangleleft G_k$$

is a composition series for G_k and we have

$$\{K_1/K_0, \ldots, K_t/K_{t-1}, G_k/K\} = \{G_1/G_0, \ldots, G_k/G_{k-1}\}$$

It follows that

$$\{H_1/H_0, \ldots, H_k/H_{k-1}\} = \{G_1/G_0, \ldots, G_k/G_{k-1}\}$$

and the conclusion follows from $G/G_k \cong H_k$ and $G/H_k \cong G_k$ (since $G = G_k \cdot H_k$). $\square$

EXAMPLE 4.9.6. If G is *abelian*, all of its composition factors are of the form $\mathbb{Z}_{p_i}$, for p_i all primes, since these *are* the simple abelian groups.

In our later work on Galois Theory (see chapter 8 on page 297) we will need a class of groups that have this property in common with abelian groups:

DEFINITION 4.9.7. A finite group, G, is defined to be *solvable* if its composition factors are all of the form $\mathbb{Z}_{p_i}$ for primes p_i. The primes that occur in the manner are called the *associated primes* of G.

Not all solvable groups are abelian. For instance S_3 has a normal subgroup generated by $(1,2,3)$ that is isomorphic to $\mathbb{Z}_3$ and $S_3/\mathbb{Z}_3 \cong \mathbb{Z}_2$.

EXAMPLE 4.9.8. On the other hand, the groups S_n for $n \geq 5$ are *not* solvable: their composition factors are $\mathbb{Z}_2$ and A_n, which are known to be *simple* (see theorem 4.5.15). Theorem 4.9.5 on page 85 implies that *all* composition series for S_n will have these factors.

EXERCISES.

1. If G is a solvable group, show that it contains a normal subgroup H such that $[G:H]$ is a prime number.

2. Show that we could've defined solvable groups as groups with a subnormal series whose quotients are abelian.

3. If G is a finite group, show that its derived series becomes constant at some point.

4. Show that a finite group, G, is solvable if and only if its derived series *terminates* after a finite number of steps, i.e. if its derived series

$$\cdots \subset G_n \subset \cdots \subset G_1 \subset G_0 = G$$

has the property that there exists an integer k such that $G_i = \{1\}$ for $i \geq k$.

5. Show that quotients and subgroups of solvable groups are solvable.

6. Let G be a group with a normal subgroup H. If H and G/H are both solvable, show that G must also be solvable.

4.10. Free Groups

4.10.1. Introduction. In this section, we will introduce groups that are "free-er" than the free abelian groups defined in section 4.6 on page 58. They can be used to define *presentations* of groups — ways of describing them using "coordinates" and "equations". This gives rise to an area of group theory called *combinatorial group theory*.

Combinatorial group theory originally arose in algebraic topology, with the study of knot-groups. One of the first purely group-theoretic problems in the field was

PROBLEM 4.10.1 (Burnside's Problem). If $n > 1$ is an integer and G is a finitely generated group that satisfies the equation

$$x^n = 1$$

for *all* $x \in G$, is G *finite*?

Burnside showed that the answer was "yes" for $n = 2, 3, 4, 6$. The breakthrough in Burnside's problem was achieved by Pyotr Novikov and Sergei Adian in 1968 (see [6]). Using a combinatorial argument, they showed that for every odd number n with $n > 4381$, there exist infinite, finitely generated groups where every element satisfies $x^n = 1$.

DEFINITION 4.10.2. If $X = \{x_1, \ldots\}$ is a set of symbols, the *free group* on X, F_X consists of equivalence classes of all possible finite strings
$$x_{i_1}^{\alpha_1} \cdots x_{i_n}^{\alpha_n}$$
where the $x_i \in X$, the $\alpha_i \in \mathbb{Z}$, and the product of two such strings is their *concatenation*. Two strings are *equivalent* or *freely equal* if one can be transformed into the other via the following operations or their inverses:

Identity: 1 is equivalent to the *empty string*,

Consolidation: whenever we have $\cdots x_j^\alpha \cdot x_j^\beta \cdots$ in a string, we may replace it by $\cdots x_j^{\alpha+\beta} \cdots$ to get an equivalent string,

Culling: whenever a factor in a string has exponent 0, we may *remove* it.

REMARK. Sometimes, we'll abuse the terminology and say things like "These strings are freely equal as elements of the free group, F_X". This is abuse because the elements of F_X are *equivalence classes* of strings, where strings are equivalent if they're freely equal."

Deciding when two strings are equivalent is fairly easy:

DEFINITION 4.10.3. If $x_{i_1}^{\alpha_1} \cdots x_{i_n}^{\alpha_n} \in F_X$ is a string in a free group, the *reduced form* of $x_{i_1}^{\alpha_1} \cdots x_{i_n}^{\alpha_n}$ is the result of applying the operations in definition 4.10.2 as many times as possible to *minimize* the number of factors in $x_{i_1}^{\alpha_1} \cdots x_{i_n}^{\alpha_n}$.

REMARK. A string, $x_{i_1}^{\alpha_1} \cdots x_{i_n}^{\alpha_n}$, is reduced if and only if

(1) $x_{i_j} \neq x_{i_{j+1}}$ for all j, and
(2) $\alpha_i \neq 0$ for all i

Since reduced forms of words are unique and equivalent to the original words, we have

PROPOSITION 4.10.4. *Two words $w_1, w_2 \in F_X$ are equivalent if and only if their reduced forms are equal.*

Note that the inverse of $x_{i_1}^{\alpha_1} \cdots x_{i_n}^{\alpha_n}$ is $x_{i_n}^{-\alpha_n} \cdots x_{i_1}^{-\alpha_1}$ — we must *negate* the exponents and *reverse* the order of the factors.

Also notice that the relations that words satisfy (in definition 4.10.2) are relations that *all* elements in *all* groups must satisfy. This has several consequences:

THEOREM 4.10.5. *Let F_X be a free group on a set of generators, X, and let G be a group. Any function*
$$f: X \to G$$
extends uniquely to a group-homomorphism
$$\bar{f}: F_X \to G$$

REMARK. This can be regarded as the *defining property* of a free group.

PROOF. Simply define
$$\bar{f}(x_{i_1}^{\alpha_1}\cdots x_{i_n}^{\alpha_n}) = f(x_{i_1})^{\alpha_1}\cdots f(x_{i_n})^{\alpha_n}$$
□

This immediately implies:

PROPOSITION 4.10.6. *If G is any group, there exists a free group F_X and a surjective homomorphism*
$$f\colon F_X \to G$$
If G is finitely generated, then X can be finite.

REMARK. Compare this to proposition 4.6.7 on page 62. Just as every *abelian* group is a quotient of a free *abelian* group, every *group* (including abelian groups) is a quotient of a *free* group.

PROOF. Let F_X be a free group on the set, X. If G is finitely generated, then the elements of X are in a 1-1 correspondence with its generators. Otherwise it is in a 1-1 correspondence with the non-identity *elements* of G, so $x_i \mapsto g_i$ and define f by
$$f(x_{i_1}^{\alpha_1}\cdots x_{i_n}^{\alpha_n}) = g_{i_1}^{\alpha_1}\cdots g_{i_n}^{\alpha_n} \in G$$
□

4.10.2. Combinatorial group theory. This leads to the topic of *presentations* of groups:

DEFINITION 4.10.7. If $X = \{x_1,\ldots\}$ is a set of symbols and $R = \{r_1,\ldots\}$ is a set of elements of F_X the symbol, called a *presentation of a group*, G,,

(4.10.1) $$\langle X|R\rangle$$

is an isomorphism
$$G = \frac{F_X}{\langle R\rangle^{F_X}}$$

where $\langle R\rangle \subset F_X$ is the subgroup generated by the elements of R and $\langle R\rangle^{F_X}$ is the *normal closure* of this subgroup — see definition 4.4.11 on page 46. The elements of X are called the *generators* of G and the elements of R are called the *relators* or *relations* of G.

REMARK. The group being presented is G, which is generated by images of the x_i under the projection
$$F_X \to \frac{F_X}{\langle R\rangle^{F_X}}$$
and the $r_i = 1$ are "equations" that these images satisfy.

For instance
$$\mathbb{Z}_n = \langle x | x^n \rangle$$
is a presentation of $\mathbb{Z}_n$.

Sometimes we write the relators as *equations*, so
$$\mathbb{Z} \oplus \mathbb{Z} = \langle x, y | xy = yx \rangle = \langle x, y | xyx^{-1}y^{-1} \rangle$$
are presentations. When a relation is not written as an equation, it is assumed to equal 1.

Proposition 4.5.17 on page 57 shows that

(4.10.2) $$D_{2n} = \langle x, y | x^n, yxy = x^{-1} \rangle$$

is a presentation of the dihedral group.

A given group can have many different presentations that may not resemble each other. For instance, it turns out that
$$S_3 = \langle a, b | a^3, b^2, ab = ba^2 \rangle = \langle x, y | x^2, y^2, (xy)^3 \rangle$$

This leads to Max Dehn's three fundamental problems in combinatorial group theory:

Given a presentation of a group G like that in equation 4.10.1 on the preceding page, we have

(1) *The word problem:* Determine whether a word in the generators, like $x_{i_1}^{\alpha_1} \cdots x_{i_n}^{\alpha_n}$, is the *identity element* in a finite number of steps.
(2) *The conjugacy problem:* Given two words in the generators, determine whether they represent *conjugate elements* of G in a finite number of steps.
(3) *The isomorphism problem:* Given two group-presentations, in a finite number of steps determine whether they represent isomorphic groups.

If a *free* group has the standard presentation, $\langle X | \rangle$, proposition 4.10.4 on page 88 solves the word problem: a word is the identity element if and only if its reduced form is *empty*.

Notice that these problems refer to *presentations* of groups, not the groups themselves. Solutions to, say, the word problem will be very different for different presentations of a group.

Max Dehn (1878 – 1952) was a German-born American mathematician and student of David Hilbert. His most famous work was in the fields of geometry, topology, and geometric group theory. He is also known for being the first to resolve one of Hilbert's famous 23 problems. Dehn solved Hilbert's third problem:

> Given two polyhedra of equal volume, is it possible to cut the first into finitely many pieces that can be reassembled to give the second?

Dehn produced a counterexample.

If our group is the dihedral group with presentation

(4.10.3) $$D_{2n} = \langle x, y | x^n, y^2, yxy = x^{-1} \rangle$$

proposition 4.5.17 on page 57 shows that the word-problem can be solved via

Given a string of x's and y's, repeat the following two operations until the string is of the form $x^i y^j$

▷ reduce all exponents of x modulo n and all exponents of y modulo 2.
▷ replace each occurrence of yx by $x^{-1}y = x^{n-1}y$

The string represents the identity element if and only if, at the end, $i = j = 0$.

This is the solution of the word problem for the presentation in equation 4.10.3 on the preceding page of the dihedral group.

In 1955, Novikov (see [85]) showed that there are finite group-presentations whose word problems are unsolvable. In 1958, Boone (see [15]) gave examples of such presentations with 2 generators and 32 relations.

Given a presentation for a group

$$G = \langle X | R \rangle$$

Tietze showed that it is possible to get any other presentation for G using the *Tietze transformations*

DEFINITION 4.10.8. The *Tietze transformations* are:

T1: If words $w_1, \ldots, w_k$ are derivable from the relators in R (i.e. they are products of conjugates of the $r_i \in R$), they can be *added* to R.
T2: If some of the relators $r_{i_1}, \ldots, r_{i_k}$ are derivable from the others, they may be *deleted* from R.
T3: If $w_1, \ldots, w_m$ are words in the symbols in X, we may add generators $y_1, \ldots, y_m$ to X if we also add relations $y_i = w_i$ to R.
T4: If we have a relation $x_i = w$, where w is a word in $X \setminus x_i$, we may delete the relation from R and the symbol x_i from X. We must also replace every occurrence of x_i in the remaining relations by w.

REMARK. The even numbered transformations are the inverses of the odd-numbered ones. Some authors list only two transformations.

THEOREM 4.10.9. *Tietze transformations replace one presentation of a group by another. If a group has two different presentations*

(4.10.4) $$G = \langle \{x_i\} | \{r_j\} \rangle$$
$$= \langle \{y_k\} | \{s_\ell\} \rangle$$

one can be transformed into the other by Tietze transformations.

REMARK. This does *not* solve the isomorphism problem. It gives no *algorithm* for finding a sequence of Tietze transformations to go from one presentation to another.

PROOF. It is not hard to see these transformations preserve the group, G. T1 asserts relations already known to be true in G, and T2 deletes them.

T3 adds new elements to G and equates them to elements already present, and T4 deletes them.

Given the two presentations for G in equation 4.10.4 on the preceding page, we have surjections

$$\begin{aligned} f\colon F_X &\to G \\ g\colon F_Y &\to G \end{aligned}$$

where $X = \{x_i\}$ and $Y = \{y_k\}$. We start with presentation $\langle\{x_i\} \mid \{r_j\}\rangle$. Let $u_i \in F_X$ have the property that $f(u_i) = g(y_i)$. Then applications of T3 convert the presentation to

$$G = \langle\{x_i\},\{y_j\} \mid \{r_j\},\{y_j = u_j\}\rangle$$

If we plug the u_j into the relations, s_ℓ and map them via f we get

$$\begin{aligned} f\left(s_\ell(u_1,\ldots,u_m)\right) &= s_\ell(f(u_1),\ldots,f(u_m)) \\ &= s_\ell(g(y_1),\ldots,g(y_m)) \\ &= g\left(s_\ell(y_1,\ldots,y_m)\right) \\ &= 1 \end{aligned}$$

which means that $s_\ell(u_1,\ldots,u_m)$ follows from the set of relations $\{r_j\}$. The relations $\{y_j = u_j\}$ imply that $s_\ell(y_1,\ldots,y_m)$ follows from the relations $\{r_j\} \cup \{y_j = u_j\}$. We can, consequently use T1 multiple times to get another presentation of G:

$$G = \langle\{x_i\},\{y_j\} \mid \{r_j\},\{y_j = u_j\},\{s_\ell\}\rangle$$

with map

$$f \cup g\colon F_{X \cup Y} \to G$$

Now let $v_i \in F_Y$ have the property that $g(v_i) = f(x_i)$. Since $g(v_i) = f(x_i)$, the words $v_i x_i^{-1}$ map to $1 \in G$ under this new map, so that $v_i x_i^{-1}$ are consequences of the relations $\{r_j\} \cup \{y_j = u_j\}$. We can use T1 to add them to the relations, getting

$$G = \langle\{x_i\},\{y_j\} \mid \{r_j\},\{y_j = u_j\},\{s_\ell\},\{v_i = x_i\}\rangle$$

Now we can use T4 several times to delete the x's from the generators

$$G = \langle\{y_j\} \mid \{r_j(v_i)\},\{y_j = u_j\},\{s_\ell\}\rangle$$

where the $r_j(v_i)$ are relators with the variables x_i replaced by the words v_i that are written in the y_j. Note that

$$\begin{aligned} g(r_j(v_i)) &= r_j(g(v_i)) \\ &= r_j(f(x_i)) \\ &= f\left(r_j(x_i)\right) \\ &= 1 \end{aligned}$$

so that the $r_j(v_i)$ are implied by the relators, s_ℓ. We can use T2 to delete them and get the presentation

$$G = \langle\{y_j\} \mid \{s_\ell\}\rangle$$

EXAMPLE 4.10.10. Suppose we start with the presentation
$$\langle a, b | aba = bab \rangle$$
Now we introduce new generators $x = ab$, and $y = aba$. Two applications of T3 give
$$\langle a, b, x, y | aba = bab, x = ab, y = aba \rangle$$
Now $x^{-1}y = a$, so we can eliminate a using T4 to get
$$\langle b, x, y | x^{-1}ybx^{-1}y = bx^{-1}yb, x = x^{-1}yb, y = x^{-1}ybx^{-1}y \rangle$$
or
$$\langle b, x, y | x^{-1}ybx^{-1}y = bx^{-1}yb, x^2 = yb, 1 = x^{-1}ybx^{-1} \rangle$$
or
$$\langle b, x, y | x^{-1}ybx^{-1}y = bx^{-1}yb, x^2 = yb, x^2 = yb \rangle$$
We use T2 to eliminate the last relation and note that $b = y^{-1}x^2$. Apply T4 again to eliminate b and get
$$\langle b, x, y | x^{-1}yy^{-1}x^2x^{-1}y = y^{-1}x^2x^{-1}yy^{-1}x^2, x^2 = yy^{-1}x^2 \rangle$$
or
$$\langle x, y | y = y^{-1}x^3, x^2 = x^2 \rangle$$
or
$$\langle x, y | y^2 = x^3, x^2 = x^2 \rangle$$
and we use T2 to eliminate the last relation to get
$$\langle x, y | y^2 = x^3 \rangle$$
This is a very different-appearing presentation for the same group.

EXERCISES.

1. Let G be a group with normal subgroup $A \triangleleft G$ and quotient
$$F = \frac{G}{A}$$
a free group. Show that $G = A \rtimes_\alpha F$, where $\alpha: F \to \text{Aut}(A)$ is defined by conjugation.

2. Show that
$$G = \langle x_n, n \in \mathbb{Z}^+ | x_n^n = x_{n-1}, n \in \{2, \dots\} \rangle$$
is a presentation for $(\mathbb{Q}, +)$, the rational numbers under addition.

3. Show, using Tietze transformations, that
$$\left\langle a,b,c \mid b(abc^{-1})^2 a, c(abc^{-1})^3 \right\rangle$$
is the free group on two generators.

Hint: In $F_{\{x,y\}}$ adjoin the new generators $a = xy$, $b = y^{-1}x$, and $c = x^3$. Then solve for x and y in terms of a, b, c.

4. Show that the Prüfer group (see example 4.1.10 on page 38), $\mathbb{Z}/p^\infty$, can be presented as
$$\mathbb{Z}/p^\infty = \left\langle x_1, x_2, x_3, \ldots \mid x_1^p = 1, x_2^p = x_1, x_3^p = x_2, \ldots \right\rangle$$

4.10.3. Quotients and subgroups. Given a group-presentation
$$G = \langle X | R \rangle$$
let H be a normal subgroup generated by the conjugates of element $g_1, \ldots, g_n \in G$. If we find words $w_1, \ldots, w_n$ in F_X representing these elements, we get a presentation of G/H:
$$G/S = \langle X | R, w_1, \ldots, w_n \rangle$$
This follows immediately from the fact that a group-presentation defines a quotient of a free group.

A related but harder (and more interesting!) problem is how we can get a presentation of a *subgroup* of G.

We begin with

DEFINITION 4.10.11. If $H \subset G = \langle X | R \rangle$ is a subgroup generated by words $w_1, \ldots, w_k \in F_X$, a *rewriting process* for H *with respect to the words* w_i is a function
$$\xi: U \to V$$
where U is the set of words of F_X that define elements of H and $V = F_S$, where $s_{w_1}, \ldots, s_{w_k}$ are symbols in a one-to-one correspondence with the w_i. If $w \in F_X$ defines an element of H, we require that
$$\xi(w)_{s_{w_1} \to w_1, \ldots, s_{w_k} \to w_k}$$
defines the same element of H as w. Here $\xi(w)_{s_{w_1} \to w_1, \ldots, s_{w_k} \to w_k}$ denotes the result of taking the word $\xi(w) \in F_S$ — a word in the s_j — and substituting w_i for s_{w_i} for all i, obtaining a word in F_X.

REMARK. In other words, ξ simply rewrites a word representing an element of H into one in the *generators* of H.

Here's an example:

EXAMPLE 4.10.12. Let
$$G = \langle x, y | \rangle$$
and let H be the normal subgroup generated by y and its conjugates. This means H is generated by elements
$$s_k = x^k y x^{-k}$$

Note that
$$s_k^\ell = \underbrace{x^k y x^{-k} \cdots x^k y x^{-k}}_{\ell \text{ times}}$$
$$= x^k y^\ell x^{-k}$$

If
$$w = x^{\alpha_1} y^{\beta_1} \cdots x^{\alpha_t} y^{\beta_t}$$
is a word with
$$\sum_{i=1}^{t} \alpha_i = 0$$
we can rewrite it as

(4.10.5) $\quad w = \left(x^{\alpha_1} y x^{-\alpha_1}\right)^{\beta_1} \cdot \left(x^{\alpha_1+\alpha_2} y x^{-\alpha_1-\alpha_2}\right)^{\beta_2} \cdots \left(x^{\alpha_1+\cdots+\alpha_t} y x^{-\alpha_1-\cdots-\alpha_t}\right)^{\beta_t}$
$\quad\quad\quad\ = s_{\alpha_1}^{\beta_1} \cdots s_{\alpha_1+\cdots+\alpha_t}^{\beta_t}$

so that $w \in H$ and equation 4.10.5 constitutes a rewriting process for H with respect to the generators $S = \{s_j\}$, i.e.
$$\xi(w) = s_{\alpha_1}^{\beta_1} \cdots s_{\alpha_1+\cdots+\alpha_t}^{\beta_t}$$

The *idea* of finding a presentation for H is fairly simple:

> Since the relators in the presentation of G represent its *identity* element, they represent an element of H as well, so rewrite the relators of G using the rewriting process and we're done!

Unfortunately, "the devil's in the details". Handling them gives us a (rather useless) presentation for H:

THEOREM 4.10.13. *If*

(4.10.6) $\quad\quad\quad\quad\quad\quad\quad\quad G = \langle X | R \rangle$

is a presentation of a group, $H = \langle g_1, \ldots, g_n \rangle$ is a subgroup, where g_i is represented by a word $w_i \in F_X$, and
$$\xi \colon \langle w_1, \ldots, w_n \rangle \to F_S$$
is a rewriting process for H with respect to the $\{w_i\}$, then the following is a presentation for H:

▷ *generators $S = \{s_1, \ldots, s_n\}$ in a one-to-one correspondence with the words $\{w_1, \ldots, w_n\}$,*
▷ *relations*
  (1) $s_i = \xi(w_i)$ *for all i.*
  (2) $\xi(w) = \xi(w')$ *for all pairs w, w' of words representing elements of H that are freely equal[6] in F_X,*
  (3) $\xi(u \cdot v) = \xi(u) \cdot \xi(v)$ *for all pairs of words u, v representing elements of H,*
  (4) $\xi(w \cdot r \cdot w^{-1}) = 1$ *where w runs over all words in F_X and r runs over all defining relations, R, in equation 4.10.6.*

REMARK. Note that this presentation for H usually has an infinite number of relations! We will work at simplifying it.

---

[6] See definition 4.10.2 on page 88, for the definition of 'freely equal.'

PROOF. The statement about generators is clear. It is also not hard to see that the stated relations are reasonable — i.e., they are *valid* in H. For instance, in statement 1 on the previous page, s_i and w_i and $\xi(w_i)$ represent the same element of H so it is valid to equate them.

If two words are freely equal in F_X, they represent the same element of G, therefore also of H and statement 2 on the preceding page makes sense. Statements 3 on the previous page and 4 on the preceding page are also clear.

The hard part of this proof is showing that *all* relations in H can be derived from these.

We begin by deriving some consequences of the four sets of relations in 1 on the previous page through 4 on the preceding page:

Statement 3 on the previous page implies that $\xi(1) \cdot \xi(1) = \xi(1)$ or $\xi(1) = 1$. If we set $v = u^{-1}$, we get

$$\xi(u) \cdot \xi(u^{-1}) = \xi(1) = 1$$

so that $\xi(u^{-1}) = \xi(u)^{-1}$.

A simple induction shows that, if $b_1, \ldots, b_k \in F_X$ are words representing elements of H, then

(4.10.7) $$\xi(b_1^{\alpha_1} \cdots b_k^{\alpha_k}) = \xi(b_1)^{\alpha_1} \cdots \xi(b_k)^{\alpha_k}$$

and this is a *direct consequence* of statements 2 on the preceding page and 3 on the previous page.

Suppose

$$s_1^{\alpha_1} \cdots s_r^{\alpha_r}$$

is a word in F_S that maps to 1 in H — i.e., is a relator in presentation of H. We will show that

$$s_1^{\alpha_1} \cdots s_r^{\alpha_r} = 1$$

is a consequence of statements 1 on the preceding page through 4 on the previous page. Let $q_i \in F_X$ be a word representing s_i for all i. Then statement 1 on the preceding page and equation 4.10.7 imply that

$$s_1^{\alpha_1} \cdots s_r^{\alpha_r} = \xi(q_1)^{\alpha_1} \cdots \xi(q_r)^{\alpha_r} = \xi(q_1^{\alpha_1} \cdots q_r^{\alpha_r})$$

Since $q_1^{\alpha_1} \cdots q_r^{\alpha_r} \in F_X$ maps to 1 in G, it is in the normal closure of the *relations* of G, namely the elements of R. It follows that $q_1^{\alpha_1} \cdots q_r^{\alpha_r}$ is freely equal to

$$r_1^{y_1} \cdots r_\ell^{y_\ell}$$

where the $r_i \in R$ and the $y_i \in F_X$ are some words. We conclude that

$$s_1^{\alpha_1} \cdots s_r^{\alpha_r} = \xi(q_1)^{\alpha_1} \cdots \xi(q_r)^{\alpha_r}$$
$$= \xi(q_1^{\alpha_1} \cdots q_r^{\alpha_r})$$
$$= \xi(r_1^{y_1} \cdots r_\ell^{y_\ell})$$
$$= \xi(r_1^{y_1}) \cdots \xi(r_\ell^{y_\ell}) = 1 \cdots 1 = 1$$

so that $s_1^{\alpha_1} \cdots s_r^{\alpha_r} = 1$ is a consequence of statements 1 through 4 on the preceding page. □

We will explore alternate rewriting processes that *automatically* satisfy many of the conditions in theorem 4.10.13 on the previous page, resulting in a simpler presentation.

DEFINITION 4.10.14. If

$$G = \langle X | R \rangle$$

is a presentation of a group and H is a subgroup let

$$C = \{1, e(g_1), \ldots\}$$

be a set of *distinct* right coset-representatives — i.e. for each *coset Hg*, we select a *single element* $e(g) \in Hg$, to represent it. For the coset of 1, namely H, we select the element 1.

A *right coset representative function* is a mapping

$$\eta \colon F_X \to F_X$$

defined as follows:

If w represents the element $g \in G$, $\eta(w) \in F_X$ is a word representing the chosen representative for the coset Hg, namely $e(g)$.

The notation $C_\eta = \{1, w_1, \ldots\}$ denotes the set of words representing the coset representatives, so $|C_\eta| = [G \colon H]$.

REMARK. Several arbitrary choices were made in defining η:

(1) the elements g selected to represent a coset Hg
(2) the words, $\eta(*)$, selected to represent these elements in F_X

Another common term for a coset representative function is *transversal*.

In general, we want to pick words that are as short and simple as possible.

EXAMPLE 4.10.15. We return to the group and subgroup in example 4.10.12 on page 94. Since H is a *normal* subgroup, the cosets of H in G are in a 1-1 correspondence with elements of $G/H = \langle x | \, \rangle = \mathbb{Z}$. For the coset Hx^k, we pick the representative x^k — and we also pick this *word* in $F_{\{x,y\}}$ so our coset representative function is

$$\eta(x^{\alpha_1} y^{\beta_1} \cdots x^{\alpha_n} y^{\beta_n}) = x^{\sum_{i=1}^n \alpha_i}$$

In this case $C_\eta = \{x^n, \forall n \in \mathbb{Z}\}$.

LEMMA 4.10.16. *Under the conditions of definition 4.10.14 on the preceding page,*

(1) *If* $w \in H$, *then*

(4.10.8)
$$\eta(w) = 1$$

(2) *If* $v, w \in F_X$ *are freely equal, then* $\eta(v) = \eta(w)$,
(3) *If* $w \in F_X$, *then* $\eta(\eta(w)) = \eta(w)$
(4) *If* $v, w \in F_X$, *then* $\eta(vw) = \eta(\eta(v)w)$
(5)

(4.10.9)
$$\left(k' x \eta(k'x)^{-1}\right)^{-1} = kx^{-1} \eta\left(kx^{-1}\right)^{-1}$$

where $k \in C_\eta$, $x \in X$, *and* $k' = \eta(kx^{-1})$.

REMARK. If we write

$$t_{k,x^\epsilon} = kx^\epsilon \eta(kx^\epsilon)^{-1}$$

with $\epsilon = \pm 1$, then equation 4 implies that

(4.10.10)
$$t_{k,x^{-1}} = t^{-1}_{\eta(kx^{-1}),x}$$

PROOF. Most of these are immediate from definition 4.10.14 on page 96. To show the final statement, set $k' = \eta(kx^{-1})$. Then

$$\left(k'x\eta(k'x)^{-1}\right) \cdot \left(kx^{-1}\eta\left(kx^{-1}\right)^{-1}\right) = \eta(kx^{-1})x\eta(\eta(kx^{-1})x)^{-1}$$
$$\cdot kx^{-1}\eta\left(kx^{-1}\right)^{-1}$$
$$= \eta(kx^{-1})x\eta(kx^{-1}x)^{-1}$$
$$\cdot kx^{-1}\eta\left(kx^{-1}\right)^{-1}$$
$$= \left(\eta(kx^{-1})x\eta(k)^{-1}\right) \cdot \left(kx^{-1}\eta\left(kx^{-1}\right)^{-1}\right)$$
$$= \left(\eta(kx^{-1})xk^{-1}\right) \cdot \left(kx^{-1}\eta\left(kx^{-1}\right)^{-1}\right)$$
$$= \eta(kx^{-1})x \cdot x^{-1} \cdot \eta\left(kx^{-1}\right)^{-1}$$
$$= 1$$

□

Given a coset representative function, we can define a rewriting process that *automatically* satisfies some of the conditions in theorem 4.10.13 on page 95 — so we can simplify the relations in the presentation.

If $G = \langle X|R \rangle$ is a presentation of a group and H is a subgroup with a coset representative function $\eta \colon F_X \to F_X$, consider elements of the form

$$k_i x_j \eta(k_i x_j)^{-1}$$

where $k_i \in C_\eta$ and $x_j \in X$. This is an element of H since $\eta(k_i x_k)$ is in the same right-coset as $k_i x_j$.

THEOREM 4.10.17. *If $G = \langle X|R \rangle$ is a presentation of a group and H is a subgroup with a coset representative function $\eta \colon F_X \to F_X$, define words*

(4.10.11) $$t_{k,x^\epsilon} = kx^\epsilon \eta(k_i x^\epsilon)^{-1}$$

where k runs over the elements of C_η and x_j runs over X. These words generate H. In fact, if

$$x_1^{\alpha_1} \cdots x_n^{\alpha_n}$$

is a word representing an element of H with all exponents equal to ± 1[7], then

(4.10.12) $$x_1^{\alpha_1} \cdots x_n^{\alpha_n} = \left(t_{1,x_1^{\alpha_1}}\right) \cdot \left(t_{\eta(x_1^{\alpha_1}),x_2^{\alpha_2}}\right) \cdots \left(t_{\eta(x_1^{\alpha_1} \cdots x_{n-1}^{\alpha_{n-1}}),x_n^{\alpha_n}}\right)$$

REMARK. This first appeared in [93], in connection with algebraic topology (knot theory).

---

[7] So x^3 is written as xxx, and x^{-2} as $x^{-1}x^{-1}$.

PROOF. Just fill in what the *t*-symbols equal (from equation 4.10.11 on the preceding page):

$$t_{1,x_1^{\alpha_1}} = x_1^{\alpha_1} \eta(x_1^{\alpha_1})^{-1}$$

$$t_{\eta(x_1^{\alpha_1}),x_2^{\alpha_2}} = \eta(x_1^{\alpha_1}) x_2^{\alpha_2} \eta(\eta(x_1^{\alpha_1}) x_2^{\alpha_2})^{-1}$$

$$= \eta(x_1^{\alpha_1}) x_2^{\alpha_2} \cdot \eta(x_1^{\alpha_1} x_2^{\alpha_2})^{-1} \text{ by lemma 4.10.16 on page 97}$$

$$t_{\eta(x_1^{\alpha_1} x_2^{\alpha_2}), x_3^{\alpha_3}} = \eta(x_1^{\alpha_1} x_2^{\alpha_2}) x_3^{\alpha_3} \cdot \eta(x_1^{\alpha_1} x_2^{\alpha_2} x_3^{\alpha_3})^{-1}$$

$$\vdots$$

$$t_{\eta(x_1^{\alpha_1} \cdots x_{n-1}^{\alpha_{n-1}}), x_n^{\alpha_n}} = \eta(x_1^{\alpha_1} \cdots x_{n-1}^{\alpha_{n-1}}) x_n^{\alpha_n} \eta(x_1^{\alpha_1} \cdots x_{n-1}^{\alpha_{n-1}} x_n^{\alpha_n})^{-1}$$

$$= \eta(x_1^{\alpha_1} \cdots x_{n-1}^{\alpha_{n-1}}) x_n^{\alpha_n}$$

because $\eta(x_1^{\alpha_1} \cdots x_{n-1}^{\alpha_{n-1}} x_n^{\alpha_n}) = 1$

since $x_1^{\alpha_1} \cdots x_{n-1}^{\alpha_{n-1}} x_n^{\alpha_n} \in H$

We get a telescoping product in which *all* of the $\eta(*)$-factors cancel, and we are left with our original word,

$$x_1^{\alpha_1} \cdots x_n^{\alpha_n}$$

□

Kurt Werner Friedrich Reidemeister (1893 – 1971) was a mathematician born in Braunschweig (Brunswick), Germany. Reidemeister's interests were mainly in combinatorial group theory, combinatorial topology, geometric group theory, and the foundations of geometry.

We can use equation 4.10.12 on the facing page to define a rewriting process for H:

DEFINITION 4.10.18. If $G = \langle X|R \rangle$ is a presentation of a group and H is a subgroup with a coset representative function $\eta: F_X \to F_X$, define words

(4.10.13) $$s_{k,x} = kx\eta(kx)^{-1}$$

where k runs over the elements of C_η and x runs over X. If

$$w = x_1^{\alpha_1} \cdots x_n^{\alpha_n}$$

represents an element of H, we take the string in equation 4.10.12 on the preceding page and replace *t*-symbols by *s*-symbols via:

$$t_{k,x} \to s_{k,x}$$

$$t_{k,x^{-1}} \to s_{\eta(kx^{-1}),x}^{-1}$$

This defines a function

$$F_X \to F_S$$

where S is the set of all possible *s*-symbols defined in equation 4.10.13. It is called a *Reidemeister rewriting process based on the coset function, η*.

REMARK. The *s*-symbols, $s_{k,x}$, are nothing but *t*-symbols, t_{k,x^ϵ} for which ϵ is *required* to be $+1$. If we allow ϵ to be *negative*, equation 4.10.10 on page 97 shows a simple relationship between *t*-symbols, t_{k,x^ϵ} No such relationship exists between the *s*-symbols — indeed, example 4.10.24 on page 103 shows that these symbols can be *completely independent*.

Theorem 4.10.17 on page 98 and equation 4.10.10 on page 97 implies that the rewritten string (in *s*-symbols) will still represent the same element of H as w.

A Reidemeister rewriting process has several interesting properties that simplify the presentation in theorem 4.10.13 on page 95:

LEMMA 4.10.19. *If $G = \langle X | R \rangle$ is a presentation of a group and H is a subgroup with a coset representative function $\eta \colon F_X \to F_X$ and associated Reidemeister rewriting process*

$$\zeta \colon F_X \to F_S$$

then

(1) *if w_1, w_2 are freely equal words in F_X representing elements of H, then $\zeta(w_1)$ is freely equal to $\zeta(w_2)$ in F_S.*
(2) *if $w_1, w_2 \in F_X$ are two words that represent elements of H, then*

$$\zeta(w_1 \cdot w_2) = \zeta(w_1) \cdot \zeta(w_2)$$

PROOF. We start with the first statement. Suppose $w_1 = uv$, where $u, v \in F_X$. It will suffice to prove the statement in the case where $w_2 = ux^\epsilon x^{-\epsilon} v$, where $\epsilon = \pm 1$. Let $\zeta(w_1) = r_1 \cdot t_1$ and $\zeta(w_2) = r_2 \cdot g \cdot t_2$ where $r_i = \zeta(u_i)$ for $i = 1, 2$, and t_i is the portion of $\zeta(w_i)$ derived from v. It is clear from equation 4.10.12 on page 98 that $r_2 = r_1$.

Since $\eta(ux^\epsilon x^{-\epsilon}) = \eta(u)$, it follows that $t_2 = t_1$, and so we must analyze the *s*-factors that come from x^ϵ and $x^{-\epsilon}$:

If $\epsilon = +1$

$$\begin{aligned} g &= s_{\eta(u), x} \cdot s^{-1}_{\eta(uxx^{-1}), x} \\ &= s_{\eta(u), x} \cdot s^{-1}_{\eta(u), x} \end{aligned}$$

If $\epsilon = -1$

$$g = s^{-1}_{\eta(ux^{-1}), x} s_{\eta(ux^{-1}), x}$$

so $g \in F_S$ is freely equal to the empty string in both cases.

Now we examine the second statement: Let $\zeta(w_1 w_2) = r \cdot t$, where $r = \zeta(w_1)$, the *s*-symbols, $s_{\eta(u), x_i}$, used to compute t will be the same as those used to compute w_2, except that the *u*-term in the subscript will be concatenated with w_1. The conclusion follows from the fact that $w_1 \in H$ so $\eta(w_1 w) = \eta(w)$, for *any* word $w \in F_X$. It follows that $t = \zeta(w_2)$. □

With these tools in hand, we are ready to give Reidemeister's simplified presentation of a subgroup.

THEOREM 4.10.20 (Reidemeister). *If $G = \langle X | R \rangle$ is a presentation of a group and H is a subgroup with a coset representative function $\eta \colon F_X \to F_X$, with associated Reidemeister rewriting process*

$$\zeta \colon F_X \to F_S$$

then

(4.10.14) $$H = \left\langle \left\{ s_{k_i, x_j} \right\} \mid \left\{ s_{k_i, x_j} = \zeta \left(k_i x_j \eta(k_i x_j)^{-1} \right) \right\}, \left\{ \zeta(k_i r_\ell k_i^{-1}) \right\} \right\rangle$$

where

(1) *the k_i run over the elements of C_η*
(2) *the x_j run over the elements of X, and*
(3) *the r_ℓ run over the elements of R*

REMARK. The curly brackets denote *sets* of terms.
Note that this is *finite* if X, R, and C_η are.

PROOF. We use lemma 4.10.19 on the preceding page to show that many of the conditions in theorem 4.10.13 on page 95 are automatically satisfied.

Statement 1 on the preceding page implies that statement 2 in theorem 4.10.13 on page 95 is automatically satisfied, and statement 2 on the preceding page implies the same for 3 on page 95.

We are left with statements 1 on page 95 and 4. Statement 1 imposes a number of conditions limited by the number of generators of H.

We can simplify statement 4 on page 95 (and reduce it to a smaller and, possibly finite, number of conditions) by noting that, for any word $w \in F_X$, and any relator $r \in R$, we can write
$$w = w\eta(w)^{-1}\eta(w)$$
where $h = w\eta(w)^{-1} \in H$. It follows that

(4.10.15) $$\zeta(w \cdot r \cdot w^{-1}) = h\zeta\left(\eta(w)r\eta(w)^{-1}\right)h^{-1}$$

where $\eta(w) \in C_\eta$. Since $h \in H$ the relation $\zeta(\eta(w)r\eta(w)^{-1}) = 1$ implies that quantities in equation 4.10.15 are equal to 1.

It follows that the (infinite) number of relations $\zeta(w \cdot r \cdot w^{-1}) = 1$ in statement 4 on page 95 are derivable from
$$\zeta(k_i r_\ell k_i^{-1}) = 1$$
where the k_i run over the (probably smaller) set C_η. They can be removed, using the T2 Tietze transformation (see definition 4.10.8 on page 91). □

We can immediately conclude:

THEOREM 4.10.21. *If G is a finitely presented group (i.e., it has a presentation with a finite number of generators and relations) and H is a subgroup of finite index, then H is also finitely presented.*

We can simplify the presentation in theorem 4.10.20 on the facing page further. We start with:

DEFINITION 4.10.22. *If $G = \langle X|R \rangle$ is a presentation of a group and H is a subgroup, a coset representative function*
$$\eta: F_X \to F_X$$
will be called Schreier if every initial segment of a word in C_η is also in C_η. A Reidemeister rewriting process based on a Schreier coset function is called a Reidemeister-Schreier rewriting process.

REMARK. The initial segment requirement means that if $xyxxy \in C_\eta$, then $xyxx$, xyx, xy, and x are also in C_η. Note that the coset representative function in example 4.10.15 on page 97 is Schreier.

> Otto Schreier (1901–1929) was an Austrian mathematician who made contributions to combinatorial group theory and the topology of Lie groups.

THEOREM 4.10.23. *If $G = \langle X|R \rangle$ is a presentation of a group and H is a subgroup with a Schreier coset representative function $\eta: F_X \to F_X$, with associated Reidemeister-Schreier rewriting process*
$$\zeta: F_X \to F_S$$

then
$$H = \left\langle \left\{s_{k_i,x_j}\right\} \mid \left\{s_{m_e,x_\ell}\right\}, \left\{\xi(k_i r_\ell k_i^{-1})\right\} \right\rangle$$
where

(1) the k_i run over the elements of C_η
(2) the x_j run over the elements of X, and
(3) the r_ℓ run over the elements of R
(4) the pairs $(m_e, x_\ell) \in S \times X$, have the property that $m_e x_\ell$ is freely equal to $\eta(m_e x_\ell)$ in F_X.

REMARK. This is a vast simplification of the presentation in equation 4.10.14 on page 100 because we have eliminated all the relations of the form
$$s_{k_i,x_j} = \xi\left(k_i x_j \eta(k_i x_j)^{-1}\right)$$
The relations $s_{m_e,x_\ell} = 1$ effectively *deletes* these s-symbols from the list of generators *and* the other relations.

PROOF. If $m_e x_\ell$ is freely equal to $\eta(m_e x_\ell)$ in F_X, then $s_{m_e,x_\ell} = m_e x_\ell \eta(m_e x_\ell)^{-1}$ is freely equal to the empty string, so the relations $s_{m_e,x_\ell} = 1$ make sense. Now we compute
$$\xi\left(kx\eta(kx)^{-1}\right)$$
for $k \in C_\eta$, and $x \in X$. Equation 2 on page 100 implies that
$$\xi\left(kx\eta(kx)^{-1}\right) = \xi(kx)\xi(\eta(kx))^{-1}$$
Suppose
$$k = u_1^{\epsilon_1} \ldots u_n^{\epsilon_n}$$
where $\epsilon_i = \pm 1$. Suppose, in computing $\xi\left(kx\eta(kx)^{-1}\right)$ we are computing the ith s-symbol, where $i < n$. If $\epsilon_i = +1$, the symbol will be
$$s_{\eta(u_1^{\epsilon_1}\ldots u_i^{\epsilon_i}), u_{i+1}}$$
and the Schreier property implies that $\eta(u_1^{\epsilon_1} \ldots u_i^{\epsilon_i}) u_{i+1} = u_1^{\epsilon_1} \ldots u_i^{\epsilon_i} u_{i+1} = \eta(u_1^{\epsilon_1} \ldots u_i^{\epsilon_i} u_{i+1})$ exactly. It follows that
$$s_{\eta(u_1^{\epsilon_1}\ldots u_i^{\epsilon_i}), u_{i+1}} = s_{m, u_{i+1}}$$
for a suitable pair (m, u_{i+1}).

If $\epsilon_i = -1$, the symbol will be
$$s^{-1}_{\eta(u_1^{\epsilon_1}\ldots u_i^{\epsilon_i} u_{i+1}^{-1}), u_{i+1}}$$
and $\eta(u_1^{\epsilon_1} \ldots u_i^{\epsilon_i} u_{i+1}^{-1}) u_{i+1} = u_1^{\epsilon_1} \ldots u_i^{\epsilon_i} u_{i+1}^{-1} u_{i+1} = u_1^{\epsilon_1} \ldots u_i^{\epsilon_i} = \eta(u_1^{\epsilon_1} \ldots u_i^{\epsilon_i})$, exactly, which implies that
$$s^{-1}_{\eta(u_1^{\epsilon_1}\ldots u_i^{\epsilon_i} u_{i+1}^{-1}), u_{i+1}} = s^{-1}_{m', u_{i+1}}$$
for another suitable pair (m', u_{i+1}).

When we reach $i = n$, we get
$$s_{k,x}$$
For $i > n$, reasoning like the above implies that all of the s-symbols of $\xi(\eta(kx))$ are of the form
$$s^{\epsilon}_{m'',v}$$
with $\epsilon = \pm 1$. If follows that
$$s_{k_i,x_j} = \xi\left(k_i x_j \eta(k_i x_j)^{-1}\right)$$

becomes

$$s_{k_i,x_j} = \underbrace{s_{m_1,u_1}^{\epsilon_1} \cdots s_{m_{n-1},u_{n-1}}^{\epsilon_{n-1}}}_{s_{m,x}\text{-factors}} \cdot s_{k_i,x_j} \cdot \underbrace{s_{m_{n+1},u_1}^{\epsilon_{n+1}} \cdots s_{m_{n+q-1},u_{n+q-1}}^{\epsilon_{n+q-1}}}_{s_{m,x}\text{-factors}}$$

Since this is implied by

$$s_{k_i,x_j} = s_{k_i,x_j}$$

and the relations $s_{m_i,u_i} = 1$, we can use Tietze transformations T2 to eliminate them. □

Now we will apply this to the group and subgroup in example 4.10.12 on page 94.

EXAMPLE 4.10.24. Let $G = \langle x, y | \, \rangle$ and let $H \triangleleft G$ be the normal subgroup generated by elements $\{x^i y x^{-i}\}$, with the coset representative function defined in 4.10.15 on page 97. It is not hard to see that the rewriting process, ζ, defined in example 4.10.12 on page 94 is Reidemeister, defined by this coset representative function, which is also Schreier since the coset-representatives are

$$\{x^i, i \in \mathbb{Z}\}$$

Theorem 4.10.23 on page 101 implies that

$$H = \left\langle s_{x^i,y}, s_{x^i,x}, i \in \mathbb{Z} | s_{x^i,x}, i \in \mathbb{Z} \right\rangle = \left\langle s_{x^i,y}, i \in \mathbb{Z} | \, \right\rangle$$

We conclude that the free group on *two* letters contains a subgroup isomorphic to a free group on a *(countable) infinity* of letters!

Compare this with proposition 4.6.6 on page 61, which shows what a difference *abelian-ness* makes!

The following result shows that we can (in principle) always find a Reidemeister-Schreier rewriting process for a subgroup.

THEOREM 4.10.25. *If $G = \langle X|R \rangle$ is a presentation of a group and H is a subgroup, then a Schreier right-coset representative function*

$$\eta \colon F_X \to F_X$$

exists for H.

REMARK. In the literature, η is often called a *Schreier transversal*. The proof explicitly constructs η, although the method used isn't particularly practical.

PROOF. For each coset Hg of H in F_X let $\ell(Hg)$ denote its length: the length of the *shortest reduced word* in F_X representing elements of the coset. We have

$$\ell(H \cdot 1) = 0$$
$$\ell(H \cdot x) = 1 \text{ for an } x \in X$$

Define $\eta(H) = 1$ and, among all the cosets of length 1 (i.e., of the form $H \cdot x$ for some $x \in X$) (arbitrarily) pick a word, u, of length 1 and define $\eta(H \cdot x) = u$. Every word of length 1 is in one of these cosets.

For each coset, C, of length 2, (arbitrarily) pick a minimal-length word, $u_1 u_2$, and define $\eta(C) = \eta(u_1) u_2$.

For each coset, C, of length 3, pick a word of length 3 like $u_1 u_2 u_3$ and define $\eta(C) = \eta(u_1 u_2) u_3$. We continue this process until we have defined η for all of the cosets of H.

Since every initial segment of a word in C_η is also in C_η, this defines a Schreier right-coset representative function. □

This immediately implies

COROLLARY 4.10.26. *Every subgroup of a free group is free.*

REMARK. This result is due to Schreier in [97].

PROOF. Given any subgroup, we can always find a Schreier right-coset representative function and apply theorem 4.10.23 on page 101. Since the original group was free, there are no relations — and the same is true for the subgroup. □

This concludes our discussion of combinatorial group theory. The interested reader is referred to [72].

EXERCISES.

5. Let F be the free group on two letters, x, and y and let H be the subgroup generated by words $\{x^i y^i\}$ for $i \in \mathbb{Z}$. Show that H is a free group on an infinite number of letters.

4.11. Groups of small order

Using the Sylow theorems and other considerations, it is possible to completely characterize small groups. We already know that groups of order p are cyclic, if p is a *prime*.

Order 4: Since $4 = 2^2$, exercise 1 on page 80 implies that the group is abelian. At this point, theorem 4.6.10 on page 65 implies that we have two groups $\mathbb{Z}_4$ and $\mathbb{Z}_2 \oplus \mathbb{Z}_2$.

Order 6: If G is a group of order 6, it must contain subgroups of order 2 and order 3, by the Sylow theorems. These subgroups are cyclic and isomorphic to $\mathbb{Z}_2$ and $\mathbb{Z}_3$. We get an abelian group $\mathbb{Z}_2 \oplus \mathbb{Z}_3 \cong \mathbb{Z}_6$. If assume G is nonabelian and a generates the copy of $\mathbb{Z}_2$ and b generates the copy of $\mathbb{Z}_3$, we get strings 1, a, b, b^2, ab, ba, where $ab \neq ba$. All of these strings must represent distinct elements of G because $b^2 = ab$ or ba implies that b is the identity. Consider the string aba. It must equal one of the strings listed above, but cannot equal a, b, ab, or ba (since these equalities would imply that a or b is the identity or one is an inverse of the other). It follows that $aba = b^2$. The mapping

$$a \mapsto (1,2)$$
$$b \mapsto (1,2,3)$$

defines an isomorphism between G and S_3.

To summarize, we have two groups $\mathbb{Z}_6$ and S_3, one *abelian* and one *nonabelian*.

Order 8: If $|G| = 8$, theorem 4.6.10 on page 65 gives us three abelian groups

$$\mathbb{Z}_2 \oplus \mathbb{Z}_2 \oplus \mathbb{Z}_2$$
$$\mathbb{Z}_2 \oplus \mathbb{Z}_4$$
$$\mathbb{Z}_8$$

Suppose G is not abelian. The order of elements must divide 8. If all elements are of order 2, let a, b, c be three distinct elements of order 2. Then the elements $\{1, a, b, c, ab, bc, ac, abc\}$ must all be distinct (equating any two of them implies that two other elements are the same), hence they constitute all of G. Consider the element ba. We have $baba = 1$ and multiplying on the left by b and the right by a gives $ab = ba$. We conclude that a, b, c commute with each other and G is abelian.

It follows that G has an element of order 4. Call this element a and denote the subgroup it generates by $H = \{1, a, a^2, a^3\}$. If $b \notin H$

$$G = H \cup bH$$

Now suppose the coset bH has an element of order 2 — we will call it b. Then

$$G = \{1, a, a^2, a^3, b, ab, a^2b, a^3b\}$$

The element ba cannot equal a or b (or one of the generators would be 1), so it must equal ab, a^2b, or a^3b. If it equals ab, then G is abelian. If

(4.11.1) $$ba = a^2b$$

then multiplying on the left by b gives

$$\begin{aligned}
b^2 a &= ba^2 b \\
&= baab \\
&= a^2 bab && \text{applying } ba = a^2b \text{ on the left} \\
&= a^4 b^2 && \text{applying it with } ba \text{ in the middle}
\end{aligned}$$

Since $b^2 = 1$, we conclude that $a = a^4$ or $a^3 = 1$, which is impossible. It follows that $ba = a^3b$ so proposition 4.5.17 on page 57 implies that $G = D_8$, the dihedral group.

Now assume that all elements of $bH = \{b, ab, a^2b, a^3b\}$ have order 4. Since b^2 has order 2 and cannot equal any of the elements of bH, it follows that $b^2 = a^2$, the only element of order 2.

Again, $ba = ab$ implies that G is abelian (and leads to other impossible conclusions). If $ba = a^2b$, then the same argument used above implies that $b^2a = a^4b^2$. Since a has order 4, it follows that $b^2a = b^2$ or $a = 1$, a contradiction. It follows that we must have $ba = a^3b$. The group G has the presentation

$$G = \left\langle a, b \mid a^4, b^4, a^2 = b^2, ba = a^3b \right\rangle$$

This is called the *quaternion group,* and denoted Q. This is the group of unit-quaternions with integer coefficients — see section 9 on page 323.

Order 9: In this case, exercise 1 on page 80 implies that the group is abelian. It follows from theorem 4.6.10 on page 65 implies that we have two groups $\mathbb{Z}_9$ and $\mathbb{Z}_3 \oplus \mathbb{Z}_3$.

CHAPTER 5

The Theory of Rings

"In the broad light of day, mathematicians check their equations and their proofs, leaving no stone unturned in their search for rigour. But at night, under the full moon, they dream, they float among the stars and wonder at the miracle of the heavens. They are inspired.
Without dreams there is no art, no mathematics, no life."
— Sir Michael Atiyah, *Notices of the AMS*, January 2010, page 8.

5.1. Basic concepts

Rings are mathematical structures with more features than groups — *two* operations, written as addition and multiplication.

DEFINITION 5.1.1. A *ring*, R, is a set equipped with two binary operations, denoted $+$ and multiplication, $\cdot$, such that, for all $r_1, r_2, r_2 \in R$,
(1) $(r_1 + r_2) + r_3 = r_1 + (r_2 + r_3)$
(2) $(r_1 \cdot r_2) \cdot r_3 = r_1 \cdot (r_2 \cdot r_3)$
(3) $r_1 \cdot (r_2 + r_3) = r_1 \cdot r_2 + r_1 \cdot r_3$
(4) $(r_1 + r_2) \cdot r_3 = r_1 \cdot r_3 + r_1 \cdot r_3$
(5) there exists elements $0, 1 \in R$ such that $r + 0 = 0 + r = r$ and $r \cdot 1 = 1 \cdot r = r$ for all $r \in R$.
(6) For every $r \in R$, there exists an element $s \in R$ such that $r + s = 0$.

The ring R will be called *commutative* if $r_1 \cdot r_2 = r_2 \cdot r_1$ for all $r_1, r_2 \in R$.

A *division ring* is one in which every nonzero element has a multiplicative inverse.

A *subring* $S \subset R$ is a subset of R that is also a ring under the operations $+$ and $\cdot$.

REMARK. We will also regard the set containing only the number 0 as a ring with $0 + 0 = 0 = 0 \cdot 0$ — the *trivial ring* (the multiplicative and additive identities are the same). When an operation is written with a '$+$' sign it is implicitly assumed to be commutative.

We have seen (and worked with) *many* examples of rings before, $\mathbb{Z}$, $\mathbb{Z}_m$, $\mathbb{Q}$, $\mathbb{R}$, and $\mathbb{C}$. The rings $\mathbb{Q}$, $\mathbb{R}$, and $\mathbb{C}$ are *commutative* division rings — also known as *fields*. For an example of a *noncommutative* division ring, see section 9.2 on page 325.

We can classify elements of a ring by certain basic properties:

DEFINITION 5.1.2. An element $u \in R$ of a ring will be called a *unit* if there exists another element $v \in R$ such that $u \cdot v = v \cdot u = 1$. The set of units of a ring, R, form a group, denoted $R^\times$. A commutative ring in which every nonzero element is a unit is called a *field*.

An element $u \in R$ is called a *zero-divisor* if it is nonzero and if there exists a nonzero element $v \in R$ such that $u \cdot v = 0$.

EXAMPLE. Perhaps the simplest example of a ring is the integers, $\mathbb{Z}$. This is simple in terms of familiarity to the reader but a detailed analysis of the integers is a very deep field of mathematics in itself (number theory). Its only units are ± 1, and it has no zero-divisors.

We can use the integers to construct:

EXAMPLE. If m is an integer, the numbers modulo m, $\mathbb{Z}_m$ is a ring under addition and multiplication modulo m. In $\mathbb{Z}_6$, the elements 2 and 3 are zero-divisors because $2 \cdot 3 = 0 \in \mathbb{Z}_6$.

EXAMPLE 5.1.3. The rational numbers, $\mathbb{Q}$, are an example of a field. Other examples: the real numbers, $\mathbb{R}$, and the complex numbers, $\mathbb{C}$.

We also have *polynomial* rings:

DEFINITION 5.1.4. If R is a ring, rings of polynomials $R[X]$ is the ring of polynomials where addition and multiplication are defined

$$\left(\sum_{i=0}^{n} a_i X^i\right) + \left(\sum_{i=0}^{m} b_i X^i\right) = \sum_{i=0}^{\max(n,m)} (a_i + b_i) X^i$$

$$\left(\sum_{i=0}^{n} a_i X^i\right)\left(\sum_{j=0}^{m} b_j X^j\right) = \sum_{k=0}^{n+m} \left(\sum_{i+j=k} a_i b_j\right) X^k$$

with $a_i, b_j \in R$ and $a_i = 0$ if $i > n$ and $b_i = 0$ if $i > m$.

More formally, one can define $R[X]$ as the set of infinite sequences

(5.1.1) $$(r_0, \ldots, r_i, \ldots)$$

with the property that all but a *finite* number of the r_i vanish, and with addition defined by

$$(r_0, \ldots, r_i, \ldots) + (s_0, \ldots, s_i, \ldots) = (r_0 + s_0, \ldots, r_i + s_i, \ldots)$$

and multiplication defined by

$$(r_0, \ldots, r_i, \ldots)(s_0, \ldots, s_i, \ldots) = (t_0, \ldots, t_i, \ldots)$$

with

$$t_n = \sum_{\substack{i+j=n \\ i \geq 0, j \geq 0}} r_i s_j$$

In this case,

$$\sum_{i=0}^{k} r_i X^i$$

becomes the *notation* for the sequence $(r_0, \ldots, r_i, \ldots, r_k, 0 \cdots)$.

EXAMPLE. $\mathbb{Z}[X]$ is the ring of polynomials with integer coefficients.

We can also discuss *noncommutative* rings:

EXAMPLE 5.1.5. The set of $n \times n$ real matrices, $M_n(\mathbb{R})$ under matrix-multiplication (see definition 6.2.13 on page 171) and addition is a ring — see exercise 2 on page 171.

A common ring used in algebraic topology and group-representation theory is:

DEFINITION 5.1.6. If G is a group and R is a commutative ring, the *group-ring*, RG, is defined as the set of all finite formal linear combinations of elements of G with coefficients in R:

$$\left(\sum_{i=1}^{n} r_i g_i\right)\left(\sum_{j=1}^{m} s_j h_j\right) = \sum_{i=1, j=1}^{n,m} r_i s_j (g_i \cdot h_j)$$

where the $r_i, s_j \in R$, $g_i, h_j \in G$ and the products $g_i \cdot h_j$ are computed using the product-operation in G. These are heavily used in group-representation theory (see chapter 11 on page 387) and algebraic topology. If G is not abelian, this will be a non-commutative ring.

We can also define *power-series* rings

DEFINITION 5.1.7. If R is a ring, the *ring of power-series* $R[[X]]$ over R is the ring of formal power series

$$\sum_{i=1}^{\infty} a_i X^i$$

with addition and multiplication defined as for $R[X]$. As with polynomial-rings, one can formally define the elements of $R[[X]]$ as infinite sequences like those in 5.1.1 on the preceding page where we allow an *infinite* number of the r_i to be *nonzero*.

REMARK. Note that these power-series are like infinite polynomials. If we impose a metric on R the ring of power-series that *converge* with respect to that metric can be very different from $R[[X]]$.

CLAIM 5.1.8. We can define a metric on $R[[X]]$ that makes power-series convergent in the usual sense.

Let $p, q \in R[[X]]$ and define the distance between them by

$$d(p, q) = \left(\frac{1}{2}\right)^{v(p-q)}$$

where $X^{v(p-q)} \mid (p-q)$ but $X^{v(p-q)+1} \nmid (p-q)$, i.e. the function $v(x)$ is equal to the degree of the lowest-degree term of x. In this metric all formal power-series series converge and we can define Cauchy-sequences, etc.

Power series rings can have very different properties than polynomial rings. For instance

PROPOSITION 5.1.9. *In the ring* $R[[X]]$, *any element*

$$\alpha = \sum_{k=0}^{\infty} a_k X^k$$

where $a_0 \in R$ *is a unit (see definition 5.1.2 on page 107) has a multiplicative inverse.*

REMARK. The multiplicative inverse of α is
$$\frac{1}{a_0} - \frac{a_1}{a_0^2}X + \frac{a_0a_2 - a_1^2}{a_0^3}X^2 + \cdots$$

PROOF. Suppose the inverse is
$$\sum_{j=0}^{\infty} b_j X^j$$
and multiply α by this to get
$$\sum_{n=0}^{\infty} c_n X^n$$
with
$$c_n = \sum_{j=0}^{n} a_j b_{n-j}$$
$$c_0 = a_0 b_0$$
$$b_0 = a_0^{-1}$$
In general, we get a recursive equation
$$b_n = -a_0^{-1} \sum_{k=0}^{n-1} b_k a_{n-k}$$
that computes b_n for any n. $\square$

We also have extension rings

DEFINITION 5.1.10. Suppose we have an embedding of rings $R \subset \Omega$ and $\alpha \in \Omega$ is some element. Then $R[\alpha] \subset \Omega$ is the subring of all possible polynomials
$$\sum_{i=1}^{n} c_i \alpha^i$$
with $c_i \in R$.

EXAMPLE. In the extension $\mathbb{Q}[\sqrt{2}]$, the fact that $(\sqrt{2})^2 \in \mathbb{Q}$ implies that all elements of $\mathbb{Q}[\sqrt{2}]$ will actually be of the form $a + b\sqrt{2}$, with $a, b \in \mathbb{Q}$.

EXERCISES.

1. In the group-ring, $\mathbb{Z}S_3$, compute the product
$$(2(1,2,3) - 6(1,2))(2(1,3) - 4)$$

5.2. Homomorphisms and ideals

Now that we have defined rings, we can define mappings of them:

DEFINITION 5.2.1. Given two rings, R and S, a function $f: R \to S$ is called a *homomorphism* if, for all $r_1, r_2 \in R$:
  (1) $f(r_1 + r_2) = f(r_1) + f(r_2) \in S$
  (2) $f(r_1 \cdot r_2) = f(r_1) \cdot f(r_2) \in S$ and $f(1) = 1$.

The set of elements $r \in R$ with the property that $f(r) = 0$ is called the *kernel* of the homomorphism, or $\ker f$. If the homomorphism is *surjective* and its kernel vanishes, it is called an *isomorphism*. An isomorphism from a ring to itself is called an *automorphism*.

REMARK. Compare this to 4.2.3 on page 40 for *groups*. It will turn out that many of the concepts in ring theory correspond to similar ones in group theory. As one might imagine, this means that they are special cases of more general concepts. Category theory (see chapter 10 on page 339) attempts to find these more general concepts.

PROPOSITION 5.2.2. *Let K be the kernel of a homomorphism $f: R \to S$ of rings. If $k \in K$ and $r \in R$, then $r \cdot k, k \cdot r \in K$.*

PROOF. The defining property of a homomorphism implies that $f(r \cdot k) = f(r) \cdot f(k) = f(r) \cdot 0 = 0$. □

We abstract out the important property of the kernel of a homomorphism with:

DEFINITION 5.2.3. If R is a ring, a *left-ideal* $\mathfrak{J} \subset R$ is a subset closed under addition with the property that, for all $x \in \mathfrak{J}, r \in R, r \cdot x \in \mathfrak{J}$. A *right-ideal* is a subset closed under addition $\mathfrak{J} \subset R$ such that $x \cdot r \in \mathfrak{J}$ for all $r \in R$. An ideal that is both left and right is called a *two-sided ideal*, or just an *ideal*.
  (1) An ideal, $\mathfrak{J} \subset R$ is *prime* if $a \cdot b \in \mathfrak{J}$ implies that $a \in \mathfrak{J}$ or $b \in \mathfrak{J}$ (or both).
  (2) The *ideal generated by* $\alpha_1, \ldots, \alpha_n \in R$, denoted $(\alpha_1, \ldots \alpha_n) \subseteq R$, is the set of all linear combinations
  $$\sum_{k=1}^{n} r_k \cdot \alpha_k \cdot s_k$$
  where the r_i and s_i run over all elements of R. The element 0 is an ideal, as well as the whole ring. The set $\alpha_1, \ldots, \alpha_n \in R$ is called a *basis* for the ideal $(\alpha_1, \ldots \alpha_n)$.
  (3) An ideal $\mathfrak{J} \subset R$ is *maximal* if $\mathfrak{J} \subset \mathfrak{K}$, where $\mathfrak{K}$ is an ideal, implies that $\mathfrak{K} = R$. This is equivalent to saying that for any $r \in R$ with $r \notin \mathfrak{J}$,
  $$\mathfrak{J} + (r) = R$$
  (4) An ideal generated by a *single element* of R is called a *principal ideal*.
  (5) Given two ideals $\mathfrak{a}$ and $\mathfrak{b}$, their *product* is the ideal generated by all products $\{(a \cdot b) | \forall a \in \mathfrak{a}, b \in \mathfrak{b}\}$.

REMARK. Following a convention in algebraic geometry, we will usually denote ideals by *Fraktur letters*.

If R is commutative, all ideals are two-sided.

EXAMPLE. We claim that the ideals of $\mathbb{Z}$ are just the sets

$$
\begin{aligned}
(0) &= \{0\} \\
(2) &= \{\ldots, -4, -2, 0, 2, 4, 6, 8, \ldots\} \\
(3) &= \{\ldots, -6, -3, 0, 3, 6, 9, 12, \ldots\} \\
&\vdots \\
(n) &= \{n \cdot \mathbb{Z}\}
\end{aligned}
$$

for various values of n. Proposition 3.1.5 on page 14 shows that $(n, m) = (\gcd(m, n))$ and a simple induction shows every ideal of $\mathbb{Z}$ is generated by a single element. Note that the ideal $(1) = \mathbb{Z}$. An ideal $(n) \subset \mathbb{Z}$ is *prime* if and only if n is a prime number.

Maximal ideals are prime:

PROPOSITION 5.2.4. *If R is a commutative ring with maximal ideal $\mathfrak{J}$, then $\mathfrak{J}$ is also prime.*

PROOF. This is similar to the proof of proposition 3.1.7 on page 15. Suppose $r, s \in R$, $r \cdot s \in \mathfrak{J}$ but $r \notin \mathfrak{J}$. Then $\mathfrak{J} + (r) = R$ so that there exists a $t \in R$ such that
$$a + t \cdot r = 1$$
where $a \in \mathfrak{J}$. If we multiply this by s, we get
$$a \cdot s + t \cdot r \cdot s = s$$
Since both terms on the left are in $\mathfrak{J}$, it follows that $s \in \mathfrak{J}$. □

Proposition 5.2.2 on the previous page shows that the kernel of a homomorphism is an ideal. The following is a converse to that:

PROPOSITION 5.2.5. *Let R be a ring and let $\mathfrak{J} \subset R$ be an ideal. For all $r_1, r_2 \in R$ define*
$$r_1 \equiv r_2 \pmod{\mathfrak{J}}$$
if $r_1 - r_2 \in \mathfrak{J}$. Then $\equiv$ is an equivalence relation. If we denote the set of equivalence-classes by $R/\mathfrak{J}$, then the ring-operations of R induce corresponding operations on $R/\mathfrak{J}$ making it into a ring (called the quotient ring of R by $\mathfrak{J}$). The canonical map
$$R \to R/\mathfrak{J}$$
that sends an element to its equivalence class is a homomorphism with kernel $\mathfrak{J}$.

REMARK 5.2.6. We can also think of the elements of $R/\mathfrak{J}$ as disjoint sets of elements of R, namely sets of the form
$$r + \mathfrak{J}$$
These are all of the elements of R equivalent to $r \in R$.

PROOF. It is not hard to see that
$$r_1 \equiv r_2 \pmod{\mathfrak{I}}$$
and
$$s_1 \equiv s_2 \pmod{\mathfrak{I}}$$
implies that
$$r_1 + s_1 \equiv r_2 + s_2 \pmod{\mathfrak{I}}$$
so that addition is well-defined in $R/\mathfrak{I}$. To see that multiplication is also well-defined note that
$$r_1 s_1 - r_2 s_2 = (r_1 - r_2) s_1 + r_2 (s_1 - s_2) \in \mathfrak{I}$$
due to the closure property in definition 5.2.3 on page 111. The final statement follows from the fact that $\mathfrak{I}$ is just the set of elements of R equivalent to 0. $\square$

EXAMPLE. Here are examples of quotient rings:
(1) For instance, $\mathbb{Z}/(n) = \mathbb{Z}_n$, the integers modulo n, where $\mathbb{Z}/(1)$ is the trivial ring.
(2) In the example given earlier, $\mathbb{Q}[X,Y]/(X) = \mathbb{Q}[Y]$ and $\mathbb{Q}[X,Y]/(X,Y) = \mathbb{Q}$.
(3) We can think of the ring $\mathbb{Q}[\sqrt{2}]$ two ways: as an extension or as a quotient
$$\mathbb{Q}[X]/(X^2 - 2)$$
There's a homomorphism
$$\mathbb{Q}[X] \to \mathbb{Q}[\sqrt{2}]$$
$$X \mapsto \sqrt{2}$$
whose kernel is exactly $(X^2 - 2)$. This induces an isomorphism $\mathbb{Q}[X]/(X^2 - 2) \cong \mathbb{Q}[\sqrt{2}]$.

Complementing the concept of kernel, we have the cokernel:

DEFINITION 5.2.7. If $f: R \to S$ is a homomorphism of rings and if $f(R) \subset S$ is an (two-sided) ideal in S, the quotient
$$\frac{S}{f(R)}$$
is called the *cokernel* of f.

REMARK. Cokernels for homomorphisms of rings do not always exist because one cannot "divide" a ring by an arbitrary subring.

DEFINITION 5.2.8. A ring R is called a *local ring* if it has a *unique* maximal ideal.

For instance, let R be the subring of $\mathbb{Q}$ of fractions
$$\frac{p}{q}$$
where q is an odd number. Then $2 \cdot R \subset R$ is the only ideal not equal to all of R. It follows that R is a local ring.

We could also have defined R by
$$R = \mathbb{Z}[\tfrac{1}{3}, \tfrac{1}{5}, \ldots, \tfrac{1}{p}, \ldots]$$
where p runs over all odd primes.

Here is how the projection to a quotient ring affects ideals:

LEMMA 5.2.9. *Let R be a commutative ring and let $\mathfrak{a} \subset R$ be an ideal and let*
$$p: R \to R/\mathfrak{a}$$
Then p induces a one-to-one correspondence between ideals of $R/\mathfrak{a}$ and ideals $\mathfrak{b} \subset R$ that contain $\mathfrak{a}$. In addition,

▷ *$p(\mathfrak{b})$ is prime or maximal in $R/\mathfrak{a}$ if and only if $\mathfrak{b}$ is prime or maximal in R*
▷ *$p^{-1}(\mathfrak{c})$ is prime or maximal in R if and only if $\mathfrak{c}$ is prime or maximal in $R/\mathfrak{a}$.*

PROOF. Let $\mathfrak{b} \subset R$ be an ideal containing $\mathfrak{a}$ and let $y \in R$ with $p(y) = x \in R/\mathfrak{a}$. Then $x \cdot p(\mathfrak{b}) = p(y \cdot \mathfrak{b}) \subset p(\mathfrak{b})$ so that $p(\mathfrak{b}) \subset R/\mathfrak{a}$ is an ideal.

Suppose $\mathfrak{b}$ is maximal in R. Then $(x) + p(\mathfrak{b}) = p((y) + \mathfrak{b}) = p(R) = R/\mathfrak{a}$ so $p(\mathfrak{b})$ is maximal in $R/\mathfrak{a}$.

If $\mathfrak{b}$ is prime, $X_1 \cdot x_2 \in p(\mathfrak{b})$ implies that $y_1 \cdot y_2 \in \mathfrak{b}$, where $p(y_i) = x_i$, and either $y_1 \in \mathfrak{b}$ or $y_2 \in \mathfrak{b}$, which implies that $x_1 \in p(\mathfrak{b})$ or $x_2 \in \mathfrak{b}$. This means that $p(\mathfrak{b}) \subset R/\mathfrak{a}$ is prime.

Now suppose $\mathfrak{c} \subset R/\mathfrak{a}$. Then $\mathfrak{a} \subset p^{-1}(\mathfrak{a})$ (since $\mathfrak{a} = p^{-1}(0)$). If $x \in R$, then $x \cdot p^{-1}(\mathfrak{c})$ has the property that its image under p is equal to $\mathfrak{c}$, i.e., it is contained in $p^{-1}(\mathfrak{c})$. It follows that $p^{-1}(\mathfrak{c})$ is an ideal of R.

Suppose $\mathfrak{c}$ is maximal in $R/\mathfrak{a}$, and suppose that $x \in R$ has the property that $x \notin p^{-1}(\mathfrak{c})$. Then $p(x) \notin \mathfrak{c}$ and $\mathfrak{I} = (x) + p^{-1}(\mathfrak{c})$ is an ideal of R that has the property that $p(\mathfrak{I}) = (p(x)) + \mathfrak{c} = R/\mathfrak{a}$. So $\mathfrak{I} = R$ and $p^{-1}(\mathfrak{c})$ is maximal.

We leave the final statement that p^{-1} of a prime ideal is prime as an exercise. □

We will also need to know the effect of multiple quotients:

LEMMA 5.2.10. *Let R be a ring with ideals $\mathfrak{a} \subset \mathfrak{b} \subset R$. Let*
(1) $f: R \to R/\mathfrak{a}$,
(2) $g: R \to R/\mathfrak{b}$ and
(3) $h: R/\mathfrak{a} \to (R/\mathfrak{a})/f(\mathfrak{b})$

be projections to the quotients. Then $(R/\mathfrak{a})/f(\mathfrak{b}) = R/\mathfrak{b}$ and the diagram

$$\begin{array}{ccc} R & \xrightarrow{f} & R/\mathfrak{a} \\ {\scriptstyle g}\downarrow & & \downarrow{\scriptstyle h} \\ R/\mathfrak{b} & = & (R/\mathfrak{a})/f(\mathfrak{b}) \end{array}$$

commutes.

PROOF. Elements of $R/\mathfrak{a}$ are equivalence classes of the equivalence relation
$$r_1 \sim_\mathfrak{a} r_2 \text{ if } r_1 - r_2 \in \mathfrak{a}$$
or *sets* of the form (see remark 5.2.6 on page 112)
$$r + \mathfrak{a} \subset R$$
and elements of $R/\mathfrak{b}$ are sets of the form
$$r + \mathfrak{b} \subset R$$
Elements of $(R/\mathfrak{a})/f(\mathfrak{b})$ are sets of the form
$$q + f(\mathfrak{b})$$
where $q \in R/\mathfrak{a}$, or sets of the form
$$r + \mathfrak{a} + \mathfrak{b} = r + \mathfrak{b}$$
This shows that $(R/\mathfrak{a})/f(\mathfrak{b}) = R/\mathfrak{b}$. The commutativity of the diagram follows from the fact that the image of $r \in R$ under the maps going down either side of the diagram is the set $r + \mathfrak{b}$. $\square$

PROPOSITION 5.2.11. *If $\mathfrak{I} \subset R$ is a proper (i.e., $1 \notin \mathfrak{I}$) ideal in a ring, then there exists a proper maximal ideal $\mathfrak{M} \subset R$ such that*
$$\mathfrak{I} \subset \mathfrak{M}$$

REMARK. 'Maximal' means no other proper ideal contains it.

PROOF. The ideals of R that contain $\mathfrak{I}$ can be ordered by inclusion. Every ascending chain of such ideals has an upper bound, namely the union. Zorn's Lemma implies (lemma 14.2.12 on page 465) that there is a maximal such ideal. $\square$

EXERCISES.

1. Show that
$$\mathbb{C} \cong \frac{\mathbb{R}[X]}{(X^2 + 1)}$$

2. If $x, y \in R$ are two elements with the property that $(x, y) = R$, show that $(x^n, y^m) = R$ for positive integers n, m.

3. Show that the converse of proposition 5.1.9 on page 109 is also true: if
$$\alpha = \sum_{i=0}^{\infty} a_i X^i \in R[[X]]$$
is a unit, so is a_0.

4. If $\mathfrak{a}$ and $\mathfrak{b}$ are ideals in a ring, show that $\mathfrak{a} \cdot \mathfrak{b} \subset \mathfrak{a} \cap \mathfrak{b}$.

5. Suppose $\mathfrak{a}, \mathfrak{b}, \mathfrak{p} \subset R$ are ideals in a commutative ring. If $\mathfrak{p}$ is a prime ideal and
$$\mathfrak{a} \cdot \mathfrak{b} \subset \mathfrak{p}$$
(for instance, if $\mathfrak{a} \cap \mathfrak{b} \subset \mathfrak{p}$) prove that either $\mathfrak{a} \subset \mathfrak{p}$ or $\mathfrak{b} \subset \mathfrak{p}$.

6. If
$$\mathfrak{p}_1 \supset \mathfrak{p}_2 \supset \cdots$$
is a decreasing sequence of prime ideals in a ring, show that
$$\mathfrak{p} = \bigcap \mathfrak{p}_i$$
is also a prime ideal.

7. In the ring $R = \mathbb{Q}[X, Y]$, show that the ideal (X) is prime but not maximal.

8. In the ring $R = \mathbb{Q}[\sqrt{2}]$, show that the map that leaves $\mathbb{Q}$ fixed and is defined by
$$\begin{aligned} f: \mathbb{Q}[\sqrt{2}] &\to \mathbb{Q}[\sqrt{2}] \\ \sqrt{2} &\mapsto -\sqrt{2} \end{aligned}$$
is an isomorphism of rings (so it is an *automorphism* of $\mathbb{Q}[\sqrt{2}]$).

9. Show that the ring $R = \mathbb{Q}[\sqrt{2}]$ is a field by finding a multiplicative inverse for any nonzero element.

10. Suppose R is a ring and $\mathfrak{J}$ is the intersection of all *maximal ideals* of R, i.e.
$$\mathfrak{J} = \bigcap_{\mathfrak{m} \text{ maximal in } R} \mathfrak{m}$$
If $r \in R$ has the property that $r \equiv 1 \pmod{\mathfrak{J}}$, show that r is a *unit* (i.e., has a multiplicative inverse).

11. If $\mathfrak{a}_1, \ldots, \mathfrak{a}_n \subset R$ are distinct ideals with the property that $\mathfrak{a}_i + \mathfrak{a}_j = R$ for any $i \neq j$, show that
$$\mathfrak{a}_i + \prod_{j \neq i} \mathfrak{a}_j = R$$
for any i where the product is take over all the integers $1, \ldots, n$ except i.

12. If $\mathfrak{a}_1, \ldots, \mathfrak{a}_n \subset R$ are distinct ideals with the property that $\mathfrak{a}_i + \mathfrak{a}_j = R$ for any $i \neq j$, and
$$\mathfrak{a} = \bigcap_{i=1}^{n} \mathfrak{a}_i$$
show that
$$\frac{R}{\mathfrak{a}} = \prod_{i=1}^{n} \frac{R}{\mathfrak{a}_i}$$
This is a generalization of theorem 3.3.5 on page 24, the Chinese Remainder Theorem.

5.3. Integral domains and Euclidean Rings

In this section, all rings will be assumed to be *commutative*.

Now we are in a position to define classes of rings with properties like those of the integers. An integral domain is a ring without zero-divisors (see definition 5.1.2 on page 107), and a Euclidean ring is one in which a version of the division algorithm (proposition 3.1.1 on page 13) applies.

DEFINITION 5.3.1. Let R be a commutative ring. Then R is an *integral domain* (or just a *domain*) if, for all $r_1, r_2 \in R$, $r_1 \cdot r_2 = 0$ implies that at least one of r_1 or r_2 is 0.

An element, x, of an integral domain is called *irreducible* if $x = a \cdot b$ implies that $x = u \cdot a$ or $x = u \cdot b$ where u is some unit of the ring (see definition 5.1.2 on page 107).

An element, x, is called *prime* if the principal ideal, (x), is prime (see definition 5.2.3 on page 111).

REMARK. For instance, $\mathbb{Z}$ is an integral domain but $\mathbb{Z}_6$ is not since $2 \cdot 3 \equiv 0 \pmod{6}$.

When we discussed the integers, we defined prime numbers as positive. In a general ring, the concept of "> 0" is not well-defined so we have to define irreducible elements "up to multiplication by a unit." It is as if we regarded 2 and -2 as essentially the same prime.

LEMMA 5.3.2. *Let $\mathfrak{a} \subset R$ be an ideal in a commutative ring. Then:*

(1) $\mathfrak{a}$ is prime if and only if $R/\mathfrak{a}$ is an integral domain.
(2) $\mathfrak{a}$ is maximal if and only if $R/\mathfrak{a}$ is a field.

PROOF. Let $a, b \in R/\mathfrak{a}$ be the images of $x, y \in R$ under the standard projection

$$R \to R/\mathfrak{a}$$

(see proposition 5.2.5 on page 112) Then $a \cdot b = 0 \in R/\mathfrak{a}$ if and only if

$$x \cdot y = 0 \pmod{\mathfrak{a}}$$

which is equivalent to saying that $x \cdot y \in \mathfrak{a}$. If $\mathfrak{a}$ is prime, $x \cdot y \in \mathfrak{a}$ implies that $x \in \mathfrak{a}$ or $y \in \mathfrak{a}$, which means that $a = 0$ or $b = 0$. Conversely, if $a \cdot b = 0 \in R/\mathfrak{a}$ always implies $a = 0$ or $b = 0$, then $x \cdot y \in \mathfrak{a}$ would always imply that $x \in \mathfrak{a}$ or $y \in \mathfrak{a}$.

If $\mathfrak{a}$ is maximal, then it is also prime (see proposition 5.2.4 on page 112) so we know that $R/\mathfrak{a}$ is an integral domain. Suppose $x \in R$ projects to $a \neq 0 \in R/\mathfrak{a}$. Since $a \neq 0$, we know that $x \notin \mathfrak{a}$, and since $\mathfrak{a}$ is maximal,

$$\mathfrak{a} + (x) = R$$

so $1 \in \mathfrak{a} + (x)$ and

$$y \cdot x + z = 1$$

for some $z \in \mathfrak{a}$ and $y \cdot x = 1 \pmod{\mathfrak{a}}$ so the image of y in $R/\mathfrak{a}$ is a multiplicative inverse of a.

The converse is left to the reader as an exercise. □

DEFINITION 5.3.3. A Euclidean domain, R, is an integral domain that has a function called the *norm*, $N\colon R \to \mathbb{N}$ that measures the "size" of an element, and such that a version of the division algorithm holds (see proposition 3.1.1 on page 13):

Given elements $a,b \in R$ with $b \nmid a$, there exist elements $q, r \in R$ such that
$$a = b \cdot q + r$$
with $r \neq 0$ and $N(r) < N(b)$.

REMARK. The term "norm" has at least two unrelated meanings in commutative algebra: the meaning above (which is like the degree of a polynomial) and norms of *field extensions* in section 7.3.1 on page 273.

EXAMPLE 5.3.4. If $\mathbb{F}$ is any field, the ring of polynomials with coefficients in $\mathbb{F}$, $\mathbb{F}[X]$ is a Euclidean domain, where the norm is the degree of a polynomial. Any irreducible polynomial generates a prime ideal.

Many basic properties of the integers immediately carry over to Euclidean rings — for instance, we have Bézout's Identity (that he originally proved for the Euclidean ring $\mathbb{R}[X]$):

PROPOSITION 5.3.5. *If R is a Euclidean ring and $a, b \in R$, and we define the greatest common divisor, $\gcd(a,b)$ of a and b to be the largest in terms of the norm, then there exist elements $u, v \in R$ such that*
$$\gcd(a,b) = u \cdot a + v \cdot b$$
If a and b have no common divisors (other than 1) then we can find $u, v \in R$ such that
$$1 = u \cdot a + v \cdot b$$

PROOF. Exactly the same as the proof of lemma 3.1.5 on page 14, but we replace every occurrence of "minimal" with "nonzero elements with minimal $N(*)$". $\square$

In fact, we can also prove this for a principal ideal domain:

PROPOSITION 5.3.6. *If R is a principal ideal domain, the concept of greatest common divisor is well-defined and, for any two elements $x, y \in R$, there exist elements $u, v \in R$ such that*
$$\gcd(x,y) = u \cdot x + v \cdot y$$

PROOF. If $x, y \in R$, then the ideal $(x,y) \subset R$ is generated by a single element (g), i.e. $(x,y) = (g)$. It follows that $g | x$ and $g | y$ — and $g = u \cdot x + v \cdot y$, which implies that any common divisor of x and y must divide g. We define g to be the greatest common divisor of x and y. $\square$

In rings with greatest common divisor, we can prove:

COROLLARY 5.3.7. *Let R be a Euclidean domain or a principal ideal domain, let $r \in R$ be some element, and let*
$$\begin{aligned} r &= p_1^{\alpha_1} \cdot \ldots \cdot p_k^{\alpha_k} \\ &= q_1^{\beta_1} \cdot \ldots \cdot q_\ell^{\beta_\ell} \end{aligned}$$

be factorizations into powers of irreducible elements. Then $k = \ell$ and there is a reordering of indices $f\colon \{1,\ldots,k\} \to \{1,\ldots,k\}$ such that $q_i = u_{f(i)} \cdot p_{f(i)}$ for some units, $u_{f(i)}$, and $\beta_i = \alpha_{f(i)}$ for all i from 1 to k.

PROOF. Simply repeat the proof of proposition 3.1.7 on page 15. □

Note that all ideals of R are principal, i.e., generated by a single element. We will be interested in general rings that share this property:

DEFINITION 5.3.8. A *principal ideal domain* is an integral domain in which all ideals are principal.

PROPOSITION 5.3.9. *All Euclidean domains are principal ideal domains.*

PROOF. Let R be a Euclidean domain with norm $N\colon R \to \mathbb{Z}$ and $\mathfrak{a} \subset R$ be an ideal. If $\mathfrak{a}' = \mathfrak{a} \setminus \{0\}$, let $x \in \mathfrak{a}'$ be a minimal element in the sense that there does not exist any element $y \in \mathfrak{a}'$ with $N(y) < N(x)$. We claim that $\mathfrak{a} = (x)$. If $y \in \mathfrak{a}$ is not a multiple of x, then we can divide y by x to get

$$y = x \cdot q + r$$

Because $\mathfrak{a}$ is an ideal, $x \cdot q \in \mathfrak{a}$. Since $y \in \mathfrak{a}$, it follows that $r \in \mathfrak{a}'$ and $N(r) < N(x)$, which contradicts the minimality of x. □

COROLLARY 5.3.10. *Let F be a field and let $F[X]$ be the ring of polynomials over F. Then $F[X]$ is a principal ideal domain.*

PROOF. It's easy to see that $F[X]$ is an integral domain. We claim that it is a Euclidean domain as well — see definition 5.3.3 on page 118. This is because we can divide polynomials as we do integers: given two polynomials $p(X), q(X)$ we can write

$$p(X) = a(X)q(X) + r(X)$$

with $a(X)$ as the quotient and $r(X)$ as the remainder where $\deg r(X) < \deg q(X)$. So the conclusion follows from proposition 5.3.9. □

Another important class of rings are unique factorization domains:

DEFINITION 5.3.11. A ring, R, is a *unique factorization domain* if it is a domain whose elements satisfy the conclusion of corollary 5.3.7 on page 118, i.e., if factorization of elements into irreducibles is unique up to units.

REMARK 5.3.12. Since Bézout's identity was used to prove unique factorization of integers (see proposition 3.1.7 on page 15), it follows that any principal ideal domain has unique factorization.

We have already seen several examples of unique factorization domains: the integers, polynomials over the rational numbers.

It is useful to give an example of a ring that is *not* a unique factorization domain. It shows that such examples are fairly common:

EXAMPLE 5.3.13. Consider the extension ring $\mathbb{Z}[\sqrt{-5}] \subset \mathbb{C}$. It is the set of all numbers

$$a + b\sqrt{-5}$$

with $a, b \in \mathbb{Z}$. These elements satisfy the multiplication law
$$(5.3.1) \quad (a_1 + b_1\sqrt{-5}) \cdot (a_2 + b_2\sqrt{-5}) = a_1 a_2 - 5 b_1 b_2 + (a_1 b_2 + a_2 b_1)\sqrt{-5}$$
It is not hard to see that the map $f\colon \mathbb{Z}[\sqrt{-5}] \to \mathbb{Z}[\sqrt{-5}]$ that sends $\sqrt{-5}$ to $-\sqrt{-5}$ is an automorphism (see definition 5.2.1 on page 111) — just plug it into equation 5.3.1.

If $x = a + b\sqrt{-5} \in \mathbb{Z}[\sqrt{-5}]$, then define
$$N(x) = x \cdot f(x) = a^2 + 5b^2 \in \mathbb{Z}$$
and

(1) $N(x) = 0$ if and only if $x = 0$.
(2) for all $x, y \in \mathbb{Z}[\sqrt{-5}]$,
$$N(x \cdot y) = x \cdot y \cdot f(x \cdot y) = x \cdot y \cdot f(x) \cdot f(y) = N(x) \cdot N(y)$$
since f is a homomorphism. This means that $a | b \in \mathbb{Z}[\sqrt{-5}]$ implies that $N(a) | N(b) \in \mathbb{Z}$.

Now note that $N(2) = 4$ and $N(3) = 9$. The only elements $z = a + b\sqrt{-5}$ with $N(z) \leq 9$ are $1 \pm \sqrt{-5}$. Both have $N(z) = 6$ which does not divide 4 or 9. It follows that the four elements $2, 3, 1 \pm \sqrt{-5} \in \mathbb{Z}[\sqrt{-5}]$ are *irreducible* — i.e., primes.

The formula
$$6 = 2 \cdot 3 = (1 - \sqrt{-5}) \cdot (1 + \sqrt{-5})$$
gives an example of non-unique factorization. So the ring $\mathbb{Z}[\sqrt{-5}]$ is not a unique factorization domain. The function, N, is an example of a *norm of a field-extension*, a topic covered in more detail in section 7.3.1 on page 273.

We conclude this section with an application to number theory. Consider the polynomial ring $\mathbb{Q}[X]$, polynomials with rational coefficients. Given any element $\alpha \in \mathbb{C}$, we can define a unique homomorphism of rings that sends X to α
$$f_\alpha \colon \mathbb{Q}[X] \to \mathbb{C}$$
$$p(X) \mapsto p(\alpha)$$
Since $\mathbb{Q}[X]$ is a PID, the *kernel* of this homomorphism will be a principal ideal of the form $(m_\alpha(X)) \subset \mathbb{Q}[X]$. We have two possibilities

1. $m_\alpha(X) \neq 0$, in which case, α is a root of a polynomial with rational coefficients and is called an *algebraic number*. In this case, $m_\alpha(X)$ is called the *minimal polynomial* of α. Example: $\sqrt{2}$ is an algebraic number with minimal polynomial $X^2 - 2$.

2. $m_\alpha(X) = 0$, in which case α is *not* the root of any polynomial with rational coefficients and is called a *transcendental number*. In this case, $\mathbb{Q}[\alpha] \cong \mathbb{Q}[X]$. Examples: e and π (although the proofs are not at all obvious).

EXERCISES.

1. If $\mathbb{F}$ is a field, show that the equation $x^n = 1$ in $\mathbb{F}$ has at most n solutions.

2. Let $C[0,1]$ be the ring of all real-valued continuous functions on the unit interval, $[0,1]$. If $a \in [0,1]$, let $\mathfrak{f}_a = \{f \in C[0,1] | f(a) = 0\}$. Show that $\mathfrak{f}_a \subset C[0,1]$ is a maximal ideal.

3. Find the greatest common divisor of
$$a(X) = X^4 + 3X^3 - 2X^2 + X + 1$$
and
$$b(X) = X^5 - X^3 + X + 5$$
in $\mathbb{Q}[X]$.

4. Show that there exists integral domains with pairs of elements that have no greatest common divisor. Hint: consider the subring $R \subset \mathbb{Q}[X]$ of polynomials with *no linear term* — i.e., polynomials of the form
$$f(x) = a_0 + a_2 X^2 + \cdots$$
and consider the monomials X^5 and X^6.

5. Let $f: R \to S$ be a homomorphism of rings and let $\mathfrak{p} \subset S$ be a prime ideal. Show that $f^{-1}(\mathfrak{p}) \subset R$ is also a prime ideal.

5.4. Noetherian rings

We add to our menagerie of ring-types (see figure 5.4.1 on page 123) with

DEFINITION 5.4.1. A ring R is *noetherian* if all of its ideals are finitely generated.

REMARK. This is a generalization of principal ideal domain. The term 'noetherian' is in honor of the mathematician Emmy Noether.

Emmy Noether (1882-1935) was a German-Jewish mathematician noted for her contributions to abstract algebra and theoretical physics. She was described by Pavel Alexandrov, Albert Einstein, Jean Dieudonné, Hermann Weyl, and Norbert Wiener as the most important woman in the history of mathematics. As one of the leading mathematicians of her time, she developed the theories of rings, fields, and algebras.

At the University of Göttingen, she proved the theorem now known as *Noether's theorem*, which shows that a conservation law is associated with any differentiable symmetry of a physical system.

After she fled Nazi Germany, she spent the last two years of her life at Bryn Mawr College and the Institute for Advanced Studies.

PROPOSITION 5.4.2. *The definition given above is equivalent to the statement:*

All increasing sequences of ideals in R eventually become constant, i.e., if
$$\mathfrak{a}_1 \subseteq \mathfrak{a}_2 \subseteq \cdots$$
then there exists a number n such that $\mathfrak{a}_i = \mathfrak{a}_{i+1}$ for all $i \geq n$. This is called the ascending chain condition *or ACC.*

PROOF. Consider the ideal
$$\mathfrak{a} = \bigcup_{i=1}^{\infty} \mathfrak{a}_i$$
If $\mathfrak{a} = (r_1, \ldots, r_n)$ for finite n, each of the r_i would occur in one of the $\mathfrak{a}_j$, say $\mathfrak{a}_{j(i)}$. If $k = \max(j(1), \ldots, j(n))$, then *all* of the $r_i \in \mathfrak{a}_K$ and
$$\mathfrak{a}_k = \mathfrak{a}_{k+1} = \cdots = \mathfrak{a}$$
On the other hand, if all ascending chains stabilize after a finite number of terms, let
$$\mathfrak{b} = (r_1, \ldots)$$
be an ideal generated by an infinite number of elements and define
$$\mathfrak{b}_n = (r_1, \ldots, r_n)$$
and consider the ascending chain of ideals
$$\mathfrak{b}_1 \subseteq \mathfrak{b}_2 \subseteq \cdots \subseteq \mathfrak{b}_k = \mathfrak{b}_{k+1} = \cdots$$
It follows that $\mathfrak{b} = (r_1, \ldots, r_k)$, which is finitely generated. □

REMARK. The similar-looking *descending* chain condition leads to a class of rings called Artinian rings — see definition 5.8.1 on page 158.

The following result (due to Emmy Noether — see [82]) shows that noetherian rings are extremely common:

LEMMA 5.4.3. *If R is noetherian, then so is $R[X]$.*

PROOF. Recall that, for a polynomial
$$f(X) = a_k X^k + \cdots + a_0$$
k is called the degree and a_k is called the leading coefficients. If $\mathfrak{a} \subseteq R[X]$ is an ideal, let $\mathfrak{c}_i$ be the set of all leading coefficients of polynomials in $\mathfrak{a}$ of degree $\leq i$.

Then $\mathfrak{c}_i \subseteq R$ is an ideal and
$$\mathfrak{c}_1 \subseteq \mathfrak{c}_2 \subseteq \cdots \subseteq \mathfrak{c}_i \subseteq \cdots$$

Because R is noetherian, this sequence eventually becomes constant, say $\mathfrak{c}_d = \mathfrak{c}_{d+1} = \cdots$. For each $i \leq d$, let
$$\mathfrak{c}_i = (a_{i,1}, \ldots, a_{i,n(i)}) \subset R$$
and let $f_{i,j} \in \mathfrak{a} \subset R[X]$ be a polynomial whose leading coefficient is $a_{i,j}$. If $f \in \mathfrak{a}$, we will show by induction on the degree of f that it lies in the ideal generated by the (finite) set of $f_{i,j}$.

When f has degree 0, the result is clear. If f has degree $s < d$ then
$$f = aX^s + \cdots$$
with $a \in \mathfrak{c}_s$, and
$$a = \sum_{j=1}^{n(s)} b_j \cdot a_{s,j}$$
for some $b_j \in R$, so
$$f - \sum_{j=1}^{n(s)} b_j \cdot f_{s,j}$$
is a polynomial of degree $s - 1$ and induction implies the conclusion.

If f has degree $s \geq d$, then
$$f = aX^s + \cdots$$
with $a \in \mathfrak{c}_d$. It follows that
$$a = \sum b_j \cdot a_{d,j}$$
for some $b_j \in R$ and that
$$f - \sum_j b_j \cdot f_{d,j} X^{s-d}$$
has degree $< \deg f$, and so lies in the ideal generated by the $\{f_{i,j}\}$ (by induction). $\square$

Some relations between classes of rings is illustrated in figure 5.4.1.

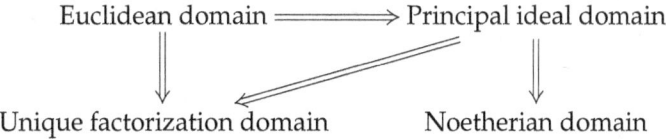

FIGURE 5.4.1.   Relations between classes of rings

THEOREM 5.4.4 (Hilbert Basis Theorem). *If R is noetherian, then so is $R[X_1, \ldots, X_n]$, i.e., every ideal is finitely generated.*

REMARK. Technically this is Noether's generalization of the Hilbert Basis Theorem. Hilbert originally proved it for R a field.

PROOF. Since R is noetherian, and
$$R[X_1, \ldots, X_n] = R[X_1, \ldots, X_{n-1}][X_n]$$
the theorem follows by an easy induction from lemma 5.4.3 on the preceding page. $\square$

 A variation of this argument even shows that *power-series rings* are noetherian.

LEMMA 5.4.5. *If R is a noetherian ring, so is $R[[X]]$.*

REMARK. As above, a simple induction shows that
$$R[[X_1, \ldots, X_k]]$$
is noetherian for any finite k.

PROOF. Given a power series $s \in R[[X]]$, let $\ell(s)$ denote its lowest nonzero coefficient.

Let $\mathfrak{a} \subset R[[X]]$ be an ideal whose elements are
$$v = \sum_{i=0}^{\infty} c(v)_i X^i$$
with $c(V)_i \in R$. Define $\mathfrak{h}_n \subset R$ to be composed of the nth coefficients of elements of $\mathfrak{a}$ whose lower coefficients vanish, i.e.
$$\mathfrak{h}_n = \{c(v)_n | v \in \mathfrak{a}, c(v)_0 = \cdots = c(v)_{n-1} = 0\}$$
We claim that the $\mathfrak{h}_n$ are all ideals of R (from the way elements of $R[[X]]$ are multiplied). Now set
$$\mathfrak{H}_n = \bigcup_{i=0}^{n} \mathfrak{h}_n$$
Then we get an ascending chain of ideals in R
$$\mathfrak{H}_0 \subset \cdots$$
which must eventually become constant with some finitely generated ideal (since R is noetherian)
$$\mathfrak{H}_{m+1} = \mathfrak{H}_m = (r_0, \ldots, r_t)$$
So the $\{r_i\}$ generate all of the coefficients of all elements of $\mathfrak{a}$. Each of the r_i is the lowest nonzero coefficient of some power series $f_i \in \mathfrak{a}$. We claim that
$$\mathfrak{a} = (f_0, \ldots, f_t)$$
Given $z = \sum_{i=0}^{\infty} c_i X^i \in \mathfrak{a}$, we must show that there exist power-series $d_i = \sum_{j=0}^{\infty} d_{i,j} X^j$ such that
(5.4.1) $$z = d_1 f_1 + \cdots + d_t f_t$$

If s is the highest degree of the lowest nonzero term that occurs in the f_i, we can subtract R-linear combinations of the f_i from z that will kill off all of its terms of degree $\leq s$ — giving z_1. This gives the *constant terms* of the d_i, i.e. $\{d_{i,0}\}$.

To cancel the lowest term of z_1, we know that its coefficient, c_{s+1}, is also a linear combination of the r_i. We must multiply suitable f_i by X to reach it, thus defining the $d_{1,i}$. Subtracting this linear combination gives z_2.

Continuing this indefinitely results in equation 5.4.1. Despite the seeming "infinite complexity" of z, we express it as a finite linear combination because we have infinite series available to us as coefficients. □

We conclude this section with a result due to Emmy Noether:

LEMMA 5.4.6. *Let $\mathfrak{J} \subset R$ be an ideal in a noetherian ring. Then:*

(1) in the set of prime ideals $\mathfrak{p}$ such that $\mathfrak{J} \subset \mathfrak{p}$, there is a minimal element
(2) the set of minimal prime ideals containing $\mathfrak{J}$ is finite.

PROOF. The first statement follows from:

(1) every ideal is contained in a maximal ideal (see proposition 5.2.11 on page 115),
(2) maximal ideals are prime (see proposition 5.2.4 on page 112) so every ideal is contained in at least one prime ideal,

(3) the intersection of a decreasing sequence of prime ideals is prime (see exercise 6 on page 116).

We prove the second statement by contradiction. Let $\mathscr{I}$ denote the set of ideals with an infinite number of minimal primes that contain them. Every ascending chain of ideals in $\mathscr{I}$

$$\mathfrak{J}_1 \subset \mathfrak{J}_2 \subset \cdots$$

has an upper bound since the sequence stabilizes after a finite number of terms (this is the only place where we use the noetherian property of R). Zorn's Lemma (14.2.12 on page 465) implies that $\mathscr{I}$ has a maximal member, $\mathfrak{M}$.

Clearly, $\mathfrak{M}$ is not prime because it would be the (one and only) minimal prime containing it. It follows that

(1) there exist $a, b \in R$ such that $a \cdot b \in \mathfrak{M}$ and $a \notin \mathfrak{M}$ and $b \notin \mathfrak{M}$.
(2) if $\mathfrak{A} = (a, \mathfrak{M})$ and $\mathfrak{B} = (b, \mathfrak{M})$, then $\mathfrak{M} \subset \mathfrak{A}$, $\mathfrak{M} \subset \mathfrak{B}$, and $\mathfrak{A} \cdot \mathfrak{B} \subset \mathfrak{M}$

If $\{\mathfrak{p}_i\}$ is the infinite set of minimal primes that contain $\mathfrak{J}$, exercise 5 on page 115 implies that, for each i, $\mathfrak{A} \subset \mathfrak{p}_i$ or $\mathfrak{B} \subset \mathfrak{p}_i$. It follows that $\mathfrak{A}$ or $\mathfrak{B}$ (or both) is contained in an infinite number of the $\mathfrak{p}_i$ — without loss of generality, we will say it is $\mathfrak{A}$. Since $\mathfrak{M} \subsetneq \mathfrak{A}$, it follows that $\mathfrak{A}$ can only have a *finite* number of minimal primes containing it. This is the contradiction. □

EXERCISES.

1. Show that every finite integral domain is a field.

2. Use the proposition 5.4.2 on page 122 to show that any quotient of a noetherian ring is noetherian.

3. Show that $\mathbb{Q}[X, Y]$ is *not* a Euclidean domain.

4. Find all the maximal ideals of $\mathbb{Z}[X]$.

5. Show that an element of $R[[X_1, \ldots, X_n]]$ is a unit if and only if its constant term is a unit in R.

6. If R is a noetherian ring, show that the nilradical is nilpotent, i.e. that there exists a integer $k > 0$ such that $\mathfrak{N}(R)^k = 0$.

7. Suppose $\mathfrak{p} \subset R$ is a minimal prime ideal in a noetherian (commutative) ring. Show that all of the elements of $\mathfrak{p}$ are zero-divisors.

5.5. Polynomial rings

Polynomial rings are important in many areas of algebra and algebraic geometry. Problems in important application-areas can often be reduced to computations in polynomial rings.

5.5.1. Ideals and systems of equations. Suppose
$$W = \mathbb{C}[X_1, \ldots, X_n]$$
suppose $f_1, \ldots, f_s \in W$, and suppose we want to solve the system of algebraic equations

(5.5.1)
$$f_1(X_1, \ldots, X_n) = 0$$
$$\vdots$$
$$f_s(X_1, \ldots, X_n) = 0$$

If $g_1, \ldots, g_t \in W$ is a set of polynomials with the property that
$$(f_1, \ldots, f_s) = (g_1, \ldots, g_t) = \mathfrak{B}$$
— i.e., the g_j are another basis for the ideal generated by the f_i, then the equations in 5.5.1 are *equivalent* to
$$g_1(X_1, \ldots, X_n) = 0$$
$$\vdots$$
$$g_t(X_1, \ldots, X_n) = 0$$

To see that, note that, since the f_i are a *basis* for $\mathfrak{B}$ and the $g_i \in \mathfrak{B}$, we have equations
$$g_i = \sum_{j=1}^{s} a_{i,j} f_j$$
where $a_{i,j} \in W$ for all i and j. It follows that $f_1 = \cdots = f_s = 0$ implies that $g_1 = \cdots = g_t = 0$. Since the g_j are *also* a basis for $\mathfrak{B}$, the *reverse* implication is also true.

EXAMPLE 5.5.1. Suppose we want to find solutions to the system of algebraic equations
$$xy = z^2$$
$$xz = 1$$
$$x^2 + y^2 = 3$$

We first make these into equations set to zero
$$xy - z^2 = 0$$
$$xz - 1 = 0$$
$$x^2 + y^2 - 3 = 0$$

and find another basis for the *ideal* these polynomials generate. It *turns out*[1] that
$$(xy - z^2, xz - 1, x^2 + y^2 - 3) = (z^8 - 3z^2 + 1, y - z^3, z^7 - 3z + x)$$

---

[1] This is not at all obvious! Later, we will look at an algorithm for coming to this conclusion.

So our original equations are equivalent to the equations

$$z^8 - 3z^2 + 1 = 0$$
$$y - z^3 = 0$$
$$z^7 - 3z + x = 0$$

or

$$z^8 - 3z^2 + 1 = 0$$
$$y = z^3$$
$$x = 3z - z^7$$

so that it follows that our original set of equations had *eight* solutions: find 8 roots of the polynomial in z and plug them into the equations for x and y.

It follows that there are applications to finding "simplified" or "improved" bases for ideals in polynomial rings.

5.5.2. Gröbner bases. One of the most powerful technique for computations in polynomial rings use a special basis for an ideal, called a *Gröbner basis*. Gröbner bases were discovered by Bruno Buchberger (in his thesis, [20]) and named after his teacher, Wolfgang Gröbner. He refined this construction in subsequent papers — see [21, 22].

One key idea in the theory of Gröbner bases involves imposing an *ordering* on the monomials in a polynomial ring:

DEFINITION 5.5.2. If $\mathbb{F}$ is a field, define an ordering on the elements of $\mathbb{N}^n$ and an induced ordering on the monomials of $\mathbb{F}[X_1, \ldots, X_n]$ by $\alpha = (a_1, \ldots, a_n) \succ \beta = (b_1, \ldots, b_n)$ implies that

$$\prod X_i^{a_i} \succ \prod X_i^{b_i}$$

The ordering of $\mathbb{N}^n$ must satisfy the conditions:
(1) if $\alpha \succ \beta$ and $\gamma \in \mathbb{N}^n$, then $\alpha + \gamma \succ \beta + \gamma$
(2) $\succ$ is a *well-ordering:* every set of elements of $\mathbb{N}^n$ has a *minimal* element.

For any polynomial $f \in \mathbb{F}[X_1, \ldots, X_n]$, let $\text{LT}(f)$ denote its *leading term* in this ordering — the polynomial's highest-ordered monomial with its coefficient.

REMARK. Condition 1 implies that the corresponding ordering of monomials is preserved by multiplication by a monomial. Condition 2 implies that there are no infinite descending sequences of monomials.

DEFINITION 5.5.3. Suppose $\mathbb{F}$ is a field and an ordering has been chosen for the monomials of $\mathbb{F}[X_1, \ldots, X_n]$. If $\mathfrak{a} \in \mathbb{F}[X_1, \ldots, X_n]$ is an ideal, let $\text{LT}(\mathfrak{a})$ denote the ideal generated by the leading terms of the polynomials in $\mathfrak{a}$.

(1) If $\mathfrak{a} = (f_1, \ldots, f_t)$, then $\{f_1, \ldots, f_t\}$ is a *Gröbner basis* for $\mathfrak{a}$ if

$$\text{LT}(\mathfrak{a}) = (\text{LT}(f_1), \ldots, \text{LT}(f_t))$$

(2) A Gröbner basis $\{f_1, \ldots, f_t\}$ is *minimal* if the leading coefficient of each f_i is 1 and for each i
$$\text{LT}(f_i) \notin (\text{LT}(f_1), \ldots, \text{LT}(f_{i-1}), \text{LT}(f_{i+1}), \ldots \text{LT}(f_t))$$
(3) A Gröbner basis $\{f_1, \ldots, f_t\}$ is *reduced* if the leading coefficient of each f_i is 1 and for each i and *no monomial* of f_i is contained in
$$(\text{LT}(f_1), \ldots, \text{LT}(f_{i-1}), \text{LT}(f_{i+1}), \ldots \text{LT}(f_t))$$

REMARK. There are many different types of orderings that can be used and a Gröbner basis with respect to one ordering will generally not be one with respect to another.

DEFINITION 5.5.4. The two most common orderings used are:
(1) *Lexicographic ordering.* Let $\alpha = (a_1, \ldots, a_n)$, $\beta = (b_1, \ldots, b_n) \in \mathbb{N}^n$. Then $\alpha > \beta \in \mathbb{N}^n$ if, in the vector difference $\alpha - \beta \in \mathbb{Z}^n$, the leftmost nonzero entry is positive — and we define
$$\prod X_i^{a_i} \succ \prod X_i^{b_i}$$
so
$$XY^2 \succ Y^3 Z^4$$
(2) *Graded reverse lexicographic order.* Here, monomials are first ordered by *total degree* — i.e., the sum of the exponents. Ties are resolved lexicographically (in reverse — higher lexicographic order represents a lower monomial).

REMARK. In Graded Reverse Lexicographic order, we get
$$X^4 Y^4 Z^7 \succ X^5 Y^5 Z^4$$
since the total degree is greater. As remarked above, Gröbner bases depend on the ordering, $\succ$: different orderings give different bases and even different *numbers* of basis elements.

Gröbner bases give an algorithmic procedure (detailed later) for deciding whether a polynomial is contained in an ideal and whether two ideals are equal.

To describe Buchberger's algorithm for finding a Gröbner (or standard) basis, we need something called the *division algorithm*. This is a generalization of the usual division algorithm for polynomials of a single variable:

ALGORITHM 5.5.5 (Division Algorithm). *Let $\succ$ be an ordering on the monomials of $\mathbb{F}[X_1, \ldots, X_n]$, where $\mathbb{F}$ is some field, and let $A = \{f_1, \ldots, f_k\}$ be a set of polynomials. If $f \in \mathbb{F}[X_1, \ldots, X_n]$ is some polynomial, the division algorithm computes polynomials $a_1, \ldots, a_s$ such that*

(5.5.2) $$f = a_1 f_1 + \cdots + a_k f_k + R$$

where $R = 0$ or no monomial in R is divisible by $\text{LT}(f_i)$ for any i.

In general, we will be more interested in the remainder, R, than the "quotients" a_i. We will use the notation
$$f \to_A R$$
to express the fact that the remainder has a certain value ("f reduces to R"). The algorithm is:

function DIVISION$(f, f_1, \ldots, f_k)$
    $a_i \leftarrow 0$
    $R \leftarrow 0$
    $g \leftarrow f$
    while $g \neq 0$ **do**
        Matched $\leftarrow$ **False**
        for $i = 1, \ldots, k$ **do**
            if $\mathrm{LT}(f_i) \,|\, \mathrm{LT}(g)$ **then**
                $h \leftarrow \frac{\mathrm{LT}(g)}{\mathrm{LT}(f_i)}$
                $a_i \leftarrow a_i + h$
                $g \leftarrow g - f_i \cdot h$
                Matched $\leftarrow$ **True** ▷ $\mathrm{LT}(g)$ *was* divisible by one of the $\mathrm{LT}(f_i)$
                Break   ▷ Leave the **for**-loop and continue the **While**-loop
            end if
        end for
        if not Matched **then** ▷ $\mathrm{LT}(g)$ was not divisible by *any* of the $\mathrm{LT}(f_i)$
            $R \leftarrow R + \mathrm{LT}(g)$ ▷ so put it into the remainder
            $g \leftarrow g - \mathrm{LT}(g)$ ▷ Subtract it from f
        end if
    end while
    return $f = a_1 f_1 + \cdots + a_k f_k + R$
    ▷ where the monomials of R are not divisible by the leading terms of *any* of the f_i
end function

REMARK. As is usual in describing algorithms, $a \leftarrow b$ represents *assignment*, i.e. "take the value in b and *plug* it into a" (the symbol '=' merely states that two quantities are equal). The symbol ▷ denotes a *comment* — on how the computation is proceeding.

It should be noted that:

PROPOSITION 5.5.6. *The division algorithm terminates in a finite number of steps and, in equation 5.5.2 on the facing page,*

(5.5.3) $$\mathrm{LT}(f) \succeq \mathrm{LT}(a_i f_i)$$

for $i = 1, \ldots, k$.

PROOF. The algorithm requires a finite number of steps because $\succeq$ is a *well-ordering*: any decreasing sequence of monomials must terminate in a finite number of steps (see the remark following definition 5.5.2 on page 127). In each iteration of the **While**-loop g (initially equal to f) loses a monomial and may gain others — which are ordered *lower* than the one it lost.

If there existed a term $a_i f_i$ with $\mathrm{LT}(a_i f_i) \succ \mathrm{LT}(f)$, there would have to be *cancellation* among leading terms of the $a_i f_i$, since their sum is f. The method used to construct the a_i guarantees that

$$\mathrm{LT}(a_i f_i) \neq \mathrm{LT}(a_j f_j)$$

for $i \neq j$, so *no* cancellation can occur among the leading terms in equation 5.5.2 on the facing page. □

EXAMPLE 5.5.7. Let $f = X^2Y + XY^2 + Y^2$, and let $f_1 = XY - 1$ and $f_2 = Y^2 - 1$.

Assume lexicographic ordering with $X \succ Y$. Then $\operatorname{LT}(f_1) \mid \operatorname{LT}(f)$ and we get

$$\begin{aligned} h &\leftarrow X \\ a_1 &\leftarrow X \\ g &\leftarrow g - X \cdot (XY - 1) \\ &= XY^2 + Y^2 + X \end{aligned}$$

In the second iteration of the **While**-loop, $\operatorname{LT}(f_1) \mid \operatorname{LT}(g)$ and

$$\begin{aligned} h &\leftarrow Y \\ a_1 &\leftarrow a_1 + Y \\ &= X + Y \\ g &\leftarrow g - Y \cdot (XY - 1) \\ &= Y^2 + X + Y \end{aligned}$$

In the third iteration of the **While**-loop, we have $\operatorname{LT}(f_1) \nmid \operatorname{LT}(g)$ and $\operatorname{LT}(f_2) \nmid \operatorname{LT}(g)$ so

$$\begin{aligned} R &\leftarrow X \\ g &\leftarrow g - X \\ &= Y^2 + Y \end{aligned}$$

In the fourth iteration of the **While**-loop, we have $\operatorname{LT}(f_1) \nmid \operatorname{LT}(g)$ but $\operatorname{LT}(f_2) \mid \operatorname{LT}(g)$ so

$$\begin{aligned} h &\leftarrow 1 \\ a_2 &\leftarrow 1 \\ g &\leftarrow g - 1 \cdot (Y^2 - 1) \\ &= Y + 1 \end{aligned}$$

Since neither Y nor 1 are divisible by the leading terms of the f_i they are thrown into the remainder and we get

$$f = (X + Y) \cdot f_1 + 1 \cdot f_2 + X + Y + 1$$

Note that our remainder depends on the *order* of the polynomials. If we set $f_1 = Y^2 - 1$ and $f_2 = XY - 1$ we get

$$f = (X + 1) \cdot f_1 + X \cdot f_2 + 2X + 1$$

It turns out that the remainder can vanish with one ordering and not another!

With the Division Algorithm in hand, we can discuss some of the more important properties of Gröbner bases:

PROPOSITION 5.5.8 (Division Property). *Let $\succ$ be an ordering of monomials in $\mathbb{F}[X_1, \ldots, X_n]$ where $\mathbb{F}$ is a field, and let $\mathfrak{a} = (g_1, \ldots, g_k) \subset \mathbb{F}[X_1, \ldots, X_n]$ be an ideal with $G = \{g_1, \ldots, g_k\}$ a Gröbner basis. If $f \in \mathbb{F}[X_1, \ldots, X_n]$, then $f \in \mathfrak{a}$ if and only if*

$$f \to_G 0$$

PROOF. If $f \to_G 0$, then $f \in \mathfrak{a}$. Conversely, suppose $f \in \mathfrak{a}$ and $f \to_G R$. If $R \neq 0$ then
$$R = f - \sum_{i=1}^{t} a_i g_i$$
so that $R \in \mathfrak{a}$ and $\mathrm{LT}(R) \in \mathrm{LT}(\mathfrak{a})$ (since G is a Gröbner basis). This contradicts the fact that the leading term of R is not divisible by the leading terms of the g_i. □

This immediately implies that

COROLLARY 5.5.9. *If $\mathbb{F}$ is a field, $\mathfrak{a} \subset \mathbb{F}[X_1, \ldots, X_n]$ is an ideal, and B is a minimal Gröbner basis then $\mathfrak{a} = (1)$ if and only if $B = \{1\}$.*

PROOF. If $1 \in \mathfrak{a}$, then
$$1 \to_B 0$$
which can only happen if $1 \in B$. Since B is minimal, $B = \{1\}$. □

5.5.3. Buchberger's Algorithm. We begin by proving a property that Gröbner bases have.

DEFINITION 5.5.10. If $\mathbb{F}$ is a field, $\succeq$ is some ordering on the monomials of $\mathbb{F}[X_1, \ldots, X_n]$, and $\mathfrak{a} = (f_1, \cdots, f_t)$ is an ideal, let

(5.5.4) $$g_{i,j} = \frac{\mathrm{LT}(f_i)}{\gcd(\mathrm{LT}(f_i), \mathrm{LT}(f_j))}$$

and define the *S-polynomial*
$$S_{i,j} = g_{j,i} \cdot f_i - g_{i,j} \cdot f_j$$

REMARK. Note that $\mathrm{LT}(g_{j,i} \cdot f_i) = \mathrm{LT}(g_{i,j} \cdot f_j)$ so that they *cancel out* in $S_{i,j}$.

Buchberger's Theorem states that the S-polynomials give a criterion for a basis being Gröbner. It quickly leads to an *algorithm* for computing Gröbner bases.

THEOREM 5.5.11. *Let $\mathbb{F}$ be a field, $F = \{f_1, \ldots, f_t\} \in \mathbb{F}[X_1, \ldots, X_n]$ be a set of polynomials and let $\succ$ be an ordering on the monomials of $\mathbb{F}[X_1, \ldots, X_n]$. Then F is a Gröbner basis of the ideal $\mathfrak{a} = (f_1, \ldots, f_t)$ if and only if*
$$S_{i,j} \to_F 0$$
for every S-polynomial one can form from the polynomials in F.

PROOF. If F is a Gröbner basis, then the division property (proposition 5.5.8 on the preceding page) implies
$$S_{i,j} \to_F 0$$
since $S_{i,j} \in \mathfrak{a}$.

On the other hand, suppose all S-polynomials reduce to 0. Then there exist expressions

(5.5.5) $$S_{i,j} = \sum_{\ell=1}^{t} a_{i,j}^{\ell} f_{\ell}$$

for all $1 \leq i < j \leq t$ and $1 \leq \ell \leq t$ such that

$$\mathrm{LT}(S_{i,j}) \succeq \mathrm{LT}(a_{i,j}^\ell f_i)$$

(5.5.6)
$$\mathrm{LT}(s_{i,j} f_j) \succ \mathrm{LT}(a_{i,j}^\ell f_i)$$

Suppose that F is *not* a Gröbner basis — i.e. $\mathrm{LT}(\mathfrak{a}) \neq (\mathrm{LT}(f_1), \ldots, \mathrm{LT}(f_t))$. Then there exists an element $f \in \mathfrak{a}$ with

(5.5.7)
$$f = \sum_{i=1}^t b_i f_i$$

such that $\mathrm{LT}(f_i) \nmid \mathrm{LT}(f)$ for all $i = 1, \ldots, t$. The only way this can happen is if the leading terms of *two* of the terms in equation 5.5.7 *cancel*. Suppose m is the *highest* (in the ordering) monomial of $\{\mathrm{LT}(b_i f_i)\}$ for $i = 1, \ldots, t$, suppose f has been chosen to make m minimal, and so that m occurs a minimal number of times.

Without loss of generality, suppose that $\bar{b}_1 \mathrm{LT}(f_1) = \mathrm{LT}(b_1 f_1)$ and $\bar{b}_2 \mathrm{LT}(f_2) = \mathrm{LT}(b_2 f_2)$ are equal to m, up to multiplication by an element of k. If we divide both of these by $\gcd(\mathrm{LT}(f_1), \mathrm{LT}(f_2))$, we get

$$k_1 \bar{b}_1 s_{1,2} = \bar{b}_2 \cdot s_{2,1}$$

where $s_{i,j}$ is as in equation 5.5.4 on the previous page. Since the $s_{i,j}$ have no common factors, we conclude that $s_{2,1} | \bar{b}_1$ or $\bar{b}_1 = c \cdot s_{2,1}$, for some monomial c, so $\bar{b}_1 \mathrm{LT}(f_1) = c \cdot s_{2,1} \cdot \mathrm{LT}(f_1)$. Now form the quantity

$$f' = f - c \left(S_{1,2} - \sum_{\ell=1}^t a_{1,2}^\ell f_\ell \right)$$

(where $S_{1,2}$ is as in definition 5.5.10 on the preceding page).

Our hypothesis (equation 5.5.5 on the previous page) implies that $f' = f$. On the other hand, the term $-c \cdot s_{2,1} \cdot f_1$ in $-c S_{1,2}$ cancels out $\bar{b}_1 \mathrm{LT}(f_1)$ and the term $+c \cdot s_{1,2} \cdot f_2$ combines with the term $b_2 f_2$ so that the number of occurrences of the monomial m decreases by at least 1. Equation 5.5.6 on page 132 implies that the terms $\{a_{1,2}^\ell f_\ell\}$ cannot affect this outcome, so we have a contradiction to the fact that m occurred a minimal number of times in f. We conclude that F must have been a Gröbner basis. $\square$

This result immediately leads to an algorithm for computing a Gröbner basis:

ALGORITHM 5.5.12 (Buchberger's Algorithm). *Given a set of polynomials* $F = \{f_1, \ldots, f_t\} \in \mathbb{F}[X_1, \ldots, X_n]$,
  (1) *for each pair* (i,j) *with* $1 \leq i < j \leq t$, *compute* $S_{i,j}$ *as in definition 5.5.10 on the preceding page,*
  (2) *compute* $S_{i,j} \to_F h_{i,j}$, *using the Division Algorithm (5.5.5 on page 128),*
  (3) *if* $h_{i,j} \neq 0$, *set*

$$F = F \cup \{h_{i,j}\}$$

and return to step 1.

The algorithm terminates when all of the $\{h_{i,j}\}$ *found in step 3 are 0.*

REMARK. The Hilbert Basis theorem (5.4.4 on page 123) implies that this process will terminate in a finite number of steps (since we are appending generators of $\text{LT}(\mathfrak{a})$).

To get a *minimal* Gröbner basis, simply throw away unnecessary elements. To get a *reduced* basis, apply the Division Algorithm to each member of the output of this algorithm with respect to the other members.

Unfortunately, Buchberger's algorithm can have *exponential* time-complexity — for graded-reverse lexicographic ordering — and *doubly-exponential* (e^{e^n}) complexity for lexicographic ordering (see [73]). This, incidentally, is why we discussed resultants of polynomials: the complexity of computing Gröbner bases (especially with lexicographic ordering, which leads to the Elimination Property) can easily overwhelm powerful computers. Computing resultants is relatively simple (they boil down to computing determinants).

In practice it seems to have a reasonable running time. In special cases, we have:
 (1) For a system of *linear* polynomials, Buchberger's Algorithm become *Gaussian Elimination* (see 6.2.30 on page 178) for putting a matrix in upper triangular form.
 (2) For polynomials over a single variable, it becomes *Euclid's algorithm* for finding the greatest common divisor for two polynomials (see 3.1.12 on page 17).

Here is an example:

EXAMPLE 5.5.13. Let $f_1 = XY + Y^2$ and $f_2 = X^2$ in $\mathbb{F}[X, Y]$ and we compute a Gröbner basis using lexicographical ordering with

$$X \succ Y$$

We have $\text{LT}(f_1) = XY$ and $\text{LT}(f_2) = X^2$. Neither is a multiple of the other and their greatest common divisor is X. Our first S-polynomial is

$$S_{1,2} = \frac{\text{LT}(f_2)}{X} f_1 - \frac{\text{LT}(f_1)}{X} f_2 = XY^2$$

The remainder after applying the Division Algorithm is $-Y^3$ so we set $f_3 = Y^3$. We compute

$$S_{1,3} = \frac{\text{LT}(f_3)}{Y} f_1 - \frac{\text{LT}(f_1)}{Y} f_3 = Y^4$$
$$S_{2,3} = \frac{\text{LT}(f_3)}{1} f_2 - \frac{\text{LT}(f_2)}{1} f_3 = 0$$

Since both of these are in the ideal generated by $\{f_1, f_2, f_3\}$, we are done.

Gröbner bases have an interesting history. In 1899, Gordon gave a new proof of the Hilbert Basis theorem[2] (theorem 5.4.4 on page 123) that demonstrated the existence of a finite Gröbner basis (with lexicographic ordering) but gave no algorithm for computing it. See [46].

---

[2]He felt that Hilbert's proof was too abstract and gave a *constructive* proof.

In 1920, Janet (see [58]) gave an algorithm for computing "involutive bases" of linear systems of partial differential equations, that can be translated into Buchberger's algorithm in a certain case. Given a system of differential equations that are linear combinations of products of partial derivatives of $\psi(x_1,\ldots,x_n)$ (with constant coefficients), one can substitute

$$\psi = e^{\sum \alpha_i x_i}$$

and get systems of *polynomials* in the α_i whose solution leads to solutions of the differential equations.

In 1950, Gröbner published a paper ([50]) that explored an algorithm for computing Gröbner bases, but could not prove that it ever terminated. One of Buchberger's signal contributions were the introduction of S-polynomials and theorem 5.5.11 on page 131.

Teo Mora (see [75, 76]) extended much of the theory of Gröbner bases to some non-polynomial rings, including local rings and power series rings.

At this point, we can prove another interesting property of Gröbner bases, when computed in a certain way:

PROPOSITION 5.5.14 (Elimination Property). *Suppose $\mathbb{F}$ is a field and $\{g_1,\ldots,g_j\}$ is a Gröbner basis for the ideal $\mathfrak{a} \in \mathbb{F}[X_1,\ldots,X_n]$, computed using lexicographic ordering with*

$$X_1 \succ X_2 \succ \cdots \succ X_n$$

If $1 \leq t \leq n$, then

$$\mathfrak{a} \cap \mathbb{F}[X_t,\ldots,X_n]$$

has a Gröbner basis that is

$$\{g_1,\ldots,g_j\} \cap \mathbb{F}[X_t,\ldots,X_n]$$

REMARK. This is particularly important in using Gröbner bases to solve systems of algebraic equations. Here, we want to eliminate variables if possible and isolate other variables. In example 5.5.1 on page 126, we have the ideal

$$\mathfrak{B} = (xy - z^2, xz - 1, x^2 + y^2 - 3)$$

and can find a Gröbner basis for it with lexicographic ordering with $x \succ y \succ z$ of

$$(z^8 - 3z^2 + 1, y - z^3, z^7 - 3z + x)$$

Here, the basis element $z^8 - 3z^2 + 1$ is an element of $\mathfrak{B} \cap \mathbb{R}[z] \subset \mathbb{R}[x,y,z]$ and the variables x and y have been *eliminated* from it. It follows that z, alone, must satisfy

$$z^8 - 3z^2 + 1 = 0$$

and we can solve for x and y in terms of z.

PROOF. Suppose $f \in \mathfrak{a} \cap \mathbb{F}[X_t,\ldots,X_n]$ and its expansion using the Division Algorithm (5.5.5 on page 128) is

$$f = \sum q_i \cdot g_i$$

with $\mathrm{LT}(q_i \cdot g_i) \preceq \mathrm{LT}(f)$ for all i (see equation 5.3.3). Lexicographic ordering implies that, if $X_1,\ldots,X_{t-1}$ occur *anywhere* in $q_i \cdot g_i$ then these variables will be in the *leading term* of $q_i \cdot g_i$ and $\text{LT}(q_i \cdot g_1) \succ \text{LT}(f)$ — a contradiction. It follows that, for all i such that g_i contains variables $X_1, \ldots, X_{t-1}$, the corresponding $q_i = 0$. Since f is a linear combination of polynomials $\{g_1, \ldots, g_j\} \cap k[X_t, \ldots, X_n]$, they *generate* $\mathfrak{a} \cap k[X_t, \ldots, X_n]$.

Since
$$S_{i,i'} \to_{G'} 0$$
whenever $g_i, g_{i'} \in \{g_1, \ldots, g_j\} \cap \mathbb{F}[X_t, \ldots, X_n]$, theorem 5.5.11 on page 131 implies that $G' = \{g_1, \ldots, g_j\} \cap \mathbb{F}[X_t, \ldots, X_n]$ is a *Gröbner* basis. □

We already know how to test *membership* of a polynomial in an ideal via the Division Algorithm and proposition 5.5.8 on page 130. This algorithm also tells us when one ideal is contained in another since $(f_1, \ldots, f_j) \subseteq (g_1, \ldots, g_\ell)$ if and only if $f_i \in (g_1, \ldots, g_\ell)$ for $i = 1, \ldots, j$.

We can use Gröbner bases to compute *intersections* of ideals:

PROPOSITION 5.5.15 (Intersections of ideals). *Let* $\mathfrak{a} = (f_1, \ldots, f_j)$, $\mathfrak{b} = (g_1, \ldots, g_\ell)$ *be ideals in* $\mathbb{F}[X_1, \ldots, X_n]$. *If we introduce a new variable,* T, *and compute the Gröbner basis of*
$$(Tf_1, \ldots, Tf_j, (1-T)g_1, \ldots (1-T)g_\ell)$$
using lexicographic ordering and ordering T *higher than the other variables, the Gröbner basis elements that do not contain* T *will be a Gröbner basis for* $\mathfrak{a} \cap \mathfrak{b}$.

REMARK. If $f, g \in \mathbb{F}[X_1, \ldots, X_n]$, this allows us to compute the least common multiple, z, of f and g since
$$(z) = (f) \cap (g)$$
and the greatest common divisor — even if we don't know how to factor polynomials! If $n > 1$, $\mathbb{F}[X_1, \ldots, X_n]$ is *not* a Euclidean domain.

PROOF. Let $\mathfrak{J} \in \mathbb{F}[T, X_1, \ldots, X_n]$ denote the big ideal defined above. We claim that
$$\mathfrak{J} \cap \mathbb{F}[X_1, \ldots, X_n] = \mathfrak{a} \cap \mathfrak{b}$$
Suppose $f \in \mathfrak{J} \cap \mathbb{F}[X_1, \ldots, X_n]$. Then
$$\begin{aligned} f &= \sum a_i T f_i + \sum b_j (1-T) g_j \\ &= T \left(\sum a_i f_i - \sum b_j g_j \right) + \sum b_j g_j \end{aligned}$$
so
$$\sum a_i f_i - \sum b_j g_j = 0$$
It follows that $f = \sum b_j g_j$ and that $\sum a_i f_i = \sum b_j g_j$ so that $f \in \mathfrak{a} \cap \mathfrak{b}$. The conclusion now follows from the Elimination Property. □

5.5.4. Mathematical software. All of the more commonly used systems of mathematical software are able to compute Gröbner bases and implement the Division Algorithm. Among commercial systems, Maple and Mathematica have very nice user-interfaces. Free software that can do this includes Maxima and Macaulay 2, and CoCoa. See [34] for much more information.

To use Macaulay 2, start it (in Unix-type systems) by typing `M2` (the command is actually capitalized). The default output format is rather awful, so you should change it by typing

`compactMatrixForm = false`

Now define a polynomial ring over $\mathbb{Q}$

`R = QQ[a..f,MonomialOrder=>Lex]`

Note that ordering is also specified here. Now define an ideal:

```
i3 : I = ideal(a*b*c-d*e*f,a*c*e-b*d*f,
a*d*f-b*c*e)
o3 = ideal (a*b*c - d*e*f, a*c*e - b*d*f,
a*d*f - b*c*e)
o3 : Ideal of R
```

To get a Gröbner basis, type:
`gens gb I`
You need to make the window wide enough to contain the entire output expression. Subscripted variables can be defined via
`x_2=3`
The ring above could have been defined via
`R = QQ[x_1..x_6,MonomialOrder=>Lex]`

In Maple the procedure is somewhat different: First, load the library via `'with(Groebner);'`. The library `PolynomialIdeals` is also very useful. Enter an ideal by simply enclosing its generators in square brackets. The command `Basis` computes a Gröbner basis:

`Basis([a*b*c-d*e*f, a*c*e-b*d*f, a*d*f-b*c*e], plex(a, b, c, d, e, f))`

The output is nicely formatted.

The expression `plex` implies lexicographic ordering, and you must explicitly give the order of variables. For instance `plex(a, b, c, d, e, f)` means

$$a \succ b \succ c \succ d \succ e \succ f$$

Maple also supports graded lexicographic ordering with the command `grlex(a,b,c,d,e,f)` or graded reverse lexicographic order via `tdeg(a,b,c,d,e,f)`.

To reduce a polynomial using the Division Algorithm (5.5.5 on page 128), the Maple command is

`NormalForm(list_polys,basis,monomial_order)` where the basis need not be Gröbner. It returns a list of remainders of the polynomials in the list.

Maxima has a package that computes Gröbner bases using lexicographic ordering (at present, no other ordering is available). To load it, type `load(grobner)`. The main commands are `poly_grobner(poly-list,var-list)`, and `poly_reduced_grobner(poly-list,var-list)`. For example:
`poly_grobner([x^2+y^2,x^3-y^4],[x,y]);` returns

$$(x^2 + y^2, x^3 - y^4, x^4 + xy^2, y^6 + y^4)$$

— the Gröbner basis with lexicographic order: $x \succ y$.

Another very powerful and free system is called Sage (it aims to "take over the world" of computer algebra systems!). It is available for all common computer systems and can even be used online (i.e., without installing it on your computer) at HTTP://www.sagemath.org/.

Here's a small example:
The command:
   R.<a,b,c,d> = PolynomialRing(QQ, 4, order='lex')
defines a polynomial ring, R, over $\mathbb{Q}$ with 4 indeterminates: a, b, c, and d. The statement order='lex' defines lexicographic ordering on monomials. The command:
   I = ideal(a+b+c+d, a*b+a*d+b*c+c*d,
      a*b*c+a*b*d+a*c*d+b*c*d, a*b*c*d-1);
defines an ideal in R. Now the command:
   B = I.groebner_basis()
computes the Gröbner basis with respect to the given ordering. Just typing the name B prints out the basis:
```
[a + b + c + d,
b^2 + 2*b*d + d^2, b*c - b*d + c^2*d^4 + c*d - 2*d^2,
b*d^4 - b + d^5 - d, c^3*d^2 + c^2*d^3 - c - d,
c^2*d^6 - c^2*d^2 - d^4 + 1]
```

5.5.5. Motion planning. Here's an application of Gröbner bases to the robotics problem in section 6.2.9 on page 203:

EXAMPLE 5.5.16. We set the lengths of the robot arms to 1. The system of equations 6.2.37 on page 205 gives rise to the ideal

$$r = (a_1 a_2 - b_1 b_2 + a_1 - x, a_2 b_1 + a_1 b_2 + b_1 - y, a_1^2 + b_1^2 - 1, a_2^2 + b_2^2 - 1)$$

in $\mathbb{C}[a_1, a_2, b_1, b_2]$. If we set $x = 1$ and $y = 1/2$, the Gröbner basis of r (using the command 'Basis(r,plex(a_1,b_1,a_2,b_2))' in Maple) is

$$(-55 + 64 b_2^2, 8 a_2 + 3, 16 b_2 - 5 + 20 b_1, -5 - 4 b_2 + 10 a_1)$$

from which we deduce that $a_2 = -3/8$ and b_2 can be either $+\sqrt{55}/8$ in which case

$$a_1 = 1/2 + \sqrt{55}/20$$
$$b_1 = 1/4 - \sqrt{55}/10$$

or $-\sqrt{55}/8$ in which case

$$a_1 = 1/2 - \sqrt{55}/20$$
$$b_1 = 1/4 + \sqrt{55}/10$$

It follows that there are precisely *two* settings that allow the robot arm in figure 6.2.2 on page 204 to reach the point $(1, 1/2)$. It is straightforward to compute the angles involved in figure 6.2.2 on page 204: in the first case,

$$\theta = -29.44710523°$$
$$\phi = 112.024312°$$

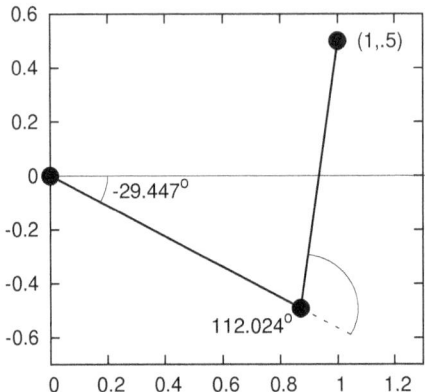

FIGURE 5.5.1. Reaching a point

as in figure 5.5.1 and in the second
$$\theta = 82.57720759°$$
$$\phi = -112.024312°$$

Another question we might ask is:

For what values of x are points on the line $y = 1 - 2x$ reachable?

In this case, we start with the ideal
$$r = (a_1 a_2 - b_1 b_2 + a_1 - x, a_2 b_1 + a_1 b_2 + b_1 + 2x - 1,$$
$$a_1^2 + b_1^2 - 1, a_2^2 + b_2^2 - 1)$$

and get the Gröbner basis (using the Maple command 'Basis(r,plex(a_1,b_1,a_2,b_2,x))'

$$(-3 + 8x + 6x^2 + 4b_2^2 - 40x^3 + 25x^4, -5x^2 + 1 + 4x + 2a_2,$$
$$-1 + 6x - 13x^2 + 2xb_2 + 10x^3 + 2b_1 - 8xb_1 + 10x^2 b_1,$$
$$3x - 2b_2 + 4x^2 + 4xb_2 - 5x^3 + 4b_1 b_2,$$
$$-1 + 4x - b_2 - 5x^2 + 2b_1 - 5xb_1 + a_1)$$

The first monomial
$$-3 + 8x + 6x^2 + 4b_2^2 - 40x^3 + 25x^4$$
is significant: When all variables are real, $4b_2^2 \geq 0$, which requires
$$-3 + 8x + 6x^2 - 40x^3 + 25x^4 \leq 0$$
— since the basis elements are assumed to be set to 0. This only happens if
$$x \in \left[\frac{2 - \sqrt{19}}{5}, \frac{2 + \sqrt{19}}{5} \right]$$
— so those are the only points on the line $y = 1 - 2x$ that the robot-arm can reach.

5.5. POLYNOMIAL RINGS

We can also analyze the Puma-type robot-arm in figure 6.2.3 on page 205:

EXAMPLE 5.5.17. If we set $\ell_1 = \ell_2 = 1$, equation 6.2.38 on page 207 implies that the endpoint of the robot-arm are solutions to the system

$$
\begin{aligned}
a_5 a_4 a_3 - a_5 b_4 b_3 + a_5 a_4 - x &= 0 \\
b_5 a_4 a_3 - b_5 b_4 b_3 + b_5 a_4 - y &= 0 \\
b_4 a_3 + a_4 b_3 + b_4 - z &= 0 \\
a_3^2 + b_3^2 - 1 &= 0 \\
a_4^2 + b_4^2 - 1 &= 0 \\
a_5^2 + b_5^2 - 1 &= 0
\end{aligned}
$$
(5.5.8)

If we want to know which points it can reach with the hand pointing in the direction

$$\begin{bmatrix} 1/\sqrt{3} \\ 1/\sqrt{3} \\ 1/\sqrt{3} \end{bmatrix}$$

use equation 6.2.39 on page 207 to get

$$
\begin{aligned}
(a_5 a_4 a_3 - a_5 b_4 b_3) a_2 + (-a_5 a_4 b_3 - a_5 b_4 a_3) b_2 - 1/\sqrt{3} &= 0 \\
(b_5 a_4 a_3 - b_5 b_4 b_3) a_2 + (-b_5 a_4 b_3 - b_5 b_4 a_3) b_2 - 1/\sqrt{3} &= 0 \\
(b_4 a_3 + a_4 b_3) a_2 + (a_4 a_3 - b_4 b_3) b_2 - 1/\sqrt{3} &= 0 \\
a_2^2 + b_2^2 - 1 &= 0
\end{aligned}
$$
(5.5.9)

We regard these terms (in equations 5.5.8 and 5.5.9 as generators of an ideal, $\mathfrak{P}$. The variety $\mathcal{V}(\mathfrak{P}) \subset \mathbb{R}^{10}$ is called the *variety of the movement problem*. Its (real-valued) points correspond to possible configurations of the robot-arm.

To understand $\mathcal{V}(\mathfrak{P})$, we compute a Gröbner basis of $\mathfrak{P}$ with lexicographic ordering — giving the lowest weight to x, y, z — to get

(5.5.10) $\mathfrak{P} = (4y^2x^2 - 4z^2 + z^4 + 2z^2x^2 + 2y^2z^2,$

$$-1 + 2b_5^2, -b_5 + a_5, 2zb_4 - z^2 - 2yx,$$

$$-4z + 4yb_4x + z^3 - 2xzy + 2y^2z + 2x^2z,$$

$$-2 - 2yx + 2b_4^2 + y^2 + x^2,$$

$$a_4 - b_5 y + b_5 x,$$

$$-b_5 yz + b_5 xz + b_3 + 2b_5 yb_4,$$

$$2 + 2a_3 - 2y^2 - z^2,$$

$$2b_5 z\sqrt{3} - \sqrt{3}b_5 y - \sqrt{3}b_5 x - 2b_4\sqrt{3}b_5 + 3b_2,$$

$$3a_2 - y\sqrt{3} - x\sqrt{3} - z\sqrt{3} + b_4\sqrt{3})$$

It follows that a point (x, y, z) is reachable (with the hand oriented as stated) *only* if it lies on the surface

$$4y^2x^2 - 4z^2 + z^4 + 2z^2x^2 + 2y^2z^2 = 0$$

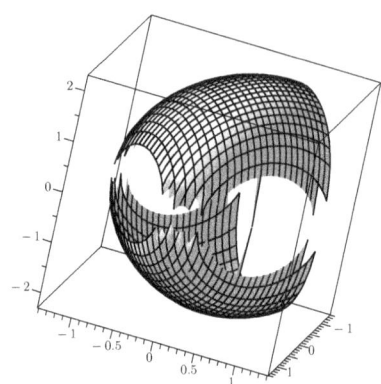

FIGURE 5.5.2.   Points reachable by the second robot arm

Solving for z^2 gives

(5.5.11) $$z^2 = 2 - y^2 - x^2 \pm \sqrt{(2-(x-y)^2)(2-(x+y)^2)}$$

The fourth expression from the top in equation 5.5.10 on the preceding page is

$$\begin{aligned} -2 - 2yx + 2b_4{}^2 + y^2 + x^2 &= 0 \\ 2b_4^2 &= 2 - 2xy - x^2 - y^2 \end{aligned}$$

which implies that

$$2 - (x-y)^2 \geq 0$$

and gives the additional constraint on (x, y, z):

$$(x-y)^2 \leq 2$$

It follows that $2 - (x-y)^2 \geq 0$ so that the square root in equation 5.5.11 is only well-defined if $2 - (x+y)^2 \geq 0$ and we get an *additional* constraint on x and y.

The requirement that $z^2 \geq 0$ implies that the only case worth considering is

$$z^2 = 2 - y^2 - x^2 + \sqrt{(2-(x-y)^2)(2-(x+y)^2)}$$

and figure 5.5.2 shows the set of points that are reachable.

EXERCISES.

1. The point $(1/2, 1/2, 1+\sqrt{2}/2)$ lies on the "reachability surface" in example 5.5.17 on page 139 that can be reached by the robot arm with its hand pointed in the direction

$$\frac{1}{\sqrt{3}}\begin{bmatrix} 1 \\ 1 \\ 1 \end{bmatrix}$$

Find the angles $\phi_1, \theta_1, \theta_2, \theta_3$ that accomplish this.

2. Find the reachability surface of the robot arm in example 5.5.17 on page 139 when we require the "hand" to point in the direction

$$\begin{bmatrix} 1 \\ 0 \\ 0 \end{bmatrix}$$

3. Find the least common multiple of the polynomials

$$-X^3 - 2YX^2 - XY^2 + 2X$$

and

$$4 - 4X^2 - 4Y^2 + X^4 - 2Y^2X^2 + Y^4$$

in $\mathbb{F}[X, Y]$.

4. Consider the ideal $\mathfrak{a} = (Y^3, X - Y)$.
Is $X + Y \in \mathfrak{a}$?

5. If $\mathfrak{a} = (Y^6, -3Y^5 + 5XY^4, X^2 - 2XY + Y^2)$ is a Gröbner basis for an ideal, what is the lowest power of $X + Y$ that is contained in $\mathfrak{a}$?

6. Find the least common multiple of the polynomials

$$-X^3 - 2YX^2 - XY^2 + 2X$$

and

$$4 - 4X^2 - 4Y^2 + X^4 - 2Y^2X^2 + Y^4$$

in $\mathbb{F}[X, Y]$.

7. Consider the ideal $\mathfrak{a} = (Y^3, X - Y)$.
Is $X + Y \in \mathfrak{a}$?

8. If $\mathfrak{a} = (Y^6, -3Y^5 + 5XY^4, X^2 - 2XY + Y^2)$ is a Gröbner basis for an ideal, what is the lowest power of $X + Y$ that is contained in $\mathfrak{a}$?

5.5.6. Elementary symmetric functions. If R is a commutative ring, consider the polynomial ring
$$P = R[X_1, \ldots, X_n]$$
The symmetric group, S_n, acts on this ring by permuting the variables. Each such permutation of the variables defines an automorphism of P so the set of elements
$$S = R[X_1, \ldots, X_n]^{S_n}$$
fixed by the action of S_n is a *subring* of P. It is interesting that the structure of this subring is completely understood — and was in the time of Isaac Newton. The actual description of this subring will be important in the sequel and is used in several areas of algebra and algebraic geometry.

In order to give this description, we will need to define the elementary symmetric functions. The quickest (if not the simplest) way to describe them is to consider a polynomial in the ring $P[T]$ where T is a new indeterminate:

(5.5.12) $$\prod_{i=1}^{n}(T - X_i) = q(T)$$
$$= T^n - \sigma_1 \cdot T^{n-1} + \cdots + (-1)^n \sigma_n$$

Since $q(T)$ is unchanged when the X_i's are permuted, the coefficients of $q(T)$ must be functions of the X_i that are also unchanged by permuting the X_i. They are

$$\sigma_0(X_1, \ldots, X_n) = 1$$
$$\sigma_1(X_1, \ldots, X_n) = \sum_{i=1}^{n} X_i$$
$$\sigma_2(X_1, \ldots, X_n) = \sum_{1 \leq i < j \leq n} X_i X_j$$
$$\vdots$$

(5.5.13) $$\sigma_n(X_1, \ldots, X_n) = \prod_{i=1}^{n} X_i$$

where $\sigma_i(X_1, \ldots, X_n)$ is $(-1)^i \times$ the coefficient of t^{n-i}.

If we consider the ring $R[\sigma_1, \ldots, \sigma_n]$ of polynomials of the σ_i, it is clear that
$$R[\sigma_1, \ldots, \sigma_n] \subset R[X_1, \ldots, X_n]$$
and even that
$$R[\sigma_1, \ldots, \sigma_n] \subset R[X_1, \ldots, X_n]^{S_n} = S$$
since the σ_i are unchanged by permutations of the X_i. It is remarkable that:

THEOREM 5.5.18. *The subring of polynomials of*
$$R[X_1, \ldots, X_n]$$
that are invariant under all permutations of the X_i is precisely the polynomial ring of elementary symmetric functions, i.e.

$$R[\sigma_1,\ldots,\sigma_n] = R[X_1,\ldots,X_n]^{S_n}$$

PROOF. Let $p(X_1,\ldots,X_n) \in R[X_1,\ldots,X_n]^{S_n}$. We will express this as a polynomial of the elementary symmetric functions. Suppose

$$m = r \cdot X_1^{\alpha_1} \cdots X_n^{\alpha_n}$$

is a monomial of p, where $r \in R$. Since p is invariant under permutations of the X_i, p also contains an ordered-monomial that is equivalent to m under the action of S_n, where an ordered monomial is of the form

$$r \cdot X_1^{\beta_1} \cdots X_n^{\beta_n}$$

where $\beta_1 \geq \beta_2 \geq \cdots \geq \beta_n$ where the β_i's are some permutation of the α_i's. We may focus our attention entirely on ordered-monomials of this type, since every monomial will be equivalent to one of these. The (unique) ordered monomial of $\sigma_i(X_1,\ldots,X_n)$ is

(5.5.14) $$X_1 \cdots X_i$$

Now we order the ordered-monomials of $p(X_1,\ldots,X_n)$ lexicographically by exponents, i.e. so

$$X_1^{\alpha_1} \cdots X_n^{\alpha_n} \succ X_1^{\beta_1} \cdots X_n^{\beta_n}$$

if $\alpha_j > \beta_j$ and $\alpha_i = \beta_i$, for $i = 1\ldots j-1$.

The polynomial p will contain a unique maximal ordered-monomial, say

$$r \cdot X_1^{\beta_1} \cdots X_n^{\beta_n}$$

and this agrees with the unique maximal ordered monomial of

(5.5.15) $$r \cdot \sigma_n^{\beta_n} \cdot \sigma_{n-1}^{\beta_{n-1}-\beta_n} \cdots \sigma_1^{\beta_1-\beta_2}$$

by equation 5.5.14. It follows that the unique maximal ordered monomial of

$$p - r \cdot \sigma_n^{\beta_n} \cdot \sigma_{n-1}^{\beta_{n-1}-\beta_n} \cdots \sigma_1^{\beta_1-\beta_2}$$

is *strictly* $\prec r \cdot X_1^{\beta_1} \cdots X_n^{\beta_n}$. Since there are only a finite number of monomials $\prec r \cdot X_1^{\beta_1} \cdots X_n^{\beta_n}$, repeating this procedure over and over again must terminate after a finite number of steps. The polynomial p is equal to the sum of the symmetric polynomials we subtracted from p. □

The proof gives us an algorithm for computing the expression of symmetric polynomials in terms of symmetric functions:

EXAMPLE. Consider

$$X^2 + Y^2 \in \mathbb{Q}[X,Y]$$

The maximal ordered monomial of this is X^2 — which corresponds to σ_1^2 in equation 5.5.15. The difference is

$$\begin{aligned} X^2 + Y^2 - \sigma_1^2 &= X^2 + Y^2 - (X+Y)^2 \\ &= -2XY \end{aligned}$$

which is equal to $-2\sigma_2$. So we get
$$X^2 + Y^2 = \sigma_1^2 - 2\sigma_2$$

An interesting consequence of formula 5.5.12 on page 142 and theorem 5.5.18 on page 142 is:

PROPOSITION 5.5.19. *Let $X^n + a_{n-1}X^{n-1} + \cdots + a_0 = p(X) \in \mathbb{Q}[X]$ and suppose $q(X_1, \ldots, X_n) \in \mathbb{Q}[X_1, \ldots, X_n]$ is invariant under permutations of the X_i. If $\mu_1, \ldots, u_n \in \mathbb{C}$ are the roots of $p(X)$, then there exists a polynomial $z(X_1, \ldots, X_n)$ such that*
$$q(\mu_1, \ldots, \mu_n) = z(a_0, \ldots, a_{n-1})$$

PROOF. Theorem 5.5.18 on page 142 implies that $q(X_1, \ldots, X_n) = z(\sigma_1, \ldots, \sigma_n)$. Equation 5.5.12 on page 142 shows that $\sigma_i(\mu_1, \ldots, \mu_n) = (-1)^i a_{n-i}$ and the result follows. □

This has an interesting application in the definition of discriminants of polynomials:

DEFINITION 5.5.20. Let $p(x) \in \mathbb{Q}[x]$ be of degree n with roots $\alpha_1, \ldots, \alpha_n \in \mathbb{C}$. The *discriminant, D,* of $p(X)$ is defined to be
$$D = \prod_{1 \leq i < j \leq n} (\alpha_i - \alpha_j)^2$$

REMARK. The discriminant is nonzero if and only if $p(X)$ has n distinct roots (so it *discriminates* between roots).

Since the discriminant is unchanged by a permutation of the roots, proposition 5.5.19 implies that

COROLLARY 5.5.21. *If*
$$p(X) = X^n + a_{n-1}X^{n-1} + \cdots + a_0$$
there is a polynomial function $z(a_0, \ldots a_{n-1})$ equal to the discriminant of $p(X)$.

For instance, the discriminant of $X^2 + aX + b$ is
$$\begin{aligned}(\alpha_1 - \alpha_2)^2 &= \alpha_1^2 - 2\alpha_1\alpha_2 + \alpha_2^2 \\ &= \sigma_1^2(\alpha_1, \alpha_2) - 4\alpha_1\alpha_2 \\ &= \sigma_1^2(\alpha_1, \alpha_2) - 4\sigma_2(\alpha_1, \alpha_2) \\ &= a^2 - 4b\end{aligned}$$

A lengthy calculation shows that the discriminant of $X^3 + aX^2 + bX + c$ is
$$D = a^2b^2 - 4b^3 - 4a^3c - 27c^2 + 18abc$$

EXERCISES.

9. Express $X^3 + Y^3 + Z^3 \in \mathbb{Q}[X,Y,Z]$ in terms of elementary symmetric functions.

10. Show that the discriminant of $X^n - 1$ is $\pm n^n$.

11. Let $p_1, p_2 \in \mathbb{Q}[t]$ are monic polynomials with roots $\alpha_1, \ldots, \alpha_n \in \mathbb{C}$ and $\beta_1, \ldots, \beta_m \in \mathbb{C}$, respectively. Let
$$\Delta = \prod (\alpha_i - \beta_j)$$
with i running from 1 to n and j running from 1 to m. Show that Δ is a polynomial function of the coefficients of p_1 and p_2.

12. Another way of defining the discriminant involves using the Vandermonde matrix. Given elements $\alpha_1, \ldots, \alpha_n$, we define the corresponding *Vandermonde matrix* as
$$V = \begin{bmatrix} 1 & \alpha_1 & \alpha_1^2 & \cdots & \alpha_1^{n-1} \\ 1 & \alpha_2 & \alpha_2^2 & \cdots & \alpha_2^{n-1} \\ 1 & \alpha_3 & \alpha_3^2 & \cdots & \alpha_3^{n-1} \\ \vdots & \vdots & \vdots & \ddots & \vdots \\ 1 & \alpha_n & \alpha_n^2 & \cdots & \alpha_n^{n-1} \end{bmatrix}$$
Show that
$$\det V = \prod_{1 \leq i < j \leq n} (\alpha_j - \alpha_i)$$

13. Suppose $\mathbb{F}$ is a field and
$$\mathfrak{I} \subset \mathbb{F}[X_1, \ldots, X_n]$$
is an ideal. If $f(X_1, \ldots, X_n) \in \mathbb{F}[X_1, \ldots, X_n]$ is any polynomial, show that
$$\frac{\mathbb{F}[X_1, \ldots, X_n]}{\mathfrak{I}} \cong \frac{\mathbb{F}[X_1, \ldots, X_{n+1}]}{\mathfrak{I} + (X_{n+1} - f(X_1, \ldots, X_n))}$$
In other words, show that a variable, like X_{n+1}, that can be expressed in terms of the others is superfluous.

5.6. Unique factorization domains

5.6.1. Introduction. This section is one of the most complex in the chapter on commutative algebra and most of it is flagged with a dangerous bend symbol. We study the important question of when a ring has unique factorization. Aside from any inherent interest, unique factorization has geometric implications that will become apparent later.

A great deal of this material is due to Gauss in his groundbreaking [41], and Weierstrass (in [107]) in his research on complex analysis in several variables.

> Johann Carl Friedrich Gauss (1777 – 1855) was a German mathematician and scientist who contributed to many fields, including number theory, analysis, statistics (the normal distribution curve), differential geometry (he essentially invented it), geophysics, electrostatics, astronomy, and optics.

We can characterize unique factorization domains by

LEMMA 5.6.1. *If R is a ring, the following three statements are equivalent*
 (1) R is a unique factorization domain
 (2) For any $r, p, q \in R$ such that

$$r \mid p \cdot q$$

with r irreducible

$$r \nmid p \implies r \mid q$$

(3) For any $r, p, q \in R$ such that

$$r \mid p \cdot q$$

and r and p have no common factors

$$r \nmid p \implies r \mid q$$

(4) For any irreducible element $r \in R$, the principal ideal $(r) \subset R$ is prime.

REMARK. Prime ideals play an important part in algebraic geometry, and statement 4 implies that they can have a particularly simple structure.

PROOF. If statement 1 is true, then R is a unique factorization domain by reasoning like that used in lemma 3.1.9 on page 3.1.9. Conversely, if R is a unique factorization domain, then

$$p = u_1 \prod_{i=1}^{m} p_i^{\beta_i}$$

(5.6.1)
$$q = u_2 \prod_{i=1}^{m} p_i^{\gamma_i}$$

where $u_1, u_2 \in R$ are units, the $p_i \in R$ are irreducible elements and the $\alpha_i, \beta_i, \gamma_i \in \mathbb{Z}$ are all ≥ 0. If $r \mid p \cdot q$, then r must equal one of the p_i, say p_j and $\alpha_j + \beta_j \geq 1$. Since $r \nmid p$, we get $\alpha_j = 0$, which implies that $\beta_j \geq 1$ and this proves the conclusion.

Statement 3 implies statement 2. Conversely, if statement 2 is true, R is a unique factorization domain and equation 5.6.1 holds as well as

$$r = u_0 \prod_{i=1}^{m} p_i^{\alpha_i}$$

Since r has no common factors with p, $\alpha_i > 0$ implies that $\beta_i = 0$, hence $\alpha_i \leq \gamma_i$ for all $i = 1, \ldots, n$. This implies that $r \mid q$.

To see statement 4, suppose $a, b \in U$ and $a \cdot b \in (r)$ or $ra \cdot b$. The previous statement implies that $r \mid a$ or $r \mid b$ which means $a \in (r)$ or $b \in (r)$. The implication clearly goes in the opposite direction too. □

One of the easiest results in this section is:

LEMMA 5.6.2. *If $\mathbb{F}$ is a field, then $\mathbb{F}[X]$ is a unique factorization domain.*

PROOF. It is a Euclidean domain in the sense of definition 5.3.3 on page 118 so it is also a unique factorization domain by corollary 5.3.7 on page 118. □

Even though the ideals in a unique factorization might not all be principal, we have:

PROPOSITION 5.6.3. *In a unique factorization domain, the concept of greatest common divisor and least common multiple are well-defined.*

REMARK. In general, we have no analogue of the Euclidean Algorithm, so it may be impossible to have a formula like equation 3.1.1 on page 14.

PROOF. If U is a unique factorization domain with elements x and y, then they have factorizations, unique up to multiplication by units. Let $\{p_1, \ldots, p_k\}$ be all of the irreducible factors that occur in their factorizations:

$$x = u \prod_{i=1}^{k} p_i^{\alpha_i}$$

$$y = u' \prod_{i=1}^{k} p_i^{\beta_i}$$

Now we can define

$$\gcd(x,y) = \prod_{i=1}^{k} p_i^{\min(\alpha_i, \beta_i)}$$

$$\operatorname{lcm}(x,y) = \prod_{i=1}^{k} p_i^{\max(\alpha_i, \beta_i)}$$

□

5.6.2. Polynomial rings. Throughout the rest of this section, we *fix* a unique factorization domain, U. We will show that $U[X]$ is also a unique factorization domain. We use a trick to prove this: embed $U[X]$ in $F[X]$, where F is the field of fractions of U and uniquely factor elements there.

The proof involves several steps.

DEFINITION 5.6.4. A polynomial $a_n X^n + \cdots + a_0 \in U[X]$ will be called *primitive*, if the greatest common divisor of its coefficients is 1.

Note that if $f \in U[X]$ is a polynomial, we can write $f = u \cdot f'$ where u is the greatest common divisor of the coefficients of f and f' is primitive.

The following result is called Gauss's Lemma:

LEMMA 5.6.5. *If $f, g \in U[X]$ are primitive polynomials, then so is fg.*

REMARK. This and the following lemma were proved by Carl Friedrich Gauss in his treatise [41].

PROOF. Suppose
$$f = a_n X^n + \cdots + a_0$$
$$g = b_m X^m + \cdots + b_0$$
$$fg = c_{n+m} X^{n+m} + \cdots + c_0$$

If $d \in U$ is irreducible, it suffices to prove that d does not divide all of the c_i. Let a_i and b_j be the first coefficients (i.e. with the lowest subscripts) not divisible by d. We claim that c_{i+j} is not divisible by d. Note that

$$c_{i+j} = \underbrace{a_0 b_{i+j} + \cdots + a_{i-1} b_{j+1}}_{\text{Group 1}} + a_i b_j + \underbrace{a_{i+1} b_{j-1} + \cdots + a_{i+j} b_0}_{\text{Group 2}}$$

By construction, d divides all of the terms in Group 1 and Group 2. Since U is a unique factorization domain, $d | c_{i+1}$ if and only if $d | a_i b_j$. But, the fact that U is a unique factorization domain also implies that $d | a_i b_j$ if and only if $d | a_i$ or $d | b_j$. □

This leads to the following:

LEMMA 5.6.6. *Let $f \in U[X]$ be primitive. Then f is irreducible in $U[X]$ if and only if it is irreducible in $F[X]$, where F is the field of fractions of U.*

PROOF. Suppose f is irreducible in $U[X]$, and that $f = gh \in F[X]$. By clearing denominators, we can assume

$$g = u_1^{-1} \bar{g}$$
$$h = u_2^{-1} \bar{h}$$

where $\bar{g}, \bar{h}$ are primitive polynomials of $U[X]$, and $u_1, u_2 \in U$. We conclude that
$$u_1 u_2 f = \bar{g} \bar{h}$$
where $\bar{g}\bar{h}$ is primitive by 5.6.5 on the preceding page. Since f is also primitive, the factor $u_1 u_2 \in U$ must be a unit. Since f is irreducible in $U[X]$, $\bar{g}$ or $\bar{h}$ must be a unit in $U[X]$ and also in $F[X]$.

On the other hand, suppose $f \in U[X]$ is irreducible in $F[X]$ and assume $f = gh \in U[X]$. Since f is irreducible in $F[X]$, either g or h must be a unit in $F[X]$, i.e. a constant polynomial. If g is a constant polynomial then the formula $f = gh$ with f primitive implies that $g \in R$ is a unit. □

We are finally ready to prove:

THEOREM 5.6.7. *If U be a unique factorization domain, then so is $U[X]$.*

REMARK. This is interesting because U doesn't have to be a euclidean domain or a principal ideal domain.

PROOF. We use a trick to prove this: embed $U[X]$ in $F[X]$, where F is the field of fractions of U and uniquely factor elements there. Suppose $r, f, g \in U[X]$ are polynomials and
(1) r is irreducible.
(2) $r \mid fg$

We will show that $r\,|\,f$ or $r\,|\,g$ and lemma 5.6.1 on page 146 will show that $U[x]$ has unique factorization.

Lemma 5.6.2 on page 147 implies that $F[X]$ is a unique factorization domain because it is Euclidean.

Write
$$r = ur'$$
$$f = u_1 f'$$
$$g = u_2 g'$$
where $u, u_1, u_2 \in U$ are, respectively, the greatest common divisors of the coefficients of r, f, g and $r', f', g' \in U[x]$ are primitive. Since r is irreducible, we can assume $u \in U$ is a unit (otherwise $r = ur'$ would be a nontrivial factorization of r).

Lemma 5.6.6 on the preceding page implies that r' is irreducible in $F[X]$ so, in $F[X]$, $r'|f'$ or $r'|g'$ in $F[x]$. Without loss of generality, assume $r'|f'$, so that
$$f' = a \cdot r'$$
where $a \in F[X]$. We can write $a = v^{-1}a'$, where $v \in U$ and $a' \in U[X]$ is primitive. We get
$$v \cdot f' = a' \cdot r'$$
in $U[X]$. Since f' and $a' \cdot r'$ are both primitive (by lemma 5.6.5 on page 147), $v \in U$ must be a unit and we get $r|f$. □

REMARK. Actually *finding* factorizations in these rings can be challenging.

To actually find a factorization of a polynomial, it is helpful to have a criterion for irreducibility. The following is called Eisenstein's Criterion:

THEOREM 5.6.8. *Let U be a unique factorization domain and let*
$$f(X) = a_n X^n + \cdots + a_0 \in U[X]$$
be a primitive polynomial and let $p \in U$ be irreducible. If $p\,|\,a_i$ for $0 \leq i \leq n-1$, $p \nmid a_n$, and $p^2 \nmid a_0$ then $f(X)$ is irreducible in $U[X]$.

REMARK. If F is the field of fractions of U, lemma 5.6.6 on the preceding page shows that this criterion works for polynomials over $F[X]$ too, after clearing the denominators of the coefficients.

Eisenstein originally proved this for $\mathbb{Z}[X]$ but it works for any unique factorization domain.

PROOF. We will reason by contradiction. Suppose there exist polynomials
$$p(X) = b_s X^s + \cdots + b_0$$
$$q(X) = c_t X^t + \cdots + c_0$$
such that $f(X) = p(X) \cdot q(X)$ and $s \geq 1$ and $t \geq 1$. Since $p^2 \nmid a_0$ we must have $p \nmid b_0$ or $p \nmid c_0$. Assume that $p|c_0$ and $p \nmid b_0$. Since f is primitive, not

all the c_i are divisible by p. Suppose c_k is the first that is not divisible by p. Then
$$a_k = b_k c_0 + \cdots + b_0 c_k$$
By assumption, $p|a_k$ and $p|c_i$ for $0 \le i < k$, which implies that $p|b_0 c_k$ and this implies that $p|b_0$, which is a contradiction. $\square$

EXAMPLE. The polynomial $X^3 + 3X^2 + 3X + 1 \in \mathbb{Q}[X]$ is irreducible by Eisenstein's Criterion with respect to the prime $p = 3$.

EXAMPLE 5.6.9. In some cases, one must first transform the polynomial a bit to use Eisenstein's Criterion. For instance, in the polynomial
$$f(X) = X^2 + X + 1 \in \mathbb{Q}[X]$$
there are no primes that divide any of the coefficients. After substituting $X = U + 1$, $f(X)$ becomes
$$g(U) = U^2 + 3U + 3$$
which satisfies Eisenstein's Criterion with respect to the prime $p = 3$. Since $X \to U + 1$ defines an isomorphism
$$\mathbb{Q}[X] \to \mathbb{Q}[U]$$
$f(X)$ is irreducible if and only if $g(U)$ is.

5.6.3. Power-series rings. Next, we tackle the question of unique factorization in *power series* rings. This appears daunting at first glance because power series seem to have "infinite complexity". For instance, it is *not* true that whenever U is a unique factorization domain, $U[[X]]$ is also — see [96].

It is gratifying to see that, in some cases, factorization is actually *easier* in power series rings. The point is that factorizations are only well-defined up to multiplication by a unit — and the power series ring $\mathbb{F}[[X]]$ (for $\mathbb{F}$ a field) has *many units:* Proposition 5.1.9 on page 109 shows that any power series
$$z = \sum_{n=0}^{\infty} c_n X^n$$
with $c_0 \ne 0$ is a unit. If the lowest nonzero coefficient in z is c_r then our unique factorization of z is
$$(5.6.2) \qquad X^r \cdot (c_r + c_{r+1} X + \cdots)$$
In other words, X is our "only prime" in $\mathbb{F}[[X]]$, and an arbitrary element of $\mathbb{F}[[X]]$ is the product of a *polynomial in X* and a *unit*.

This will turn out to be true in *general*: the Weierstrass Preparation Theorem will show that certain elements (every element can be transformed into one of these — see lemma 5.6.15 on page 153) of
$$\mathbb{F}[[X_1, \ldots, X_n]]$$
will equal units $\times$ polynomials in
$$\mathbb{F}[[X_1, \ldots, X_{n-1}]][X_n]$$

i. e., polynomials in X_n with coefficients in $\mathbb{F}[[X_1,\ldots,X_{n-1}]]$. Unique factorization in $\mathbb{F}[[X_1,\ldots,X_{n-1}]]$ and $\mathbb{F}[[X_1,\ldots,X_{n-1}]][X_n]$ will imply it in $\mathbb{F}[[X_1,\ldots,X_n]]$ (via lemma 5.6.1 on page 146).

We will fix the following notation throughout the rest of this section: $P_n = \mathbb{F}[[X_1,\ldots,X_n]]$, *where* $\mathbb{F}$ *is a field.*

We need to develop some properties of power-series rings.

PROPOSITION 5.6.10. *An element $p \in P_n$ is a unit if and only if its constant term is nonzero.*

PROOF. Straightforward induction using proposition 5.1.9 on page 109 and exercise 3 on page 115. □

DEFINITION 5.6.11. An element $p(X_1,\ldots,X_n) \in P_n$ will be called X_n-*general* if $p(0,\ldots,0,X_n) \neq 0$. If $X_n^d \mid p(0,\ldots,0,X_n)$ and $X_n^{d+1} \nmid p(0,\ldots,0,X_n)$ for some integer $d > 0$, we say that p is X_n-*general of degree d.*

REMARK. A power series is X_n-general if it has a term that only involves X_n. For instance $X_1 + X_2$ is X_2-general but $X_1 X_2$ is not.

Next, we have a kind of division algorithm for power-series rings (even though these rings are not Euclidean):

THEOREM 5.6.12 (Weierstrass Division Theorem). *Let $p \in P_n$ be X_n-general power-series of degree d that is not a unit of P_n. For every power series $g \in P_n$, there exists a power series $u \in P_n$ and a polynomial $r \in P_{n-1}[X_n]$ of degree $d - 1$ such that*

(5.6.3) $$g = u \cdot p + r$$

The power-series u and polynomial r are uniquely determined.

REMARK. A shorter way to say this is that

$$\frac{P_n}{(p)} = P_{n-1} \oplus X_n \cdot P_{n-1} \oplus \cdots \oplus X_n^{d-1} P_{n-1}$$

or that it is a module over P_{n-1} generated by $\{1,\ldots,X_n^{d-1}\}$.

PROOF. We will explicitly construct u and r.

For every $f \in P_n$, let $r(f)$ equal the set of terms, T, such that $X_n^d \nmid T$, and let $h(f)$ be the factor of X_n^d in $f - r(f)$. Then

$$f = r(f) + X_n^d h(f)$$

for all power series in P_n. So $r(f), h(f) \in P_n$ and $r(f)$ is a polynomial in $P_{n-1}[X_n]$ of degree $< d$. Note that, regarding P_n as a vector space (see definition 6.2.1 on page 165) over k, both $r(*)$ and $h(*)$ are linear maps.

CLAIM 5.6.13. In addition, $h(p)$ is a unit (since its constant term is the element of k multiplying X_n^d, and $r(f)$ has no constant terms since f is not a unit.

We claim that equation 5.6.3 is equivalent to

(5.6.4) $$h(g) = h(u \cdot p)$$

for some $u \in P_n$. If equation 5.6.3 holds then $h(g - u \cdot p) = 0$ and equation 5.6.4 is true. Conversely, if equation 5.6.4 on page 151is true, then $h(g - u \cdot p) = 0$ and $g - u \cdot p = r(q - u \cdot p)$, a degree $d - 1$ polynomial in $P_{n-1}[X_n]$.

Since $p = r(p) + X_n^d \cdot h(p)$, equation 5.6.4 on the previous page is equivalent to

(5.6.5) $$u \cdot p = u \cdot r(p) + X_n^d \cdot u \cdot h(p)$$

Since $h(p)$ is a unit (see the claim above), it suffices to compute the power-series $v = u \cdot h(p)$. Set
$$m = -r(f) \cdot h(p)^{-1}$$
Then $u \cdot r(p) = -m \cdot v$ and we can rewrite equation 5.6.5 to the equivalent equation

(5.6.6) $$h(g) = -h(m \cdot v) + v$$

or

(5.6.7) $$v = h(g) + s(v)$$

where, for any power series, $f \in P_n$, we have defined $s(f) = h(m \cdot f)$. Note that s is a linear operation on power-series.

Let $\mathfrak{m} = (X_1, \ldots, X_{n-1}) \subset P_{n-1}$ be the maximal ideal. Note that $r(p) \in \mathfrak{m}[X_n] \subset P_{n-1}[X_n]$ since it is not a unit (so it has vanishing constant term). This means that, if the coefficients of $f \in P_n = P_{n-1}[[X_n]]$ lie in $\mathfrak{m}^j$, then the coefficients of $s(f)$ will lie in $\mathfrak{m}^{j+1}$.

Now we plug equation 5.6.7 into itself to get
$$\begin{aligned} v &= h(g) + s(h(g) + s(v)) \\ &= h(g) + s(h(g)) + s^2(v) \end{aligned}$$

We can iterate this any number of times:
$$v = \sum_{j=0}^{t} s^j(h(g)) + s^{t+1}(v)$$

or
$$v - \sum_{j=0}^{t} s^j(h(g)) \in \mathfrak{m}^{t+1}[[X_n]] \subset P_{n-1}[[X_n]]$$

Since lemma 6.3.34 on page 246 implies that
$$\bigcap_{j=1}^{\infty} \mathfrak{m}^j = (0)$$

we claim that
$$v = \sum_{j=0}^{\infty} s^j(h(g))$$

is the unique solution to our problem. It is easy to verify that it satisfies equation 5.6.7. Now all we have to do is set
$$u = v \cdot h(p)^{-1}$$
and $r = r(q - u \cdot p)$. □

The Weierstrass Preparation Theorem is a simple corollary:

THEOREM 5.6.14 (Weierstrass Preparation Theorem). *Let $p \in P_n$ be X_n-general power-series of degree d that is not a unit of P_n. Then there exists a unit $u \in P_n$ and a monic polynomial $w \in P_{n-1}[X_n]$ of degree d such that*

(5.6.8) $$p = u \cdot w$$

and u and w are uniquely determined by p.

REMARK. This is the general analogue of 5.6.2 on page 150 for power-series of n variables. Weierstrass originally proved it for convergent power series using the methods of complex analysis. It gives valuable information on the behavior of the zero-sets of analytic functions of several complex variables (besides implying that the ring of such functions has unique factorization).

Our proof of the Division Theorem is the "combinatorial" or "algebraic" form — that does not use contour integrals.

The polynomial $w \in P_{n-1}[X_n]$ is called the Weierstrass Polynomial of p.

Case 1. Apply the division theorem (5.6.12 on page 151) to $g = X_n^d$. It gives q and r such that
$$X_n^d = q \cdot p + r$$
so we get
$$q \cdot p = X_n^d - r = w \in P_{n-1}[X_n]$$
We claim that q must be a unit since the lowest X_n term in p is X_n^d. The only way the product could contain X_n^d is for u to have a nonvanishing constant term. Now set $u = q^{-1}$, so that $p = u \cdot w$.

If A is an $n \times n$ invertible matrix whose entries are in k, then A induces an automorphism
$$\begin{array}{rcl} A^*\colon P_n & \to & P_n \\ p(X_1,\ldots,X_n) & \mapsto & p^A = p(A^{-1} \cdot (X_1,\ldots,X_n)) \end{array}$$
The inverse is given by the inverse of A.

LEMMA 5.6.15. *Let $p \in P_n$ be a power series that is not a unit. If the field $\mathbb{F}$ is infinite, then there exists a matrix, A, such that p^A is X_n-general.*

PROOF. Let L be the leading term of p — this consists of the terms of lowest total degree in the power series and will be a homogeneous polynomial in $X_1,\ldots,X_n$. Let
$$(k_1,\ldots,k_n) \in \mathbb{F}^n$$
be a set of values on which L is nonvanishing. Such a set of values exists because we can plug in 1 for all of the X_i except one, and the resulting polynomial of one variable vanishes at a finite number of values. Since the field $\mathbb{F}$ is infinite, we can find a value for the remaining variable that makes $L \neq 0$ vanish. Let A be an invertible matrix that transforms this point
$$(k_1,\ldots,k_n)$$
to $(0,\ldots,0,1)$. The conclusion follows. □

It is easy (and necessary) to generalize this a bit:

COROLLARY 5.6.16. *Let $p_1,\ldots,p_t \in P_n$ be a finite set of power series that are nonunits. Then there exists an invertible matrix A such that $p_1^A,\ldots,p_t^A \in P_n$ are all X_n-regular.*

PROOF. Simply apply lemma 5.6.15 to the product $p_1 \cdots p_t$. □

We are finally ready to prove the main result:

THEOREM 5.6.17. *If $\mathbb{F}$ is an infinite field, then the ring $P_n = \mathbb{F}[[X_1,\ldots,X_n]]$ is a unique factorization domain.*

REMARK. The requirement that $\mathbb{F}$ be an infinite field is not really necessary but it simplifies the proof of lemma 5.6.15 on the previous page — and $\mathbb{F}$ will be infinite in all of our applications of this result.

Weierstrass originally proved this for $\mathbb{F} = \mathbb{C}$ and $P_n = \mathbb{C}\{X_1,\ldots,X_n\}$ — the ring of *convergent* power-series. This is essentially the ring of complex-analytic functions. See [56].

PROOF. We prove this by induction on n, the number of indeterminates. The result is almost trivial for $n = 1$ — see 5.6.2 on page 150.

Let $p_1 \in P_n$ be irreducible and suppose $p_1 \mid p_2 \cdot p_2$ and $p_1 \nmid p_2$. We will show that this forces $p_1 \mid p_3$. Use corollary 5.6.16 on the previous page to transform p_1, p_2, p_3 to X_n-regular power series of degrees d_1, d_2, d_3, respectively. Then the Weierstrass Preparation Theorem (theorem 5.6.14 at page 152) implies that

$$
\begin{aligned}
p_1^A &= u_1 \cdot w_1 \\
p_2^A &= u_2 \cdot w_2 \\
p_3^A &= u_3 \cdot w_3
\end{aligned}
$$

where $u_1, u_2, u_3 \in P_n$ are units and $w_1, w_2, w_3 \in P_{n-1}[X_n]$ are the Weierstrass polynomials of the p_i. We claim that the polynomial $w_1 \in P_{n-1}[X_n]$ is irreducible. This is because a nontrivial factorization of it would give a nontrivial factorization of p_1, since A induces an automorphism. Since P_{n-1} is a unique factorization domain by induction and $P_{n-1}[X_n]$ is one by theorem 5.6.7 on page 148, we must have

$$w_1 \mid w_3$$

which implies that

$$p_1^A \mid p_3^A$$

and

$$p_1 \mid p$$

which means P_n is a unique factorization domain, by lemma 5.6.1 on page 146. □

5.7. The Jacobson radical and Jacobson rings

In this section, we give a very brief treatment of a construct similar to the nilradical.

DEFINITION 5.7.1. If R is a commutative ring, the *Jacobson radical*, $\mathfrak{J}(R)$, of R is defined by

$$\mathfrak{J}(R) = \bigcap_{\text{maximal } \mathfrak{m} \subset R} \mathfrak{m}$$

— the intersection of all of the maximal ideals of R.

REMARK. Since the nilradical is the intersection of all prime ideals and maximal ideals are prime, it is easy to see that

$$\mathfrak{N}(R) \subset \mathfrak{J}(R)$$

is always true.

DEFINITION 5.7.2. A commutative ring, R, is called a *Jacobson ring* if for any ideal $\mathfrak{J} \subset R$

$$\sqrt{\mathfrak{J}} = \bigcap_{\mathfrak{J} \subset \mathfrak{m}} \mathfrak{m}$$

where the intersection is taken over all maximal ideals containing $\mathfrak{J}$.

REMARK. The term *Jacobson ring* was coined by Krull in [**64**] in honor of the notable American mathematician, Nathan Jacobson (1910–1999). Krull used Jacobson rings to generalize Hilbert's famous Nullstellensatz in algebraic geometry. Because of their relation to the Nullstellensatz, they are sometimes called Hilbert rings or Jacobson-Hilbert rings.

Theorem 12.2.8 on page 421 shows that $\sqrt{\mathfrak{J}}$ is the intersection of all prime ideals containing $\mathfrak{J}$. In a Jacobson ring, there are "enough" maximal ideals so the corresponding statement is true for the primes that are maximal.

We can characterize Jacobson rings by how prime ideals behave:

PROPOSITION 5.7.3. *The following statements are equivalent*

(1) *R is a Jacobson ring*
(2) *every prime ideal $\mathfrak{p} \subset R$ satisfies*

(5.7.1) $$\mathfrak{p} = \bigcap_{\mathfrak{p} \subset \mathfrak{m}} \mathfrak{m}$$

where the intersections is taken over all maximal ideals containing $\mathfrak{p}$.
(3) $\mathfrak{J}(R') = 0$ *for every quotient, R', of R that is an integral domain.*

PROOF. If R is Jacobson, the statement is clearly true because prime ideals are radical, so 1 $\implies$ 2. Conversely, if $\mathfrak{J} \subset R$ is any ideal, theorem 12.2.8 on page 421 implies that $\sqrt{\mathfrak{J}}$ is the intersection of all prime ideals that contain $\mathfrak{J}$ and equation 5.7.1 implies that each of these is the intersection of all the maximal ideals that contain it. It follows that the condition in definition 5.7.2 on page 154 is satisfied, so 2 $\implies$ 1. Statement 2 is equivalent to statement 2 because $R' = R/\mathfrak{p}$ for some prime ideal and lemma 5.2.9 on page 114 implies that the maximal ideals of $R/\mathfrak{p}$ are in a one to one correspondence with the maximal ideals of R containing $\mathfrak{p}$. □

This immediately implies:

COROLLARY 5.7.4. *Every quotient of a Jacobson ring is Jacobson.*

It is not hard to find examples of Jacobson rings:

PROPOSITION 5.7.5. *A principle ideal domain is a Jacobson ring if and only if it has an infinite number of prime ideals.*

REMARK. We immediately conclude that
(1) any field is a Jacobson ring,
(2) $\mathbb{Z}$ is a Jacobson ring,
(3) $k[X]$ is a Jacobson ring, where k is any field. An argument like that used in number theory implies that $k[X]$ has an infinite number of primes.

PROOF. We use the characterization of Jacobson rings in proposition 5.7.3. Let R denote the ring in question — this is a unique factorization domain (see remark 5.3.12 on page 119). *All* of the prime ideals of R are *maximal* except for (0). It follows that all prime ideals are equal to the intersection of maximal ideals that contain them, with the *possible* exception of (0).

If there are only a finite number of prime ideals, $(x_1), \ldots, (x_k)$ then

$$(x_1) \cap \cdots \cap (x_k) = (x_1 \cdots x_k) \neq (0)$$

so R fails to be Jacobson.

If there are an infinite number of prime ideals and $x \neq 0 \in R$ is an arbitrary element, then x factors as a finite product of primes. It follows that there exists a prime *not* in this factorization so that $x \notin \mathfrak{J}(R)$ — since nonzero prime ideals are maximal. It follows that the intersection of all maximal ideals that contain (0) *is* (0) and the ring is Jacobson. □

It is well-known that Jacobson rings are *polynomially-closed*: if J is a Jacobson ring, so is $J[X]$. To prove this, we need what is widely known as the Rabinowich Trick (which first appeared in [**92**]):

LEMMA 5.7.6. *The following statements are equivalent:*

(1) *the ring R is Jacobson*
(2) *if $\mathfrak{p} \subset R$ is any prime ideal and $S = R/\mathfrak{p}$ has an element $t \in S$ such that $S[t^{-1}]$ is a field, then S is a field.*

PROOF. If R is Jacobson, so is S. The prime ideals of $S[t^{-1}]$ are those of S that do not contain t. Since $S[t^{-1}]$ is a field, it follows that t is contained in every nonzero prime ideal. If any nonzero prime ideals existed in S, t would be contained in them. Since R is Jacobson, so is S and $\mathfrak{J}(R) = 0$ (see proposition 5.7.3 on page 155), so there cannot exist any nonzero prime ideals, and S must be a field.

Conversely, suppose the hypotheses are satisfied and $\mathfrak{p} \subset R$ is a prime ideal with
$$\mathfrak{p} \subsetneq \bigcap_{\mathfrak{p} \subset \mathfrak{m}} \mathfrak{m}$$
where the intersection is taken over all maximal ideals containing $\mathfrak{p}$. We will derive a contradiction.

If $t \in \bigcap_{\mathfrak{p} \subset \mathfrak{m}} \mathfrak{m} \setminus \mathfrak{p}$, the set of prime ideals, $\mathfrak{q}$, with $t \notin \mathfrak{q}$ has a maximal element (by Zorn's Lemma — 14.2.12 on page 465), $\mathfrak{Q}$. This ideal is not maximal since t is contained in all maximal ideals, so $R/\mathfrak{Q}$ is not a field. On the other hand $\mathfrak{Q}$ generates a maximal ideal of $R[t^{-1}]$ so
$$R[t^{-1}]/\mathfrak{Q} \cdot R[t^{-1}] = (R/\mathfrak{Q})[t^{-1}]$$
(see lemma 10.6.19 on page 373) is a field. The hypotheses imply that $R/\mathfrak{Q}$ is also a field — which is a contradiction. □

We need one more lemma to prove our main result:

LEMMA 5.7.7. *Let R be a Jacobson domain and let S be an algebra over R generated by a single element, i.e. $S = R[\alpha]$ and an integral domain. If there exists an element $t \in S$ such that $S[t^{-1}]$ is a field, then R and S are both fields, and S is a finite extension of R.*

PROOF. Let F be the field of fractions of R. We have $S = R[X]/\mathfrak{p}$ where $\mathfrak{p} \subset R[X]$ is a prime ideal and X maps to α under projection to the quotient. We claim that $\mathfrak{p} \neq (0)$. Otherwise, there would exist an element $t \in R[X]$ that makes $R[X][t^{-1}]$ a field. Since $R[X][t^{-1}] = F[X][t^{-1}]$, the fact that $F[X]$ is known to be Jacobson (by proposition 5.7.5 on page 155) and lemma 5.7.6 imply that $F[X]$ is *also* a field, which is a contradiction.

Since $\mathfrak{p} \neq 0$, let $p(X) \in \mathfrak{p}$ be any nonzero polynomial
$$p_n X^n + \cdots + p_0$$
that vanishes in S. In $S[p_n^{-1}]$ we may divide by p_n to get a monic polynomial — showing that α is integral over $R[p_n^{-1}]$ (see definition 6.5.1 on page 252) so corollary 6.5.6 on page 254 implies that $S[p_n^{-1}]$ is integral over $R[p_n^{-1}]$.

Let

(5.7.2) $$c_n t^n + \cdots + c_0 = 0$$

be a polynomial that t satisfies in S (factor off copies of t to guarantee that $c_0 \neq 0$). Now, invert $p_n c_0$ in R and S, so we get $S[(p_n c_0)^{-1}]$ integral over $R[(p_n c_0)^{-1}]$.

After doing this, we can divide equation 5.7.2 on the preceding page by $c_0 t^n$ in $S[(c_0 p_n)^{-1}, t^{-1}]$ to get a monic polynomial for t^{-1}

$$t^{-n} + \left(\frac{c_1}{c_0}\right) t^{-(n-1)} + \cdots + \frac{c_n}{c_0} = 0$$

It follows that $S[(c_0 p_n)^{-1}, t^{-1}]$ is integral over $R[(p_n c_0)^{-1}]$. Since $S[t^{-1}]$ is a field, so is $S[(c_0 p_n)^{-1}, t^{-1}]$ (the same field) and proposition 6.5.7 on page 254 implies that $R[(p_n c_0)^{-1}]$ is *also* a field. The fact that R is Jacobson, and lemma 5.7.6 on the facing page implies that R is also a field. So $R = R[p_n^{-1}]$ and $R[\alpha] = S$ is integral over R. Proposition 6.5.7 on page 254 applied a second time implies that S is *also* a field and the conclusion follows. □

We are now ready to prove the main result:

THEOREM 5.7.8. *If R is a Jacobson ring, any finitely generated algebra over R is also a Jacobson ring.*

REMARK. This result provides a huge number of Jacobson rings:

▷ $\mathbb{Z}[X_1, \ldots, X_n]$

PROOF. We start with $S = R[\alpha]$. The general case follows by a simple induction. If $\mathfrak{p} \subset S$ is a prime ideal, then $S/\mathfrak{p}$ will be an integral domain and the image of R in $S/\mathfrak{p}$ will be a Jacobson domain. If there exists $t \in S/\mathfrak{p}$ such that $(S/\mathfrak{p})[t^{-1}]$ is a field, lemma 5.7.7 implies that $S/\mathfrak{p}$ (and, for that matter $R/R \cap \mathfrak{p}$) is *also* a field — satisfying the conditions of lemma 5.7.7 on the preceding page. It follows that S is Jacobson. □

It is also easy to find *non*-Jacobson rings:

EXAMPLE 5.7.9. If $\mathfrak{t} = 2 \cdot \mathbb{Z} \subset \mathbb{Z}$, then $\mathbb{Z}_\mathfrak{t}$, is the ring[3] of rational numbers with odd denominators. This is a local ring with a unique maximal ideal, $2 \cdot \mathbb{Z}_{(2)}$ so $\mathfrak{J}(\mathbb{Z}_\mathfrak{t}) = 2 \cdot \mathbb{Z}_\mathfrak{t}$ but $\mathfrak{N}(\mathbb{Z}_\mathfrak{t}) = 0$, since it is an integral domain.

This example induces many more

EXAMPLE 5.7.10. Let $R = \mathbb{Z}_\mathfrak{t}[X_1, \ldots, X_n]$ be a polynomial ring over $\mathbb{Z}_\mathfrak{t}$ from example 5.7.9 above. The maximal ideals of $\mathbb{Q}[X_1, \ldots, X_n]$ are of the form $(X_1 - q_1, \ldots, X_n - q_n)$. If we restrict the q_i to be in $\mathbb{Z}_\mathfrak{t}$, we get ideals of R that are no longer maximal because the quotient of R by them is $\mathbb{Z}_\mathfrak{t}$, which is not a field. We can make these ideal maximal by adding one additional element. The ideals

$$\mathfrak{L}(q_1, \ldots, q_n) = (2, X_1 - q_1, \ldots, X_n - q_n)$$

are maximal because the quotient of R by them is $\mathbb{Z}_2$. The intersection of the ideals $\mathfrak{L}(q_1, \ldots, q_n)$ contains (at least) (2) or $2 \cdot R$. Since R is an integral domain, $\mathfrak{N}(R) = 0$ but $(2) \subset \mathfrak{J}(R)$. So R is not Jacobson, either.

5.8. Artinian rings

Artinian rings are an example of the effect of slightly changing the defining property of noetherian rings. It turns out (theorem 5.8.5 on page 159) that Artinian rings *are* noetherian rings with a special property.

---

[3] We do not use the notation $\mathbb{Z}_{(2)}$ because that would conflict with the notation for 2-adic integers (see example 10.4.14 on page 359).

In algebraic geometry, Artinian rings are used to understand the geometric properties of finite maps — see chapter 2 of [99].

> Emil Artin, (1898 – 1962) was an Austrian mathematician of Armenian descent. He is best known for his work on algebraic number theory, contributing largely to class field theory and a new construction of L-functions. He also contributed to the pure theories of rings, groups and fields. His work on Artinian rings appears in [3] and [2].

DEFINITION 5.8.1. A ring, R, will be called *Artinian* if every *descending* sequence of ideals becomes constant from some finite point on — i.e., if

$$\mathfrak{a}_1 \supseteq \mathfrak{a}_2 \supseteq \cdots$$

is a descending chain of ideals, there exists an integer n such that $\mathfrak{a}_i = \mathfrak{a}_{i+1}$ for all $i \geq n$.

REMARK. Emil Artin (1898-1962) introduced these rings in the papers [3] and [2]. At first glance, this definition appears very similar to the definition of noetherian ring in definition 5.4.1 on page 121 (at least if you look at the remark following the definition).

For instance $\mathbb{Z}$ is noetherian but not Artinian since we have an infinite descending sequence of ideals that does *not* become constant

$$(2) \supset (4) \supset \cdots \supset (2^k) \supset \cdots$$

Artinian rings have some unusual properties:

LEMMA 5.8.2. *If R is an Artinian ring:*
  (1) every quotient of R is Artinian
  (2) if R is an integral domain, it is a field
  (3) every prime ideal of R is maximal
  (4) the number of maximal ideals in R is finite.

REMARK. Statement 3 implies that all Artinian rings are *Jacobson* rings — see definition 5.7.2 on page 154.

PROOF. The first statement follows immediately from the definition of Artinian ring and lemma 5.2.9 on page 114.

To prove the second statement, suppose R is an integral domain and $x \neq 0 \in R$. Then the descending chain of ideals

$$(x) \supset (x^2) \subset \cdots \supset (x^n) \supset \cdots$$

must stabilize after a finite number of steps, so $(x^t) = (x^{t+1})$ and $x^t = r \cdot x^{t+1}$ for some $r \in R$, or $x^t - r \cdot x^{t+1} = 0$. Since R is an integral domain $x^t \cdot (1 - r \cdot x) = 0$ implies $1 - r \cdot x = 0$ so $r = x^{-1}$.

The third statement follows from the first two: if $\mathfrak{p} \subset R$ is a prime ideal, then $R/\mathfrak{p}$ is an Artinian integral domain, hence a field. This implies that $\mathfrak{p}$ is maximal.

Suppose we have an infinite set of distinct maximal ideals, $\{\mathfrak{m}_i\}$ and consider the following descending sequence of ideals

$$\mathfrak{m}_1 \supset \mathfrak{m}_1 \cdot \mathfrak{m}_2 \supset \cdots \supset \mathfrak{m}_1 \cdots \mathfrak{m}_k \supset \cdots$$

The Artinian property implies that this becomes constant at some point, i.e.,

$$\mathfrak{m}_1 \cdots \mathfrak{m}_n \subset \mathfrak{m}_1 \cdots \mathfrak{m}_{n+1} \subset \mathfrak{m}_{n+1}$$

The fact that maximal ideals are prime (see proposition 5.2.4 on page 112) and exercise 5 on page 115 implies that either

$$\mathfrak{m}_1 \subset \mathfrak{m}_{n+1}$$

a contradiction, *or*

$$\mathfrak{m}_2 \cdots \mathfrak{m}_n \subset \mathfrak{m}_{n+1}$$

In the latter case, a simple induction shows that *one* of the $\mathfrak{m}_i \subset \mathfrak{m}_{n+1}$, so a contradiction cannot be avoided. □

We can completely characterize Artinian rings. The first step to doing this is:

LEMMA 5.8.3. *Let R be a ring in which there exists a finite product of maximal ideals equal to zero, i.e.*

$$\mathfrak{m}_1 \cdots \mathfrak{m}_k = 0$$

Then R is Artinian if and only if it is noetherian.

PROOF. We have a descending chain of ideals

$$R \supset \mathfrak{m}_1 \supset \mathfrak{m}_1\mathfrak{m}_2 \supset \cdots \supset \mathfrak{m}_1 \cdots \mathfrak{m}_{k-1} \supset \mathfrak{m}_1 \cdots \mathfrak{m}_k = 0$$

Let for $1 \le i \le k$, let $M_i = \mathfrak{m}_1 \cdots \mathfrak{m}_{i-1}/\mathfrak{m}_1 \cdots \mathfrak{m}_i$, a module over $R/\mathfrak{m}_i$ i.e., a vector space over $R/\mathfrak{m}_i$. Then M_i is Artinian if and only if it is noetherian — if and only if it is finite-dimensional. The conclusion follows from induction on k, proposition 6.3.12 on page 226 and the short exact sequences

$$0 \to \mathfrak{m}_1 \cdots \mathfrak{m}_i \to \mathfrak{m}_1 \cdots \mathfrak{m}_{i-1} \to M_i \to 0$$

□

PROPOSITION 5.8.4. *If R is an Artinian ring, the nilradical, $\mathfrak{N}(R)$, is nilpotent, i.e. there exists an integer k such that $\mathfrak{N}(R)^k = 0$.*

REMARK. We have already seen this for noetherian rings — see exercise 6 on page 125.

PROOF. Since R is Artinian, the sequence of ideals

$$\mathfrak{N}(R) \supset \mathfrak{N}(R)^2 \supset \cdots$$

becomes constant after a finite number of steps. Suppose $\mathfrak{n} = \mathfrak{N}(R)^k = \mathfrak{N}(R)^{k+1}$. We claim that $\mathfrak{n} = 0$.

If not, consider the set, $\mathscr{I}$ of ideals, $\mathfrak{a}$, in R such that $\mathfrak{a} \cdot \mathfrak{n} \ne 0$. Since all descending sequences of such ideals have a lower bound (because R is Artinian), Zorn's Lemma (14.2.12 on page 465) implies that $\mathscr{I}$ has a minimal element, $\mathfrak{b}$. There exists an element $x \in \mathfrak{b}$ such that $x \cdot \mathfrak{n} \ne 0$, and the minimality of $\mathfrak{b}$ implies that $\mathfrak{b} = (x)$. The fact that $\mathfrak{n}^2 = \mathfrak{n}$ implies that $(x \cdot \mathfrak{n}) \cdot \mathfrak{n} = x \cdot \mathfrak{n}^2 = x \cdot \mathfrak{n}$ so $x \cdot \mathfrak{n} \subset (x)$. The minimality of $\mathfrak{b} = (x)$ implies that $x \cdot \mathfrak{n} = (x)$ so that there is an element $y \in \mathfrak{n}$ such that

$$x \cdot y = x = x \cdot y^2 = \cdots = x \cdot y^m$$

Since $y \in \mathfrak{N}(R)$, we have $y^n = 0$ for some $n > 0$, which implies that $x = 0$, which in turn contradicts the requirement that $x \cdot \mathfrak{n} \ne 0$. This contradiction is the result of assuming that $\mathfrak{n} \ne 0$. □

We are finally ready to characterize Artinian rings:

THEOREM 5.8.5. *A ring is Artinian if and only if it is noetherian and all of its prime ideals are maximal.*

REMARK. The reader may wonder whether this contradicts our statement that $\mathbb{Z}$ is not Artinian. After all, all of its prime ideals of the form (p) for a prime number $p \in \mathbb{Z}$ are maximal. The one exception is $(0) \subset (p)$ which is a prime ideal that is proper subset of another prime ideal.

PROOF. If R is Artinian, then lemma 5.8.2 on page 158 implies that all of its prime ideals are maximal. Proposition 5.8.4 on the previous page implies that $\mathfrak{N}(R)^k = 0$ for some $k > 0$ and the proof of 12.2.8 on page 421 implies that

$$\mathfrak{m}_1 \cap \cdots \cap \mathfrak{m}_n \subset \mathfrak{N}(R)$$

where $\mathfrak{m}_1, \ldots, \mathfrak{m}_n$ are the finite set of maximal ideals of R. Since

$$\mathfrak{m}_1 \cdots \mathfrak{m}_n \subset \mathfrak{m}_1 \cap \cdots \cap \mathfrak{m}_n$$

it follows that a finite product of maximal ideals is equal to 0. Lemma 5.8.3 on the previous page then implies that R is noetherian.

Conversely, if R is noetherian and all of its prime ideals are maximal, lemma 5.4.6 on page 124 implies that the number of these will be finite. Since the nilradical is nilpotent (see exercise 6 on page 125), the argument above implies that R is Artinian. □

Another interesting property of Artinian rings is:

THEOREM 5.8.6. *An Artinian ring decomposes (uniquely) into a product of finitely many local Artinian rings.*

REMARK. Recall that a *local* ring is one that has a *unique* maximal ideal — see definition 5.2.8 on page 113.

PROOF. Let A be an Artinian ring with maximal ideals $\{\mathfrak{m}_1, \ldots, \mathfrak{m}_n\}$. Then

$$\mathfrak{N}(R) = \mathfrak{m}_1 \cap \cdots \cap \mathfrak{m}_n = \mathfrak{m}_1 \cdots \mathfrak{m}_n$$

Let k be a value for which $\mathfrak{N}(R)^k = 0$ (this exists by proposition 5.8.4 on the previous page). Then

$$(\mathfrak{m}_1 \cdots \mathfrak{m}_n)^k = \mathfrak{m}_1^k \cdots \mathfrak{m}_n^k$$

and the Chinese Remainder Theorem (see exercise 12 on page 116) implies that

$$R = \frac{R}{(\mathfrak{m}_1 \cdots \mathfrak{m}_n)^k} = \prod_{i=1}^{n} \frac{R}{\mathfrak{m}_i^k}$$

Each of the quotients $R/\mathfrak{m}_i^k$ has a unique maximal ideal, namely the image of $\mathfrak{m}_i$ so it is a local ring.

Suppose we have an expression

$$A = A_1 \times \cdots \times A_t$$

where the A_i are Artinian local rings. Then every ideal, $\mathfrak{I} \subset A$ is of the form

$$\mathfrak{I} = \mathfrak{I}_1 \times \cdots \times \mathfrak{I}_t$$

and the maximal ideals of A are of the form

$$\mathfrak{m}_i = A_1 \times \cdots \times \mathfrak{M}_i \times \cdots \times A_t$$

where $\mathfrak{M}_i \subset A_i$ is a maximal ideal. This implies that $t = n$ — i.e., the number of factors is uniquely determined by A. We also conclude that

$$\mathfrak{m}_1^k \cdots \mathfrak{m}_n^k = \mathfrak{M}_1^k \times \cdots \times \mathfrak{M}_n^k = 0$$

so that $\mathfrak{m}_i^k = 0$ for all i. We finally note that
$$\frac{R}{\mathfrak{m}_i^k} = \frac{A_1 \times \cdots \times A_n}{A_1 \times \cdots \times A_{i-1} \times 0 \times A_{i+1} \times \cdots \times A_n} = A_i$$
so the decomposition is unique. □

CHAPTER 6

Modules and Vector Spaces

"Every mathematician worthy of the name has experienced, if only rarely, the state of lucid exaltation in which one thought succeeds another as if miraculously, and in which the unconscious (however one interprets that word) seems to play a role."
— André Weil.

6.1. Introduction

Linear algebra includes two main themes: *vectors* and systems of *linear equations*[1].

Vectors in $\mathbb{R}^n$ have a geometric significance, and were first studied in the mid-19th century. They were conceived as objects with *magnitude* and *direction* and often represented as arrows:

where the length of the arrow is the magnitude. One is to imagine that this vector displaces the universe from its tail to its head. Although they have magnitude and direction, they do not have a location, so the two vectors

are the same.

The *sum* of two vectors is the result of applying one displacement after the other

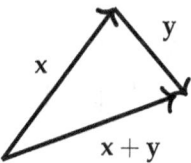

---

[1]This is not in historical order. The study of linear equations precedes that of vectors by at least a century.

Of course, there is only so much one can do with pictures. The need to quantify vectors quickly became apparent and vectors were given a numerical representation, where

$$\mathbf{v} = \begin{bmatrix} 1 \\ 2 \\ 3 \end{bmatrix}$$

represents the vector that displaces the *origin* to the point $(1,2,3)$. Addition of vectors is easy to formulate

$$\begin{bmatrix} a \\ b \\ c \end{bmatrix} + \begin{bmatrix} u \\ v \\ w \end{bmatrix} = \begin{bmatrix} a+u \\ b+v \\ c+w \end{bmatrix}$$

Linear equations are problems like

$$2x + 3y = 5$$
(6.1.1)
$$6x - y = 2$$

where we must solve for x and y.

The obstacle to solving these equations easily is that they both have two variables. We must, somehow, isolate one of the variables — and solve for it. Getting the second variable is straightforward.

In this case it is not hard to see that we can subtract $3\times$ the first equation from the second to give

$$2x + 3y = 5$$
(6.1.2)
$$6x - 6x - 10y = -13$$

so $y = 13/10$. Now we substitute this into the first equation to get

$$2x + 3 \cdot 13/10 = 5$$

or $x = 11/20$.

Mathematicians quickly developed a shorthand for a set of variables or quantities by writing them in tables

$$\begin{bmatrix} x \\ y \end{bmatrix} \text{ and } \begin{bmatrix} 5 \\ 2 \end{bmatrix}$$

called *vectors*, and writing the coefficients in a table

(6.1.3)
$$\begin{bmatrix} 2 & 3 \\ 6 & -1 \end{bmatrix}$$

called a *matrix*. Strictly speaking, a vector is a matrix with one column. The original problem is rewritten as

$$\begin{bmatrix} 2 & 3 \\ 6 & -1 \end{bmatrix} \cdot \begin{bmatrix} x \\ y \end{bmatrix} = \begin{bmatrix} 5 \\ 2 \end{bmatrix}$$

where we, somehow, define multiplication of a vector by a matrix (see equations 6.2.6 on page 170 and 6.2.7 on page 170) in a way that reproduces equation 6.1.1.

The reader might protest that we haven't simplified the process of solving equation 6.1.1 at all. We have, however, clarified the most important aspect of the solution: subtracting $3\times$ the first equation from the second *only* involves the *coefficients* — i.e. the *matrix* in equation 6.1.3 on the preceding page — and *not* the variables x and y.

We define a vector space to be the mathematical structure where vectors "live."

6.2. Vector spaces

6.2.1. Basic properties. Vector-spaces and maps between them are the main objects of study of linear algebra.

DEFINITION 6.2.1. If $\mathbb{F}$ is a field, a *vector space*, V, over $\mathbb{F}$ is an abelian group with an *action* of $\mathbb{F}$ on V

$$\bullet\colon \mathbb{F} \times V \to V$$

— i.e. multiplication of elements of V by elements of $\mathbb{F}$ is well-defined. In addition
  (1) if $x \in \mathbb{F}$ and $v_1, v_2 \in V$, then $x \cdot v_1 + x \cdot v_2 = x \cdot (v_1 + v_2)$,
  (2) if $x, y \in \mathbb{F}$ and $v \in V$, then $x \cdot v + y \cdot v = (x+y) \cdot v$, and $x \cdot (y \cdot v) = (xy) \cdot v$,
  (3) $0 \cdot v = 0 \in V$, and $1 \cdot v = v$ for all $v \in V$.

REMARK. It is easy to find examples of vector spaces:
  ▷ $\mathbb{R}^n$ is a vector space over $\mathbb{R}$
  ▷ $C^\infty([0,1])$ — the set of infinitely differentiable real-valued functions on the interval $[0,1]$ — is a vector space over $\mathbb{R}$.
  ▷ a field is a vector space over any of its subfields.
  ▷ $M_{n,m}(k)$ — the set of $n \times m$ matrices with entries in a field, k, is a vector space.

DEFINITION 6.2.2. If V is a vector space over a field, k, and $s = \{v_1, \ldots, v_n\}$ is a set with $v_i \in V$ for $i = 1, \ldots, n$ then s is *linearly dependent* if there exists an equation

$$(6.2.1) \qquad \sum_{i=1}^{n} \alpha_i v_i = 0$$

with $\alpha_i \in k$ and at least one $\alpha_i \neq 0$. If s is not linearly dependent, it is said to be *linearly independent*.

An *infinite* set of vectors is linearly independent if all of its *finite* subsets are.

REMARK. If s is linearly *dependent*, then one of its elements is a linear combination of the others: if $\alpha_k \neq 0$ in equation 6.2.1, then

$$v_k = -\alpha_k^{-1} \sum_{i=1, i \neq k}^{n} \alpha_i v_i$$

Note that linear independence is a property of the *set* of vectors that is inherited by every subset:

A single vector is a linearly independent set if it is nonzero.

Two vectors are linearly independent if neither is a multiple of the other. This also means both of the vectors are nonzero.

DEFINITION 6.2.3. If V is a vector-space over a field k and $s = \{v_1, \ldots, v_n\}$ are elements of V, their span, $\mathrm{Span}(s)$ is the set of all possible linear combinations

$$\sum_{i=1}^{n} \alpha_i v_i$$

for $\alpha_i \in k$.

REMARK. It's not hard to see that the span of a set of vectors is always a vector space.

PROPOSITION 6.2.4. *Let V be a vector space and let $s = \{v_1, \ldots, v_n\}$ be a linearly independent set of vectors. If $v \in V$ and $v \notin \mathrm{Span}(s)$, then the set*

$$s \cup \{v\}$$

is also linearly independent.

PROOF. Suppose

$$\alpha v + \sum_{i=1}^{n} \alpha_i v_i = 0$$

with some nonzero coefficients (making the set $s \cup \{v\}$ linearly *dependent*). If $\alpha = 0$, then we get

$$\sum_{i=1}^{n} \alpha_i v_i = 0$$

which forces all of the α_i to be zero, a *contradiction*. It follows that $\alpha \neq 0$ so that

$$v = -\alpha^{-1} \sum_{i=1}^{n} \alpha_i v_i$$

which contradicts the hypothesis that $v \notin \mathrm{Span}(s)$. □

DEFINITION 6.2.5. A *basis* for a vector space, V, is a linearly independent set of vectors $s = \{v_1, \ldots\}$ such that

$$V = \mathrm{Span}(s)$$

Now we can outline an inductive procedure for finding a basis of a vector space.
  (1) Start with a nonzero vector, $v_1 \in V$, and set $s_1 = \{v_1\}$,
  (2) In step n, $s_n = \{v_1, \ldots, v_n\}$. If $V = \mathrm{Span}(s)$, we are done and s_n is a basis. If not, there exists $v_{n+1} \in V \setminus \mathrm{Span}(s)$, and set $s_{n+1} = s_n \cup \{v_{n+1}\}$. This will be linearly independent, by proposition 6.2.4.

If this terminates in a finite number of steps, it proves a basis exists and explicitly constructs one. Unfortunately, this is unlikely to happen in many cases — for instance if $V = C^{\infty}([0,1])$.

In light of this, we have

THEOREM 6.2.6. *If V is a vector-space and $s = \{v_1, \ldots, v_k\}$ is a linearly independent set of vectors, then V has a basis containing s.*

PROOF. We will assume that the procedure described above never terminates, since we would be done in that case. Let S be the set of *all* linearly independent *sets* of vectors of V containing s.

We begin by claiming that every *ascending chain* of elements in S

$$A_1 \subset A_2 \subset \cdots$$

has a *maximum*, namely

$$A = \bigcup_{i=1}^{\infty} A_i$$

This involves proving that $A = \{w_1, \dots\}$ is a linearly independent set of vectors, where $w_i = v_i$ for $i \leq k$. Any linear *dependence* equation

$$\sum_{i=1}^{n} \alpha_i w_i = 0$$

involves a *finite* set of vectors of A, hence would involve the vectors of A_j for some j. This contradicts the fact that each A_j is a linearly independent set of vectors.

Zorn's lemma (14.2.12 on page 465) implies that S contains a *maximal* element, T. This is linearly independent, and we claim that it spans V. If not, there exists $v \in V \setminus \text{Span}(T)$ and

$$T \cup \{v\} \supset T$$

is linearly independent (by proposition 6.2.4 on the preceding page). This contradicts the maximality of T. □

Bases for vector spaces are interesting because they constitute coordinate systems:

PROPOSITION 6.2.7. *If V is a vector space over a field k with a basis $B = \{b_1, \dots\}$ and $v \in V$ is any vector, then there exists a unique expression*

$$v = \sum_{i=1}^{n} \alpha_i b_{j_i}$$

PROOF. We know that such an expression exists because $V = \text{Span}(B)$. If we had two distinct expressions for v, i.e.

$$v = \sum_{i=1}^{n} \alpha_i b_{j_i} = \sum_{i=1}^{n} \beta_i b_{j_i}$$

then

$$\sum_{i=1}^{n} (\alpha_i - \beta_i) b_{j_i} = 0$$

and the linear independence of B implies that $\alpha_i - \beta_i = 0$ for all i. □

It turns out that the *size* of a basis is uniquely determined

PROPOSITION 6.2.8. *If V is a vector space over a field k and $B = \{b_1, \dots, b_n\}$ is a basis, then every other basis has precisely n elements. This value n is called the dimension of V.*

REMARK. It turns out that this result is true even if the basis-sets are *infinite*, but one must use transfinite induction and the axiom of choice (14.2.11 on page 465) to prove it.

PROOF. Let $F = \{f_1, \ldots, f_m\}$ be another basis for V.

If $n = 1$, the fact that B is a basis implies that $f_i = \alpha_i b_1$, so that $\{f_1, \ldots, f_k\}$ is linearly *dependent* unless $k = 1$.

Now we consider the case where $n > 1$.

We will show that $m \leq n$ and the conclusion will follow by symmetry. Since F is linearly independent, $f_1 \neq 0$. Since B is a basis, we have

$$(6.2.2) \qquad f_1 = \alpha_1 b_1 + \cdots + \alpha_n b_n$$

Without loss of generality, assume $\alpha_1 \neq 0$ (if necessary, rearrange the b_i to make this happen) so we can solve for b_1:

$$(6.2.3) \qquad b_1 = \alpha_1^{-1}(f_1 - \alpha_2 b_2 - \cdots - \alpha_n b_n)$$

We claim that $\{f_1, b_2, \ldots, b_n\}$ is a basis for V. If $v \in V$, there is a formula

$$v = \beta_1 b_1 + \cdots + \beta_n b_n$$

and plugging equation 6.2.3 into it gives a formula for v in terms of $\{f_1, b_2, \ldots, b_n\}$. It follows that $\{f_1, b_2, \ldots, b_n\}$ spans V.

If

$$(6.2.4) \qquad \gamma_1 f_1 + \sum_{i=2}^{n} \gamma_i b_i = 0$$

is a linear dependence relation (i.e., some of the $\gamma_i \neq 0$), we claim that $\gamma_1 \neq 0$. If $\gamma_1 = 0$, we get a linear dependence equation

$$\sum_{i=2}^{n} \gamma_i b_i = 0$$

which contradicts the fact that B is a linearly independent set. Now we plug equation 6.2.2 into equation 6.2.4 to get

$$\gamma_1(\alpha_1 b_1 + \cdots + \alpha_n b_n) + \gamma_2 b_2 + \cdots + \gamma_n b_n = 0$$
$$\gamma_1 \alpha_1 b_1 + (\gamma_1 \alpha_2 + \gamma_2) b_2 + \cdots + (\gamma_1 \alpha_n + \gamma_n) = 0$$

Since we know $\gamma_1 \alpha_1 \neq 0$, this gives a nontrivial linear dependence relation between the b_i, a contradiction.

Assume $m > n$. We have established that we can replace *one* of the b_i by *one* of the f_i and still have a basis. We do this *repeatedly* until we have a basis $\{f_1, \ldots f_n\}$. If $m > n$ the f_i with $i > n$ will be linear combinations of $\{f_1, \ldots f_n\}$ and the original set $\{f_1, \ldots, f_m\}$ cannot be linearly independent: a contradiction.

It follows that $m \leq n$ and symmetry (i.e., starting with the F-basis and successively replacing basis elements by the b_i) shows that $n \leq m$, so they are equal. $\square$

DEFINITION 6.2.9. If a vector-space V over field k has a finite basis $B = \{b_1, \ldots, b_n\}$ and

$$v = \sum_{i=1}^{n} \alpha_i b_i$$

is the unique expression for $v \in V$, we use the notation

(6.2.5)
$$v = \begin{bmatrix} \alpha_1 \\ \vdots \\ \alpha_n \end{bmatrix}_B$$

which we will call a *matrix* with one column and n rows.

REMARK. Note that the elements α_i in equation 6.2.5 depend on the basis used to compute them.

6.2.2. Linear transformations. Having defined vector spaces, we turn to the question of maps between them:

DEFINITION 6.2.10. If V_1 and V_2 are vector spaces over the field k, a *homomorphism* or *linear transformation*

$$f\colon V_1 \to V_2$$

is a function with the properties
 (1) for all $v, w \in V_1$, we have $f(v+w) = f(v) + f(w)$
 (2) for all $v \in V_1$, $x \in k$, we have $f(x \cdot v) = x \cdot f(v)$

The set of vectors $v \in V_1$ with the property that $f(v) = 0$ is called the *nullspace* of f, denoted $\text{Null}(f)$.

REMARK. The nullspace is just the kernel of f, regarded as a map of abelian groups (for instance). It is clearly a subspace of V_1.

Suppose we have a linear transformation

$$f\colon V \to W$$

and $B = \{b_1, \ldots, b_n\}$ is a basis for B and $G = \{g_1, \ldots, g_m\}$ is a basis for W. If $v \in V$ is given by

$$v = \sum_{i=1}^n \alpha_i b_i$$

then the definition of a linear transformation implies that

$$f(v) = \sum_{i=1}^n \alpha_i f(b_i)$$

If

$$f(b_i) = \begin{bmatrix} \beta_{1,i} \\ \vdots \\ \beta_{n,i} \end{bmatrix}_G$$

then

$$f(v) = \begin{bmatrix} \beta_{1,1} & \cdots & \beta_{1,n} \\ \vdots & \ddots & \vdots \\ \beta_{m,1} & \cdots & \beta_{m,n} \end{bmatrix} \begin{bmatrix} \alpha_1 \\ \vdots \\ \alpha_n \end{bmatrix}$$

in the G-basis. The quantity

$$M = \begin{bmatrix} \beta_{1,1} & \cdots & \beta_{1,n} \\ \vdots & \ddots & \vdots \\ \beta_{m,1} & \cdots & \beta_{m,n} \end{bmatrix}$$

is called an $n \times m$ matrix, and the *product* is defined by

$$(6.2.6) \quad \begin{bmatrix} \beta_{1,1} & \cdots & \beta_{1,n} \\ \vdots & \ddots & \vdots \\ \beta_{m,1} & \cdots & \beta_{m,n} \end{bmatrix} \begin{bmatrix} \alpha_1 \\ \vdots \\ \alpha_n \end{bmatrix} = \alpha_1 \begin{bmatrix} \beta_{1,1} \\ \vdots \\ \beta_{m,1} \end{bmatrix} + \cdots + \alpha_n \begin{bmatrix} \beta_{1,n} \\ \vdots \\ \beta_{m,n} \end{bmatrix}$$

or, if we regard the columns of M as $m \times 1$ matrices, so

$$(6.2.7) \quad M = \begin{bmatrix} C_1 & \cdots & C_n \end{bmatrix}$$

where the C_i are called *column-vectors* of M, we get

$$f(v) = \sum_{i=1}^{n} \alpha_i C_i$$

or

$$f(v)_i = \sum_{j=1}^{n} \beta_{i,j} \alpha_j$$

so that a matrix is a convenient notation for representing a linear transformation. We will also use the notation

$$(6.2.8) \quad M = \begin{bmatrix} R_1 \\ \vdots \\ R_m \end{bmatrix}$$

where the R_i are the *rows* of M, regarded as $1 \times n$ matrices.

By abuse of notation, we will often refer to the nullspace (see definition 6.2.10 on the previous page) of a matrix.

Note that the matrix-representation of a linear transformation depends in a crucial way on the *bases* used.

DEFINITION 6.2.11. If $n \geq 1$ is an integer, define the *identity matrix*

$$I_n = \begin{bmatrix} 1 & 0 & \cdots & 0 \\ 0 & 1 & \cdots & 0 \\ 0 & 0 & \ddots & \vdots \\ 0 & 0 & \cdots & 1 \end{bmatrix}$$

or

$$I_{i,j} = \begin{cases} 1 & \text{if } i = j \\ 0 & \text{otherwise} \end{cases}$$

REMARK. It is not hard to see that

$$I_n \begin{bmatrix} \alpha_1 \\ \vdots \\ \alpha_n \end{bmatrix} = \begin{bmatrix} \alpha_1 \\ \vdots \\ \alpha_n \end{bmatrix}$$

which motivates the term identity matrix. It represents the identity *map* regarded as a linear transformation.

DEFINITION 6.2.12. If A is an $n \times m$ matrix, its *transpose*, A^t is an $m \times n$ matrix defined by

$$(A^t_{i,j}) = A_{j,i}$$

6.2. VECTOR SPACES

Composites of linear transformations give rise to *matrix products*:

DEFINITION 6.2.13. If A is an $n \times m$ matrix and B is an $m \times k$ matrix with

$$B = \begin{bmatrix} C_1 & \cdots & C_k \end{bmatrix}$$

where the C_i are the column-vectors, then

$$AB = \begin{bmatrix} AC_1 & \cdots & AC_k \end{bmatrix}$$

or

$$(AB)_{i,j} = \sum_{\ell=1}^{m} A_{i,\ell} B_{\ell,j}$$

— an $n \times k$ matrix. If A and B are $n \times n$ matrices and $AB = I$, A is said to be *invertible* and B is its *inverse*.

Note that any *isomorphism* of vector spaces is represented by a square matrix since the size of a basis is uniquely determined.

EXERCISES.

1. Which of the following functions define linear transformations?
    a. $f(x,y,z) = xy + z$
    b. $f(x,y,z) = 3x + 2y - z$
    c. $f(x,y,z) = x + y + z + 1$
    d. $f(x,y,z) = z - x + 5$

2. Show that matrix-multiplication is distributive over addition and associative, so that the set of all $n \times n$ matrices over a ring, R, forms a (non-commutative) ring $M_n(R)$.

3. Show that *powers* of a matrix commute with each other, i.e. for any matrix, A, over a ring, R, and any $n, m \geq 0$, show that

$$A^n \cdot A^m = A^m \cdot A^n$$

4. If R is a commutative ring, $M_n(R)$ is the ring of $n \times n$ matrices, and $A \in M_n(R)$ is a specific matrix, show that there exists a ring-homomorphism

$$R[X] \to M_n(R)$$
$$X \mapsto A$$

so that the matrix A generates a commutative subring $R[A] \subset M_n(R)$.

6.2.3. Determinants.

DEFINITION 6.2.14. If M is an $n \times n$ matrix, its *determinant*, $\det(M)$ is defined by

$$(6.2.9) \qquad \det(M) = \sum_{\sigma \in S_n} \wp(\sigma) M_{1,\sigma(1)} \cdots M_{n,\sigma(n)}$$

where the sum is taken over all $n!$ permutations in S_n. If $\tau \in S_n$ is *any* transposition and $A_n \triangleleft S_n$ is the *alternating group* (see definition 4.5.10 on page 54), then τ projects to the nontrivial element of $S_n/A_n = \mathbb{Z}_2$ so

$$S_n = A_n \sqcup \tau \cdot A_n$$

as *sets*. We can rewrite formula 6.2.9 as

$$(6.2.10) \quad \det(M) = \sum_{\sigma \in A_n} M_{1,\sigma(1)} \cdots M_{n,\sigma(n)} - \sum_{\sigma \in A_n} M_{1,\tau\sigma(1)} \cdots M_{n,\tau\sigma(n)}$$

$$(6.2.11) \qquad = \sum_{\sigma \in A_n} \left(M_{1,\sigma(1)} \cdots M_{n,\sigma(n)} - M_{1,\tau\sigma(1)} \cdots M_{n,\tau\sigma(n)} \right)$$

since all the permutations in A_n are even.

REMARK. Equation 6.2.9 is due to Euler. It is not particularly suited to computation of the determinant since it is a sum of $n!$ terms.

In a few simple cases, one can write out the determinant explicitly:

$$(6.2.12) \qquad \det \begin{bmatrix} a & b \\ c & d \end{bmatrix} = ad - bc$$

and

$$(6.2.13) \quad \det \begin{bmatrix} a_{11} & a_{12} & a_{13} \\ a_{21} & a_{22} & a_{23} \\ a_{31} & a_{32} & a_{33} \end{bmatrix} = a_{11}a_{22}a_{33} + a_{11}a_{23}a_{31} + a_{13}a_{32}a_{21}$$
$$- a_{13}a_{22}a_{31} - a_{23}a_{32}a_{11} - a_{33}a_{21}a_{12}$$

DEFINITION 6.2.15. An $n \times n$ matrix, A, is called *upper-triangular* if $A_{i,j} = 0$ whenever $i > j$. The matrix, A, is lower triangular if $A_{i,j} = 0$ whenever $j > i$.

REMARK. The term "upper-triangular" comes from the fact that A looks like

$$\begin{bmatrix} A_{1,1} & A_{1,2} & \cdots & A_{1,n-1} & A_{1,n} \\ 0 & A_{2,2} & \ddots & A_{2,n-1} & A_{2,n} \\ 0 & 0 & \ddots & \vdots & \vdots \\ \vdots & \vdots & \ddots & A_{n-1,n-1} & A_{n-1,n} \\ 0 & 0 & \cdots & 0 & A_{n,n} \end{bmatrix}$$

We explore a closely related concept:

DEFINITION 6.2.16. Let A be an $n \times n$ upper-triangular matrix. In any given row, the *pivot element* is the first *nonzero* entry encountered, scanning from left to right. Suppose the pivot element in row i is $A_{i,p(i)}$, so that the pivot-column for row i is $p(i)$. We define the matrix A to be in *echelon form* if

(1) $p(j) > p(i)$ whenever $j > i$.
(2) rows of all zeroes lie below rows that have nonzero entries

REMARK. Oddly enough, the term echelon form comes from the military, referring to a formation of soldiers. Here's an example of a matrix in echelon form

$$\begin{bmatrix} \mathbf{2} & 0 & 1 & 0 & 3 \\ 0 & \mathbf{1} & 0 & 4 & 0 \\ 0 & 0 & 0 & \mathbf{1} & 5 \\ 0 & 0 & 0 & 0 & \mathbf{3} \\ 0 & 0 & 0 & 0 & 0 \end{bmatrix}$$

with its pivots highlighted. The point of echelon form is that the pivot element in every row is to the *right* of the pivots in the rows *above* it.

Here's an example of a upper-triangular matrix that is *not* in echelon form

$$\begin{bmatrix} \mathbf{2} & 0 & 1 & 0 & 3 \\ 0 & \mathbf{1} & 0 & 4 & 0 \\ 0 & 0 & 0 & \mathbf{1} & 5 \\ 0 & 0 & 0 & 0 & \mathbf{3} \\ 0 & 0 & 0 & 0 & \mathbf{2} \end{bmatrix}$$

The problem is that it has two pivots in the same column. Another example of an upper-triangular matrix that is not echelon is

$$\begin{bmatrix} 2 & 0 & 1 & 0 & 3 \\ 0 & 0 & 0 & 0 & 0 \\ 0 & 0 & 2 & 1 & 5 \\ 0 & 0 & 0 & 7 & 3 \\ 0 & 0 & 0 & 0 & 2 \end{bmatrix}$$

Here, the problem is that there is a row of zeroes above rows that have nonzero entries.

In the next few results we will investigate the properties of determinants.

PROPOSITION 6.2.17. *If M is an $n \times n$ matrix, then*

$$\det(M^t) = \det(M)$$

PROOF. Definitions 6.2.14 on the preceding page and 6.2.12 on page 170 imply that

$$\begin{aligned} \det(M^t) &= \sum_{\sigma \in S_n} \wp(\sigma) M_{\sigma(1),1} \cdots M_{\sigma(n),n} \\ &= \sum_{\sigma \in S_n} \wp(\sigma) M_{1,\sigma^{-1}(1)} \cdots M_{n,\sigma^{-1}(n)} \quad \text{sorting by left subscripts} \\ &= \sum_{\sigma \in S_n} \wp(\sigma^{-1}) M_{1,\sigma^{-1}(1)} \cdots M_{n,\sigma^{-1}(n)} \quad \text{since } \wp(\sigma^{-1}) = \wp(\sigma) \\ &= \det(M) \end{aligned}$$

since, if σ runs over *all* permutations in S_n, so does σ^{-1}. □

PROPOSITION 6.2.18. *If M is an $n \times n$ upper- or lower-triangular matrix*
$$\det M = \prod_{i=1}^{n} M_{i,i}$$

PROOF. We'll do the upper-triangular case — the other case follows from proposition 6.2.17 on the previous page. In equation 6.2.9 on page 172, all terms with $\sigma(i) < i$ will have a factor of 0, so we must have $\sigma(i) \geq i$ for all i. For all permutations, σ
$$\sum_{i=1}^{n}(i - \sigma(i)) = 0$$
If $\sigma \neq 1$, some terms will have $(j - \sigma(j)) > 0$ and must be balanced by other terms $(j' - \sigma(j')) < 0$. If follows that $\sigma(i) \geq i$ for all i implies that σ is the identity. □

COROLLARY 6.2.19. *If M is an $n \times n$ matrix in echelon form and $\det M = 0$, then the bottom row of M consists of zeros.*

PROOF. Since $\det M = 0$, proposition 6.2.15 on page 172 implies that one of the diagonal entries of M is 0. This implies that there is a pivot element ≥ 2 columns to the right of the pivot of the previous row. Since M has the same number of rows and columns, the pivot elements must reach the *right* side of M *before* they reach the bottom, so the bottom row consists of zeroes. □

PROPOSITION 6.2.20. *If M is an $n \times n$ matrix and M' is the result of interchanging rows i and $j \neq i$ or columns i and j of M, then*
$$\det M' = -\det M$$

PROOF. We will prove this for interchanged columns; the conclusion for interchanged rows follows from 6.2.17 on the preceding page.

Let $\tau = (i, j)$, the transposition. Then equation 6.2.10 on page 172 (and the fact that $\tau^2 = 1$) implies that
$$\det(M') = \sum_{\sigma \in A_n} M'_{1,\sigma(1)} \cdots M'_{n,\sigma(n)} - \sum_{\sigma \in A_n} M'_{1,\tau\sigma(1)} \cdots M'_{n,\tau\sigma(n)}$$
$$= \sum_{\sigma \in A_n} M_{1,\tau\sigma(1)} \cdots M_{n,\tau\sigma(n)} - \sum_{\sigma \in A_n} M_{1,\sigma(1)} \cdots M_{n,\sigma(n)}$$
$$= -\det(M)$$
□

We have the related result:

PROPOSITION 6.2.21. *If M is an $n \times n$ matrix and M has two rows or two columns that are the same, then $\det(M) = 0$.*

REMARK. If we are working over an integral domain whose characteristic is $\neq 2$, this follows immediately from proposition 6.2.20 since it shows that $\det(M) = -\det(M)$.

PROOF. We'll suppose columns i and $j \neq i$ are identical; the corresponding result for rows follows from proposition 6.2.17 on page 173.

Let $\tau \in S_n$ be the transposition (i,j). Then

$$M_{1,\sigma(1)} \cdots M_{n,\sigma(n)} = M_{1,\tau\sigma(1)} \cdots M_{n,\tau\sigma(n)}$$

for all $\sigma \in A_n$ and equation 6.2.10 on page 172 implies that

$$\det(M) = \sum_{\sigma \in A_n} \left(M_{1,\sigma(1)} \cdots M_{n,\sigma(n)} - M_{1,\tau\sigma(1)} \cdots M_{n,\tau\sigma(n)} \right)$$
$$= 0$$

□

The next two results follow immediately from equation 6.2.9 on page 172; their proofs are left as exercises for the reader.

PROPOSITION 6.2.22. *If*

$$A = \begin{bmatrix} R_1 \\ \vdots \\ R_i \\ \vdots \\ R_n \end{bmatrix}$$

is an $n \times n$ matrix and

$$B = \begin{bmatrix} R_1 \\ \vdots \\ k \cdot R_i \\ \vdots \\ R_n \end{bmatrix}$$

where k is a constant, then $\det B = k \cdot \det A$.

PROPOSITION 6.2.23. *If*

$$A = \begin{bmatrix} R_1 \\ \vdots \\ R_{i-1} \\ R_i \\ R_{i+1} \\ \vdots \\ R_n \end{bmatrix} \text{ and } B = \begin{bmatrix} R_1 \\ \vdots \\ R_{i-1} \\ S_i \\ R_{i+1} \\ \vdots \\ R_n \end{bmatrix}$$

are $n \times n$ matrices and

$$C = \begin{bmatrix} R_1 \\ \vdots \\ R_{i-1} \\ R_i + S_i \\ R_{i+1} \\ \vdots \\ R_n \end{bmatrix}$$

then

$$\det C = \det A + \det B$$

DEFINITION 6.2.24. If A is an $m \times n$ matrix, the act of adding a multiple of one row to another is called a *type 1 elementary row operation*. If I_m is an $m \times m$ identity matrix, the result of performing an elementary row operation on I_m is called an *type 1 elementary matrix*.

REMARK 6.2.25. It is not hard to see that performing a type 1 elementary row operation on A is the same as forming the product EA, where E is an $m \times m$ type 1 elementary matrix.

There are type 2 and three elementary row operations, too:

Type 2: involves swapping two rows of a matrix
Type 3: involves multiplying a row of a matrix by a nonzero constant.

PROPOSITION 6.2.26. *Elementary matrices are invertible and their inverses are other elementary matrices.*

REMARK. For instance

$$\begin{bmatrix} 1 & 0 & 0 \\ 0 & 1 & 0 \\ \alpha & 0 & 1 \end{bmatrix} \cdot \begin{bmatrix} 1 & 0 & 0 \\ 0 & 1 & 0 \\ -\alpha & 0 & 1 \end{bmatrix} = \begin{bmatrix} 1 & 0 & 0 \\ 0 & 1 & 0 \\ 0 & 0 & 1 \end{bmatrix}$$

PROOF. The inverse of adding $\alpha \times$ row i to row j is adding $-\alpha \times$ row i to row j. $\square$

COROLLARY 6.2.27. *If A is an $n \times n$ matrix and we perform a type 1 elementary row operation on A, the determinant is unchanged.*

PROOF. Suppose we add a multiple of row i to row k with $j > i$ and suppose

$$A = \begin{bmatrix} R_1 \\ \vdots \\ R_i \\ \vdots \\ R_j \\ \vdots \\ R_n \end{bmatrix}, B = \begin{bmatrix} R_1 \\ \vdots \\ R_i \\ \vdots \\ R_i \\ \vdots \\ R_n \end{bmatrix}, C = \begin{bmatrix} R_1 \\ \vdots \\ R_i \\ \vdots \\ k \cdot R_i \\ \vdots \\ R_n \end{bmatrix} \text{ and } D = \begin{bmatrix} R_1 \\ \vdots \\ R_i \\ \vdots \\ k \cdot R_i + R_j \\ \vdots \\ R_n \end{bmatrix}$$

where k is a constant. Then $\det C = 0$, by corollary 6.2.21 on page 174 and proposition 6.2.22 on page 175. Proposition 6.2.23 on page 175 implies that
$$\det D = \det A + \det C = \det A$$

□

Proposition 6.2.17 on page 173, coupled with these results, immediately implies

PROPOSITION 6.2.28. *Let* $A = \begin{bmatrix} C_1 & \cdots & C_n \end{bmatrix}$ *be an $n \times n$ matrix with column-vectors C_i.*
  (1) *if $C_i = C_j$ for $i \neq j$, then $\det A = 0$,*
  (2) *if $A' = \begin{bmatrix} C_1 \cdots & C_{i-1} & k \cdot C_i & C_{i+1} & \cdots & C_n \end{bmatrix}$ — the ith column has been multiplied by a constant k — then*
$$\det A' = k \cdot \det A$$
  (3) *If*
$$B = \begin{bmatrix} C_1 \cdots & C_{i-1} & D_i & C_{i+1} & \cdots & C_n \end{bmatrix}$$
  and
$$C = \begin{bmatrix} C_1 \cdots & C_{i-1} & C_i + D_i & C_{i+1} & \cdots & C_n \end{bmatrix}$$
  — i.e., their columns are the same as those of A except for the ith — then $\det C = \det A + \det B$

This leads to an *application* for determinants[2]:

THEOREM 6.2.29 (Cramer's Rule). *Let*
$$A = \begin{bmatrix} C_1 & \cdots & C_n \end{bmatrix} = \begin{bmatrix} A_{1,1} & \cdots & A_{1,n} \\ \vdots & \ddots & \vdots \\ A_{n,1} & \cdots & A_{n,n} \end{bmatrix}, \text{ and } B = \begin{bmatrix} b_1 \\ \vdots \\ b_n \end{bmatrix}$$
and let $\bar{A}_i = \begin{bmatrix} C_1 & \cdots & C_{i-1} & B & C_{i+1} & \cdots & C_n \end{bmatrix}$ *— the result of replacing the ith column of A by B. If*
$$A_{1,1}x_1 + \cdots + A_{1,n}x_n = b_1$$
$$\vdots$$
(6.2.14)
$$A_{n,1}x_1 + \cdots + A_{n,n}x_n = b_n$$
is a system of linear equations, then

(6.2.15)
$$(\det A) \cdot x_i = \det \bar{A}_i$$

If $\det A \neq 0$, then
$$x_i = \frac{\det \bar{A}_i}{\det A}$$

REMARK. Note that we have not used the fact that we are working over a field until the last step. It follows that equation 6.2.15 is valid over any commutative ring. This is used in a proof of the Cayley-Hamilton Theorem (6.2.57 on page 199) in Example 6.3.18 on page 229.

---

[2] The reader has, no doubt, wondered what they are "good for."

PROOF. Just plug both sides of equation 6.2.14 on the previous page into the ith column of A and take determinants. If
$$E = \begin{bmatrix} A_{1,1}x_1 + \cdots + A_{1,n}x_n \\ \vdots \\ A_{n,1}x_1 + \cdots + A_{n,n}x_n \end{bmatrix}$$
then equations 6.2.14 on the preceding page become
$$(6.2.16) \qquad \det \begin{bmatrix} C_1 \cdots & C_{i-1} & E & C_{i+1} & \cdots & C_n \end{bmatrix} = \det \bar{A}_i$$
On the other hand
$$(6.2.17) \quad \det \begin{bmatrix} C_1 \cdots & C_{i-1} & E & C_{i+1} & \cdots & C_n \end{bmatrix}$$
$$= \det \begin{bmatrix} C_1 \cdots & C_{i-1} & x_1 \cdot C_1 & C_{i+1} & \cdots & C_n \end{bmatrix} +$$
$$\cdots + \det \begin{bmatrix} C_1 \cdots & C_{i-1} & x_i \cdot C_i & C_{i+1} & \cdots & C_n \end{bmatrix} +$$
$$\cdots + \det \begin{bmatrix} C_1 \cdots & C_{i-1} & x_n \cdot C_n & C_{i+1} & \cdots & C_n \end{bmatrix}$$
$$= x_1 \cdot \det \begin{bmatrix} C_1 \cdots & C_{i-1} & C_1 & C_{i+1} & \cdots & C_n \end{bmatrix} +$$
$$\cdots + x_i \cdot \det \begin{bmatrix} C_1 \cdots & C_{i-1} & C_i & C_{i+1} & \cdots & C_n \end{bmatrix} +$$
$$\cdots + x_n \cdot \det \begin{bmatrix} C_1 \cdots & C_{i-1} & C_n & C_{i+1} & \cdots & C_n \end{bmatrix}$$
Now, note that
$$\det \begin{bmatrix} C_1 \cdots & C_{i-1} & C_j & C_{i+1} & \cdots & C_n \end{bmatrix}$$
$$= \begin{cases} \det A & \text{if } i = j \\ 0 & \text{if } i \neq j \text{ because column } j \text{ is duplicated} \end{cases}$$
Plugging this into equation 6.2.17 gives
$$\det \begin{bmatrix} C_1 \cdots & C_{i-1} & E & C_{i+1} & \cdots & C_n \end{bmatrix} = x_i \cdot \det A$$
and equation 6.2.15 on the preceding page follows. □

> Gabriel Cramer (1704-1752), was a Swiss mathematician born in Geneva who worked in geometry and analysis. Cramer used his rule in the book, [25], to solve a system of five equations in five unknowns. The rule had been used before by other mathematicians. Although Cramer's Rule is not the most efficient method of solving systems of equations, it has theoretic applications.

We can develop an efficient algorithm for computing determinants using propositions 6.2.27 on page 176 and 6.2.20 on page 174:

PROPOSITION 6.2.30 (Gaussian Elimination). *If A is an $n \times n$ matrix, the following algorithm computes an echelon matrix, $\bar{A}$, with $d \cdot \det \bar{A} = \det A$ (where d is defined below):*

$d \leftarrow 1$
$c \leftarrow 1$
for $r = 1, \ldots, n-2$ **do** ▷ Main for-loop
    if $A_{r,i} = 0$ for all $c \leq i \leq n$ **then**
        Move row r to the last row ▷ This is a row of zeroes.

        end if
        if $A_{r,c} = 0$ **then**
            if $\exists$ row k with $r \leq k \leq n$ and $A_{k,c} \neq 0$ **then**
                Swap row r with row k
                $d \leftarrow -d$
            else
                $c \leftarrow c + 1$ ▷ The new pivot is more than one column
                                                       ▷ further to the right
                Perform next iteration of the Main for-loop
            end if
        end if
        for $i = r+1, \ldots, n$ **do**   ▷ For all lower rows
            for $j = c, \ldots, n$ **do**   ▷ do an type 1 elementary row-operation
                $A_{i,j} \leftarrow A_{i,j} - \frac{A_{i,c}}{A_{r,c}} \cdot A_{r,j}$
            end for
        end for
        $c \leftarrow c + 1$ ▷ Advance the pivot-column
    end for

REMARK. This algorithm is reasonably fast (proportional to n^3 steps rather than $n!$) and can easily be turned into a computer program.

This algorithm first appeared in Europe in the notes of Isaac Newton. In the 1950's it was named after Gauss due to confusion about its history. Forms of it (with no proofs) appeared in Chapter 8: Rectangular Arrays of the ancient Chinese treatise, *The Nine Chapters on the Mathematical Art* (see [26]).

PROOF. The variable d keeps track of how many times we swap rows — because proposition 6.2.20 on page 174 implies that each such swap multiplies the determinant by -1. The only *other* thing we do with A is elementary row operations, which leave the determinant unchanged by corollary 6.2.27 on page 176. In the end, the matrix is echelon and proposition 6.2.18 on page 174 implies that the determinant is the product of diagonal entries. □

Here are some examples:
Start with

$$A = \begin{bmatrix} 1 & 2 & 0 & 6 & 3 \\ 1 & 1 & 1 & 0 & 2 \\ 2 & 0 & 1 & 2 & 3 \\ 0 & 5 & 1 & 1 & 1 \\ 1 & 3 & 0 & 5 & 1 \end{bmatrix}$$

The first pivot column is 1 and we start with the first row. We subtract the first row from the second, $2\times$ the first row from the third and the first row from the last, to get

$$A_1 = \begin{bmatrix} 1 & 2 & 0 & 6 & 3 \\ 0 & -1 & 1 & -6 & -1 \\ 0 & -4 & 1 & -10 & -3 \\ 0 & 5 & 1 & 1 & 1 \\ 0 & 1 & 0 & -1 & -2 \end{bmatrix}$$

The new pivot-column is 2 and we subtract 4× the second row from the third and so on, to get

$$A_2 = \begin{bmatrix} 1 & 2 & 0 & 6 & 3 \\ 0 & -1 & 1 & -6 & -1 \\ 0 & 0 & -3 & 14 & 1 \\ 0 & 0 & 6 & -29 & -4 \\ 0 & 0 & 1 & -7 & -3 \end{bmatrix}$$

Now we add 2× the third row to the fourth, and so on, to get

$$A_4 = \begin{bmatrix} 1 & 2 & 0 & 6 & 3 \\ 0 & -1 & 1 & -6 & -1 \\ 0 & 0 & -3 & 14 & 1 \\ 0 & 0 & 0 & -1 & -2 \\ 0 & 0 & 0 & -7/3 & -8/3 \end{bmatrix}$$

We finally get

$$A_5 = \begin{bmatrix} 1 & 2 & 0 & 6 & 3 \\ 0 & -1 & 1 & -6 & -1 \\ 0 & 0 & -3 & 14 & 1 \\ 0 & 0 & 0 & -1 & -2 \\ 0 & 0 & 0 & 0 & 2 \end{bmatrix}$$

Since this is in echelon-form, the determinant of the original matrix is

$$\det A = 1 \cdot (-1) \cdot (-3) \cdot (-1) \cdot 2 = -6$$

It is not hard to see that

PROPOSITION 6.2.31. *If the matrix A in proposition 6.2.30 on page 178 was lower-triangular with nonzero diagonal entries, then $\bar{A}$ is a diagonal matrix with the same diagonal entries as A.*

PROOF. We use induction on n. It's clearly true for $n = 1$. For a larger value of n, simply use column 1 as the pivot column. Since all the other elements in the first row are 0, subtracting multiples of the first row from the lower rows will have no effect on columns larger than 1. Columns 2 through n and the rows below the first constitute an $(n-1) \times (n-1)$ lower-triangular matrix and the inductive hypotheses implies the conclusion. □

PROPOSITION 6.2.32. *If A and $\bar{A}$ are $n \times n$ matrices as in proposition 6.2.30 on page 178 and*

$$\mathbf{x} = \begin{bmatrix} x_1 \\ \vdots \\ x_n \end{bmatrix}$$

then the solution-sets of the equations $A\mathbf{x} = 0$ and $\bar{A}\mathbf{x} = 0$ are the same. It follows that, $\det A = 0$ implies that $A\mathbf{x} = 0$ has nonzero solutions.

PROOF. Both $Ax = 0$ and $\bar{A}x = 0$ represent systems of n equations in the n unknowns, $x_1, \ldots, x_n$. Rearranging rows of A simply renumbers the equations, and has no effect on the solution-set. An elementary row-operation consists in adding *one* equation to *another* and (since the left sides are equated to 0) *also* does not effect the solution-set.

The final statement follows from corollary 6.2.19 on page 174, which implies that the bottom row (at least!) of $\bar{A}$ consists of zeroes when $\det A = 0$. That implies that the unknown x_n can take on *arbitrary* values and the other equations express the *other* variables in terms of x_n. □

Here's an example of this:

$$A = \begin{bmatrix} 1 & 2 & 3 \\ 2 & 1 & 6 \\ -1 & 1 & -3 \end{bmatrix}$$

In the first step, the pivot column is 1 and subtract $2\times$ the first row from the second and add the first row to the third to get

$$A_1 = \begin{bmatrix} 1 & 2 & 3 \\ 0 & -3 & 0 \\ 0 & 3 & 0 \end{bmatrix}$$

Now the pivot column is 2 and we add the second row to the third to get

$$\bar{A} = A_2 = \begin{bmatrix} 1 & 2 & 3 \\ 0 & -3 & 0 \\ 0 & 0 & 0 \end{bmatrix}$$

so that $\det A = 0$. The equation

$$\bar{A}\mathbf{x} = 0$$

has nontrivial solutions, because this matrix equation represents the linear equations

$$x_1 + 2x_2 + 3x_3 = 0$$
$$-3x_2 = 0$$
$$0x_3 = 0$$

so that

(1) x_3 can be *arbitrary*
(2) $x_2 = 0$
(3) $x_1 = -3x_3$

and the *general* solution can be written as

$$\mathbf{x} = x_3 \begin{bmatrix} -3 \\ 0 \\ 1 \end{bmatrix}$$

It is easy to check that $A\mathbf{x} = 0$, so *that* equation has nonzero solutions (an infinite number of them, in fact).

COROLLARY 6.2.33. *If A is an $n \times n$ matrix, the following statements are equivalent:*

(1) $\det A \neq 0$
(2) *if*

$$\mathbf{x} = \begin{bmatrix} x_1 \\ \vdots \\ x_n \end{bmatrix}$$

the equation $A\mathbf{x} = 0$ has $\mathbf{x} = 0$ as its only solution.
(3) *the rows of A are linearly independent,*
(4) *the columns of A are linearly independent*
(5) *A has an inverse*

PROOF. Cramer's Rule (theorem 6.2.29 on page 177) also implies that $A\mathbf{x} = 0$ has $\mathbf{x} = 0$ as its only solution.

Suppose

$$A = \begin{bmatrix} C_1 & \cdots & C_n \end{bmatrix}$$

and

$$\alpha_1 C + \cdots + \alpha_n C_n = 0$$

is a linear dependence equation. If $\det A \neq 0$, then theorem 6.2.29 on page 177 implies

$$\alpha_1 = \cdots = \alpha_n = 0$$

so the columns of A are linearly independent.

Taking the transpose of A and invoking proposition 6.2.17 on page 173 implies that the *rows* are linearly independent.

Now suppose $\det A = 0$. Then proposition 6.2.32 on the preceding page implies that the equation $A\mathbf{x} = 0$ has nontrivial solutions and

$$A\mathbf{x} = x_1 C_1 + \cdots + x_n C_n = 0$$

is a nontrivial dependence relation between columns. Since A annihilates nonzero vectors (the nontrivial solutions of $A\mathbf{x} = 0$) it cannot have an inverse (even as a *function*). Cramer's Rule (theorem 6.2.29 on page 177) implies that the equations

$$A\mathbf{x}_i = \begin{bmatrix} 0 \\ \vdots \\ 0 \\ 1 \\ 0 \\ \vdots \\ 0 \end{bmatrix}$$

have unique solutions, where the 1 on the right occurs in the ith row. The matrix
$$\begin{bmatrix} \mathbf{x}_1 & \cdots & \mathbf{x}_n \end{bmatrix}$$
is easily seen to be A^{-1}. □

This, remark 6.2.25 on page 176, and proposition 6.2.26 on page 176 imply that:

COROLLARY 6.2.34. *If A is an $n \times n$ matrix, there exists an equation*
$$S_1 \cdots S_k \cdot A = U$$
where the S_i are either lower triangular elementary matrices or transpositions and U is upper-triangular. It follows that
$$(6.2.18) \qquad A = T_n \cdots T_1 \cdot U$$
where $T_i = S_i^{-1}$ are also lower triangular elementary matrices. We also have a decomposition
$$A = L \cdot R_1 \cdots R_k$$
where L is lower-triangular and the R_i are upper-triangular elementary matrices or transpositions. If $\det A \neq 0$, we have a decomposition
$$(6.2.19) \qquad A = T_n \cdots T_1 \cdot D \cdot R_1 \cdots R_k$$
where D is a diagonal matrix.

PROOF. To prove the second statement, simply take the transpose of A and apply equation 6.2.18:
$$A^t = T_k \cdots T_1 \cdot U$$
and take the transpose
$$A = U^t \cdot T_1^t \cdots T_k^t$$
The transpose of an upper-triangular matrix is lower-triangular and the transpose of an elementary matrix is elementary.

The final statement follows from proposition 6.2.31 on page 180, applied to L. □

The determinant has an interesting property:

THEOREM 6.2.35. *If A and B are $n \times n$ matrices, then*
$$\det(AB) = \det A \cdot \det B$$

PROOF. We divide this into cases.
(1) $\det B = 0$. In this case, the equation $B\mathbf{x} = 0$ has nontrivial solutions, by proposition 6.2.32 on page 181. It follows that $AB\mathbf{x} = 0$ as well, and corollary 6.2.33 on the preceding page implies that $\det AB = 0$.
(2) From now on, $\det B \neq 0$. If B is an elementary matrix, $\det B = 1$. We have $(AB)^t = B^t A^t$ and $B^t A^t$ is the result of performing an elementary row operation on A^t so corollary 6.2.27 on page 176 implies that $\det B^t A^t = \det A^t$. Since proposition 6.2.17 on page 173 implies that $\det M = \det M^t$ for any $n \times n$ matrix, M, it follows that $\det AB = \det A$ and the conclusion follows.

(3) In this case, use corollary 6.2.34 on the preceding page to represent B via
$$B = T_n \cdots T_1 \cdot D \cdot R_1 \cdots R_k$$
and the T's and R's are elementary matrices. The product of the diagonal entries in D give the determinant of B, i.e. $\det B = \prod_{i=1}^n D_{i,i}$. Then

$$\begin{aligned}
\det(AB) &= \det(A \cdot T_n \cdots T_1 \cdot D \cdot R_1 \cdots R_k) \\
&= \det(A' \cdot D) \quad \text{where } \det A' = \det A \\
&= \det A' \cdot \det B \\
&= \det A \cdot \det B
\end{aligned}$$

since $A'D$ is the result of multiplying row i of A' by $D_{i,i}$ and each such multiplication multiplies $\det A' = \det A$ by $D_{i,i}$, by proposition 6.2.23 on page 175.

□

DEFINITION 6.2.36. If A is an $n \times n$ matrix and $1 \leq i, j \leq n$, then the $(i,j)^{\text{th}}$ minor of A, denoted $M_{i,j}(A)$ is the determinant of the $(n-1) \times (n-1)$ matrix formed from A by deleting its ith row and its jth column.

The use of minors leads to another way to compute determinants:

PROPOSITION 6.2.37. *If A is an $n \times n$ matrix, and $k = 1, \ldots, n$ is fixed, then*

(6.2.20) $$\det A = \sum_{i=1}^n (-1)^{i+j} A_{k,i} \cdot M_{k,i}(A)$$

Furthermore, if $j \neq k$, we have

$$A_{j,1} \cdot M_{k,1}(A) - A_{j,2} \cdot M_{k,2}(A) + \cdots + (-1)^{n+1} A_{j,n} \cdot M_{k,n}(A) = 0$$

PROOF. We begin by proving the claim for $k = 1$. Recall the equation of the determinant

(6.2.21) $$\det(A) = \sum_{\sigma \in S_n} \wp(\sigma) A_{1,\sigma(1)} \cdots A_{n,\sigma(n)}$$

Now note that $A_{1,i} \cdot M_{1,i}(A)$ is *all terms* of equation 6.2.21 for which $\sigma(1) = i$ and

$$M_{1,i}(A) = \sum_{\tau \in S_{n-1}} \wp(\tau) A_{2,f(\tau(1))} \cdots A_{n,f(\tau(n-1))}$$

where
$$f(k) = \begin{cases} k & \text{if } k < i \\ k+1 & \text{if } k \geq i \end{cases}$$

Write $\tau' = \tau$, extended to S_n by setting $\tau'(n) = n$. Then $\wp(\tau') = \wp(\tau)$. Converting from τ' to σ involves

(1) decreasing all indices by 1 and mapping 1 to n (which τ' fixes)
(2) applying τ'
(3) increasing all indices $\geq i$ by 1 and mapping n to i

In other words
$$\sigma = (i, i+1, \ldots, n) \circ \tau' \circ (n, \ldots, 1)$$
which implies that
$$\wp(\tau') = (-1)^{(n+1)+(n-i)} \wp(\sigma) = (-1)^{2n+i+1} \wp(\sigma) = (-1)^{i+1} \wp(\sigma)$$

The sum of all the terms of equation 6.2.20 on the preceding page accounts for *all* of the terms in equation 6.2.21 on the facing page.

The statement for $k \neq 1$ follows from constructing a matrix, V, via

$$V = \begin{bmatrix} A_{k,1} & \cdots & A_{k,n} \\ A_{1,1} & \cdots & A_{1,n} \\ \vdots & \ddots & \vdots \\ A_{k-1,1} & \cdots & A_{k-1,n} \\ A_{k+1,1} & \cdots & A_{k+1,n} \\ \vdots & \ddots & \vdots \\ A_{n,1} & \cdots & A_{n,n} \end{bmatrix}$$

where the kth row has been "shuffled" to the top via $k-1$ *swaps* of rows — in other words, the kth row has been moved to the top in a way that preserves the relative order of the *other* rows. It follows that
$$\det V = (-1)^{k+1} \det A$$
and the conclusion follows from the fact that $M_{1,i}(V) = M_{k,i}(A)$, $V_{1,i} = A_{k,i}$ for $i = 1, \ldots, n$.

The final statement follows from the fact that we are computing the determinant of a matrix whose j^{th} and k^{th} rows are the *same*. □

We can define the *adjoint* or *adjugate* of a matrix:

DEFINITION 6.2.38. If A is an $n \times n$ matrix, we define the *adjoint* or *adjugate* of A, denoted $\text{adj}(A)$ via
$$\text{adj}(A)_{i,j} = (-1)^{i+j} M_{j,i}(A)$$

Proposition 6.2.37 on the preceding page immediately implies that

PROPOSITION 6.2.39. *If A is an $n \times n$ matrix*
$$A \cdot \text{adj}(A) = \det A \cdot I$$
where I is the $n \times n$ identity matrix. If $\det A \neq 0$, we get
$$A^{-1} = \frac{1}{\det A} \cdot \text{adj}(A)$$

REMARK. Note that A commutes with its adjugate because the latter is a scalar multiple of the inverse of A.

6.2.4. Application of determinants: Resultants of polynomials.

Resultants of polynomials answer many of the same questions as Gröbner bases, but are computationally more tractable.

We begin by trying to answer the question:

Given polynomials

(6.2.22) $$f(x) = a_n x^n + \cdots + a_0$$
(6.2.23) $$g(x) = b_m x^m + \cdots + b_0$$

when do they have a common root?

An initial (but not very helpful) answer is provided by:

LEMMA 6.2.40. *If $f(x)$ is a nonzero degree n polynomial and $g(x)$ is a nonzero degree m polynomial, they have a common root if and only if there exist nonzero polynomials $r(x)$ of degree $\leq m - 1$ and $s(x)$ of degree $\leq n - 1$ such that*

(6.2.24) $$r(x)f(x) + s(x)g(x) = 0$$

REMARK. Note that the conditions on the degrees of $r(x)$ and $s(x)$ are important. Without them, we could just write

$$\begin{aligned} r(x) &= g(x) \\ s(x) &= -f(x) \end{aligned}$$

and *always* satisfy equation 6.2.24.

PROOF. Suppose $f(x), g(x)$ have a common root, α. Then we can set

$$\begin{aligned} r(x) &= g(x)/(x - \alpha) \\ s(x) &= -f(x)/(x - \alpha) \end{aligned}$$

and satisfy equation 6.2.24.

On the other hand, if equation 6.2.24 is satisfied it follows that $r(x)f(x)$ and $s(x)g(x)$ are degree $t \leq n + m - 1$ polynomials that have the *same t* factors

$$x - \alpha_1, \ldots, x - \alpha_t$$

since they cancel each other out. This set (of factors) of size t includes the n factors of $f(x)$ and the m factors of $g(x)$. The pigeonhole principal implies that at least 1 of these factors must be common to $f(x)$ and $g(x)$. And this common factor implies the existence of a common root. □

Suppose

$$\begin{aligned} r(x) &= u_{m-1} x^{m-1} + \cdots + u_0 \\ s(x) &= v_{n-1} x^{n-1} + \cdots + v_0 \end{aligned}$$

Then

(6.2.25) $$r(x) \cdot f(x) = \sum_{i=0}^{n+m-1} x^i c_i$$

$$s(x) \cdot g(x) = \sum_{i=0}^{n+m-1} x^i d_i$$

where $c_i = \sum_{j+k=i} u_j a_k$. We can compute the coefficients $\{c_i\}$ by matrix products

$$[u_{m-1}, \ldots, u_0] \begin{bmatrix} a_n \\ 0 \\ \vdots \\ 0 \end{bmatrix} = c_{n+m-1}$$

and
$$[u_{m-1},\ldots,u_0]\begin{bmatrix} a_{n-1} \\ a_n \\ 0 \\ \vdots \\ 0 \end{bmatrix} = c_{n+m-2}$$

or, combining the two,
$$[u_{m-1},\ldots,u_0]\begin{bmatrix} a_n & a_{n-1} \\ 0 & a_n \\ 0 & 0 \\ \vdots & \vdots \\ 0 & 0 \end{bmatrix} = [c_{n+m-1}, c_{n+m-2}]$$

where the subscripts of a_k increase from top to bottom and those of the u_j increase from left to right.

On the other end of the scale
$$[u_{m-1},\ldots,u_0]\begin{bmatrix} 0 \\ \vdots \\ 0 \\ a_0 \end{bmatrix} = c_0$$

and
$$[u_{m-1},\ldots,u_0]\begin{bmatrix} 0 \\ \vdots \\ 0 \\ a_0 \\ a_1 \end{bmatrix} = c_1$$

so we get
$$[u_{m-1},\ldots,u_0]\begin{bmatrix} 0 & 0 \\ \vdots & \vdots \\ 0 & 0 \\ a_0 & 0 \\ a_1 & a_0 \end{bmatrix} = [c_1, c_0]$$

This suggests creating a matrix
$$M_1 = \begin{bmatrix} a_n & a_{n-1} & \cdots & a_0 & 0 & \cdots & 0 \\ 0 & a_n & \cdots & a_1 & \ddots & \cdots & 0 \\ \vdots & \ddots & \ddots & \vdots & \ddots & \ddots & \vdots \\ 0 & \cdots & 0 & a_n & \cdots & a_1 & a_0 \end{bmatrix}$$

of m rows and $n+m$ columns. The top row contains the coefficients of $f(x)$ followed by $m-1$ zeros and each successive row is the one above shifted to the right. We stop when a_0 reaches the rightmost column. Then

$$\begin{bmatrix} u_{m-1} & \cdots & u_0 \end{bmatrix} M_1 = \begin{bmatrix} c_{n+m-1} & \cdots & c_0 \end{bmatrix} = [\mathbf{c}]$$

so we get the coefficients of $r(x)f(x)$. In like fashion, we can define a matrix with n rows and $n+m$ columns

$$M_2 = \begin{bmatrix} b_m & b_{m-1} & \cdots & b_0 & 0 & \cdots & 0 \\ 0 & b_m & \cdots & b_1 & & \ddots & 0 \\ \vdots & \ddots & \ddots & \vdots & \ddots & \ddots & \vdots \\ 0 & \cdots & 0 & b_m & \cdots & b_1 & b_0 \end{bmatrix}$$

whose top row is the coefficients of $g(x)$ followed by $n-1$ zeros and each successive row is shifted one position to the right, with b_0 on the right in the bottom row. Then

$$\begin{bmatrix} v_{n-1} & \cdots & v_0 \end{bmatrix} M_2 = [d_{n+m-1}, \ldots, d_0] = [\mathbf{d}]$$

— a vector of the coefficients of $s(x)g(x)$. If we combine the two together, we get an $(n+m) \times (n+m)$-matrix

$$S = \begin{bmatrix} M_1 \\ M_2 \end{bmatrix}$$

with the property that

(6.2.26) $$\begin{bmatrix} \mathbf{u} & \mathbf{v} \end{bmatrix} S = [\mathbf{c} + \mathbf{d}]$$

— an $n+m$ dimensional vector of the coefficients of $r(x)f(x) + s(x)g(x)$, where

(6.2.27) $$\begin{aligned} \mathbf{u} &= \begin{bmatrix} u_{m-1} & \cdots & u_0 \end{bmatrix} \\ \mathbf{v} &= \begin{bmatrix} v_{n-1} & \cdots & v_0 \end{bmatrix} \end{aligned}$$

It follows that S reduces the question of the existence of a common root of $f(x)$ and $g(x)$ to *linear algebra:* The equation

(6.2.28) $$\begin{bmatrix} \mathbf{u} & \mathbf{v} \end{bmatrix} S = [0]$$

has a nontrivial solution if and only if $\det(S) = 0$ by corollary 6.2.33 on page 182.

DEFINITION 6.2.41. If

$$f(x) = a_n x^n + \cdots + a_0$$
$$g(x) = b_m x^m + \cdots + b_0$$

are two polynomials, their *Sylvester Matrix* is the $(n+m) \times (n+m)$-matrix

$$S(f, g, x) = \begin{bmatrix} a_n & a_{n-1} & \cdots & a_0 & 0 & \cdots & 0 \\ 0 & a_n & \cdots & a_1 & & \ddots & 0 \\ \vdots & \ddots & \ddots & \vdots & \ddots & \ddots & \vdots \\ 0 & \cdots & 0 & a_n & \cdots & a_1 & a_0 \\ b_m & b_{m-1} & \cdots & b_0 & 0 & \cdots & 0 \\ 0 & b_m & \cdots & b_1 & & \ddots & 0 \\ \vdots & \ddots & \ddots & \vdots & \ddots & \ddots & \vdots \\ 0 & \cdots & 0 & b_m & \cdots & b_1 & b_0 \end{bmatrix}$$

and its determinant $\det(S(f, g, x)) = \text{Res}(f, g, x)$ is called the *resultant* of f and g.

> James Joseph Sylvester (1814–1897) was an English mathematician who made important contributions to matrix theory, invariant theory, number theory and other fields.

The reasoning above shows that:

PROPOSITION 6.2.42. *The polynomials $f(x)$ and $g(x)$ have a common root if and only if $\text{Res}(f,g,x) = 0$.*

PROOF. Equations 6.2.23 on page 186 and 6.2.26 on the preceding page imply that the hypothesis of lemma 6.2.40 on page 186 is satisfied if and only if $\det(S(f,g,x)) = 0$. □

EXAMPLE. For instance, suppose

$$\begin{aligned} f(x) &= x^2 - 2x + 5 \\ g(x) &= x^3 + x - 3 \end{aligned}$$

Then the Sylvester matrix is

$$M = \begin{bmatrix} 1 & -2 & 5 & 0 & 0 \\ 0 & 1 & -2 & 5 & 0 \\ 0 & 0 & 1 & -2 & 5 \\ 1 & 0 & 1 & -3 & 0 \\ 0 & 1 & 0 & 1 & -3 \end{bmatrix}$$

and the resultant is 169, so these two polynomials have no common roots.

There are many interesting applications of the resultant. Suppose we are given parametric equations for a curve

$$\begin{aligned} x &= \frac{f_1(t)}{g_1(t)} \\ y &= \frac{f_2(t)}{g_2(t)} \end{aligned}$$

where f_i and g_i are polynomials, and want an implicit equation for that curve, i.e. one of the form

$$F(x,y) = 0$$

This is equivalent to finding x, y such that the polynomials

$$\begin{aligned} f_1(t) - xg_1(t) &= 0 \\ f_2(t) - yg_2(t) &= 0 \end{aligned}$$

have a common root (in t). So the condition is

$$\text{Res}(f_1(t) - xg_1(t), f_2(t) - yg_2(t), t) = 0$$

This resultant will be a polynomial in x and y. We have *eliminated* the variable t — in a direct generalization of Gaussian elimination (see 6.2.30 on page 178) — and the study of such algebraic techniques is the basis of Elimination Theory.

For example, let

$$\begin{aligned} x &= t^2 \\ y &= t^2(t+1) \end{aligned}$$

Then the Sylvester matrix is

$$\begin{bmatrix} 1 & 0 & -x & 0 & 0 \\ 0 & 1 & 0 & -x & 0 \\ 0 & 0 & 1 & 0 & -x \\ 1 & 1 & 0 & -y & 0 \\ 0 & 1 & 1 & 0 & -y \end{bmatrix}$$

and the resultant is
$$\operatorname{Res}(t^2 - x, t^2(t+1) - y, t) = -x^3 + y^2 - 2yx + x^2$$
and it is not hard to verify that
$$-x^3 + y^2 - 2yx + x^2 = 0$$
after plugging in the parametric equations for x and y.

EXERCISES.

5. Compute an implicit equation for the curve defined parametrically by
$$\begin{aligned} x &= t/(1+t^2) \\ y &= t^2/(1-t) \end{aligned}$$

6. Compute an implicit equation for the curve
$$\begin{aligned} x &= t/(1-t^2) \\ y &= t/(1+t^2) \end{aligned}$$

7. Compute an implicit equation for the curve
$$\begin{aligned} x &= (1-t)/(1+t) \\ y &= t^2/(1+t^2) \end{aligned}$$

8. Solve the equations
$$\begin{aligned} x^2 + y^2 &= 1 \\ x + 2y - y^2 &= 1 \end{aligned}$$
by computing a suitable resultant to eliminate y.

9. Find implicit equations for x, y, and z if
$$\begin{aligned} x &= s+t \\ y &= s^2 - t^2 \\ z &= 2s - 3t^2 \end{aligned}$$

Hint: Compute resultants to eliminate s from every pair of equations and then eliminate t from the resultants.

6.2.5. A geometric property of determinants. Consider the effect of operating on the unit cube, $[0,1]^n$, in $\mathbb{R}^n$ via a matrix. In $\mathbb{R}^2$ the unit square at the origin becomes a *parallelogram,* as in figure 6.2.1 on the facing page

We give a heuristic argument that an elementary matrix does not alter the n-dimensional volume of a region of $\mathbb{R}^n$:

PROPOSITION 6.2.43. *In* $\mathbb{R}^2$, *an elementary matrix leaves the area of the square* $[x, x+1] \times [y, y+1]$ *unchanged.*

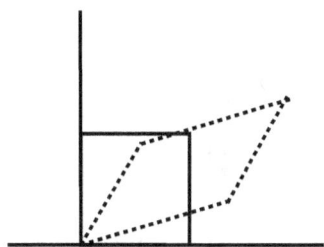

FIGURE 6.2.1.   Image of a unit square under a matrix

PROOF. An elementary matrix
$$\begin{bmatrix} 1 & a \\ 0 & 1 \end{bmatrix}$$
maps $[x, x+1] \times [y, y+1]$ to the parallelogram spanning the four points
$$(x+ay, y)$$
$$(x+ay+a, y+1)$$
$$(x+1+ay, y)$$
$$(x+1+ay+a, y+1)$$
so its base and height are still 1 and its area is still 1. A similar argument holds for the elementary matrix
$$\begin{bmatrix} 1 & 0 \\ a & 1 \end{bmatrix}$$
□

COROLLARY 6.2.44. *An elementary $n \times n$ matrix leaves the volume of a unit cube unchanged.*

PROOF. If the elementary matrix E adds a multiple of row i to row j, then it only alters the copy of $\mathbb{R}^2 \subset \mathbb{R}^n$ spanned by coordinates x_i and x_j, where it behaves like a 2×2 elementary matrix. The volume of $E(\text{Cube})$ is $1^{n-2} \times A$ where A is the area of the parallelogram cross-section of $E(\text{Cube})$ in the x_i-x_j plane, which is 1, by 6.2.43 on the preceding page. □

DEFINITION 6.2.45. If $B = \prod_{i=1}^{n}[a_i, b_i]$ is a box in $\mathbb{R}^n$, its *volume*, $\text{vol}(B)$ is defined by
$$\text{vol}(B) = \prod_{i=1}^{n}(b_i - a_i)$$

REMARK. This is used to define the Lebesgue measure:

DEFINITION 6.2.46. *If $R \subset \mathbb{R}^n$ is a region, its outer Lebesgue measure is defined by*

$\lambda(R) =$

$$\inf \left\{ \sum_{B \in \mathcal{C}} \text{vol}(B) : \mathcal{C} \text{ is a countable set of boxes whose union covers } R \right\}$$

PROPOSITION 6.2.47. *Let E be an $n \times n$ elementary matrix and let $R \subset \mathbb{R}^n$ be a region. Then*
$$\lambda(E(R)) = \lambda(R)$$

PROOF. The matrix E does not change the volume of the boxes covering R so the Lebesgue measure will also be unchanged. □

PROPOSITION 6.2.48. *If D is an $n \times n$ diagonal matrix and $R \subset \mathbb{R}^n$ is a region, then*
$$\lambda(D(R)) = \lambda(R) \cdot \prod_{i=1}^{n} |D_{i,i}|$$

PROOF. The matrix D dilates the ith coordinate of $\mathbb{R}^n$ by $D_{i,i}$, thereby multiplying the volume of all boxes by that factor. □

This is all building up to

THEOREM 6.2.49. *If A is an $n \times n$ matrix, $R \subset \mathbb{R}^n$, then*
$$\lambda(A(R)) = |\det A| \cdot \lambda(R)$$

REMARK. So the determinant gives the effect of a linear transformation on *volumes*. Analytic geometry and manifold theory considers volumes to have *signs*, in which case we do not take the *absolute value* of the determinant.

PROOF. If $\det A \neq 0$, this follows immediately from equation 6.2.19 on page 183, proposition 6.2.47 and proposition 6.2.48.

If $\det A = 0$, then there are linear dependency relations between the rows of A (see corollary 6.2.33 on page 182) so the image of A is a linear subspace of $\mathbb{R}^n$ of dimension $< n$, and the n-dimensional volume of it is 0. □

6.2.6. Changes of basis. Suppose we have a vector-space with basis $\{e_i\}, i = 1, \ldots, n$ and we are given a new basis $\{b_i\}$. If

$$\begin{bmatrix} x_1 \\ \vdots \\ x_n \end{bmatrix}$$

is a vector in this new basis, then

$$x_1 b_1 + \cdots + x_n b_n = \begin{bmatrix} b_1 & \cdots & b_n \end{bmatrix} \cdot \begin{bmatrix} x_1 \\ \vdots \\ x_n \end{bmatrix}$$

is the same vector in the old basis, where $P = \begin{bmatrix} b_1 & \cdots & b_n \end{bmatrix}$ is an $n \times n$ matrix whose columns are the basis-vectors. Since the basis-vectors are linearly independent, corollary 6.2.33 on page 182 implies that P is invertible. Since P converts from the new basis to the old one, P^{-1} performs the reverse transformation.

For instance, suppose $\mathbb{R}^3$ has the standard basis and we have a new basis

$$b_1 = \begin{bmatrix} 28 \\ -25 \\ 7 \end{bmatrix}, b_2 = \begin{bmatrix} 8 \\ -7 \\ 2 \end{bmatrix}, b_3 = \begin{bmatrix} 3 \\ -4 \\ 1 \end{bmatrix}$$

We form a matrix from these columns:

$$P = \begin{bmatrix} 28 & 8 & 3 \\ -25 & -7 & -4 \\ 7 & 2 & 1 \end{bmatrix}$$

whose determinant is verified to be 1. The vector

$$\begin{bmatrix} 1 \\ -1 \\ 1 \end{bmatrix}$$

in the new basis is

$$b_1 - b_2 + b_3 = P \begin{bmatrix} 1 \\ -1 \\ 1 \end{bmatrix} = \begin{bmatrix} 23 \\ -36 \\ 10 \end{bmatrix}$$

If we want to convert the vector

$$\begin{bmatrix} 1 \\ 2 \\ 3 \end{bmatrix}$$

in the standard basis into the new basis, we get

$$P^{-1} \begin{bmatrix} 1 \\ 2 \\ 3 \end{bmatrix} = \begin{bmatrix} 1 & -2 & -11 \\ -3 & 7 & 37 \\ -1 & 0 & 4 \end{bmatrix} \begin{bmatrix} 1 \\ 2 \\ 3 \end{bmatrix} = \begin{bmatrix} -36 \\ 122 \\ 11 \end{bmatrix}$$

and a simple calculation shows that

$$-36 b_1 + 122 b_2 + 11 b_3 = \begin{bmatrix} 1 \\ 2 \\ 3 \end{bmatrix}$$

For matrices, changes of basis are a bit more complicated. Definition 6.2.10 on page 169 shows that a matrix depends in a crucial way on the basis used to compute it.

Suppose V is an n-dimensional vector-space and an $n \times n$ matrix, A, represents a linear transformation

$$f: V \to V$$

with respect to some basis. If $\{b_1, \ldots, b_n\}$ is a new basis for V, let

$$P = [b_1, \ldots, b_n]$$

be the matrix whose columns are the b_i. We can compute the matrix representation of f in this new basis, $\bar{A}$, via

$$\begin{array}{ccc} V_{\text{old}} & \xrightarrow{A} & V_{\text{old}} \\ P \uparrow & & \downarrow P^{-1} \\ V_{\text{new}} & \xrightarrow{\bar{A}} & V_{\text{new}} \end{array}$$

In other words, to compute a matrix representation for f in the new basis:
 (1) convert to the *old* basis (multiplication by P)
 (2) *act* via the matrix A, which represents f in the old basis
 (3) convert the result to the *new* basis (multiplication by P^{-1}).

We summarize this with

THEOREM 6.2.50. *If A is an $n \times n$ matrix representing a linear transformation*

$$f: V \to V$$

with respect to some basis $\{e_1, \ldots, e_n\}$ and we have a new basis $\{b_1, \ldots, b_n\}$ with

$$P = \begin{bmatrix} b_1 & \cdots & b_n \end{bmatrix}$$

then, in the new basis, the transformation f is represented by

$$\bar{A} = P^{-1} A P$$

EXERCISES.

10. Suppose

$$C = \left[\begin{array}{c|c} A & 0 \\ \hline 0 & B \end{array} \right]$$

where A is an $n \times n$ matrix and B is an $m \times m$ matrix and all the other entries of C are 0. Show that

$$\det C = \det A \cdot \det B$$

11. Solve the system of linear equations

$$\begin{aligned} 2x + 3y + z &= 8 \\ 4x + 7y + 5z &= 20 \\ -2y + 2z &= 0 \end{aligned}$$

12. Solve the system

$$\begin{aligned} 2x + 3y + 4z &= 0 \\ x - y - z &= 0 \\ y + 2z &= 0 \end{aligned}$$

13. If V is a 3-dimensional vector-space with a standard basis and
$$b_1 = \begin{bmatrix} 8 \\ 4 \\ 3 \end{bmatrix}, b_2 = \begin{bmatrix} -1 \\ 0 \\ -1 \end{bmatrix}, b_3 = \begin{bmatrix} 2 \\ 1 \\ 1 \end{bmatrix}$$
is a new basis, convert the matrix
$$A = \begin{bmatrix} 1 & 0 & -1 \\ 2 & 1 & 3 \\ -1 & 2 & 1 \end{bmatrix}$$
to the new basis.

6.2.7. Eigenvalues and the characteristic polynomial. Suppose V is a vector space over a field k. If $A\colon V \to V$ is a linear transformation, consider the equation

(6.2.29) $$Av = \lambda v$$

where we require $v \neq 0$ and λ to be a *scalar*. A nonzero vector, v, satisfying this equation is called an *eigenvector* of A and the value of λ that makes this work is called the corresponding *eigenvalue*.

Eigenvectors and eigenvalues are defined in terms of each other, but eigenvalues are computed first.

We rewrite equation 6.2.29 as
$$Av = \lambda I v$$
where I is the suitable identity matrix and get
$$(A - \lambda I)v = 0$$
This must have solutions for *nonzero* vectors, v. Corollary 6.2.33 on page 182 and proposition 6.2.32 on page 181 imply that this can *only* happen if
$$\det(A - \lambda I) = 0$$

DEFINITION 6.2.51. If A is an $n \times n$ matrix
$$\det(\lambda I - A) = \chi_A(\lambda)$$
is a degree-n polynomial called the *characteristic polynomial* of A. Its roots are the *eigenvalues* of A.

REMARK. Its roots are understood to lie in an algebraic closure (see section 7.5 on page 283) of the field k. Essentially, this is a larger field containing k with the property that roots of polynomials lie in it. For instance, if $k = \mathbb{R}$, the eigenvalues would lie in $\mathbb{C}$.

This is often defined as $\det(A - \lambda I)$. The present definition ensures that the coefficient of the highest power of λ is 1.

EXAMPLE 6.2.52. If
$$A = \begin{bmatrix} 1 & 2 \\ 3 & 4 \end{bmatrix}$$
its characteristic polynomial is
$$\chi_A(\lambda) = \lambda^2 - 5\lambda - 2$$
with roots
$$\frac{5 \pm \sqrt{33}}{2}$$
with corresponding *eigenvectors*
$$\begin{bmatrix} 1 \\ \frac{3-\sqrt{33}}{4} \end{bmatrix} \text{ and } \begin{bmatrix} 1 \\ \frac{3+\sqrt{33}}{4} \end{bmatrix}$$

We have another interesting invariant of a matrix:

DEFINITION 6.2.53. If A is an $n \times n$ matrix, its *trace*, $\text{Tr}(A)$ is defined by:
$$\text{Tr}(A) = \sum_{i=1}^{n} A_{i,i}$$

REMARK. The trace seems like an innocuous-enough (and easily computed!) quantity, but it has a number of interesting properties: see exercises 18 on page 201 and 6 on page 240.

The reader might wonder why we are interested in eigenvalues and eigenvectors. The answer is simple:

> Equation 6.2.29 on the preceding page shows that A behaves like a *scalar* when it acts on an eigenvector.

If we could find a basis for our vector space of *eigenvectors*, A would become a *diagonal matrix* in that basis — because it merely multiplies each basis-vector by a scalar.

Diagonal matrices are relatively easy to compute with[3]: adding or multiplying them simply involves adding or multiplying corresponding *diagonal entries*. They *almost* behave like scalars.

Finding a basis of eigenvectors is made somewhat easier by the following:

PROPOSITION 6.2.54. *Let A be an $n \times n$ matrix with eigenvalues and corresponding eigenvectors*
$$\{(\lambda_1, v_1), \ldots, (\lambda_n, v_n)\}$$
If the λ_i are all distinct, then the v_i are linearly independent.

PROOF. We prove this by contradiction. Assume the v_i are linearly dependent and let
$$\sum_{i=1}^{k} a_i v_i = 0$$

---

[3] We don't need the complex formula in definition 6.2.13 on page 171.

be a minimal linear relation between them (assume we have renumbered the eigenvectors so that the linearly dependent ones are first). If we multiply this by λ_k, we get

$$\sum_{i=1}^{k} \lambda_k a_i v_i = 0 \tag{6.2.30}$$

and if we act on this via A we get

$$\sum_{i=1}^{k} \lambda_i a_i v_i = 0 \tag{6.2.31}$$

If we subtract equation 6.2.31 from 6.2.30, we get

$$\sum_{i=1}^{k-1} (\lambda_k - \lambda_i) a_i v_i = 0$$

Since the eigenvalues are all *distinct*, $\lambda_k - \lambda_i \neq 0$. This dependency-relation is smaller than the original one, contradicting its minimality.

Consider the matrix

$$A = \begin{bmatrix} -9 & 2 & -3 \\ 8 & 1 & 2 \\ 44 & -8 & 14 \end{bmatrix} \tag{6.2.32}$$

Its characteristic polynomial is

$$-\lambda^3 + 6\lambda^2 - 11\lambda + 6$$

and its roots are $\{1, 2, 3\}$ — the eigenvalues of A. These are all distinct, so the corresponding eigenvectors are linearly independent and form a basis for $\mathbb{R}^3$. The corresponding eigenvectors are computed $\square$

$\lambda = 1$: We solve the equation

$$\left(\begin{bmatrix} -9 & 2 & -3 \\ 8 & 1 & 2 \\ 44 & -8 & 14 \end{bmatrix} - 1 \cdot \begin{bmatrix} 1 & 0 & 0 \\ 0 & 1 & 0 \\ 0 & 0 & 1 \end{bmatrix} \right) \cdot \begin{bmatrix} x \\ y \\ z \end{bmatrix}$$

$$= \begin{bmatrix} -10 & 2 & -3 \\ 8 & 0 & 2 \\ 44 & -8 & 13 \end{bmatrix} \cdot \begin{bmatrix} x \\ y \\ z \end{bmatrix} = \begin{bmatrix} 0 \\ 0 \\ 0 \end{bmatrix}$$

and we get

$$\begin{bmatrix} 1 \\ -1 \\ -4 \end{bmatrix}$$

$\lambda = 2$ We solve the equation

$$\left(\begin{bmatrix} -9 & 2 & -3 \\ 8 & 1 & 2 \\ 44 & -8 & 14 \end{bmatrix} - 2 \cdot \begin{bmatrix} 1 & 0 & 0 \\ 0 & 1 & 0 \\ 0 & 0 & 1 \end{bmatrix} \right) \cdot \begin{bmatrix} x \\ y \\ z \end{bmatrix}$$

$$= \begin{bmatrix} -11 & 2 & -3 \\ 8 & -1 & 2 \\ 44 & -8 & 12 \end{bmatrix} \cdot \begin{bmatrix} x \\ y \\ z \end{bmatrix} = \begin{bmatrix} 0 \\ 0 \\ 0 \end{bmatrix}$$

and we get
$$\begin{bmatrix} 1 \\ -2 \\ 5 \end{bmatrix}$$
When $\lambda = 3$, we get the eigenvector
$$\begin{bmatrix} 1 \\ 0 \\ -4 \end{bmatrix}$$
So our basis of eigenvectors can be assembled into a matrix
$$P = \begin{bmatrix} 1 & 1 & 1 \\ -1 & -2 & 0 \\ -4 & -5 & -4 \end{bmatrix}$$
with inverse
$$P^{-1} = \begin{bmatrix} 8 & -1 & 2 \\ -4 & 0 & -1 \\ -3 & 1 & -1 \end{bmatrix}$$
and
$$D = P^{-1}AP = \begin{bmatrix} 1 & 0 & 0 \\ 0 & 2 & 0 \\ 0 & 0 & 3 \end{bmatrix}$$
as expected. To pass back to the original basis, we compute
$$A = PDP^{-1}$$
Both A and D represent the *same* linear transformation — viewed from different bases.

Since
$$D^n = \begin{bmatrix} 1 & 0 & 0 \\ 0 & 2^n & 0 \\ 0 & 0 & 3^n \end{bmatrix}$$
we easily get a *closed form* expression for A^n:

(6.2.33) $\quad A^n = PD^n P^{-1}$
$$= \begin{bmatrix} -2^{n+2} - 3^{n+1} + 8 & 3^n - 1 & -3^n - 2^n + 2 \\ 2^{n+3} - 8 & 1 & 2^{n+1} - 2 \\ 5 \cdot 2^{n+2} + 4 \cdot 3^{n+1} - 32 & 4 - 4 \cdot 3^n & 4 \cdot 3^n + 5 \cdot 2^n - 8 \end{bmatrix}$$

This is even valid for *non-integral* values of n — something that is not evident from looking at definition 6.2.13 on page 171.

We can use this technique to compute *other* functions of matrices like e^A, $\sin A$, $\cos A$.

Even in cases where eigenvalues are *not* all distinct, it is possible to get a basis of eigenvectors for a vector space.

DEFINITION 6.2.55. Given an $n \times n$ matrix A with an eigenvalue λ, the *eigenspace* of λ is defined to be the nullspace (see 6.2.10 on page 169) of
$$A - \lambda \cdot I$$

REMARK. An eigenspace may be more than one-dimensional, in which case it is possible for eigenvectors to span a vector space even if not all eigenvalues are distinct.

EXAMPLE 6.2.56. It is also possible for eigenvectors to *not* span a vector space. Consider the matrix

$$B = \begin{bmatrix} 1 & 1 \\ 0 & 1 \end{bmatrix}$$

This has a single eigenvalue, $\lambda = 1$, and its eigenspace is one-dimensional, spanned by

$$\begin{bmatrix} 1 \\ 0 \end{bmatrix}$$

so there *doesn't exist* a basis of $\mathbb{R}^2$ of eigenvectors of B. All matrices (even those like B above) have a standardized form that is "almost" diagonal called Jordan Canonical Form — see section 6.3.3 on page 235.

For a class of matrices that can *always* be diagonalized, see corollary 6.2.95 on page 222.

We conclude this section with

THEOREM 6.2.57 (Cayley-Hamilton). *If A is an $n \times n$ matrix with characteristic polynomial*

$$\chi_A(\lambda)$$

then $\chi_A(A) = 0$.

REMARK. In other words, every matrix "satisfies" its characteristic polynomial.

The proof given here works over an arbitrary commutative ring; it doesn't require computations to be performed over a *field*.

Using the idea of *modules over a ring*, we can get a shorter proof of the theorem — see example 6.3.18 on page 229.

The Cayley-Hamilton Theorem can be useful in computing *powers* of a matrix. For instance, if the characteristic polynomial of a matrix, A, is $\lambda^2 - 5\lambda + 3$, we know that

$$A^2 = 5A - 3I$$

so *all* powers of A will be linear combinations of A and I. Since A is invertible

$$A = 5I - 3A^{-1}$$

or

$$A^{-1} = \frac{1}{3}(5I - A)$$

This can also be used to calculate other functions of a matrix. If

$$f(X)$$

is a high-order polynomial or even an infinite series, write

$$f(X) = \chi_A(X) \cdot g(X) + r(X)$$

where $r(X)$ is the remainder with $\deg r(X) < \deg \chi_A(X)$ and

$$f(A) = r(A)$$

PROOF. Consider the matrix $\lambda \cdot I - A$, where λ is an indeterminate. This matrix has an adjugate (see definition 6.2.38 on page 185)

$$B = \mathrm{adj}(\lambda \cdot I - A)$$

with the property

$$(\lambda \cdot I - A)B = \det(\lambda \cdot I - A) \cdot I = \chi_A(\lambda) \cdot I$$

Since B is also a matrix whose entries are polynomials in λ, we can gather terms of the same powers of λ

$$B = \sum_{i=0}^{n-1} \lambda^i B_i$$

where we know B is of degree $n-1$ in λ since its product with $A - \lambda \cdot I$ is of degree n.

We have

$$\begin{aligned}
\chi_A(\lambda)I &= (\lambda \cdot I - A)B \\
&= (\lambda \cdot I - A) \sum_{i=0}^{n-1} \lambda^i B_i \\
&= \sum_{i=0}^{n-1} \lambda \cdot I \lambda^i B_i - \sum_{i=0}^{n-1} \lambda^i A B_i \\
&= \sum_{i=0}^{n-1} \lambda^{i+1} B_i - \sum_{i=0}^{n-1} \lambda^i A B_i \\
&= \lambda^n B_{n-1} + \sum_{i=1}^{n-1} \lambda^i (B_{i-1} - A B_i) \\
&\quad - A B_0
\end{aligned}$$

If $\chi_A(\lambda) = \lambda^n + c_{n-1}\lambda^{n-1} + \cdots + c_0$, we equate equal powers of λ to get

$$B_{n-1} = I,\ B_{i-1} - A B_i = c_{i-1} I,\ \cdots,\ -A B_0 = c_0 I$$

If we multiply the equation of the coefficient of λ^i by A^i we get (note that $c_i A^i = A^i c_i$ because the c_i are scalars)

$$\begin{aligned}
A^n + c_{n-1}A^{n-1} + \cdots + c_0 I &= \\
& A^n B_{n-1} + A^{n-1}(B_{n-2} - A B_{n-1}) \\
& + A^{n-2}(B_{n-3} - A B_{n-2}) + \cdots + A(B_0 - A B_1) - A B_0 \\
&= \underbrace{A^n - A^n}_{} + \underbrace{A^{n-1} B_{n-2} - A^{n-1} B_{n-2}}_{} + A^{n-1} B_{n-3} - \cdots \\
& \cdots - A^2 B_1 + \underbrace{A B_0 - A B_0}_{} = 0
\end{aligned}$$

This is a telescoping sum whose terms all cancel out. □

> Sir William Rowan Hamilton, (1805 – 1865) was an Irish physicist, astronomer, and mathematician who made major contributions to mathematical physics (some had applications to quantum mechanics), optics, and algebra. He invented quaternions, a generalization of the complex numbers (see section 9 on page 323).

EXERCISES.

14. Compute a *square root* of the matrix, A, in 6.2.32 on page 197 using equation 6.2.33 on page 198.

15. Give a *simple* proof of the Cayley-Hamilton Theorem for a diagonal matrix. Generalize that to a matrix whose eigenvalues are all distinct (this gives an *intuitive* motivation for the theorem)[4].

16. Suppose
$$C = \left[\begin{array}{c|c} A & 0 \\ \hline 0 & B \end{array}\right]$$
where A is an $n \times n$ matrix and B is an $m \times m$ matrix and all the other entries of C are 0. Show that their characteristic polynomials satisfy
$$\chi_C(\lambda) = \chi_A(\lambda) \cdot \chi_B(\lambda)$$

17. Suppose A and B are $n \times n$ matrices and there exists an invertible matrix, C, such that
$$B = C^{-1}AC$$
Show that
$$\chi_A(\lambda) = \chi_B(\lambda)$$
So the characteristic polynomial of a linear transformation is independent of the basis used.

18. Suppose A and B are $n \times n$ matrices and there exists an invertible matrix, C, such that
$$B = C^{-1}AC$$
Show that
$$\text{Tr}(B) = \text{Tr}(A)$$
(see definition 6.2.53 on page 196). It follows that the trace of a matrix is an invariant of the *linear transformation* — independent of the basis used to compute it.

19. This is a partial converse to exercise 17. Suppose A and B are two diagonalizable $n \times n$ matrices with
$$\chi_A(\lambda) = \chi_B(\lambda)$$
Show that there exists an invertible matrix C such that
$$B = C^{-1}AC$$

---

[4]We claim that every matrix is "arbitrarily close" to one with distinct eigenvalues: a small perturbation of the *characteristic polynomial* has distinct roots.

6.2.8. Geometric Groups. In this section we will analyze groups that originate in geometry — groups of symmetries and motions. We will also give applications of the latter.

DEFINITION 6.2.58. If $\mathbb{F}$ is any field (see definition 7.1.1 on page 261) — for instance $\mathbb{F} = \mathbb{Q}$, $\mathbb{Z}_p$ for p prime, $\mathbb{R}$, or $\mathbb{C}$ — then the *general linear group*, $\mathrm{GL}(n,\mathbb{F})$ is the set of $n \times n$ matrices with entries in $\mathbb{F}$ whose determinants are $\neq 0$.

REMARK. Regard $\mathbb{F}$ as an abelian group under addition and let

$$\mathbb{F}^n = \underbrace{\mathbb{F} \oplus \cdots \oplus \mathbb{F}}_{n \text{ summands}}$$

In this case, $\mathrm{GL}(n,\mathbb{F})$ is the group of *automorphisms* of $\mathbb{F}^n$. If V is a vector-space, $\mathrm{GL}(V)$ is the group of *automorphisms* of V, i.e., linear transformations $f:V \to V$ that are isomorphisms.

If $\mathbb{F} = \mathbb{R}$, the abelian group $\mathbb{R}^n$ is a space in which geometry is defined and $\mathrm{GL}(n,\mathbb{F})$ represents ways of "deforming" that space in ways that preserve the origin and are invertible.

We will often be concerned with various *subgroups* of $\mathrm{GL}(n,\mathbb{F})$:

DEFINITION 6.2.59. Under the assumptions of definition 6.2.58, the *special linear group* $\mathrm{SL}(n,\mathbb{F}) \subset \mathrm{GL}(n,\mathbb{F})$ is the subgroup of matrices whose determinant is $1 \in \mathbb{F}$.

REMARK. If $\mathbb{F} = \mathbb{R}$, $\mathrm{SL}(n,\mathbb{R})$ is the set of "deformations" of $\mathbb{R}^n$ that preserve n-dimensional volume — see theorem 6.2.49 on page 192. If $A, B \in \mathrm{GL}(n,\mathbb{F})$ then

$$\det(ABA^{-1}) = \det(A)\det(B)\det(A^{-1}) = \det(B)$$

so that $\mathrm{SL}(n,\mathbb{F})$ is a *normal* subgroup of $\mathrm{GL}(n,\mathbb{F})$.

DEFINITION 6.2.60. Under the assumptions of definition 6.2.58, the *orthogonal group* $\mathrm{O}(n,\mathbb{F}) \subset \mathrm{GL}(n,\mathbb{F})$ is the group of matrices $M \in \mathrm{GL}(n,\mathbb{F})$ with the property that

$$MM^t = I$$

where M^t is the transpose of M and I is the identity matrix.

The *special orthogonal group*, $\mathrm{SO}(n,\mathbb{F}) \subset \mathrm{GL}(n,\mathbb{F})$ is given by

$$\mathrm{SO}(n,\mathbb{F}) = \mathrm{O}(n,\mathbb{F}) \cap \mathrm{SL}(n,\mathbb{F})$$

REMARK. If $\mathbb{F} = \mathbb{R}$, the orthogonal group is the set of "deformations" of $\mathbb{R}^n$ that preserve distances and angles between lines through the origin. In other words, the matrices in $\mathrm{O}(n,\mathbb{R})$ represent rotations and reflections. The group $\mathrm{SO}(n,\mathbb{R})$ eliminates the reflections.

To understand the geometry of $\mathbb{R}^n$, it is not enough to simply be able to rotate space about a fixed point (namely, the origin — which matrix-operations do). We must also be able to *move* objects through space, to *displace* them. This leads to the *affine groups*.

Regard $\mathbb{R}^n$ as the plane $x_{n+1} = 1$ in $\mathbb{R}^{n+1}$. An $(n+1) \times (n+1)$ matrix of the form

(6.2.34) $$D(a_1, \ldots, a_n) = \begin{bmatrix} 1 & 0 & \cdots & 0 & a_1 \\ 0 & 1 & \ddots & \vdots & \vdots \\ \vdots & \ddots & \ddots & 0 & a_{n-1} \\ \vdots & \ddots & 0 & 1 & a_n \\ 0 & \cdots & \cdots & 0 & 1 \end{bmatrix}$$

preserves this imbedded copy of $\mathbb{R}^n$ and displaces it so that the origin is moved to the point $(a_1, \ldots, a_n)$. A simple calculation shows that

$$D(a_1, \ldots, a_n) \cdot D(b_1, \ldots, b_n) = D(a_1 + b_1, \ldots, a_n + b_n)$$

which implies that the matrices of the form $D(a_1, \ldots, a_n) \in \mathrm{GL}(n, \mathbb{R})$ form a subgroup, $S \subset \mathrm{GL}(n+1, \mathbb{R})$ isomorphic to $\mathbb{R}^n$.

DEFINITION 6.2.61. If $n > 0$ is an integer and $G \subset \mathrm{GL}(n, \mathbb{F})$ is a subgroup, the subgroup of $\mathrm{GL}(n+1, \mathbb{F})$ generated by matrices

$$M = \begin{bmatrix} g & 0 \\ 0 & 1 \end{bmatrix}$$

for $g \in G$ and matrices of the form $D(a_1, \ldots, a_n)$ with the $a_i \in \mathbb{F}$ is called the *affine group* associated to G and denoted $\mathrm{Aff}(G)$.

6.2.9. An application of geometric groups. Now we perform computations in $\mathrm{Aff}(\mathrm{SO}(3, \mathbb{R}))$ to study problems in motion-planning.

Recall that if we want to represent rotation in $\mathbb{R}^2$ via an angle of θ in the counterclockwise direction, we can use a matrix

$$\begin{bmatrix} \cos(\theta) & -\sin(\theta) \\ \sin(\theta) & \cos(\theta) \end{bmatrix} : \mathbb{R}^2 \to \mathbb{R}^2$$

Regard $\mathbb{R}^2$ as the subspace, $z = 1$, of $\mathbb{R}^3$. The linear transformation

(6.2.35) $$f = \begin{bmatrix} \cos(\theta) & -\sin(\theta) & a \\ \sin(\theta) & \cos(\theta) & b \\ 0 & 0 & 1 \end{bmatrix} : \mathbb{R}^3 \to \mathbb{R}^3$$

in $\mathrm{Aff}(\mathrm{SO}(2, \mathbb{R}))$ sends

$$\begin{bmatrix} x \\ y \\ 1 \end{bmatrix} \in \mathbb{R}^2 \subset \mathbb{R}^3$$

to

$$\begin{bmatrix} x\cos(\theta) - y\sin(\theta) + a \\ x\sin(\theta) + y\cos(\theta) + b \\ 1 \end{bmatrix} \in \mathbb{R}^2 \subset \mathbb{R}^3$$

and represents
  (1) *rotation* by θ (in a counterclockwise direction), followed by
  (2) *displacement* by (a, b).

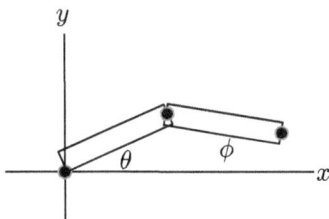

FIGURE 6.2.2.   A simple robot arm

Affine group-actions are used heavily in computer graphics: creating a scene in $\mathbb{R}^3$ is done by creating objects at the *origin* of $\mathbb{R}^3 \subset \mathbb{R}^4$ and moving them into position (and rotating them) via linear transformations in $\mathbb{R}^4$. A high-end (and not so high end) computer graphics card performs millions of affine group-operations per second.

Suppose we have a simple robot-arm with two links, as in figure 6.2.2.

If we assume that both links are of length ℓ, suppose the second link were attached to the origin rather than at the end of the second link.

Then its endpoint would be at (see equation 6.2.35 on the preceding page)

$$\begin{bmatrix} \ell \cos(\phi) \\ \ell \sin(\phi) \\ 1 \end{bmatrix} = \begin{bmatrix} \cos(\phi) & -\sin(\phi) & 0 \\ \sin(\phi) & \cos(\phi) & 0 \\ 0 & 0 & 1 \end{bmatrix} \begin{bmatrix} 1 & 0 & \ell \\ 0 & 1 & 0 \\ 0 & 0 & 1 \end{bmatrix} \begin{bmatrix} 0 \\ 0 \\ 1 \end{bmatrix}$$
$$= \begin{bmatrix} \cos(\phi) & -\sin(\phi) & \ell\cos(\phi) \\ \sin(\phi) & \cos(\phi) & \ell\sin(\phi) \\ 0 & 0 & 1 \end{bmatrix} \begin{bmatrix} 0 \\ 0 \\ 1 \end{bmatrix}$$

In other words, the effect of moving from the origin to the end of the second link (attached to the origin) is

(1) *displacement* by ℓ — so that $(0,0)$ is moved to $(\ell, 0) = (\ell, 0, 1) \in \mathbb{R}^3$.
(2) *rotation* by ϕ

This is the effect of the *second* link on all of $\mathbb{R}^2$. If we want to compute the effect of *both* links, *insert* the first link into the system — i.e. rigidly attach the second link to the first, displace by ℓ, and rotate by θ. The effect is equivalent to multiplying by

$$M_2 = \begin{bmatrix} \cos(\theta) & -\sin(\theta) & \ell\cos(\theta) \\ \sin(\theta) & \cos(\theta) & \ell\sin(\theta) \\ 0 & 0 & 1 \end{bmatrix}$$

It is clear that we can compute the endpoint of *any* number of links in this manner — always inserting new links at the *origin* and moving the rest of the chain accordingly.

At this point, the reader might wonder

> Where does *algebra* enter into all of this?

The point is that we do not have to deal with trigonometric functions until the very last step. If $a, b \in \mathbb{R}$ are numbers with the property that

(6.2.36) $$a^2 + b^2 = 1$$

there is a *unique* angle θ with $a = \cos(\theta)$ and $b = \sin(\theta)$. This enables us to replace the trigonometric functions by real numbers that satisfy equation 6.2.36 and derive purely algebraic equations for

(1) the set of points in $\mathbb{R}^2$ reachable by a robot-arm

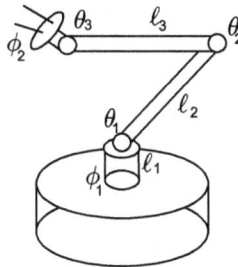

FIGURE 6.2.3.   A more complicated robot arm

(2) strategies for reaching those points (solving for explicit angles).

In the simple example above, let $a_1 = \cos(\theta)$, $b_1 = \sin(\theta)$, $a_2 = \cos(\phi)$, $b_2 = \sin(\phi)$ so that our equations for the endpoint of the second link become

$$\begin{bmatrix} x \\ y \\ 1 \end{bmatrix} = \begin{bmatrix} a_1 & -b_1 & \ell a_1 \\ b_1 & a_1 & \ell b_1 \\ 0 & 0 & 1 \end{bmatrix} \begin{bmatrix} \ell a_2 \\ \ell b_2 \\ 1 \end{bmatrix}$$

$$= \begin{bmatrix} \ell a_1 a_2 - \ell b_2 b_1 + \ell a_1 \\ \ell b_1 a_2 + \ell a_1 b_2 + \ell b_1 \\ 1 \end{bmatrix}$$

It follows that the points (x, y) reachable by this link are those for which the system of equations

$$\begin{aligned} \ell a_1 a_2 - \ell b_2 b_1 + \ell a_1 - x &= 0 \\ \ell b_1 a_2 + \ell a_1 b_2 + \ell b_1 - y &= 0 \\ a_1^2 + b_1^2 - 1 &= 0 \\ a_2^2 + b_2^2 - 1 &= 0 \end{aligned}$$

(6.2.37)

has *real* solutions (for a_i and b_i). Given values for x and y, we can solve for the set of configurations of the robot arm that will *reach* (x, y). Section 5.5.2 on page 127 develops the theory needed.

We conclude this chapter with a more complicated robot-arm in figure 6.2.3—somewhat like a Unimation Puma 560[5].

It has:
(1) A base of height ℓ_1 and motor that rotates the whole assembly by ϕ_1 — with 0 being the positive x-axis.
(2) An arm of length ℓ_2 that can be moved forward or backward by an angle of θ_1 — with 0 being straight forward (in the positive x-direction).
(3) A second arm of length ℓ_3 linked to the first by a link of angle θ_2, with 0 being when the second arm is in the same direction as the first.
(4) A little "hand" of length ℓ_4 that can be inclined from the second arm by an angle of θ_3 and rotated perpendicular to that direction by an angle ϕ_2.

---

[5]In 1985, this type of robot-arm was used to do brain-surgery! See [66].

We do our computations in $\mathbb{R}^4$, start with the "hand" and work our way back to the base. The default position of the hand is on the origin and pointing in the positive x-direction. It displaces the origin in the x-direction by ℓ_4, represented by the matrix

$$D_0 = \begin{bmatrix} 1 & 0 & 0 & \ell_4 \\ 0 & 1 & 0 & 0 \\ 0 & 0 & 1 & 0 \\ 0 & 0 & 0 & 1 \end{bmatrix}$$

The angle ϕ_2 rotates the hand in the yz-plane, and is therefore represented by

$$\begin{bmatrix} 1 & 0 & 0 & 0 \\ 0 & \cos(\phi_2) & -\sin(\phi_2) & 0 \\ 0 & \sin(\phi_2) & \cos(\phi_2) & 0 \\ 0 & 0 & 0 & 1 \end{bmatrix}$$

or

$$Z_1 = \begin{bmatrix} 1 & 0 & 0 & 0 \\ 0 & a_1 & -b_1 & 0 \\ 0 & b_1 & a_1 & 0 \\ 0 & 0 & 0 & 1 \end{bmatrix}$$

with $a_1 = \cos(\phi_2)$ and $b_1 = \sin(\phi_2)$. The "wrist" inclines the hand in the xz-plane by an angle of θ_3, given by the matrix

$$Z_2 = \begin{bmatrix} a_2 & 0 & -b_2 & 0 \\ 0 & 1 & 0 & 0 \\ b_2 & 0 & a_2 & 0 \\ 0 & 0 & 0 & 1 \end{bmatrix}$$

with $a_2 = \cos(\theta_3)$ and $b_2 = \sin(\theta_3)$ and the composite is

$$Z_2 Z_1 D_0 = \begin{bmatrix} a_2 & -b_2 b_1 & -b_2 a_1 & a_2 \ell_4 \\ 0 & a_1 & -b_1 & 0 \\ b_2 & a_2 b_1 & a_2 a_1 & b_2 \ell_4 \\ 0 & 0 & 0 & 1 \end{bmatrix}$$

The second arm displaces everything by ℓ_3 in the x-direction, giving

$$D_1 = \begin{bmatrix} 1 & 0 & 0 & \ell_3 \\ 0 & 1 & 0 & 0 \\ 0 & 0 & 1 & 0 \\ 0 & 0 & 0 & 1 \end{bmatrix}$$

so

$$D_1 Z_2 Z_1 D_0 = \begin{bmatrix} a_2 & -b_2 b_1 & -b_2 a_1 & a_2 \ell_4 + \ell_3 \\ 0 & a_1 & -b_1 & 0 \\ b_2 & a_2 b_1 & a_2 a_1 & b_2 \ell_4 \\ 0 & 0 & 0 & 1 \end{bmatrix}$$

so and then inclines it by θ_2 in the xz-plane, represented by

$$Z_3 = \begin{bmatrix} a_3 & 0 & -b_3 & 0 \\ 0 & 1 & 0 & 0 \\ b_3 & 0 & a_3 & 0 \\ 0 & 0 & 0 & 1 \end{bmatrix}$$

so that $Z_3 D_1 Z_2 Z_1 D_0$ is

$$\begin{bmatrix} a_3 a_2 - b_3 b_2 & (-a_3 b_2 - b_3 a_2) b_1 & (-a_3 b_2 - b_3 a_2) a_1 & (a_3 a_2 - b_3 b_2) \ell_4 + a_3 \ell_3 \\ 0 & a_1 & -b_1 & 0 \\ b_3 a_2 + a_3 b_2 & (a_3 a_2 - b_3 b_2) b_1 & (a_3 a_2 - b_3 b_2) a_1 & (b_3 a_2 + a_3 b_2) \ell_4 + b_3 \ell_3 \\ 0 & 0 & 0 & 1 \end{bmatrix}$$

Continuing in this fashion, we get a huge matrix, Z. To find the endpoint of the robot-arm, multiply

$$\begin{bmatrix} 0 \\ 0 \\ 0 \\ 1 \end{bmatrix}$$

(representing the origin of $\mathbb{R}^3 \subset \mathbb{R}^4$) by Z to get

(6.2.38) $\begin{bmatrix} x \\ y \\ z \\ 1 \end{bmatrix} =$

$$\begin{bmatrix} ((a_5 a_3 + b_5 b_4 b_3) a_2 + (-a_5 b_3 + b_5 b_4 a_3) b_2) \ell_4 + (a_5 a_3 + b_5 b_4 b_3) \ell_3 + a_5 \ell_2 \\ ((b_5 a_3 - a_5 b_4 b_3) a_2 + (-b_5 b_3 - a_5 b_4 a_3) b_2) \ell_4 + (b_5 a_3 - a_5 b_4 b_3) \ell_3 + b_5 \ell_2 \\ (a_4 b_3 a_2 + a_4 a_3 b_2) \ell_4 + a_4 b_3 \ell_3 + \ell_1 \\ 1 \end{bmatrix}$$

where $a_3 = \cos(\theta_2)$, $b_3 = \sin(\theta_2)$, $a_4 = \cos(\theta_1)$, $b_4 = \sin(\theta_1)$ and $a_5 = \cos(\phi_1)$, $b_5 = \sin(\phi_1)$. Note that $a_i^2 + b_i^2 = 1$ for $i = 1, \ldots, 5$. We are also interested in the *angle* that the hand makes (for instance, if we want to pick something up). To find this, compute

(6.2.39) $Z \begin{bmatrix} 1 \\ 0 \\ 0 \\ 1 \end{bmatrix} - Z \begin{bmatrix} 0 \\ 0 \\ 0 \\ 1 \end{bmatrix} = Z \begin{bmatrix} 1 \\ 0 \\ 0 \\ 0 \end{bmatrix} =$

$$\begin{bmatrix} (a_5 a_3 + b_5 b_4 b_3) a_2 + (-a_5 b_3 + b_5 b_4 a_3) b_2 \\ (b_5 a_3 - a_5 b_4 b_3) a_2 + (-b_5 b_3 - a_5 b_4 a_3) b_2 \\ a_4 b_3 a_2 + a_4 a_3 b_2 \\ 0 \end{bmatrix}$$

The numbers in the top three rows of this matrix are the *direction-cosines* of the hand's direction. We can ask what points the arm can reach with its hand aimed in a particular direction.

EXERCISES.

20. If $S \subset \text{GL}(n+1, \mathbb{F})$ is the subgroup of matrices of the form $D(a_1, \ldots, a_n)$ (see equation 6.2.34 on page 203) that is isomorphic to $\mathbb{F}^n$ and $G \subset \text{GL}(n, \mathbb{F})$, show that S is a *normal* subgroup of $\text{Aff}(G)$.

21. Under the assumptions of exercise 20 above, show that

$$\frac{\text{Aff}(G)}{S} \cong G$$

6.2.10. Geometry of Vectors in $\mathbb{R}^n$. Given vectors on $\mathbb{R}^n$, we can define a *product* that has a geometric significance.

DEFINITION 6.2.62. If $\mathbf{v}, \mathbf{w} \in \mathbb{R}^n$ are

$$\mathbf{v} = \begin{bmatrix} v_1 \\ \vdots \\ v_n \end{bmatrix}, \quad \mathbf{w} = \begin{bmatrix} w_1 \\ \vdots \\ w_n \end{bmatrix}$$

then their *dot-product*, denoted $\mathbf{v} \bullet \mathbf{w}$, is the scalar

$$\mathbf{v} \bullet \mathbf{w} = \sum_{i=1}^{n} v_i w_i \in \mathbb{R}$$

If we regard a vector as an $n \times 1$ matrix, then

$$\mathbf{v} \bullet \mathbf{w} = \mathbf{v}^t \mathbf{w} \in \mathbb{R}$$

where $\mathbf{v}^t$ is the *transpose* of $\mathbf{v}$ (a $1 \times n$ matrix) and we perform matrix-multiplication.

Give this, we can express the magnitude in terms of the dot-product:

DEFINITION 6.2.63. If $\mathbf{v} \in \mathbb{R}^n$, define $\|\mathbf{v}\| = \sqrt{\mathbf{v} \bullet \mathbf{v}}$ is the norm of $\mathbf{v}$. A *unit vector* $\mathbf{u} \in \mathbb{R}^n$ is one for which $\|\mathbf{u}\| = 1$.

The following properties of the dot-product are clear:

PROPOSITION 6.2.64. *Let* $\mathbf{x}, \mathbf{y} \in \mathbb{R}^n$. *Then:*

(1) $\mathbf{x} \bullet \mathbf{y} = \mathbf{y} \bullet \mathbf{x}$
(2) $(k \cdot \mathbf{x}) \bullet \mathbf{y} = k \cdot (\mathbf{x} \bullet \mathbf{y})$, *for* $k \in \mathbb{R}$,
(3) $(\mathbf{x} \pm \mathbf{y}) \bullet (\mathbf{x} \pm \mathbf{y}) = \mathbf{x} \bullet \mathbf{x} \pm 2\mathbf{x} \bullet \mathbf{y} + \mathbf{y} \bullet \mathbf{y}$, *or*

$$\|\mathbf{x} \pm \mathbf{y}\|^2 = \|\mathbf{x}\|^2 + \|\mathbf{y}\|^2 \pm 2\mathbf{x} \bullet \mathbf{y} \tag{6.2.40}$$

We need a generalization of Pythagoras's Theorem (proved in most precalculus books):

THEOREM 6.2.65 (Law of Cosines). *Given a triangle:*

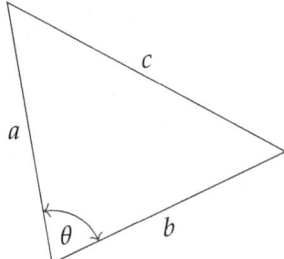

we have the formula

$$c^2 = a^2 + b^2 - 2ab\cos(\theta)$$

REMARK. If $\theta = 90° = \pi/2$, $\cos(\theta) = 0$ and we recover Pythagoras's original theorem.

The following result gives the geometric significance of dot-products

THEOREM 6.2.66. *Let $\mathbf{x}, \mathbf{y} \in \mathbb{R}^n$ be two vectors with an angle θ between them. Then*

$$\cos(\theta) = \frac{\mathbf{x} \bullet \mathbf{y}}{\|\mathbf{x}\| \cdot \|\mathbf{y}\|}$$

PROOF. Simply compare the triangle in theorem 6.2.65 on the facing page with

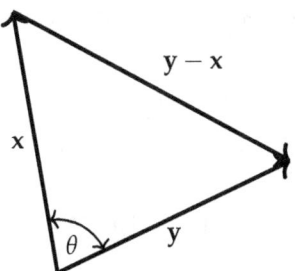

Here $a = \|\mathbf{x}\|$, $b = \|\mathbf{y}\|$, and $c = \|\mathbf{y} - \mathbf{x}\|$, so the Law of Cosines gives

$$\|\mathbf{y} - \mathbf{x}\|^2 = \|\mathbf{x}\|^2 + \|\mathbf{y}\|^2 - 2\|\mathbf{x}\| \cdot \|\mathbf{y}\| \cos(\theta)$$
$$= \|\mathbf{x}\|^2 + \|\mathbf{y}\|^2 - 2\mathbf{x} \bullet \mathbf{y} \quad \text{from equation 6.2.40 on the preceding page}$$

It follows that $2\|\mathbf{x}\| \cdot \|\mathbf{y}\| \cos(\theta) = 2\mathbf{x} \bullet \mathbf{y}$, which implies the conclusion.
□

COROLLARY 6.2.67. *Two nonzero vectors $\mathbf{x}, \mathbf{y} \in \mathbb{R}^n$ are perpendicular if and only if $\mathbf{x} \bullet \mathbf{y} = 0$.*

PROPOSITION 6.2.68. *If $M \in O(n)$, the orthogonal group defined in 6.2.60 on page 202, then M preserves all dot-products, i.e. for any vectors $\mathbf{u}, \mathbf{v} \in \mathbb{R}^n$*

$$(M\mathbf{u}) \bullet (M\mathbf{v}) = \mathbf{u} \bullet \mathbf{v}$$

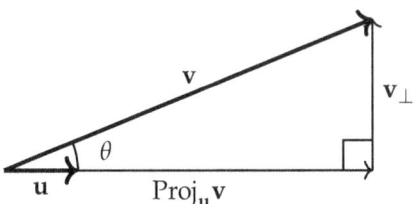

FIGURE 6.2.4.   Projection of a vector onto another

PROOF. We write
$$(M\mathbf{u}) \bullet (M\mathbf{v}) = (M\mathbf{u})^t (M\mathbf{v})$$
$$= \mathbf{u}^t M^t M \mathbf{v}$$
$$= \mathbf{u}^t \mathbf{v}$$
$$= \mathbf{u} \bullet \mathbf{v}$$
since $M^t = M^{-1}$ (see definition 6.2.60 on page 202). □

This, definition 6.2.63 on page 208 and theorem 6.2.66 on the preceding page immediately imply

COROLLARY 6.2.69. *An orthogonal matrix defines a linear transformation that preserves lengths of vectors and the angles between them.*

REMARK. It follows that orthogonal matrices define *geometric operations:* rotations and reflections.

We can use products of vectors to define other geometric concepts:

DEFINITION 6.2.70. Let $\mathbf{u} \in \mathbb{R}^n$ be a unit vector and $\mathbf{v} \in \mathbb{R}^n$ be some other vector. Define *the projection of* $\mathbf{v}$ *onto* $\mathbf{u}$ via
$$\mathrm{Proj}_{\mathbf{u}} \mathbf{v} = (\mathbf{u} \bullet \mathbf{v}) \mathbf{u}$$
Also define
$$\mathbf{v}_\perp = \mathbf{v} - \mathrm{Proj}_{\mathbf{u}} \mathbf{v}$$

REMARK. Note that $\mathrm{Proj}_{\mathbf{u}} \mathbf{v}$ is parallel to $\mathbf{u}$ with a length of $\|\mathbf{v}\| \cdot \cos \theta$, where θ is the angle between $\mathbf{u}$ and $\mathbf{v}$. Also note that
$$\mathbf{u} \bullet \mathbf{v}_\perp = \mathbf{u} \bullet (\mathbf{v} - (\mathbf{u} \bullet \mathbf{v})\mathbf{u})$$
$$= \mathbf{u} \bullet \mathbf{v} - (\mathbf{u} \bullet \mathbf{v})\mathbf{u} \bullet \mathbf{u}$$
$$= \mathbf{u} \bullet \mathbf{v} - \mathbf{u} \bullet \mathbf{v} = 0$$
so $\mathbf{v}_\perp$ is *perpendicular* to $\mathbf{u}$.

Since $\mathbf{v} = \mathrm{Proj}_{\mathbf{u}} \mathbf{v} + \mathbf{v}_\perp$, we have represented $\mathbf{v}$ as a sum of a vector *parallel* to $\mathbf{u}$ and one *perpendicular* to it. See figure 6.2.4

EXERCISES.

22. Compute the angle between the vectors
$$\begin{bmatrix} 1 \\ 2 \\ 3 \end{bmatrix} \text{ and } \begin{bmatrix} 1 \\ -2 \\ 0 \end{bmatrix}$$

23. If $\mathbf{u}_1, \ldots, \mathbf{u}_n \in \mathbb{R}^n$ are an orthonormal basis, show that the matrix
$$M = \begin{bmatrix} \mathbf{u}_1 \\ \vdots \\ \mathbf{u}_n \end{bmatrix}$$
— i.e. the matrix whose rows are the $\mathbf{u}_i$, is an orthogonal matrix — i.e. defines an element of $O(n)$ in definition 6.2.60 on page 202.

24. Suppose A is a symmetric matrix (i.e., $A = A^t$). If (λ_1, v_1) and (λ_2, v_2) are eigenvalue-eigenvector pairs with $\lambda_1 \neq \lambda_2$, show that $v_1 \bullet v_2 = 0$. Hint: If a vector is regarded as an $n \times 1$ matrix, its transpose is a $1 \times n$ matrix and $u^t v = u \bullet v$ — i.e. the matrix product is the dot-product.

6.2.11. Vectors in $\mathbb{R}^3$. Besides the dot-product, there's another way to form products of vectors — that is only well-defined in $\mathbb{R}^3$ (to see what happens in *other* dimensions, look at the discussion following corollary 10.7.13 on page 383).

DEFINITION 6.2.71. Define vectors
$$i = \begin{bmatrix} 1 \\ 0 \\ 0 \end{bmatrix}, \quad j = \begin{bmatrix} 0 \\ 1 \\ 0 \end{bmatrix}, \quad k = \begin{bmatrix} 0 \\ 0 \\ 1 \end{bmatrix}$$
so that vectors in $\mathbb{R}^3$ can be written as linear combinations of i, j, and k. Given this definition and vectors in $\mathbb{R}^3$
$$\mathbf{x} = \begin{bmatrix} x_1 \\ x_2 \\ x_3 \end{bmatrix}, \quad \mathbf{y} = \begin{bmatrix} y_1 \\ y_2 \\ y_3 \end{bmatrix}$$
set
$$(6.2.41) \qquad \mathbf{x} \times \mathbf{y} = \det \begin{bmatrix} i & j & k \\ x_1 & x_2 & x_3 \\ y_1 & y_2 & y_3 \end{bmatrix}$$
— the *cross product* of $\mathbf{x}$ and $\mathbf{y}$.

REMARK. Expansions by minors in the first row (see definitions 6.2.37 on page 184 and proposition 6.2.36 on page 184) gives
$$(6.2.42) \quad \mathbf{x} \times \mathbf{y} = i \cdot \det \begin{bmatrix} x_2 & x_3 \\ y_2 & y_3 \end{bmatrix} - j \cdot \det \begin{bmatrix} x_1 & x_3 \\ y_1 & y_3 \end{bmatrix} + k \cdot \det \begin{bmatrix} x_1 & x_2 \\ y_1 & y_2 \end{bmatrix}$$

We have the following properties of cross-products:

PROPOSITION 6.2.72. *Given vectors*
$$\mathbf{x} = \begin{bmatrix} x_1 \\ x_2 \\ x_3 \end{bmatrix}, \quad \mathbf{y} = \begin{bmatrix} y_1 \\ y_2 \\ y_3 \end{bmatrix}, \quad \mathbf{z} = \begin{bmatrix} z_1 \\ z_2 \\ z_3 \end{bmatrix}$$
then

(1) $(\mathbf{x} + \mathbf{y}) \times \mathbf{z} = \mathbf{x} \times \mathbf{z} + \mathbf{y} \times \mathbf{z}$
(2) $\mathbf{y} \times \mathbf{x} = -\mathbf{x} \times \mathbf{y}$
(3) $\mathbf{x} \times \mathbf{x} = 0$
(4)
$$\mathbf{x} \bullet (\mathbf{y} \times \mathbf{z}) = \det \begin{bmatrix} x_1 & x_2 & x_3 \\ y_1 & y_2 & y_3 \\ z_1 & z_2 & z_3 \end{bmatrix}$$
(5) *so* $\mathbf{x} \bullet (\mathbf{x} \times \mathbf{y}) = \mathbf{y} \bullet (\mathbf{x} \times \mathbf{y}) = 0$ *so* $\mathbf{x} \times \mathbf{y}$ *is perpendicular to* $\mathbf{x}$ *and* $\mathbf{y}$.

PROOF. The first statement follows from corollary 6.2.27 on page 176, proposition 6.2.23 on page 175, and corollary 6.2.17 on page 173.

Statement 2 follows from proposition 6.2.20 on page 174, and statement 3 follows from equation 6.2.42 on the preceding page, which implies that

$$\mathbf{x} \bullet (\mathbf{y} \times \mathbf{z}) = x_1 \cdot \det \begin{bmatrix} y_2 & y_3 \\ z_2 & z_3 \end{bmatrix} - x_2 \cdot \det \begin{bmatrix} y_1 & y_3 \\ z_1 & z_3 \end{bmatrix} + x_3 \cdot \det \begin{bmatrix} y_1 & y_2 \\ z_1 & z_2 \end{bmatrix}$$
$$= \det \begin{bmatrix} x_1 & x_2 & x_3 \\ y_1 & y_2 & y_3 \\ z_1 & z_2 & z_3 \end{bmatrix}$$

by proposition 6.2.37 on page 184. □

Statement 4 and theorem 6.2.49 on page 192 implies that:

COROLLARY 6.2.73. *Given vectors*
$$\mathbf{x} = \begin{bmatrix} x_1 \\ x_2 \\ x_3 \end{bmatrix}, \quad \mathbf{y} = \begin{bmatrix} y_1 \\ y_2 \\ y_3 \end{bmatrix}, \quad \mathbf{z} = \begin{bmatrix} z_1 \\ z_2 \\ z_3 \end{bmatrix}$$
the composite
$$\mathbf{x} \bullet (\mathbf{y} \times \mathbf{z})$$
is equal to the volume of the parallelepiped spanned by $\mathbf{x}$, $\mathbf{y}$, *and* $\mathbf{z}$.

This leads to a geometric interpretation of $\mathbf{x} \times \mathbf{y}$:

PROPOSITION 6.2.74. *If* $\mathbf{x}$ *and* $\mathbf{y}$ *are vectors in* $\mathbb{R}^3$ *with an angle of* θ *between them, then*
$$\|\mathbf{x} \times \mathbf{y}\| = \|\mathbf{x}\| \cdot \|\mathbf{y}\| \cdot \sin(\theta)$$
— *the area of the parallelogram generated by* $\mathbf{x}$ *and* $\mathbf{y}$.

REMARK. This almost completes a geometric description of $\mathbf{x} \times \mathbf{y}$: it is perpendicular to $\mathbf{x}$ and $\mathbf{y}$ and has the length given above. The vector has two possible directions and the one it takes is given by the Right Hand Rule: if fingers of the right hand go from $\mathbf{x}$ and $\mathbf{y}$, the thumb points in the direction of $\mathbf{x} \times \mathbf{y}$.

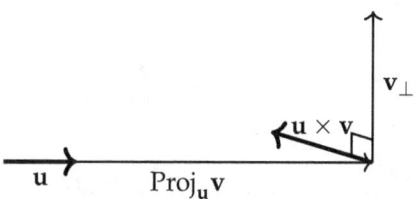

FIGURE 6.2.5.   The plane perpendicular to **u**

PROOF. Let **u** be a unit vector in the direction of $\mathbf{x} \times \mathbf{y}$. Then
$$\mathbf{u} \bullet (\mathbf{x} \times \mathbf{y}) = \|\mathbf{x} \times \mathbf{y}\|$$
is the volume of the parallelepiped spanned by **u**, **x**, and **y**. Since **u** is perpendicular to **x**, and **y** and of *unit length*, the volume of this parallelepiped is equal to the area of its base:

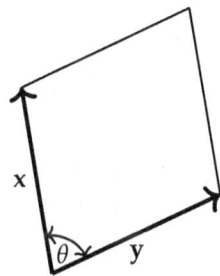

which is $\|\mathbf{x}\| \cdot \|\mathbf{y}\| \cdot \sin(\theta)$. □

Examination of figure 6.2.4 on page 210 (and the definition of the sine-function) shows that $\|\mathbf{v}_\perp\| = \|\mathbf{v}\| \cdot \sin \theta$.

If we form the cross-product $\mathbf{u} \times \mathbf{v}_\perp$, we get
$$\begin{aligned}\mathbf{u} \times \mathbf{v}_\perp &= \mathbf{u} \times (\mathbf{v} - (\mathbf{u} \bullet \mathbf{v})\mathbf{u}) \\ &= \mathbf{u} \times \mathbf{v} - (\mathbf{u} \bullet \mathbf{v})\mathbf{u} \times \mathbf{u} \\ &= \mathbf{u} \times \mathbf{v}\end{aligned}$$
producing a vector perpendicular to **u** and $\mathbf{v}_\perp$ *and* **v** — see figure 6.2.5. The two vectors, $\mathbf{v}_\perp$ and $\mathbf{u} \times \mathbf{v}$ span a plane perpendicular to **u**, and both vectors have lengths of $\|\mathbf{v}\| \cdot \sin \theta$.

Now consider the linear combination
$$\mathbf{v}_\phi = \cos \phi \cdot \mathbf{v}_\perp + \sin \phi \cdot \mathbf{u} \times \mathbf{v}$$
as ϕ runs from 0 to 2π.

We claim that $\mathbf{v}_\phi$ traces out a circle in the plane *perpendicular* to **u**, with a *radius* of $\|\mathbf{u}\| \cdot \sin \theta$ — see figure 6.2.6 on the next page.

We have our final geometric result:

THEOREM 6.2.75. *If* $\mathbf{u}, \mathbf{v} \in \mathbb{R}^3$ *are vectors with* $\|\mathbf{u}\| = 1$, *the result*, **r**, *of rotating* **v** *(counterclockwise, viewed in the direction of* **u***) by an angle of* ϕ *around*

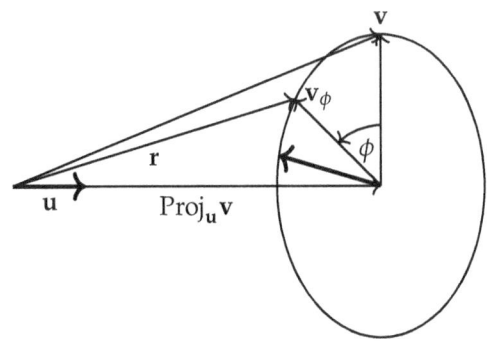

FIGURE 6.2.6.   Rotation of a vector in $\mathbb{R}^3$

the axis defined by $\mathbf{u}$ is

$$\mathbf{r} = \mathrm{Proj}_\mathbf{u}\mathbf{v} + \mathbf{v}_\phi$$
(6.2.43)
$$= (\mathbf{u}\bullet\mathbf{v})\mathbf{u} + \cos\phi \cdot (\mathbf{v} - (\mathbf{u}\bullet\mathbf{v})\mathbf{u}) + \sin\phi \cdot (\mathbf{u}\times\mathbf{v})$$

REMARK. Compare this with theorem 9.2.9 on page 329 in section 9.2 on page 325 on quaternions.

PROOF. This follows from the fact that $\mathbf{v}_\phi$ is the portion of $\mathbf{v}$ perpendicular to $\mathbf{u}$ — and that has been rotated by an angle of ϕ. The portion of $\mathbf{v}$ parallel to $\mathbf{u}$ has been unchanged. □

EXERCISES.

25. If $\mathbf{u}, \mathbf{v}, \mathbf{w} \in \mathbb{R}^3$, show that

$$\mathbf{u}\times(\mathbf{v}\times\mathbf{w}) = \mathbf{v}(\mathbf{u}\cdot\mathbf{w}) - \mathbf{w}(\mathbf{u}\cdot\mathbf{v})$$

This is called *Lagrange's formula* (one of many called this!).

26. Rotate the vector

$$\mathbf{v} = \begin{bmatrix} 1 \\ 2 \\ 3 \end{bmatrix}$$

by 30° around the vector

$$\mathbf{u} = \frac{1}{\sqrt{2}}\begin{bmatrix} 1 \\ 0 \\ -1 \end{bmatrix}$$

6.2.12. Inner-product spaces over $\mathbb{R}$ and $\mathbb{C}$. Inner product spaces are vector space equipped with an analogue to the dot product, called an inner product:

DEFINITION 6.2.76. A vector space, V, over $k = \mathbb{R}$ or $\mathbb{C}$ is called a *Hilbert space* if it has a function
$$\langle *, * \rangle : V \times V \to k$$
called its *inner product*, such that

(1) $\langle \alpha \cdot \mathbf{v}_1 + \beta \cdot \mathbf{v}_2, \mathbf{w} \rangle = \alpha \cdot \langle \mathbf{v}_1, \mathbf{w} \rangle + \beta \cdot \langle \mathbf{v}_2, \mathbf{w} \rangle$ for all $\alpha, \beta \in k$, and $\mathbf{v}_1, \mathbf{v}_2, \mathbf{w} \in V$,
(2) $\langle \mathbf{v}, \mathbf{w} \rangle = \overline{\langle \mathbf{w}, \mathbf{v} \rangle}$, the complex conjugate, for all $\mathbf{v}, \mathbf{w} \in V$. Note that this implies $\langle \mathbf{v}, \mathbf{v} \rangle = \overline{\langle \mathbf{v}, \mathbf{v} \rangle}$, or that $\langle \mathbf{v}, \mathbf{v} \rangle \in \mathbb{R}$ for all $\mathbf{v} \in V$.
(3) $\langle \mathbf{v}, \mathbf{v} \rangle = 0$ implies that $\mathbf{v} = 0$.
(4) Define $\|\mathbf{v}\| = \sqrt{\langle \mathbf{v}, \mathbf{v} \rangle}$
(5) Let $\{\mathbf{v}_i\}$ be an infinite sequence of elements of V. If the series
$$\sum_{n=1}^{\infty} \|\mathbf{v}_i\|$$
converges, then so does
$$\sum_{n=1}^{\infty} \mathbf{v}_i$$

REMARK. The first rule implies that $\langle 0, \mathbf{v} \rangle = \langle \mathbf{v}, 0 \rangle = 0$.

We have already seen an example of an inner product space over $\mathbb{R}$, namely $\mathbb{R}^n$, equipped with the *dot product* — see 6.2.62 on page 208.

It turns out that the last condition is always satisfied if V is finite-dimensional (the proof is beyond the scope of this book!). We mention it here because many interesting Hilbert spaces are infinite-dimensional.

We have a version of the last statement in proposition 6.2.64 on page 208

PROPOSITION 6.2.77. *If V is an inner product space and $\mathbf{x}, \mathbf{y} \in V$, then*
$$\|\mathbf{x} \pm \mathbf{y}\|^2 = \|\mathbf{x}\|^2 + \|\mathbf{y}\|^2 \pm \langle \mathbf{x}, \mathbf{y} \rangle \pm \langle \mathbf{y}, \mathbf{x} \rangle$$
In particular, if $\langle \mathbf{x}, \mathbf{y} \rangle = 0$ then
$$\|\mathbf{x} \pm \mathbf{y}\|^2 = \|\mathbf{x}\|^2 + \|\mathbf{y}\|^2$$

PROOF. Straightforward computation:
$$\langle \mathbf{x} \pm \mathbf{y}, \mathbf{x} \pm \mathbf{y} \rangle = \langle \mathbf{x}, \mathbf{x} \rangle \pm \langle \mathbf{x}, \mathbf{y} \rangle$$
$$\pm \langle \mathbf{y}, \mathbf{x} \rangle + (\pm 1)^2 \langle \mathbf{y}, \mathbf{y} \rangle$$

□

DEFINITION 6.2.78. Let $V = \mathbb{C}^n$ for some finite $n > 0$. For any $\mathbf{v}, \mathbf{w} \in V$ define the standard inner product via
$$\langle \mathbf{v}, \mathbf{w} \rangle = \sum_{i=1}^{n} \bar{v}_i w_i$$

Even over the real numbers, we can have spaces with "exotic" inner products:

EXAMPLE 6.2.79. Let
$$M = \begin{bmatrix} 2 & -1 & 0 \\ -1 & 2 & -1 \\ 0 & -1 & 2 \end{bmatrix}$$
and define an inner product in $\mathbb{R}^3$ via
$$\langle \mathbf{v}, \mathbf{w} \rangle = \mathbf{v}^t M \mathbf{w}$$
This is clearly bilinear (i.e., it satisfies rules 1 and 2). We must verify rule 3:
If
$$\mathbf{v} = \begin{bmatrix} x \\ y \\ z \end{bmatrix}$$
then
$$\langle \mathbf{v}, \mathbf{v} \rangle = \mathbf{v}^t M \mathbf{v} = \begin{bmatrix} x & y & z \end{bmatrix} \begin{bmatrix} 2 & -1 & 0 \\ -1 & 2 & -1 \\ 0 & -1 & 2 \end{bmatrix} \begin{bmatrix} x \\ y \\ z \end{bmatrix}$$
$$= \begin{bmatrix} 2x - y, & -x + 2y - z, & -y + 2z \end{bmatrix} \begin{bmatrix} x \\ y \\ z \end{bmatrix}$$
$$= 2x^2 - 2xy + 2y^2 - 2yz + 2z^2$$
$$= x^2 + (x-y)^2 + (y-z)^2 + z^2$$
$$\geq 0$$

If $\langle \mathbf{v}, \mathbf{v} \rangle = 0$, it follows that $x = 0 = z$ and $x = y$ so $\mathbf{v} = 0$.
If
$$e_1 = \begin{bmatrix} 1 \\ 0 \\ 0 \end{bmatrix}, \quad e_2 = \begin{bmatrix} 0 \\ 1 \\ 0 \end{bmatrix}, \quad e_3 = \begin{bmatrix} 0 \\ 0 \\ 1 \end{bmatrix}$$
is the standard basis for $\mathbb{R}^3$, $\langle e_1, e_1 \rangle = 2$ so the norm of e_1 with this new inner-product is
$$\sqrt{\langle e_1, e_1 \rangle} = \sqrt{2}$$

Since $\langle e_1, e_2 \rangle = -1$, e_1 and e_2 are not even *perpendicular* in this new inner-product.

Here's an example of an infinite dimensional Hilbert space used heavily in functional analysis:

EXAMPLE 6.2.80. If $a, b \in \mathbb{R}$ with $a < b$ define $L^2([a,b])$ to be the set of equivalence classes of complex-valued functions, f, on $[a, b]$ with
$$\int_a^b \|f(x)\|^2 dx < \infty$$
where two functions f, g are regarded as equivalent if
$$\int_a^b \|f(x) - g(x)\|^2 dx = 0$$

and with an inner product defined by

$$\langle f, g \rangle = \int_a^b f(x)\overline{g(x)}dx$$

From now on, we'll only be concerned with *finite-dimensional* Hilbert spaces like example 6.2.78 on page 215.

DEFINITION 6.2.81. Given an $n \times n$ matrix, A, with complex entries, define the *Hermitian transpose*, A^H, via

$$\left(A^H\right)_{i,j} = \bar{A}_{j,i}$$

A matrix, A, is called *Hermitian* if

$$A = A^H$$

A $n \times n$ matrix, U, is called *unitary* if

$$UU^H = U^H U = I$$

REMARK. Note that these are direct generalizations of familiar concepts to the complex domain: a real-valued Hermitian matrix is *symmetric*, and a real-valued unitary matrix is *orthogonal*.

> Charles Hermite (1822 – 1901) was a French mathematician who did research involving number theory, quadratic forms, invariant theory, orthogonal polynomials, elliptic functions, and algebra. He was the first to prove that e is a transcendental number.
> The Hermite crater near the Moon's north pole is named in his honor.

Proof of the following two statements is left as an exercise for the reader:

PROPOSITION 6.2.82. *The product of two unitary matrices is unitary, so the set of $n \times n$ unitary matrices form a group, $U(n)$. If A and B are complex-valued matrices then*

$$(AB)^H = B^H A^H$$

REMARK. Note that the standard inner product on $\mathbb{C}^n$ can be defined as the matrix-product

$$\langle \mathbf{u}, \mathbf{v} \rangle = \mathbf{u}^H \mathbf{v}$$

We have a unitary version of proposition 6.2.68 on page 209:

PROPOSITION 6.2.83. *Let $V = \mathbb{C}^n$ with the standard inner product (see definition 6.2.78 on page 215) and let U be a unitary matrix. If $\mathbf{v}, \mathbf{w} \in V$ then*

$$\langle U\mathbf{v}, U\mathbf{w} \rangle = \langle \mathbf{v}, \mathbf{w} \rangle$$

PROOF. Straight computation:

$$\begin{aligned}
\langle U\mathbf{v}, U\mathbf{w} \rangle &= (U\mathbf{v})^H U\mathbf{w} \\
&= \mathbf{v}^H U^H U\mathbf{w} \\
&= \mathbf{v}^H \mathbf{w} \\
&= \langle \mathbf{v}, \mathbf{w} \rangle
\end{aligned}$$

We start with:

DEFINITION 6.2.84. Let V be an inner product space and let $\mathbf{u}_1, \ldots, \mathbf{u}_k \in V$ be a set of vectors. This set is defined to be *orthonormal* if

$$(6.2.44) \qquad \langle \mathbf{u}_i, \mathbf{u}_j \rangle = \begin{cases} 1 & \text{if } i = j \\ 0 & \text{otherwise} \end{cases}$$

Unitary matrices are a complex analogue to orthogonal matrices (see exercise 23 on page 211):

PROPOSITION 6.2.85. *If*

$$M = \begin{bmatrix} \mathbf{v}_1 & \cdots & \mathbf{v}_n \end{bmatrix}$$

be an $n \times n$ matrix with entries in $\mathbb{C}$ and column-vectors $\mathbf{v}_1, \cdots, \mathbf{v}_n$, then M is unitary if and only if the $\mathbf{v}_i$ form an orthonormal set with respect to the standard inner product (see definition 6.2.78 on page 215).

PROOF. We have
$$M^H M = A$$
where
$$A_{i,j} = \mathbf{v}_i^H \mathbf{v}_j = \langle \mathbf{v}_i, \mathbf{v}_j \rangle$$

□

As with vector spaces over $\mathbb{R}$, we can define projections:

DEFINITION 6.2.86. If V is an inner-product space, $\mathbf{v} \in V$, and $S = \{\mathbf{u}_1, \ldots, \mathbf{u}_k\}$ is an orthonormal set of vectors with $W = \text{Span}(S)$ (see definition 6.2.3 on page 166), then define

$$\text{Proj}_W \mathbf{v} = \sum_{i=1}^{k} \langle \mathbf{u}_i, \mathbf{v} \rangle \mathbf{u}_i$$

REMARK. If V is over the complex numbers, we've lost the precise geometric significance of projection.

It is easy to express vectors in terms of an orthonormal basis

PROPOSITION 6.2.87. *Let V be an inner product space over k ($\mathbb{R}$ or $\mathbb{C}$) with an orthonormal basis $\mathbf{u}_1, \ldots, \mathbf{u}_n$ and let $\mathbf{v} \in V$ be any vector. Then*

$$\mathbf{v} = \sum_{i=1}^{n} \langle \mathbf{v}, \mathbf{u}_i \rangle \mathbf{u}_i = \text{Proj}_V \mathbf{v}$$

PROOF. Since $\mathbf{u}_1, \ldots, \mathbf{u}_n$ is a basis for V, there is a unique expression

$$\mathbf{v} = \sum_{i=1}^{n} a_i \mathbf{u}_i$$

with $a_i \in k$. If we take the inner product

$$\langle \mathbf{v}, \mathbf{u}_j \rangle = \sum_{i=1}^{n} a_i \langle \mathbf{u}_i, \mathbf{u}_j \rangle = a_j$$

due to equation 6.2.44. □

Consider an inner product space, V. If $S = \{\mathbf{u}_1,\ldots,\mathbf{u}_k\}$ is an orthonormal set of vectors that *don't* span V, they span some vector subspace $W \subset V$. Proposition 6.2.87 on the preceding page tells us that if $\mathbf{v} \in W$

$$\mathbf{v} = \text{Proj}_W \mathbf{v}$$

This raises the question:

If $\mathbf{v} \notin W$, what is the relation between $\mathbf{v}$ and $\text{Proj}_W \mathbf{v}$?

PROPOSITION 6.2.88. *If V is an inner product space, $S = \{\mathbf{u}_1,\ldots,\mathbf{u}_k\}$ is an orthonormal set of vectors that span $W \subset V$, and $\mathbf{v}$ is any other vector, then*

$$\mathbf{v}_\perp = \mathbf{v} - \text{Proj}_W \mathbf{v}$$

has the property that $\langle \mathbf{v}_\perp, \mathbf{u}_j \rangle = 0$ for all $j = 1,\ldots,k$, making it perpendicular to all of W. It follows that $\text{Proj}_W \mathbf{v}$ is the vector in W closest to $\mathbf{v}$ in the sense that

$$\|\mathbf{v} - \mathbf{w}\| > \|\mathbf{v} - \text{Proj}_W \mathbf{v}\|$$

for any $\mathbf{w} \in W$ with $\mathbf{w} \neq \text{Proj}_W \mathbf{v}$.

REMARK. This result shows the advantages of having orthonormal bases for vector spaces — in this case, W.

PROOF. If we form the inner product $\langle \mathbf{v}_\perp, \mathbf{u}_j \rangle$ for *any* $j = 1,\ldots,k$, we get

$$\begin{aligned}
\langle \mathbf{v} - \text{Proj}_W \mathbf{v}, \mathbf{u}_j \rangle &= \langle \mathbf{v}, \mathbf{u}_j \rangle - \left\langle \sum_{i=1}^k \langle \mathbf{v}, \mathbf{u}_i \rangle \mathbf{u}_i, \mathbf{u}_j \right\rangle \\
&= \langle \mathbf{v}, \mathbf{u}_j \rangle - \sum_{i=1}^k \langle \langle \mathbf{v}, \mathbf{u}_i \rangle \mathbf{u}_i, \mathbf{u}_j \rangle \\
&= \langle \mathbf{v}, \mathbf{u}_j \rangle - \langle \mathbf{v}, \mathbf{u}_j \rangle \langle \mathbf{u}_j, \mathbf{u}_j \rangle \qquad \text{by equation 6.2.44} \\
&= 0
\end{aligned}$$

It follows that, if $\mathbf{w} \in W$, then $\langle \mathbf{v}_\perp, \mathbf{w} \rangle = 0$.

Suppose $\mathbf{v}' \in W$ is a vector. Note that

$$\begin{aligned}
\mathbf{v} - \mathbf{v}' &= (\mathbf{v} - \text{Proj}_W \mathbf{v}) + (\text{Proj}_W \mathbf{v} - \mathbf{v}') \\
&= \mathbf{v}_\perp + (\text{Proj}_W \mathbf{v} - \mathbf{v}')
\end{aligned}$$

Since $\mathbf{w} = \text{Proj}_W \mathbf{v} - \mathbf{v}' \in W$, we have

$$\langle \mathbf{v}_\perp, \mathbf{w} \rangle = 0$$

Proposition 6.2.77 on page 215 implies that

$$\|\mathbf{v} - \mathbf{v}'\|^2 = \|\mathbf{v}_\perp\|^2 + \|\mathbf{w}\|^2$$

so that $\|\mathbf{v} - \mathbf{v}'\| > \|\mathbf{v} - \text{Proj}_W \mathbf{v}\| = \|\mathbf{v}_\perp\|$ for *all* $\mathbf{v}' \neq \text{Proj}_W \mathbf{v}$. □

Given the power of an orthonormal basis, it is gratifying that we can always find one:

THEOREM 6.2.89 (Gram-Schmidt orthonormalization). *Let* $\mathbf{v}_1, \ldots, \mathbf{v}_k \in V$ *be a basis for an inner-product space, V over* $k = \mathbb{R}$ *or* $\mathbb{C}$. *Then there exists an orthonormal basis*
$$\mathbf{u}_1, \ldots, \mathbf{u}_k$$
for V such that

(6.2.45) $$\mathrm{Span}(\mathbf{u}_1, \ldots, \mathbf{u}_i) = \mathrm{Span}(\mathbf{v}_1, \ldots, \mathbf{v}_i)$$

for $i = 1, \ldots, k$.

REMARK. This is named after Jørgen Pedersen Gram and Erhard Schmidt but it appeared earlier in the work of Laplace and Cauchy[6].

PROOF. This is an inductive process: We initially set
$$\mathbf{u}_1 = \frac{\mathbf{v}_1}{\|\mathbf{v}_1\|}$$
Having performed $j - 1$ steps, define $W_j = \mathrm{Span}(\mathbf{u}_1, \ldots, \mathbf{u}_{j-1}) \subset V$ and set
$$\mathbf{u}_j = \frac{\mathbf{v}_j - \mathrm{Proj}_{W_j} \mathbf{v}_j}{\|\mathbf{v}_j - \mathrm{Proj}_{W_j} \mathbf{v}_j\|}$$
and continue until $j = k$. The process *works* because $\mathbf{v}_j - \mathrm{Proj}_{W_j} \mathbf{v}_j \neq 0$ since:

(1) the $\mathbf{u}_1, \ldots, \mathbf{u}_{j-1}$ are linear combinations of the $\mathbf{v}_1, \ldots, \mathbf{v}_{j-1}$ — in a way that is easily *reversed*[7] — so the *span* of $\mathbf{u}_1, \ldots, \mathbf{u}_{j-1}$ is the *same* as that of $\mathbf{v}_1, \ldots, \mathbf{v}_{j-1}$ — i.e., equation 6.2.45.
(2) the $\mathbf{v}_1, \ldots, \mathbf{v}_k$ are linearly independent, so $\mathbf{v}_j$ is *not* in the span of $\mathbf{v}_1, \ldots, \mathbf{v}_{j-1}$ — or that of $\mathbf{u}_1, \ldots, \mathbf{u}_{j-1}$.

□

PROPOSITION 6.2.90. *The eigenvalues of a Hermitian matrix lie in* $\mathbb{R}$.

PROOF. Let A be an $n \times n$ Hermitian matrix with eigenvalue-eigenvector pair $(\lambda, \mathbf{v})$. Then

(6.2.46) $$A\mathbf{v} = \lambda \mathbf{v}$$

If we form the Hermitian transpose, we get

(6.2.47) $$\mathbf{v}^H A^H = \bar{\lambda} \mathbf{v}^H$$

Now, we left-multiply equation 6.2.46 by $\mathbf{v}^H$ and right multiply equation 6.2.47 by $\mathbf{v}$, we get
$$\mathbf{v}^H A \mathbf{v} = \mathbf{v}^H \lambda \mathbf{v} = \lambda \|\mathbf{v}\|^2$$
$$\mathbf{v}^H A^H \mathbf{v} = \bar{\lambda} \|\mathbf{v}\|^2$$
Since $A = A^H$, $\lambda \|\mathbf{v}\|^2 = \bar{\lambda} \|\mathbf{v}\|^2$, and since $\|\mathbf{v}\|^2 \neq 0$, we conclude that $\lambda = \bar{\lambda}$ or $\lambda \in \mathbb{R}$. □

Since real-valued matrices are Hermitian if and only if they are *symmetric*, we also conclude:

---

[6]Part of the mathematical tradition of naming things after people who didn't discover them!

[7]So the $\mathbf{v}_i$ can be written as linear combinations of the $\mathbf{u}_i$.

COROLLARY 6.2.91. *The eigenvalues of a real-valued symmetric matrix lie in $\mathbb{R}$.*

REMARK. It's interesting that we must go into the complex domain to prove this statement about real matrices.

Although many matrices cannot be diagonalized (see example 6.2.56 on page 199), there is an important class of them that can. We begin with:

LEMMA 6.2.92. *If A is a complex-valued $n \times n$ matrix, then there exists an $n \times n$ unitary matrix, U, such that*
$$U^{-1}AU = B$$
where B has the form
$$B = \begin{bmatrix} \lambda & * \\ 0 & A_1 \end{bmatrix}$$
where A_1 is an $(n-1) \times (n-1)$ matrix and the first column of B has zeros below the first row. If A is real-valued, so are U and B.

PROOF. Let λ be an eigenvalue of A with eigenvector $\mathbf{v}$. By adjoining other vectors, we extend $\mathbf{v}$ to a basis for $\mathbb{C}^n$ and perform Gram-Schmidt orthonormalization (theorem 6.2.89 on the preceding page) to get an orthonormal basis $\mathbf{u}_1, \ldots, \mathbf{u}_n$. We create a matrix with the $\mathbf{u}_i$ as its columns:
$$U = \begin{bmatrix} \mathbf{u}_1 & \cdots & \mathbf{u}_n \end{bmatrix}$$
This will be unitary, by proposition 6.2.85 on page 218. In the new $\mathbf{u}_i$-basis, the linear transformation that A defines will have a matrix representation that looks like B — since $\mathbf{u}_1 = \mathbf{v}/\|\mathbf{v}\|$ is still an eigenvector corresponding to the eigenvalue λ. □

COROLLARY 6.2.93. *If A is an $n \times n$ matrix over $\mathbb{C}$, then there exists an $n \times n$ unitary matrix, U, such that $U^{-1}AU$ is upper-triangular. If A is real-valued, so is U.*

PROOF. Simply apply lemma 6.2.92 to A, and then to A_1 and the smaller matrices that result. After $n-1$ such applications, we get
$$U_{n-1}^{-1} \cdots U_1^{-1} A U_1 \cdots U_{n-1}$$
is upper-triangular, where $U = U_1 \cdots U_{n-1}$. □

THEOREM 6.2.94. *If A is a Hermitian $n \times n$ matrix, then there exists an $n \times n$ unitary matrix, U, such that $U^{-1}AU$ is diagonal. If A is real-valued, so is U.*

PROOF. Let U be such that $B = U^{-1}AU$ is upper-triangular. We claim that B is also Hermitian. Note that $U^{-1} = U^H$ and take the Hermitian transpose of
$$B = U^H A U$$
We get
$$B^H = U^H A^H U = U^H A U = B$$
But a Hermitian upper-triangular matrix is diagonal. □

Since a real-valued matrix is Hermitian if and only if it is *symmetric*, and unitary if and only if it is *orthogonal*, we get:

COROLLARY 6.2.95. *If A is an $n \times n$ real-valued symmetric matrix, then there exists an $n \times n$ orthogonal matrix, U, such that $U^{-1}AU$ is diagonal.*

REMARK. This is interesting because it says not only that a symmetric matrix can be diagonalized, but that this can be done by *geometric operations*: rotations and/or reflections.

Here's an intuitive argument for why symmetric matrices can be diagonalized: Every matrix is arbitrarily "close" to one with distinct eigenvalues. For instance, we can we can perturb the matrix in example 6.2.56 on page 199 so it has two *distinct* eigenvalues:

$$\begin{bmatrix} 1-\epsilon & 1 \\ 0 & 1+\epsilon \end{bmatrix}$$

namely $1 - \epsilon$ and $1 + \epsilon$. The eigenvectors corresponding to these are, respectively

$$\begin{bmatrix} 1 \\ 0 \end{bmatrix} \text{ and } \begin{bmatrix} 1 \\ 2\epsilon \end{bmatrix}$$

As we let ϵ approach 0 the eigenvalues and eigenvectors merge, and the eigenvectors no longer span $\mathbb{R}^2$.

Exercise 24 on page 211 shows that the eigenvectors of a symmetric matrix are *orthogonal* — so they *cannot* merge, even if the *eigenvalues* do — i.e., the angle between two vectors is a continuous function of the vectors, so it cannot abruptly jump from 90° to 0°.

EXERCISES.

27. Find an orthonormal basis for $\mathbb{R}^3$, using the inner-product in example 6.2.79 on page 216 (use the Gram-Schmidt process).

6.3. Modules

6.3.1. Basic properties. Modules are like vector-spaces over *rings* rather than fields. This turns out to make the subject infinitely more complex and many basic questions have no known general answers.

DEFINITION 6.3.1. If R is a commutative ring, a *module* over R is
(1) an abelian group, A,
(2) an action of R on A, i.e. a map

$$f: R \times A \to A$$

such that

$$f(r, *): r \times A \to A$$

is a homomorphism of abelian groups, for all $r \in R$, and
$$f(r_1, f(r_2, a)) = f(r_1 r_2, a)$$
and
$$f(r_1 + r_2, a) = f(r_1, a) + f(r_2, a)$$
This action is usually written with a product-notation, i.e. $f(r, a) = r \cdot a$ (in analogy with multiplication by scalars in a vector space).

If $B \subset A$ is a subgroup with the property that $r \cdot B \subset B$ for all $r \in R$, then B is called a *submodule* of A. If R is not commutative, we can have left- or right-modules over R, and they are generally not equivalent to each other. This definition given here is for a *left* module.

EXAMPLE. We can regard a ring, R, as a module over *itself*. Its *submodules* are precisely its ideals.

If $g: R \to S$ is a homomorphism of rings, S naturally becomes a module over R by defining $r \cdot s = g(r)s$ for all $r \in R$ and $s \in S$.

EXAMPLE 6.3.2. If R is a ring and $R^n = \bigoplus_{i=1}^n R$, then R^n is a module over R with the action defined by multiplication in R. This is called the *free module of rank n over R*. An n-dimensional *vector space* over a field k is a free module of rank n over that field.

It is possible to come up with more "exotic" examples of modules:

EXAMPLE 6.3.3. Let V be an n-dimensional vector space over a field F and let M be an $n \times n$ matrix over F. Then V is a module over the polynomial ring $k[X]$, where a polynomial, $p(X) \in k[X]$ acts via
$$p(M): V \to V$$
In other words, we plug M into $p(X)$ to get a matrix and then act on V via that matrix.

Note that a vector-subspace $W \subset V$ is a *submodule* if and only if $M(W) \subset W$. It follows that the module-structure of V over $k[X]$ depends strongly on the matrix M.

DEFINITION 6.3.4. Let M_1 and M_2 be modules over the same ring, R. A *homomorphism* of modules is a map of their underlying abelian groups
$$f: M_1 \to M_2$$
such that $f(r \cdot m) = r \cdot f(m)$ for all $m \in M$ and $r \in R$. The set of elements $m \in M_1$ with $f(m) = 0$ is called *the kernel of f* and denoted $\ker f$. The set of elements $m \in M_2$ of the form $f(n)$ for some $n \in M_1$ is called the *image of f* and denoted $\operatorname{im} f$.

If $\ker f = 0$, the homomorphism f is said to be *injective*. If $\operatorname{im} f = M_2$, the homomorphism is said to be *surjective*. If f is both injective and surjective, it is called an *isomorphism*.

REMARK. Note that, if f above is injective, we can regard M_1 as a submodule of M_2. Isomorphic modules are algebraically equivalent.

The corresponding statements about abelian groups imply that

PROPOSITION 6.3.5. *Let M be a module over a ring R and let A and B be submodules of M. Then:*

(1) *we can define the quotient M/A as the set of equivalence classes of the equivalence relation*

$$m_1 \equiv m_2 \pmod{A}$$

if $m_1 - m_2 \in A$, for all $m_1, m_2 \in M$. We can also define M/A as the set of cosets $\{m + A\}$ for $m \in M$.

(2) *the map*

$$p \colon M \to M/A$$

sending an element to its equivalence class, is a homomorphism of modules.

(3) *the map p defines a 1-1 correspondence between submodules of M containing A and submodules of M/A*

(4) *there is a canonical isomorphism*

$$\frac{A+B}{A} \cong \frac{B}{A \cap B}$$

DEFINITION 6.3.6. If $f \colon M_1 \to M_2$ is a homomorphism of modules, the quotient

$$\frac{M_2}{f(M_1)}$$

is called the *cokernel* of f.

REMARK. Since one can form quotients of modules with respect to arbitrary submodules, cokernels always exist for module-homomorphisms.

DEFINITION 6.3.7. A sequence of modules and homomorphisms (all over the same ring)

$$\cdots \xrightarrow{f_{n+1}} M_{n+1} \xrightarrow{f_n} M_n \xrightarrow{f_{n-1}} M_{n-1} \to \cdots$$

is said to be *exact* if $\operatorname{im} f_{n+1} = \ker f_n$ for all n. An exact sequence with five terms like

$$0 \to A \xrightarrow{f} B \xrightarrow{g} C \to 0$$

is called a *short exact sequence*.

REMARK. In the *short* exact sequence above, the kernel of f must be 0, so A can be identified with a submodule of B, and the map g must be surjective (since the kernel of the rightmost map is all of C).

The exactness of the (long) sequence above is equivalent to saying that the short sequences

$$0 \to \operatorname{im} f_n \to M_n \to \operatorname{im} f_{n-1} \to 0$$

are exact for all n.

Exact sequences are widely used in homological algebra and algebraic topology, facilitating many types of computations.

DEFINITION 6.3.8. If M is a module over a ring R, a set of elements $S = \{m_1, \ldots\} \in M$ will be called a *generating set* if every element $m \in M$ can be expressed in terms of the elements of S

$$m = \sum_{m_i \in S} r_i \cdot m_i$$

with the $r_i \in R$.

A module is said to be *finitely generated* if it has a finite generating set.

EXAMPLE 6.3.9. As in example 6.3.3 on page 223Let V be an n-dimensional vector space over a field F and let M be the $n \times n$ permutation matrix

$$M = \begin{bmatrix} 0 & 0 & 0 & \cdots & 1 \\ 0 & 1 & 0 & \cdots & 0 \\ 0 & 0 & 1 & \ddots & 0 \\ 1 & 0 & 0 & \cdots & 0 \end{bmatrix}$$

Then, as a module over $k[X]$ (defined as in example 6.3.3 on page 223), V has a *single generator*, namely

$$g = \begin{bmatrix} 1 \\ 0 \\ \vdots \\ 0 \end{bmatrix}$$

This is because

$$Mg = \begin{bmatrix} 0 \\ 1 \\ \vdots \\ 0 \end{bmatrix}$$

and $M^k g = g_k$, the kth basis element of V. So all of the basis-elements of V are in the orbit of powers of M and of $k[X]$.

In analogy with proposition 4.6.7 on page 62, we have:

PROPOSITION 6.3.10. *Every R-module is a quotient of a free R-module.*

PROOF. If M is an R-module with generating set $\{m_1, \ldots\}$, let F be free on generators $\{e_1, \ldots\}$ — one for each of the m_i. We define a map

$$f: F \to M$$
$$\sum r_j e_j \mapsto \sum r_j m_j$$

This is clearly surjective, so that

$$\frac{F}{\ker f} \cong M$$

□

It is interesting to consider what properties of vector-spaces carry over to modules, or whether we can do a kind of "linear algebra" over a general ring. This is a deep field of mathematics that includes several areas, such as group-representations (see [**40**]), homological algebra (see [**109**]) and algebraic K-theory (see [**74**]).

Even a simple question like

"Is a submodule of a finitely generated module finitely generated?"

can have a complex answer. For instance, let $R = k[X_1, \ldots]$ — a polynomial ring over an infinite number of variables. It is finitely generated as a module over itself (generated by 1). The submodule of polynomials with vanishing constant term is *not* finitely generated since every polynomial has a finite number of variables.

We need to find a class of modules that is better-behaved.

DEFINITION 6.3.11. A module M over a ring R will be called *noetherian* if all of its submodules are finitely generated — this is equivalent to saying that all ascending chains of submodules of M

$$M_1 \subset M_2 \subset \cdots \subset M_i \subset \cdots$$

becomes constant from some finite point on, i.e. $M_t = M_{t+i}$ for all $i > 0$. A module will be said to be *Artinian* if every *descending* chain of submodules

$$M_1 \supset M_2 \supset \cdots \subset M_i \supset \cdots$$

becomes constant from some finite point on.

REMARK. A ring is noetherian if and only if it is noetherian as a module over itself.

PROPOSITION 6.3.12. *Let*

$$0 \to M_1 \xrightarrow{f} M_2 \xrightarrow{g} M_3 \to 0$$

be a short exact sequence (see definition 6.3.7 on page 224) of modules over a ring. Then M_2 is noetherian or Artinian if and only if M_1 and M_3 are both noetherian or Artinian, respectively.

PROOF. We will prove this in the noetherian case; the Artinian case is almost identical. Clearly, if M_2 is noetherian, M_1 will inherit this property since it is a submodule. Any increasing chain of submodules of M_3 will lift to one in M_2, which becomes constant from some finite point on. It follows that M_2 being noetherian implies that M_1 and M_3 are also noetherian.

Conversely, suppose that M_1 and M_3 are noetherian and

$$N_1 \subset N_2 \subset \cdots \subset N_i \subset \cdots$$

is an increasing sequence of submodules of M_2. Since M_3 is noetherian, this image of this sequence in M_3 will become constant from some finite point on, say k. Then

$$\frac{N_i}{N_i \cap M_1} = \frac{N_{i+1}}{N_{i+1} \cap M_1} \tag{6.3.1}$$

for $i > k$. Since M_1 is noetherian, the sequence
$$N_j \cap M_1 \subset N_{j+1} \cap M_2 \subset \cdots$$
will become constant from some finite point on — say $j = t$. Then, for $i > \max(k, t)$, equation 6.3.1 on the preceding page and
$$N_i \cap M_1 = N_{i+1} \cap M_1$$
imply that $N_i = N_{i+1}$. □

COROLLARY 6.3.13. *If R is a noetherian ring, then R^n is a noetherian module.*

PROOF. This is a simple induction on n. If $n = 1$, $R^1 = R$ is noetherian over itself since it is a noetherian ring. For $n > 1$ we use proposition 6.3.12 on the facing page with the short exact sequence
$$0 \to R \xrightarrow{f} R^{n+1} \xrightarrow{g} R^n \to 0$$
where
$$\begin{aligned} g(r_1, \ldots, r_{n+1}) &= (r_1, \ldots, r_n) \\ f(r) &= (0, \ldots 0, r) \end{aligned}$$
where the image of f is the $n+1^{\text{st}}$ entry in R^{n+1}. □

Now we can define a large class of well-behaved modules:

LEMMA 6.3.14. *If R is a noetherian ring, a module, M over R is noetherian if and only if it is finitely generated.*

PROOF. If M is noetherian it must be finitely generated (since its submodules, including itself, are). Suppose M is finitely generated, say by generators $(a_1, \ldots, a_n)$. Then there exists a surjective homomorphism of modules
$$\begin{aligned} R^n &\xrightarrow{f} M \\ (r_1, \ldots, r_n) &\mapsto r_1 \cdot a_1 + \cdots + r_n \cdot a_n \end{aligned}$$
This map fits into a short exact sequence
$$0 \to \ker f \to R^n \xrightarrow{f} M \to 0$$
and proposition 6.3.12 on the preceding page and corollary 6.3.13 imply the conclusion. □

Although noetherian modules are somewhat "well-behaved," they still are more complex than vector-spaces. For instance, a subspace of a vector space of dimension n must have dimension $< n$. The ring $k[X, Y, Z]$ is a module over itself with *one* generator: 1. On the other hand, the ideal $(X, Y, Z) \subset k[X, Y, Z]$, is a proper submodule that requires *three* generators.

The most "straightforward" modules are the *free* ones like R^n above. They are closely related to projective modules:

DEFINITION 6.3.15. *If R is a ring and P is an R-module, then P is said to be projective if it is a direct summand of a free module.*

REMARK. In other words, P is projective if there exists an R-module, Q, such that $P \oplus Q = R^n$ for some n. All free modules are (trivially) projective but not all projective modules are free. For instance, if $R = \mathbb{Z}_6$, note that $\mathbb{Z}_2 \oplus \mathbb{Z}_3 = \mathbb{Z}_6$ as rings so $\mathbb{Z}_2$ and $\mathbb{Z}_3$ are projective modules that are not free.

Projective modules have an interesting property that is often (usually?) used to define them:

PROPOSITION 6.3.16. *Let R be a ring and let P be a projective module over R. If $\alpha\colon M \to N$ is a surjective homomorphism of R-modules and $\beta\colon P \to N$ is any homomorphism, then a homomorphism, $\gamma\colon P \to M$ exists that makes the diagram*

(6.3.2)
$$\begin{array}{ccc} & & P \\ & {}^{\gamma}\swarrow & \downarrow{\beta} \\ M & \xrightarrow{\alpha} & N \end{array}$$

commute.

PROOF. Since P is projective, there exists a module Q such that $P \oplus Q = R^n$ for some value of n. Consider the diagram

$$\begin{array}{ccc} & & P \oplus Q \\ & & \downarrow{\beta \oplus 1} \\ M \oplus Q & \xrightarrow{\alpha \oplus 1} & N \oplus Q \end{array}$$

and note that

$$P \oplus Q = R^n = \bigoplus_{i=1}^{n} x_i \cdot R$$

where the $x_i \in R^n$ are its generators. Since α is surjective, $\alpha \oplus 1$ will also be, and we can choose $y_i \in M \oplus Q$ such that $(\alpha \oplus 1)(y_i) = (\beta \oplus 1)(x_i)$ for $i = 1, \ldots, n$. Then we can define a homomorphism

$$\begin{array}{rcl} G\colon R^n & \to & M \oplus Q \\ x_i & \mapsto & y_i \end{array}$$

making the diagram

$$\begin{array}{ccc} & & P \oplus Q \\ & {}^{G}\swarrow & \downarrow{\beta \oplus 1} \\ M \oplus Q & \xrightarrow{\alpha \oplus 1} & N \oplus Q \end{array}$$

commute. Since $(\alpha \oplus 1)(Q) = Q$, we can extend diagram this to a commutative diagram

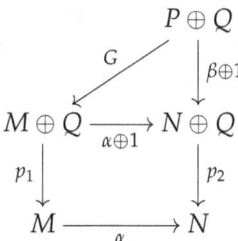

where the p_i are projections to the factors. If γ is the composite $P \hookrightarrow P \oplus Q \xrightarrow{G} M \oplus Q \xrightarrow{p_1} M$, it will have the desired properties. □

EXAMPLE 6.3.17. Let $R = \mathbb{C}[X]$ and let A be an $n \times n$ matrix over $\mathbb{C}$. If $V = \mathbb{C}^n$ is a vector-space, we can make it a module over R by defining

$$X \cdot v = Av$$

for all $v \in V$ — this is actually a module over the commutative ring $\mathbb{C}[A]$ — see exercise 4 on page 171. We know that some element of R annihilates V because the Cayley-Hamilton theorem (see theorem 6.2.57 on page 199) states that A "satisfies" its characteristic polynomial, i.e., $\chi_A(A) = 0$. Because R is a principal ideal domain (see corollary 5.3.10 on page 119), the annihilator of V is a principal ideal $(p(X))$ such that $\chi_A(X) \in (p(X))$, i.e. $p(X)|\chi_A(X)$. The polynomial, $p(X)$, is called the *minimal polynomial* of A.

In general, the minimal polynomial of a matrix is not equal to its characteristic polynomial. For instance, if $A = 3 \cdot I$, where I is the identity matrix then $p(X) = X - 3$ and $\chi_A(X) = (X - 3)^n$.

We can use this example to get a shorter, more straightforward proof of the Cayley-Hamilton Theorem (theorem 6.2.57 on page 199).

EXAMPLE 6.3.18. Let A be an $n \times n$ matrix acting on an n-dimensional vector-space, V with basis $\{x_1, \ldots, x_n\}$, and call the generator (representing A itself) of $\mathbb{C}[A]$ be γ. Then

$$\gamma \cdot x_i = \sum_{j=1}^{n} A_{i,j} x_j$$

or

$$\gamma I \begin{bmatrix} x_1 \\ \vdots \\ x_n \end{bmatrix} = A \cdot \begin{bmatrix} x_1 \\ \vdots \\ x_n \end{bmatrix}$$

and

(6.3.3) $$(\gamma I - A) \begin{bmatrix} x_1 \\ \vdots \\ x_n \end{bmatrix} = \begin{bmatrix} 0 \\ \vdots \\ 0 \end{bmatrix}$$

Now we apply Cramer's Rule (theorem 6.2.29 on page 177) to this, particularly equation 6.2.15 on page 177, to get

$$\det(\gamma I - A) \cdot x_i = \det(\bar{M}_i) = 0$$

where $M = \gamma I - A$ and $\bar{M}_i$ is the result of replacing the ith column of M by the right side of equation 6.3.3 on the previous page — zeroes. It follows that $\det(\gamma I - A) = \chi_A(\gamma)$ annihilates all of the x_i. If we get a matrix-representation (i.e., by replacing γ by A) we get a matrix that annihilates all basis-elements of V, i.e., the *zero-matrix*.

EXERCISES.

1. If $\mathfrak{A} \subset R$ is an ideal in a commutative ring, R, then $\mathfrak{A}$ is also an R-module. Show that, if $\mathfrak{A}$ is a *free* R-module, it *must* be of the form $\mathfrak{A} = (x)$ where $x \in R$ is a non-zero divisor.

2. Let $f: M_1 \to M_2$ be a homomorphism of modules (over a commutative ring, or left-modules over a noncommutative one). Suppose $A \subset M_1$ is a submodule and
$$f|A: A \to f(A)$$
Show that there is an induced map
$$\bar{f}: \frac{M_1}{A} \to \frac{M_2}{f(A)}$$

6.3.2. Modules over principal ideal domains. When a ring is a PID, it turns out one can easily classify modules over it. Throughout this section, R will denote a PID — see definition 5.3.8 on page 119.

PROPOSITION 6.3.19. *If $S \subset A$ is a submodule of a free R-module, then S is free. If $\operatorname{rank}(A) < \infty$, then*
$$\operatorname{rank}(S) \le \operatorname{rank}(A)$$

REMARK. Compare this with proposition 4.6.6 on page 61. The proof is essentially the same. If we had proved the current result first, we could've regarded proposition 4.6.6 on page 61 as a corollary with $R = \mathbb{Z}$.

EXAMPLE 6.3.20. Exercise 1 produces plenty of counterexamples if R is *not* a PID. For instance, $\mathbb{Z}_6$ is free as a $\mathbb{Z}_6$-module, but the ideal (i.e., submodule) $2\mathbb{Z}_6 = \{0, 2, 4\}$ is not because $3 \cdot 2 = 3 \cdot 4 = 0 \in \mathbb{Z}_6$ so *none* of these can be free generators of $2\mathbb{Z}_6$. The ideal $(X, Y) \subset \mathbb{C}[X, Y]$ is another counterexample.

REMARK. There's a whole subfield of algebra that studies the extent to which proposition 6.3.19 fails in general (among other things). It is called Algebraic K-theory — see [106].

PROOF. Let X be a free basis for A. By the Well Ordering Axiom of set theory (see axiom 14.2.9 on page 465), we can *order* the elements of X
$$x_1 \succ x_2 \succ \cdots$$

so that every subset of X has a *minimal* element. Note that this ordering is completely *arbitrary* — we only use the fact that such an ordering *exists*.

If $s \in S$, we have
$$s = \sum r_i x_i$$
— a unique finite sum, with $r_i \in R$. Define the *leading term*, $\lambda(s) = r_\alpha x_\alpha$ — the *highest* ordered term in the sum. Let $Y \subset X$ be the set of basis elements that occur in these leading terms of elements of S. For each $y \in Y$, there is at least one $s \in S$ whose leading term is $r \cdot y$. Let

(6.3.4) $\quad I(y) = \{0\} \cup \{r \in R | \lambda(s) = r \cdot y \text{ for some } s \in S\}$

It is not hard to see that
$$I(y) \subset R$$
is an ideal — since $\lambda(r's) = r'\lambda(s)$. Since R is a PID, it follows that $I(y) = (r_y)$, and there exists a set of elements $S_y \subset S$ whose leading terms are $r_y y$. The Axiom of Choice (see theorem 14.2.11 on page 465) implies that we can select *one* such element for each $y \in Y$ — i.e. we can define a function
$$f: Y \to S$$
where $f(y)$ is some (arbitrarily selected) element of S_y. We claim that the set
$$B = \{f(y) | y \in Y\}$$
is a basis for S. We must demonstrate two things:

(1) *the set B is linearly independent.* This follows immediately from the fact that its leading terms are multiples of *distinct* basis elements of A. Suppose we have a linear combination
$$\sum_{i=1}^{n} r_i y_i + H = 0$$
where H is the *non*-leading terms. Of those leading terms, one is ordered the *highest* — say $r_\alpha y_\alpha$ and it can *never* be canceled out by any other term. It follows that $r_\alpha = 0$ and a similar argument applied to the remaining terms eventually leads to the conclusion that *all* of the coefficients are 0.

(2) *the set B generates S.* Suppose *not* and suppose $g \in S \setminus \langle B \rangle$ has the property that $\lambda(g) = ry$ for $r \in R$ and $y \in Y$ is *minimal* (in the ordering on X — axiom 14.2.9 on page 465). Then $r \in I(g) = (r_g)$ so that $r = r' \cdot r_g$ and $r' \cdot f(y) \in \langle B \rangle$ also has leading term ry. It follows that
$$g - r' \cdot f(y) \in S \setminus \langle B \rangle$$
has a leading term *strictly lower* than that of g — a contradiction.

□

We also get a form of the Chinese Remainder Theorem:

PROPOSITION 6.3.21. *Let $\{r_1,\ldots,r_k\}$ be a set of elements of R such that $\gcd(r_i,r_j) = 1$ (see Proposition 5.3.6 on page 118) whenever $i \neq j$. Then*

$$\frac{R}{(r_1\cdots r_k)} \cong \bigoplus_{i=1}^{k} \frac{R}{(r_i)}$$

PROOF. The hypotheses imply that

$$(r_i) + (r_j) = R$$

whenever $i \neq j$ and

$$(r_1\cdots r_k) = \bigcap_{i=1}^{k}(r_i)$$

so the conclusion follows from exercise 12 on page 116. □

LEMMA 6.3.22. *Let $r,s \in R$ with $t = \gcd(r,s)$. Let $\alpha, \beta \in R$ be such that $\alpha r + \beta s = t$. Then the matrix*

$$A = \begin{bmatrix} \alpha & \beta \\ -s/t & r/t \end{bmatrix}$$

has inverse

$$\begin{bmatrix} r/t & -\beta \\ s/t & \alpha \end{bmatrix}$$

and

$$A\begin{bmatrix} r \\ s \end{bmatrix} = \begin{bmatrix} t \\ 0 \end{bmatrix}$$

If A^t is the transpose of A, then

$$\begin{bmatrix} r & s \end{bmatrix} A^t = \begin{bmatrix} t & 0 \end{bmatrix}$$

REMARK. The existence of α and β follows from the fact that $(r,s) = (t)$.

PROOF. Straightforward computation. □

THEOREM 6.3.23. *If M is a finitely generated R-module then*

(6.3.5) $$M \cong F \oplus \frac{R}{(g_1)} \oplus \cdots \oplus \frac{R}{(g_k)}$$

where F is free and $g_1 \mid g_2 \mid \cdots \mid g_k$. The free submodule, F, and the $R/(g_i)$ are uniquely determined by M and are called its invariant factors.

REMARK. The proof is very similar to that of theorem 4.6.8 on page 62 except that we do not have the Euclidean algorithm to simplify a matrix. Lemma 6.3.22 takes its place.

PROOF. Propositions 6.3.10 on page 225 and 6.3.19 on page 230 imply that

$$M \cong \frac{G}{K}$$

where G and K are both free modules of respective ranks n and $m \leq n$. If $\{x_1, \ldots, x_m\}$ is a free basis for K and $\{y_1, \ldots, y_n\}$ is one for G then

$$y_i = \sum_{j=1}^n a_{i,j} x_j$$

the $m \times n$ matrix $A = [a_{i,j}]$ expresses how K fits in G and the quotient. We consider ways of "simplifying" this matrix in such a way that the quotient G/K remains unchanged. Multiplication on the right by an invertible matrix corresponds to a change of basis for K and on the left a change of basis for G. It follows that the quotient is unchanged. As in theorem 4.6.8 on page 62, we do row-reduction, and then column-reduction.

We'll call this part of our algorithm, Diagonalization:

This consists of two phases: *row-reduction* and *column-reduction*.

By permuting rows and columns, if necessary, assume $a_{1,1} \neq 0$. If this is impossible, we have a matrix of zeros and there is no work to be done.

We will call the $(1,1)$ position in the matrix, the *pivot*.

Now we scan down column 1.

(1) If $a_{1,1} | a_{k,1}$ with $k > 1$, we subtract $(a_{1,1}/a_{k,1}) \times$ row 1 from row k, replacing $a_{k,1}$ by 0.
(2) If $a_{1,1} \nmid a_{k,1}$, perform *two* steps:
    (a) swap row k with row 2 so $a_{k,1}$ becomes 0 and $a_{2,1}$ becomes the original value of $a_{k,1}$,
    (b) apply lemma 6.3.22 on the preceding page and left-multiply the matrix A by a matrix whose first two rows and columns are

(6.3.6)
$$\begin{bmatrix} \alpha & \beta \\ -a_{2,1}/\delta & a_{1,1}/\delta \end{bmatrix}$$

and whose remaining rows are the identity matrix. Here $\delta = \gcd(a_{1,1}, a_{2,1})$, α and β are such that $\alpha a_{1,1} + \beta a_{2,1} = \delta$. Since this matrix is invertible, it represents a basis change of K and has no effect on the quotient.

The effect is[8]:

$$a_{1,1} \to \delta$$
$$a_{2,1} \to 0$$
$$a_{k,1} \to 0$$

Repeating these operations for all lower rows ultimately clears *all* elements of column 1 below $a_{1,1}$ to 0. This completes row-reduction for column 1.

Now we do *column*-reduction on row 1, ultimately clearing out all elements to the *right* of $a_{1,1}$ to 0. Here, we *right*-multiply the matrix by the transpose of a matrix like that in equation 6.3.6.

After these steps, our matrix looks like

$$\begin{bmatrix} \bar{a}_{1,1} & 0 \\ 0 & \bar{M} \end{bmatrix}$$

---

[8]Where the symbol '$\to$' means "is replaced by."

where $\bar{M}$ is an $(m-1) \times (n-1)$ matrix.

After recursively carrying out our algorithm above on this and the smaller sub-matrices that result, we get the diagonal form:

$$\begin{bmatrix} \bar{a}_{1,1} & 0 & 0 & \cdots & 0 \\ 0 & \bar{a}_{2,2} & 0 & \cdots & 0 \\ \vdots & \vdots & \ddots & \cdots & \vdots \\ 0 & 0 & 0 & \cdots & 0 \end{bmatrix}$$

where *all* off-diagonal entries are 0 and we have $n - m$ columns of 0's on the right. This matrix means that the new bases for K and F satisfy

$$y_i = \bar{a}_{i,i} \cdot x_i$$

for $1 \leq i \leq m$. This completes the phase of the algorithm called diagonalization.

Now we may run into the problem $\bar{a}_{i,i} \nmid \bar{a}_{i+1,i+1}$. This is easily resolved: just add column $i+1$ to column i to give

$$\begin{bmatrix} \bar{a}_{i,i} & 0 \\ \bar{a}_{i+1,i+1} & \bar{a}_{i+1,i+1} \end{bmatrix}$$

Left multiplication by

$$\begin{bmatrix} \alpha & \beta \\ -\bar{a}_{i+1,i+1}/\delta & \bar{a}_{i,i}/\delta \end{bmatrix}$$

where $\alpha \bar{a}_{i,i} + \beta \bar{a}_{i+1,i+1} = \delta = \gcd(\bar{a}_{i,i}, \bar{a}_{i+1,i+1})$ (as in lemma 6.3.22 on page 232), gives

$$\begin{bmatrix} \delta & \beta \bar{a}_{i+1,i+1} \\ 0 & \bar{a}_{i,i}\bar{a}_{i+1,i+1}/\delta \end{bmatrix}$$

At this point, we subtract $(\beta \bar{a}_{i+1,i+1}/\delta) \times$ column i from column $i+1$ to get

$$\begin{bmatrix} \delta & 0 \\ 0 & \bar{a}_{i,i}\bar{a}_{i+1,i+1}/\delta \end{bmatrix}$$

and we have $\delta \mid (\bar{a}_{i,i}\bar{a}_{i+1,i+1}/\delta) = \text{lcm}(\bar{a}_{i,i}, \bar{a}_{i+1,i+1})$. Notice that the product of these diagonal entries is the same as before. On the other hand, the number of distinct prime factors in δ must be strictly less than those in $\bar{a}_{i,i}$ and $\bar{a}_{i+1,i+1}$ (equality only occurs if $\bar{a}_{i,i}$ and $\bar{a}_{i+1,i+1}$ have the *same* prime factors).

Note that this operation shifts prime factors to the right and down. After a finite number of steps, $\bar{a}_{1,1}$ must *stabilize* (i.e. remain constant over iterations of this process) since it has a finite number of prime factors. One can say the same of $\bar{a}_{2,2}$ as well. Eventually the entire list of elements

$$\{\bar{a}_{1,1}, \ldots, \bar{a}_{m,m}\}$$

must stabilize, finally satisfying the condition

$$\bar{a}_{1,1} \mid \cdots \mid \bar{a}_{m,m}$$

▷ each *nonzero* entry $a_{i,i}$ results in a direct summand $R/(a_{i,i})$, if $a_{i,i}$ is not a unit, and 0 otherwise.

▷ each column of *zeros* contributes a direct summand of R to F in equation 6.3.5 on page 232.

□

EXERCISES.

3. In theorem 6.3.23 on page 232 why are F and the invariant factors uniquely determined by M?

4. If R is a PID and A is an $m \times n$ matrix over R, there exists an invertible $m \times m$ matrix, S, and an invertible $n \times n$ matrix, T, such that

$$SAT$$

is a diagonal matrix with the property that the diagonal entries $\{d_i\}$ satisfy $d_i \mid d_{i+1}$ for all i. This is called the Smith Normal Form of A and the proof of theorem 6.3.23 on page 232 gives an algorithm for computing it. The matrices S and T are computed by keeping careful track of the steps in the algorithm (i.e., adding a row to another is multiplication by a suitable elementary matrix). Find the Smith normal form of the matrix (over $\mathbb{Z}$):

$$\begin{bmatrix} 2 & 4 & 4 \\ -6 & 6 & 12 \\ 10 & -4 & -16 \end{bmatrix}$$

6.3.3. Jordan Canonical Form. In this section, we will apply a knowledge of rings and modules to find a standard form for matrices. It is well-known that not all matrices can be diagonalized — for instance

$$B = \begin{bmatrix} 1 & 1 \\ 0 & 1 \end{bmatrix}$$

has a single eigenvalue of 1 and eigenvector

$$\begin{bmatrix} 1 \\ 0 \end{bmatrix}$$

which *doesn't* span $\mathbb{C}^2$. All matrices can be put into Jordan Canonical Form, which is as close as one can get to diagonalization.

Marie Ennemond Camille Jordan (1838 – 1922) was a French mathematician known for (besides Jordan Canonical Form) the Jordan Curve Theorem and the Jordan–Hölder theorem (see theorem 4.9.5 on page 85), and his influential series of books, *Cours d'analyse*.

PROPOSITION 6.3.24. *Let V be an n-dimensional vector-space over an $\mathbb{C}$. If $A\colon V \to V$ is a linear transformation, we can define an action of $\mathbb{C}[X]$ on V as in example 6.3.17 on page 229:*
$$X \cdot v = Av$$
for all $v \in V$. Then, as $\mathbb{C}[X]$-modules

(6.3.7)
$$V \cong \frac{\mathbb{C}[X]}{(p_1(X))} \oplus \cdots \oplus \frac{\mathbb{C}[X]}{(p_k(X))}$$

where $p_1(X) \mid \cdots \mid p_k(X)$ where $p_k(X)$ is the minimal polynomial of V (see example 6.3.17 on page 229),
$$\sum_{i=1}^{k} \deg p_i(X) = n$$
and
$$\prod_{i=1}^{k} p_i(X) = \chi_A(X)$$
the characteristic polynomial of A.

PROOF. Equation 6.3.7 follows immediately from theorem 6.3.23 on page 232 since $\mathbb{C}[X]$ is a PID (see corollary 5.3.10 on page 119). That $p_k(X)$ is the minimal polynomial of V follows from the fact that it is the lowest-degree polynomial that annihilates the right side of equation 6.3.7. The statement regarding dimensions follows from
$$\dim \frac{\mathbb{C}[X]}{(p(X))} = \deg p(X)$$
To prove the final statement, note that each of the summands
$$\frac{\mathbb{C}[X]}{(p_i(X))}$$
represents an A-invariant subspace of V so there exists a basis of V for which A is a block-matrix
$$A = \begin{bmatrix} A_1 & 0 & \cdots & 0 \\ 0 & \ddots & & 0 \\ \vdots & & \ddots & \vdots \\ 0 & \cdots & 0 & A_k \end{bmatrix}$$
where the A_i are the restrictions of A to each of these invariant subspaces (in this basis!). We claim that, in the subspace
$$\frac{\mathbb{C}[X]}{(p_i(X))}$$
the characteristic polynomial of A_i is just $p_i(X)$, same as the minimum polynomial. This is because they have the same degree, and one must divide the other. The conclusion follows from exercise 16 on page 201. □

COROLLARY 6.3.25. *Under the assumptions of proposition 6.3.24 on the preceding page we have an isomorphism of $\mathbb{C}[X]$-modules*

$$V \cong \frac{\mathbb{C}[X]}{(X - \lambda_1)^{\alpha_1}} \oplus \cdots \oplus \frac{\mathbb{C}[X]}{(X - \lambda_m)^{\alpha_m}}$$

where $\{\lambda_1, \ldots, \lambda_m\}$ are the eigenvalues of A and

$$\sum_{i=1}^{m} \alpha_i = n$$

Each summand

$$\frac{\mathbb{C}[X]}{(X - \lambda_i)^{\alpha_i}}$$

is isomorphic to an A-invariant subspace, V_i, of V and there exists a basis of V in which A is a block matrix

$$A = \begin{bmatrix} A_1 & 0 & \cdots & 0 \\ 0 & \ddots & & 0 \\ \vdots & & \ddots & \vdots \\ 0 & \cdots & 0 & A_m \end{bmatrix}$$

where

$$A_i = A|V_i \subset V$$

PROOF. Simply factor the $\{p_j(X)\}$ in proposition 6.3.24 on the facing page into powers of linear factors (possible in $\mathbb{C}$). These factors are pairwise relatively prime to each other so the conclusion follows from proposition 6.3.21 on page 232. □

Our final conclusion is

THEOREM 6.3.26. *Let V be an n-dimensional vector-space over an $\mathbb{C}$. If $A: V \to V$ is a linear transformation, there exists a basis for V such that A is a block-matrix*

$$A = \begin{bmatrix} J_{\alpha_1}(\lambda_1) & 0 & \cdots & 0 \\ 0 & \ddots & & 0 \\ \vdots & & \ddots & \vdots \\ 0 & \cdots & 0 & J_{\alpha_m}(\lambda_m) \end{bmatrix}$$

where the block

(6.3.8)
$$J_\alpha(\lambda) = \begin{bmatrix} \lambda & 0 & 0 & \cdots & 0 \\ 1 & \lambda & \ddots & \ddots & \vdots \\ 0 & 1 & \ddots & \ddots & 0 \\ \vdots & \cdots & \ddots & \lambda & 0 \\ 0 & \cdots & 0 & 1 & \lambda \end{bmatrix}$$

is an $\alpha \times \alpha$ matrix with λ on the main diagonal, 1's in the diagonal below it and zeros elsewhere. It is called a Jordan Block.

REMARK. Note that $J_1(\lambda) = [\lambda]$. If all the eigenvalues are distinct, the Jordan blocks are all of size 1, and we get the usual *diagonalized* form of the matrix,

This representation is called the *Jordan Canonical Form* of the matrix A. The form given here is the *transpose* of the usual Jordan block, due to our applying the matrix A on the left.

PROOF. We need to show that
$$A_i : V_i = \frac{\mathbb{C}[X]}{(X-\lambda_i)^{\alpha_i}} \to \frac{\mathbb{C}[X]}{(X-\lambda_i)^{\alpha_i}}$$
in corollary 6.3.25 on the preceding page can be put in the form $J_{\alpha_i}(\lambda_i)$. We claim that the image of
$$\{e_1 = 1, e_2 = X - \lambda_i, \ldots, e_{\alpha_i} = (X-\lambda_i)^{\alpha_i-1}\}$$
in V_i is a basis. The $\{e_j\}$ must be linearly independent since a linear relation would imply that a polynomial of degree $\alpha_i - 1$ could annihilate V_i — contradicting the idea that $(X - \lambda_i)^{\alpha_i}$ is its minimal polynomial. The e_i also span V_i since all linear combinations of the X^i can be written in terms of them. The matrix A_i acts like multiplication by X in the quotient
$$\frac{\mathbb{C}[X]}{(X-\lambda_i)^{\alpha_i}}$$
and it is not hard to see that $X \cdot e_j = \lambda_i e_j + e_{j+1}$ for $j = 1, \ldots, \alpha_i - 1$ and $X \cdot e_{\alpha_i} = \lambda_i e_{\alpha_i}$. Written as a matrix, this is precisely equation 6.3.8 on the previous page. □

We will do some computations. Let
$$A = \begin{bmatrix} -5 & -1 & -2 & -1 & -4 \\ -1 & 3 & 2 & -1 & 0 \\ 21 & 2 & 5 & 4 & 11 \\ 32 & 4 & 8 & 7 & 18 \\ 0 & 0 & 0 & 0 & 1 \end{bmatrix}$$

The characteristic polynomial is
$$\begin{aligned} \chi_A(\lambda) &= \det(\lambda I - A) \\ &= \lambda^5 - 11\lambda^4 + 47\lambda^3 - 97\lambda^2 + 96\lambda - 36 \\ &= (\lambda - 1) \cdot (\lambda - 3)^2 \cdot (\lambda - 2)^2 \end{aligned}$$

So its eigenvalues are 1, 2, and 3.

Eigenvalue 1. We compute the nullspace of $A - I$. This turns out to be the span of
$$v_1 = \begin{bmatrix} 1 \\ -3 \\ 2 \\ -3 \\ -1 \end{bmatrix}$$

Eigenvalue 2. The nullspace of $A - 2I$ is the span of

$$\begin{bmatrix} 1 \\ -1 \\ -1 \\ -4 \\ 0 \end{bmatrix}$$

Something odd happens here: the nullspace of $(A - 2I)^2$ is *two-dimensional*:

$$\begin{bmatrix} 1 \\ 0 \\ -2 \\ -4 \\ 0 \end{bmatrix}, \begin{bmatrix} 0 \\ 1 \\ -1 \\ 0 \\ 0 \end{bmatrix}$$

This is how we detect Jordan blocks: the nullspace of $(A - \lambda I)^n$ expands with increasing n — up to the dimension of the Jordan block. If we compute the nullspace of $(A - 2I)^3$ we get the same two-dimensional subspace. This means the Jordan block for the eigenvalue 2 is two-dimensional. For a basis of this, we'll take

$$v_2 = \begin{bmatrix} 0 \\ 1 \\ -1 \\ 0 \\ 0 \end{bmatrix}$$

and

$$v_3 = (A - 2I)v_2 = \begin{bmatrix} 1 \\ -1 \\ -1 \\ -4 \\ 0 \end{bmatrix}$$

Eigenvalue 3: The nullspace of $A - 3I$ is the span of

$$\begin{bmatrix} 1 \\ -1 \\ -3/2 \\ -4 \\ 0 \end{bmatrix}$$

and the nullspace of $(A - 3I)^2$ is the span of

$$\begin{bmatrix} 1 \\ 0 \\ -9/4 \\ -9/2 \\ 0 \end{bmatrix}, \begin{bmatrix} 0 \\ 1 \\ -3/4 \\ -3/2 \\ 0 \end{bmatrix}$$

so we have another Jordan block of size 2. As its basis, we'll take

$$v_4 = \begin{bmatrix} 0 \\ 1 \\ -3/4 \\ -3/2 \\ 0 \end{bmatrix}$$

and

$$v_5 = (A - 3I)v_4 = \begin{bmatrix} 1 \\ -1 \\ -3/2 \\ -4 \\ 0 \end{bmatrix}$$

If we assemble these basis vectors into an array (as columns), we get

$$P = \begin{bmatrix} 1 & 0 & 1 & 0 & 1 \\ -3 & 1 & -1 & 1 & -1 \\ 2 & -1 & -1 & -\frac{3}{4} & -\frac{3}{2} \\ -3 & 0 & -4 & -\frac{1}{2} & -4 \\ -1 & 0 & 0 & 0 & 0 \end{bmatrix}$$

and

$$P^{-1}AP = \begin{bmatrix} 1 & 0 & 0 & 0 & 0 \\ 0 & 2 & 0 & 0 & 0 \\ 0 & 1 & 2 & 0 & 0 \\ 0 & 0 & 0 & 3 & 0 \\ 0 & 0 & 0 & 1 & 3 \end{bmatrix}$$

which is the Jordan Canonical Form for A.

EXERCISES.

5. Show, using Jordan Canonical Form, that the "simple proof" of the Cayley-Hamilton Theorem in exercise 15 on page 201 can *always* be made to work.

6. If A is an $n \times n$ matrix, use the Jordan Canonical Form to show that

$$\text{Tr}(A) = \sum_{i=1}^{n} \lambda_i$$

(see definition 6.2.53 on page 196) where the $\{\lambda_i\}$ are the (not necessarily distinct) n eigenvalues of A.

7. If $\lambda \neq 0$, show that

$$J_\alpha(\lambda)^n = \begin{bmatrix} \lambda^n & 0 & 0 & \cdots & 0 \\ n\lambda^{n-1} & \lambda^n & \ddots & \ddots & \vdots \\ 0 & n\lambda^{n-1} & \ddots & \ddots & 0 \\ \vdots & \cdots & \ddots & \lambda^n & 0 \\ 0 & \cdots & 0 & n\lambda^{n-1} & \lambda^n \end{bmatrix}$$

where $J_\alpha(\lambda)$ is defined in equation 6.3.8 on page 237. It follows that powers of Jordan blocks are not diagonal unless the block size is 1.

8. Use exercise 7 on the preceding page to get a closed-form expression for the nth power of
$$A = \begin{bmatrix} -5 & -1 & -2 & -1 & -4 \\ -1 & 3 & 2 & -1 & 0 \\ 21 & 2 & 5 & 4 & 11 \\ 32 & 4 & 8 & 7 & 18 \\ 0 & 0 & 0 & 0 & 1 \end{bmatrix}$$

9. If M is a square matrix over $\mathbb{C}$ with $M^n = I$ for some positive integer n, use theorem 6.3.26 on page 237 to show that M is diagonalizable and that its eigenvalues have absolute value 1.

6.3.4. Prime decompositions. When a ring is *not* a PID, no simple classification for modules over it like theorem 6.3.23 on page 232 exists.

It is still possible to represent modules over a noetherian ring in terms of prime ideals of that ring.

DEFINITION 6.3.27. Let $m \in M$ be an element of a module over a ring, R. Then the *annihilator* of m, denoted $\text{ann}(m)$, is defined by
$$\text{ann}(m) = \{r \in R | r \cdot m = 0\}$$
The annihilator of M is defined by
$$\text{Ann}(M) = \{r \in R | \forall_{m \in M} \, r \cdot m = 0\} = \bigcap_{m \in M} \text{ann}(m)$$

A prime ideal $\mathfrak{p} \subset R$ is *associated* to M if it annihilates an element $m \in M$. The set of associated primes is denoted $\text{Assoc}(M)$.

REMARK. It is not hard to see that $\text{ann}(m), \text{Ann}(M) \subset R$ are always ideals. The following properties are also easy to verify:
(1) If $m \in M$ then $\text{ann}(m) = R$ if and only if $m = 0$.
(2) If $m \in M$ and $s \in R$, then $\text{ann}(m) \subseteq \text{ann}(s \cdot m)$.

It is not at all obvious that *any* associated primes exist, since $\text{Ann}(M)$ is usually not prime.

In studying the structure of ideals that annihilate elements of a module, we begin with:

LEMMA 6.3.28. *Let M be a finitely generated module over a noetherian ring R. If $\mathfrak{J} \subset R$ is an ideal that is maximal among ideals of R of the form $\mathrm{ann}(m)$ for $m \neq 0 \in R$, then $\mathfrak{J}$ is prime.*

REMARK. We will construct an ascending chain of ideals — the fact that R is noetherian implies that this chain has a maximal element.

This shows that, for a finitely generated module over a noetherian ring, at least *one* associated prime exists.

PROOF. Suppose $r, s \in R$ and $rs \in \mathfrak{J}$ but $s \notin \mathfrak{J}$. Then we will show that $r \in \mathfrak{J}$. We have
$$rs \cdot m = 0$$
but $s \cdot m \neq 0$. It follows that $(r) + \mathfrak{J}$ annihilates $sm \in M$. Since $\mathfrak{J}$ is maximal among ideals that annihilate elements of M, we have $(r) + \mathfrak{J} \subseteq \mathfrak{J}$ so $r \in \mathfrak{J}$. □

We go from this to show that *many* associated primes exist:

COROLLARY 6.3.29. *Let M be a finitely generated module over a noetherian ring, R. If $Z \subset R$ is the set of elements that annihilate nonzero elements of M, then*
$$Z = \bigcup_{\mathfrak{p} \in \mathrm{Assoc}(M)} \mathfrak{p}$$

PROOF. The definition of $\mathrm{Assoc}(M)$ implies that
$$\bigcup_{\mathfrak{p} \in \mathrm{Assoc}(M)} \mathfrak{p} \subset Z$$
If $x \in Z$, then $x \cdot m = 0$ for $m \in M$, $m \neq 0$. The submodule $R \cdot m \subset M$ has an associated prime, $\mathfrak{p} = \mathrm{ann}(y \cdot m)$, by lemma 6.3.28 on page 242, which is also an associated prime to M. Since $x \cdot m = 0$, it follows that $xy \cdot m = 0$ so that $x \in \mathrm{ann}(y \cdot m) = \mathfrak{p}$. □

Our main result classifying the structure of modules is

THEOREM 6.3.30. *Let M be a finitely generated module over a noetherian ring R. Then there exist a finite filtration*
$$0 = M_0 \subsetneq M_1 \subsetneq \cdots \subsetneq M_n = M$$
such that each
$$\frac{M_{i+1}}{M_i} \cong \frac{R}{\mathfrak{p}_i}$$
for prime ideals $\mathfrak{p}_i \subset R$.

REMARK. This sequence $\{M_i\}$ is called the *prime filtration* of M, and the primes $\{\mathfrak{p}_i\}$ that occur here are called the *prime factors* of M. Note that the $\{\mathfrak{p}_i\}$ might not all be distinct — a given prime ideal may occur more than once (see examples 6.3.31 on the facing page and 6.3.32 on the next page).

The associated primes occur among the primes that appear in this decomposition, so that $\mathrm{Assoc}(M)$ is *finite* (for finitely generated modules over a noetherian ring)

PROOF. Lemma 6.3.28 on the facing page states that the maximal ideal, $\mathfrak{p}$, that annihilates an element $m \in M$ is prime. Consider the submodule, M_0, generated by this $m \in M$, i.e., $R \cdot m \subset M$. We get a homomorphism of modules

$$R \to R \cdot m$$
$$r \mapsto r \cdot m$$

Since $\mathfrak{p}$ is in the kernel, we get a homomorphism of R-modules

$$g: R/\mathfrak{p} \to R \cdot m$$

Since $\mathfrak{p}$ is *maximal*, any element in the kernel of g must lie in $\mathfrak{p}$, so

$$\frac{R}{\mathfrak{p}} \cong R \cdot m = M_1$$

Now, form the quotient M/M_1 and carry out the same argument, forming a submodule $M_2' \subset M/M_1$ and its inverse image over the projection

$$M \to M/M_1$$

is M_2. We continue this process over and over until we get to 0. It must terminate after a finite number of steps because M is finitely generated over a noetherian ring (see definition 6.3.11 on page 226). □

EXAMPLE. If $M = \mathbb{Z}_{60}$ is regarded as a module over $\mathbb{Z}$, then a maximal ideal is of the form $(p) \subset \mathbb{Z}$ for some prime p. For instance, (2) annihilates $M_1 = 30 \cdot \mathbb{Z}_{60}$ and $M/M_1 = \mathbb{Z}_{30}$. The ideal (2) annihilates $M_2 = 15 \cdot \mathbb{Z}_{30}$ and we get $M_1/M_2 = \mathbb{Z}_{15}$. The ideal (3) annihilates $5 \cdot \mathbb{Z}_{15}$ and we are done (the final quotient is $\mathbb{Z}_5$). We can lift these modules up into M to get

$$0 \subset 30 \cdot \mathbb{Z}_{60} \subset 15 \cdot \mathbb{Z}_{60} \subset 5 \cdot \mathbb{Z}_{60} \subset \mathbb{Z}_{60}$$

with prime factors, $\mathbb{Z}_2, \mathbb{Z}_2, \mathbb{Z}_3$ and $\mathbb{Z}_5$, respectively.

EXAMPLE 6.3.31. Returning to example 6.3.17 on page 229, let $\mathfrak{p} \subset R$ be a prime ideal that annihilates an element $v \in V$. Then $\mathfrak{p} = (X - \lambda)$ and λ must be an *eigenvalue*. The element, v, annihilated by $\mathfrak{p}$ is the corresponding *eigenvector*. A simple induction shows that all of the prime ideals we get in the prime decomposition of V are of the form $(X - \lambda_i)$ where the λ_i run through the eigenvalues of A.

Here's a much more detailed example:

EXAMPLE 6.3.32. Let $R = \mathbb{C}[X, Y]/\mathfrak{a}$ where

(6.3.9) $$\mathfrak{a} = (Y^3, XY + Y, X^2 + Y^2 - 1) \subset \mathbb{C}[X, Y]$$

Lemma 5.2.9 on page 114 implies that the prime ideals of R are images under the projection

$$p: \mathbb{C}[X, Y] \to \mathbb{C}[X, Y]/\mathfrak{a} = R$$

of the prime ideals of $\mathbb{C}[X, Y]$ that contain $\mathfrak{a}$. We will skip ahead and use theorem 12.2.3 on page 420 to conclude that the prime ideals of $\mathbb{C}[X, Y]$ are either of the form $(f(X, Y))$ for some irreducible polynomial, f, or of the form $(X - \alpha, Y - \beta)$ for $\alpha, \beta \in \mathbb{C}$. We reject the possibility of a principal ideal because $f(X,Y)|Y^3$ implies that $f(X,Y) = Y$ but that does not divide $X^2 + Y^2 - 1$.

If $\mathfrak{a} \subset (X - \alpha, Y - \beta)$, then equations like

$$\begin{aligned} p_1(X,Y)(X-\alpha) + q_1(X,Y)(Y-\beta) &= Y^3 \\ p_2(X,Y)(X-\alpha) + q_2(X,Y)(Y-\beta) &= XY + Y \\ p_3(X,Y)(X-\alpha) + q_3(X,Y)(Y-\beta) &= X^2 + Y^2 - 1 \end{aligned}$$

must hold, for some $p_i, q_i \in k[X,Y]$. The top equation forces $\beta = 0$ (i.e., set $X = \alpha$). This also satisfies the second equation since we can set $p_2 = 0$ and $q_2 = X + 1$. The bottom equation becomes

$$p_3(X,Y)(X-\alpha) + q_3(X,Y)Y = X^2 + Y^2 - 1$$

We simplify this by setting $q_3 = Y$ and subtracting to get

$$p_3(X,Y)(X-\alpha) = X^2 - 1 = (X-1)(X+1)$$

which implies that $\alpha = \pm 1$. It follows that $\mathfrak{P}_1 = (X-1, Y)$ and $\mathfrak{P}_2 = (X+1, Y)$ are the only two ideals of the form $(X - \alpha, Y - \beta)$ that contain $\mathfrak{a}$.

Let $x, y \in R$ be the images of X and Y, respectively under the projection, p above — they clearly generate R as a ring. Then the prime ideals of R are $\mathfrak{p}_1 = (x-1, y)$ and $\mathfrak{p}_2 = (x+1, y)$. In addition, lemma 5.2.10 on page 114 implies that

$$\frac{R}{\mathfrak{p}_i} = \frac{\mathbb{C}[X,Y]}{\mathfrak{P}_i} = \mathbb{C}$$

for $i = 1, 2$.

Now we will compute a prime filtration of R as a module over itself. Since $y^2 \cdot \mathfrak{p}_2 = 0$, we regard $y^2 \cdot R$ as a candidate for R_1. Let us compute what happens to $\{1, x, y\}$ when we multiply by y^2

$$y^2 \cdot 1 = y^2$$

$$y^2 \cdot x = y \cdot (-y) = -y^2 \quad \text{because of the relation } xy + y = 0 \text{ in equation 6.3.9}$$

$$y^2 \cdot y = 0 \quad \text{because of the relation } y^3 = 0 \text{ in equation 6.3.9}$$

It follows that $y^2 \cdot R = \mathbb{C} \cdot y^2 = \mathbb{C}$ and we get an isomorphism

$$R_1 = y^2 \cdot R = \frac{R}{\mathfrak{p}_2} = \mathbb{C} \cdot y^2$$

Following the proof of theorem 6.3.30 on page 242, we form the quotient

$$(6.3.10) \quad R' = \frac{R}{R_1} = \frac{k[X,Y]}{(Y^3, Y^2, XY+Y, X^2+Y^2-1)}$$

$$= \frac{k[X,Y]}{(Y^2, XY+Y, X^2-1)}$$

— where we eliminated Y^3 since $(Y^3) \subset (Y^2)$ and eliminated the Y^2 term from $X^2 + Y^2 - 1$. Notice that $\mathfrak{p}_2 \cdot y = 0$. This suggests using $y \cdot R'$ as our second prime quotient. As before, we enumerate the effect of $y \cdot$ on the generators of R':

$$y \cdot 1 = y$$
$$y \cdot x = -y \quad \text{because of the relation } xy + y = 0$$
$$\text{in equation 6.3.10}$$
$$y \cdot y = 0 \quad \text{because of the relation } y^2 = 0$$
$$\text{in equation 6.3.10}$$

Again, we conclude that $y \cdot R' = \mathbb{C}$, generated by y, and get
$$\frac{R}{\mathfrak{p}} = y \cdot R' = \mathbb{C} \cdot y$$
and we take the inverse image of $y \cdot R'$ over the projection $R \to R/R_1 = R'$ to get $R_2 = y \cdot R$ and a *partial* prime filtration
$$0 \subsetneq y^2 \cdot R \subsetneq y \cdot R$$
Continuing, we form the quotient again

(6.3.11) $$R'' = \frac{R'}{\mathbb{C} \cdot y} = \frac{k[X,Y]}{(Y^2, Y, XY+Y, X^2-1)} = \frac{k[X,Y]}{(Y, X^2-1)}$$

and notice that $\mathfrak{p}_2 \cdot (x-1) = 0$. Computing $(x-1) \cdot R''$ gives

$$(x-1) \cdot 1 = (x-1)$$
$$(x-1) \cdot x = x^2 - x = -(x-1) \quad \text{because of the relation}$$
$$x^2 - 1 = 0 \text{ in equation 6.3.11}$$
$$(x-1) \cdot y = 0 \quad \text{because of the relation } y = 0$$
$$\text{in equation 6.3.11}$$

so $(x-1) \cdot R'' = \mathbb{C}$, generated by $x-1$ and
$$\frac{R}{\mathfrak{p}_2} = (x-1) \cdot R'' = \mathbb{C} \cdot (x-1)$$
and this lifts to $(x-1) \cdot R \subset R$. In the final step, we get
$$R''' = \frac{R''}{(x-1) \cdot R''} = \frac{k[X,Y]}{(Y, X^2-1, X-1)} = \frac{k[X,Y]}{(Y, X-1)} = \frac{R}{\mathfrak{p}_1}$$
(since $(X^2 - 1) \subset (X - 1)$), so we get our complete prime filtration
$$0 \subsetneq y^2 \cdot R \subsetneq y \cdot R \subsetneq (x-1) \cdot R \subsetneq R$$
The prime $\mathfrak{p}_2$ occurs three times, and the last factor involves the prime $\mathfrak{p}_1$.

Another interesting and useful result is called *Nakayama's Lemma* — it has a number of applications to algebraic geometry and other areas of algebra:

LEMMA 6.3.33. *Let M be a finitely-generated module over a commutative ring, R. If $\mathfrak{a} \subset R$ is an ideal with the property that*
$$\mathfrak{a} \cdot M = M$$
then there exists an element $r \in R$ such that $r \equiv 1 \pmod{\mathfrak{a}}$ and
$$r \cdot M = 0$$

REMARK. This result is named after Tadashi Nakayama who introduced it in [78]. Special cases of it had been discovered earlier by Krull, and Azumaya had published the general case in [7] before Nakayama's paper. The version for noncommutative rings is called the Krull-Azumaya Theorem.

PROOF. Let $m_1 \ldots, m_k$ denote the generators of M over R so

$$M = R \cdot m_1 + \cdots + R \cdot m_k$$

Since $\mathfrak{a} \cdot M = M$, we have

$$m_i = \sum_{j=1}^{k} A_{i,j} m_j$$

for some $k \times k$ matrix $A = [A_{i,j}]$ with entries in $\mathfrak{a}$. Subtracting gives

$$\sum_{j=1}^{n} (\delta_{i,j} - A_{i,j}) m_j = 0$$

where $\delta_{i,j}$ is the $(i,j)^{\text{th}}$ entry of the identity matrix, or

$$\delta_{i,j} = \begin{cases} 1 & \text{if } i = j \\ 0 & \text{otherwise} \end{cases}$$

or

$$(I - A) \begin{bmatrix} m_1 \\ \vdots \\ m_k \end{bmatrix} = 0$$

Cramer's Rule (theorem 6.2.29 on page 177) implies that

$$\det(I - A) m_i = C_i = 0$$

for all i, where C_i is the determinant of the matrix one gets by replacing the ith column by 0's. So $r \in R$ in the statement of the lemma is just $\det(I - A)$.

We claim that $\det(I - A) = 1 + a$ for some $a \in \mathfrak{a}$. The determinant of $I - A$ is what one gets from the characteristic polynomial $p_A(x)$ by setting $x = 1$. Since the characteristic polynomial is monic, one term is equal to 1 and the remaining terms are linear combinations of elements of $\mathfrak{a}$. □

Here's a consequence of Nakayama's lemma (a special case of the Krull Intersection Theorem — see [63]):

LEMMA 6.3.34. *Let $\mathfrak{m} \subset R$ be a maximal ideal of a noetherian ring, R or an arbitrary ideal of a noetherian domain. Then*

$$\bigcap_{j=1}^{\infty} \mathfrak{m}^j = (0)$$

PROOF. Call this infinite intersection $\mathfrak{b}$. Since R is noetherian, $\mathfrak{b}$ is finitely generated as a module over R. Since

$$\mathfrak{m} \cdot \mathfrak{b} = \mathfrak{b}$$

Nakayama's Lemma (6.3.33 on page 245) implies that $\mathfrak{b}$ is annihilated by an element $x \in R$ such that $x \equiv 1 \pmod{\mathfrak{m}}$ and such an element is a unit so $\mathfrak{b} = (0)$.

If R is an integral domain and $\mathfrak{m}$ is an arbitrary ideal, $x \neq 0$ and $x \cdot \mathfrak{b} = 0$ implies $\mathfrak{b} = 0$ since R has no zero-divisors. □

We also get a result for local rings:

COROLLARY 6.3.35. *Let R be a local ring (see definition 5.2.8 on page 113) with unique maximal ideal $\mathfrak{m} \subset R$. If M is an R-module with the property that*

$$\mathfrak{m} \cdot M = M$$

then $M = 0$.

PROOF. Nakayama's lemma implies that there exists $r \in R$ such that $r \equiv 1 \pmod{\mathfrak{m}}$ and $r \cdot M = 0$. Since R is a *local* ring, $\mathfrak{m}$ is the *only* maximal ideal and therefore equal to the intersection of *all* maximal ideals. Exercise 10 on page 116 implies that this r is a unit, i.e., has a multiplicative inverse, $s \in R$. Consequently

$$r \cdot M = 0 \implies s \cdot r \cdot M = 0 \implies 1 \cdot M = 0$$

and the conclusion follows. □

EXERCISES.

10. If
$$0 \to U \to V \to W \to 0$$
is a short exact sequence of vector-spaces, show that
$$\dim V = \dim U + \dim W$$

11. Prove this basic result in linear algebra:

    A vector-space over an infinite field cannot be a finite union of proper subspaces.

12. Give a counterexample to statement in exercise 11 if the field of definition is *finite*.

13. If P and M are R-modules, with P projective and
$$f \colon M \to P$$
is a surjective homomorphism, show that there exists a homomorphism $g \colon P \to M$ such that $f \circ g = 1 \colon P \to P$.

14. If
$$0 \to M_1 \to M_2 \to M_3 \to 0$$
is a short exact sequences of modules over a ring R, show that
$$\mathrm{Ann}(M_1) \cdot \mathrm{Ann}(M_3) \subset \mathrm{Ann}(M_2) \subset \mathrm{Ann}(M_1) \cap \mathrm{Ann}(M_3)$$

15. Let
$$0 \to U \xrightarrow{q} V \xrightarrow{p} W \to 0$$
be a short exact sequence of modules, and suppose there exists a homomorphism
$$h \colon W \to V$$
such that $p \circ h = 1 \colon W \to W$ (such short exact sequences are said to be *split* and h is called a *splitting map*). Show that there exists an isomorphism
$$V \cong U \oplus W$$

16. Let
$$0 \to U \xrightarrow{q} V \xrightarrow{p} P \to 0$$
be a short exact sequence of modules, and suppose that P is a projective module. Show that there exists an isomorphism
$$V \cong U \oplus P$$

6.4. Rings and modules of fractions

We begin by defining multiplicative sets:

DEFINITION 6.4.1. A *multiplicative set*, S, is a set of elements of a ring, R that:
(1) contains 1
(2) is closed under multiplication.

Our main application of multiplicative sets will be in constructing rings and modules of fractions in section 6.4. We need this concept here, to prove 12.2.8 on page 421.

EXAMPLE. For instance, if $\mathfrak{p} \subset R$ is a prime ideal, then $S = R \setminus \mathfrak{p}$ is a multiplicative set.

We have a kind of converse to this:

PROPOSITION 6.4.2. *If $S \subset R$ is a multiplicative set in a commutative ring with $S^{-1}R \neq 0$, then any ideal $\mathfrak{I} \subset R$ with $\mathfrak{I} \cap S = \emptyset$ that is maximal with respect to this property is prime.*

REMARK. "Maximal with respect to this property" means that, given any other ideal $\mathfrak{J}$ with $\mathfrak{J} \cap S = \emptyset$ and $\mathfrak{I} \subset \mathfrak{J}$, then $\mathfrak{I} = \mathfrak{J}$ — i.e. $\mathfrak{I}$ is not *properly* contained in $\mathfrak{J}$.

Such a maximal ideal always exists, by Zorn's Lemma (14.2.12 on page 465).

PROOF. Let $\mathfrak{J}$ be such a maximal ideal and assume it is not prime. The there exist $a, b \in R$ such that $ab \in \mathfrak{J}$ and $a \notin \mathfrak{J}$ and $b \notin \mathfrak{J}$. Then $(a + \mathfrak{J}) \cap S \neq \emptyset$ and $(b + \mathfrak{J}) \cap S \neq \emptyset$. Let $s_1 \in (a + \mathfrak{J}) \cap S$ and $s_2 \in (b + \mathfrak{J}) \cap S$. Then

$$s_1 s_2 \in (a + \mathfrak{J})(b + \mathfrak{J}) \subset ab + a\mathfrak{J} + b\mathfrak{J} + \mathfrak{J}^2 \subset \mathfrak{J}$$

which is a contradiction. □

Given a multiplicative set, we can define the corresponding ring of fractions:

DEFINITION 6.4.3. Let M be a module over a ring, R, and let $S \subset R$ be a multiplicative set. Then the module $S^{-1}M$ consists of pairs $(s, m) \in S \times M$, usually written m/s, subject to the relation

$$\frac{m_1}{s_1} \equiv \frac{m_2}{s_2}$$

if $u \cdot (s_2 \cdot m_1 - s_1 \cdot m_2) = 0$ for some $u \in S$ and $m_1, m_2 \in M$. We make $S^{-1}M$ a module by defining:

$$\frac{m_1}{s_1} + \frac{m_2}{s_2} = \frac{s_2 \cdot m_1 + s_1 \cdot m_2}{s_1 \cdot s_2}$$

$$r \cdot \frac{m_1}{s_1} = \frac{r \cdot m_1}{s_1}$$

for all $m_1, m_2 \in M$, $s_1, s_2 \in S$, and $r \in R$.

There exists a canonical homomorphism $f \colon M \to S^{-1}M$ that sends $m \in M$ to $m/1 \in S^{-1}M$.

If $M = R$ as a module over itself, then $S^{-1}R$ is a ring with multiplication defined by

$$\frac{r_1}{s_1} \cdot \frac{r_2}{s_2} = \frac{r_1 \cdot r_2}{s_1 \cdot s_2}$$

for all $r_1, r_2 \in R$ and $s_1, s_2 \in S$.

REMARK. The kernel of the canonical map $M \to S^{-1}M$ consists of elements of M that are annihilated by elements of S. If R is an integral domain, the map $R \to S^{-1}R$ is injective.

This construction has a universal property described in proposition 10.4.8 on page 355.

PROPOSITION 6.4.4. If a multiplicative set $S \subset R$ contains elements s_1, s_2 with the property that $s_1 s_2 = 0$, then $S^{-1}M = 0$, for any R-module, M.

PROOF. Suppose $m \in M$. We claim that

$$\frac{m}{1} = \frac{0}{1} \in S^{-1}M$$

In order for this to be true, we must have

$$s(m - 0) = 0$$

for some $s \in S$. But the fact that $s_1 s_2 = 0$ implies that $0 \in S$ and we can just set

$$0(m - 0) = 0$$

□

DEFINITION 6.4.5. Let R be a ring and let $h \in R$. Then $S_h = \{1, h, h^2, \ldots\}$ is a multiplicative subset of A and we define $R_h = S_h^{-1} R$.

REMARK. Every element of R_h can be written in the form a/h^m and
$$\frac{a}{h^m} = \frac{b}{h^n} \Leftrightarrow h^J(ah^n - bh^m) = 0$$
for some integer $J \geq 0$.

LEMMA 6.4.6. *For any ring A and $h \in A$, the map*
$$\sum a_i x^i \mapsto \sum \frac{a_i}{h^i}$$
defines an isomorphism
$$A[X]/(1 - hX) \to A_h$$

PROOF. If $h = 0$, both rings are zero, so assume $h \neq 0$. In the ring $A' = A[X]/(1 - hX)$, $1 = hX$ so h is a unit. Let $\alpha \colon A \to B$ be a homomorphism of rings that $\alpha(h)$ is a unit in B.

The homomorphism
$$\sum a_i X^i \mapsto \sum \alpha(a_i)\alpha(h)^{-i} \colon A[X] \to B$$
factors through A' because $1 - hX \mapsto 1 - \alpha(a)\alpha(h)^{-1} = 0$.

Because $\alpha(h)$ is a unit in B, this is the unique extension of α to A'. Therefore A' has the same universal property as A_h so the two are uniquely isomorphic.

When $h|h'$ so $h' = hg$, there is a canonical homomorphism
$$\frac{a}{b} \mapsto \frac{ag}{h'} \colon A_h \to A_{h'}$$
so the rings A_h form a direct system indexed by the set S. □

PROPOSITION 6.4.7. *Suppose A is a ring and $S \subset A$ is a multiplicative set. Then:*
  ▷ *If $S \subset A$ and $\mathfrak{b} \subset A$ is an ideal, then $S^{-1}\mathfrak{b}$ is an ideal in $S^{-1}A$.*
  ▷ *If $\mathfrak{b}$ contains any element of S, then $S^{-1}\mathfrak{b} = S^{-1}A$.*

It follows that

COROLLARY 6.4.8. *The ideals in $S^{-1}A$ are in a 1-1 correspondence with the ideals of A that are disjoint from S.*

DEFINITION 6.4.9. If $\mathfrak{p} \subset A$ is a prime ideal, then $S = A \setminus \mathfrak{p}$ is a multiplicative set. Define $A_\mathfrak{p} = S^{-1}A$.

REMARK. Since any ideal $\mathfrak{b} \not\subseteq \mathfrak{p}$ intersects S, it follows that $S^{-1}\mathfrak{p}$ is the *unique* maximal ideal in $S^{-1}A$.

$S^{-1}A$ is, therefore, a *local ring* (a ring with a unique maximal ideal). The word "local" is motivated by algebraic geometry (where it corresponds to the geometric concept of "local").

If a ring is not an integral domain, it has no field of fractions. Nevertheless we can define a "closest approximation" to it

DEFINITION 6.4.10. If R is a ring and S is the set of non-zero-divisors of R then
$$Q(R) = S^{-1}R$$
is called the *total quotient ring* of R.

REMARK. If R is an integral domain, $Q(R)$ is just the field of fractions of R.

EXERCISES.

1. Suppose R is a ring with a multiplicative set S and $a \cdot s = 0$ for $a \in R$ and $s \in S$. Show that
$$\frac{a}{1} = 0 \in S^{-1}R$$

2. Use the results of proposition 5.4.2 on page 122 to show that if R is noetherian, so is $S^{-1}R$ for any multiplicative set S.

3. If R and S are rings, show that $Q(R \times S) = Q(R) \times Q(S)$. Here, $R \times S$ is the ring of pairs (r,s) with pairwise addition and multiplication:
$$\begin{aligned}(r_1, s_1) + (r_2, s_2) &= (r_1 + r_2, s_1 + s_2) \\ (r_1, s_1) \cdot (r_2, s_2) &= (r_1 \cdot r_2, s_1 \cdot s_2)\end{aligned}$$

4. If R is a ring and M is an R-module, show that an element $m \in M$ goes to 0 in all localizations $M_\mathfrak{a}$, where $\mathfrak{a} \subset R$ runs over the maximal ideals of R if and only if $m = 0$.

5. If R is a ring and M is an R-module, show that $M_\mathfrak{a} = 0$ for all maximal ideals $\mathfrak{a} \subset R$ if and only if $M = 0$.

6. Suppose k is a field and $R = k[[X]]$ is the ring of power-series in X (see definition 5.1.7 on page 109). If $F = k((X))$, the field of fractions of R, show that every element of F can be written in the form
$$X^\alpha \cdot r$$
for some $\alpha \in \mathbb{Z}$ and some $r \in R$.

6.5. Integral extensions of rings

The theory of integral extensions of rings is crucial to algebraic number theory and algebraic geometry. It considers the question of "generalized integers:"

> If $\mathbb{Z} \subset \mathbb{Q}$ is the subring of integers, what subring, $R \subset \mathbb{Q}[\sqrt{2}]$, is like its "ring of integers"?

DEFINITION 6.5.1. If $A \subset K$ is the inclusion of an integral domain in a field, $x \in K$ will be called *integral* over A if it satisfies an equation
$$x^j + a_1 x^{j-1} + \cdots + a_k = 0 \in A$$
with the $a_i \in A$ (i.e., is a root of a *monic* polynomial).

REMARK. For instance, consider $\mathbb{Z} \subset \mathbb{Q}$. The only integral elements over $\mathbb{Z}$ are in $\mathbb{Z}$ itself.

In the case of $\mathbb{Z} \subset \mathbb{Q}(i)$, we get integral elements $n_1 + n_2 \cdot i$ where $n_1, n_2 \in \mathbb{Z}$ — the ring of Gaussian Integers.

PROPOSITION 6.5.2. *Let $R \subset S$ be integral domains. The following statements are equivalent*
(1) *An element $s \in S$ is integral over R*
(2) *$R[s]$ is a finitely-generated R-module (see definition 6.3.8 on page 225).*
(3) *$s \in T$ for some subring of S with $R \subseteq T \subseteq S$ and T is a finitely-generated R-module.*

REMARK. Note that being finitely generated as a module is very different from being finitely generated as a *ring or field*. For instance $R[X]$ is finitely generated as a *ring* over R but, as a *module*, it is
$$\bigoplus_{n=0}^{\infty} R \cdot X^n$$

PROOF. $1 \implies 2$. If s is integral over R, then
$$s^n + a_{n-1} s^{n-1} + \cdots + a_0 = 0$$
with the $a_i \in R$, so
$$s^n = -a_{n-1} s^{n-1} - \cdots - a_0$$
This means that $R[s]$ — the ring of polynomials in s will only have polynomials of degree $< n$, so $R[s]$ will be finitely generated as a *module* over R. Compare this argument to that used in proposition 7.2.2 on page 266.

$2 \implies 3$. Just set $T = R[s]$.

$3 \implies 1$. Suppose that $t_1, \ldots, t_n \in T$ is a set of generators of T as an R-module. Then
$$s t_i = \sum_{j=1}^{n} A_{i,j} t_j$$
for some $n \times n$ matrix A, so
$$\sum_{j=1}^{n} (\delta_{i,j} s - A_{i,j}) t_j = 0$$
where
$$\delta_{i,j} = \begin{cases} 1 & \text{if } i = j \\ 0 & \text{otherwise} \end{cases}$$
Cramer's Rule (6.2.29 on page 177) implies that
$$\det(sI - A) t_i = C_i = 0$$

for all i, where C_i is the determinant of the matrix one gets by replacing the i^{th}column by 0's. It follows that s is a root of the monic polynomial
$$\det(XI - A) = 0 \in R[X]$$
□

DEFINITION 6.5.3. If $R \subseteq S$ is an inclusion of integral domains and every element of S is integral over R, then S will be said to be integral over R.

REMARK. It is not hard to see that this property is preserved in *quotients*. If $\mathfrak{a} \subset S$ is an ideal then $S/\mathfrak{a}$ will be integral over $R/\mathfrak{a} \cap R$ because the monic polynomials satisfied by every element of S over R will map to monic polynomials in the quotients.

COROLLARY 6.5.4. *Let $f\colon R \to S$ be an integral extension of integral domains. If $\mathfrak{a} \subset R$ is a proper ideal, then so is $\mathfrak{a} \cdot S \subset S$.*

PROOF. We will prove the contrapositive: If $\mathfrak{a} \cdot S = S$, then $\mathfrak{a} = R$. The statement that $\mathfrak{a} \cdot S = S$ and Nakayama's Lemma 6.3.33 on page 245 imply that there exists $r \in R$ with $r \equiv 1 \pmod{\mathfrak{a}}$ with $r \cdot S = 0$. Since S is an integral domain, we must have $r = 0$ to $0 \equiv 1 \pmod{\mathfrak{a}}$ or $1 \in \mathfrak{a}$, so $\mathfrak{a} = R$. □

PROPOSITION 6.5.5. *Suppose $R \subseteq S$ are integral domains and let $s, t \in S$. Then:*

(1) *If s and t are integral over R, so are $t \pm s$ and st. Consequently integral elements over R form a ring.*
(2) *Let T be a commutative ring with $S \subseteq T$. If T is integral over S and S is integral over R, then T is integral over R.*

PROOF. If s and t are integral over R, let
$$R[s] = Rs_1 + \cdots + Rs_k$$
$$R[t] = Rt_1 + \cdots + Rt_\ell$$
as R-modules. Then
$$(6.5.1) \quad R[s,t] = Rs_1 + \cdots + Rs_k$$
$$+ Rt_1 + \cdots + Rt_\ell$$
$$+ \sum_{i=1, j=1}^{k,\ell} Rs_i t_j$$

which contains $s \pm t$ and st and is still a finitely generated R-module. This proves the first statement.

To prove the second statement, suppose $t \in T$ satisfies the monic polynomial
$$t^k + s_{k-1} t^{k-1} + \cdots + s_0 = 0$$
with $s_i \in S$, and $S[t]$ is a finitely generated S-module. Since S is integral over R, $R[s_i]$ is a finitely-generated R-module, and so is
$$R' = R[s_0, \ldots, s_{k-1}]$$

— equation 6.5.1 on the preceding page gives some ideal of how one could obtain a finite set of generators. The element t is also monic over R', so $R'[t]$ is a finitely-generated R'-module and

$$R[s_0, \ldots, s_{k-1}, t]$$

is a finitely-generated R-module. It follows that t is integral over R. □

This immediately implies:

COROLLARY 6.5.6. *If $R \subset S$ is an inclusion of integral domains and $\alpha \in S$ is integral over R, then $R \subset R[\alpha]$ is an integral extension of rings.*

Integral extensions of rings have interesting properties where fields are concerned:

PROPOSITION 6.5.7. *If R and S are integral domains, and S is an integral extension of R, then S is a field if and only if R is a field.*

PROOF. If R is a field, and $s \in S$ is a nonzero element, then $s \in R[s]$ is a finitely generated module over R — i.e., a vector space. Since S is an integral domain, multiplication by s induces a linear transformation of $R[s]$ whose kernel is 0. This means it is an isomorphism and has an inverse.

Conversely, if S is a field and $r \in R$. Then $r^{-1} \in S$ and it satisfies a monic polynomial over r:

$$r^{-n} + a_{n-1} r^{-(n-1)} + \cdots + a_0 = 0$$

with the $a_i \in R$. If we multiply this by r^{n-1}, we get

$$r^{-1} + a_{n-1} + \cdots + a_0 r^{n-1} = 0$$

□

DEFINITION 6.5.8. If K is a field containing an integral domain, R, the ring of elements of K that are integral over R will be called the *integral closure* of R in K. If K is the field of fractions of R and its integral closure is equal to R itself, R will be called *integrally closed* or *normal*.

REMARK. Proposition 6.5.5 on the previous page shows that the set of all integral elements over R form a ring — the integral closure of R.

PROPOSITION 6.5.9. *Every unique factorization domain is integrally closed.*

REMARK. This shows that $\mathbb{Z}$ is integrally closed in $\mathbb{Q}$. It is possible for R to be normal but not integrally closed in a field *larger* than its field of fractions. For instance $\mathbb{Z}$ is integrally closed in $\mathbb{Q}$ but not in $\mathbb{Q}[\sqrt{2}]$.

PROOF. Let a/b be integral over A, with $a, b \in A$. If $a/b \notin A$ then there is an irreducible element p that divides b but not a. As a/b is integral,

$$(a/b)^n + a_1 (a/b)^{n-1} + \cdots + a_n = 0, \text{ with } a_i \in A$$

Multiplying by b^n gives

$$a^n + a_1 a^{n-1} b + \cdots + a_n b^n = 0$$

Now p divides every term of this equation except the first. This is a contradiction! □

A simple induction, using theorem 5.6.7 on page 148 shows that

COROLLARY 6.5.10. *For any $n > 0$, the rings $\mathbb{Z}[X_1,\ldots,X_n]$, $F[[X_1,\ldots,X_n]]$, and $F[X_1,\ldots,X_n]$, where F is any field, have unique factorization and are integrally closed (see definition 6.5.8 on the preceding page) in their respective fields of fractions.*

REMARK. In most of these examples, the rings are not Euclidean.

Normality of a ring is a "local" property:

PROPOSITION 6.5.11. *An integral domain, R, is normal if and only if its localizations, $R_\mathfrak{p}$, at all primes are normal.*

PROOF. If R is normal and $S \subset R$ is any multiplicative set, the solution to exercise 1 on page 259 implies that the integral closure of $S^{-1}R$ is $S^{-1}R$. The converse follows from the fact that

$$(6.5.2) \qquad R = \bigcap_{\text{all primes } \mathfrak{p} \subset R} R_\mathfrak{p} \subset F$$

□

The following result gives a test for an element being integral over a ring

LEMMA 6.5.12. *Let R be an integral domain with field of fractions, F, let $F \subset H$ be a finite extension of fields, and let $\alpha \in H$ be integral over R. Then*

(1) all conjugates of α (in the algebraic closure of H) are integral over R,
(2) all coefficients of the characteristic polynomial, $\chi_\alpha(X) \in F[X]$, are integral over R,
(3) the norm $N_{H/F}(\alpha) \in F$ is integral over R.

REMARK. If R is normal, this implies that $\chi_\alpha(X) \in R[X]$ and provides a necessary and sufficient condition for α to be integral.

For instance, $a + b\sqrt{2} \in \mathbb{Q}[\sqrt{2}]$ is integral over $\mathbb{Z}$ if and only if

$$\chi_\alpha(X) = X^2 - 2aX + a^2 - 2b^2 \in \mathbb{Z}[X]$$

This implies that all elements $a + b\sqrt{2}$ with $a, b \in \mathbb{Z}$ are integral over $\mathbb{Z}$. Since $-2a \in \mathbb{Z}$, the only other possibility is for

$$a = \frac{2n+1}{2}$$

Plugging this into

$$\frac{(2n+1)^2}{4} - 2b^2 = m \in \mathbb{Z}$$

or

$$b^2 = \frac{(2n+1)^2 - 4m}{8}$$

which is never an integer much less a square, giving a contradiction.

PROOF. Let $p(X) \in R[X]$ be a monic polynomial such that $p(\alpha) = 0$. If α' is any conjugate of α, then the isomorphism

$$F[\alpha] \to F[\alpha']$$

that leaves F and $R \subset F$ fixed implies that $p(\alpha') = 0$ as well. The statement about the characteristic polynomial follows from the fact that its coefficients are elementary symmetric functions of the conjugates of α (see equation 5.5.12 on page 142), and the fact that the set of integral elements form a ring (see proposition 6.5.5 on page 253).

The final statement about the norm follows from lemma 7.5.11 on page 286. $\square$

We conclude this section with a result on the behavior of integral closures under algebraic field extensions. To prove it, we will need the concept of *bilinear form*:

DEFINITION 6.5.13. If V is a vector-space over a field, F, a *bilinear form* on V is a function

$$b \colon V \times V \to F$$

such that

(1) $b(c \cdot v_1, v_2) = b(v_1, c \cdot v_2) = c \cdot b(v_1, v_2)$ for all $v_1, v_2 \in V$ and $c \in F$.
(2) $b(v_1 + w, v_2) = b(v_1, v_2) + b(w, v_2)$ for all $v_1, v_2, w \in V$.
(3) $b(v_1, w + v_2) = b(v_1, w) + b(v_1, v_2)$ for all $v_1, v_2, w \in V$.

A bilinear form, $b(*, *)$, is called *symmetric* if $b(v_1, v_2) = b(v_2, v_1)$ for all $v_1, v_2 \in V$. If $v = \{v_1, \ldots, v_n\}$ is a basis for V, then the *associated matrix* of b is M defined by

$$M_{i,j} = b(v_i, v_j)$$

A bilinear form, $b(*, *)$, is said to be *degenerate* if there exists a nonzero vector $v \in V$ such that $b(v, w) = 0$ for all $w \in W$.

REMARK. If M is the associated matrix of b, then we can write b as

(6.5.3) $$b(u, v) = u^t M v$$

where u and v are vectors expanded in the basis used to compute M, and u^t is the transpose.

PROPOSITION 6.5.14. *Let V be a vector space over a field, F, equipped with a bilinear form*

$$b \colon V \times V \to F$$

Then b is nondegenerate if and only if its associated matrix is invertible.

PROOF. If M is invertible, then $u^t M \neq 0$ if $u \neq 0$ and we can define $v = (u^t M)^t = M^t u$ in which case

$$b(u, v) = \|u^t M\|^2 \neq 0$$

by equation 6.5.3. If M is not invertible, it sends some nonzero vector u to 0 and

$$b(u, v) = 0$$

for all $v \in V$. $\square$

We will be interested in nondegenerate bilinear forms because:

PROPOSITION 6.5.15. *Let V be a vector-space over a field F with basis $\{u_1, \ldots, u_n\}$ and suppose that*

$$b \colon V \times V \to F$$

is a nondegenerate bilinear form. Then there exists a dual basis $\{v^1, \ldots v^n\}$ of V such that

$$b(u_i, v^j) = \begin{cases} 1 & \text{if } i = j \\ 0 & \text{otherwise} \end{cases}$$

PROOF. If M is the associated matrix (with respect to the u-basis), simply define

$$v = M^{-1}u$$

The conclusion follows from equation 6.5.3 on the preceding page. □

Now we introduce a special bilinear form significant in studying field extensions:

DEFINITION 6.5.16. Let $F \subset H$ be a finite extension of fields. Then define the *trace form* of H over F via

$$b_{H/F}(h_1, h_2) = T_{H/F}(h_1 \cdot h_2)$$

(see section 7.3.1 on page 273 for information about $T_{H/F}$).

REMARK. Lemma 7.3.3 on page 274 implies that trace form is bilinear, and it is easy to see that it is also symmetric.

It is interesting to consider what happens if the trace form is degenerate. In this case, there exists $h \in H$ such that $b_{H/F}(h, h') = 0$ for all $h' \in H$, in particular, when $h' = h^{-1}$. It follows that $b_{H/F}(h, h^{-1}) = T_{H/F}(1) = 0$. But lemma 7.5.11 on page 286 implies that

$$T_{H/F}(1) = [H\colon F] \cdot 1 = 0 \in F$$

The only way this can happen is if F has finite characteristic, p, and $p \mid [H\colon F]$. This happens when H is an inseparable extension of F (see definition 7.2.11 on page 269).

LEMMA 6.5.17. *If $F \subset H$ is a separable extension of fields, then the trace form, $b_{H/F}$, is nondegenerate.*

REMARK. Note that "separable" implies "finite."

PROOF. Since the extension is separable, theorem 7.2.13 on page 270 implies that there exists a primitive element $\alpha \in H$ such that $H = F[\alpha]$. If $[H\colon F] = n$, then $\{1, \alpha, \ldots, \alpha^{n-1}\}$ are a basis for H over F.

We have $b_{H/F}(\alpha^i, \alpha^j) = T_{H/F}(\alpha^{i+j})$ and the associated matrix to $b_{H/F}$ is given by

$$M_{i,j} = T_{H/F}(\alpha^{i-1} \cdot \alpha^{j-1}) = T_{H/F}(\alpha^{i+j-2})$$

Let $\bar{H}$ be the algebraic closure of H and let $\alpha = \alpha_1, \ldots, \alpha_n$ be the conjugates of α in $\bar{H}$ (see definition 7.5.10 on page 285). Lemma 7.5.11 on page 286 implies that

$$T_{H/F}(\alpha^j) = \sum_{i=1}^n \alpha_i^j$$

Let V be the Vandermonde matrix $V(\alpha_1,\ldots,\alpha_n)$ — see exercise 12 on page 145. It is defined by
$$V_{i,j} = \alpha_i^{j-1}$$

Now, note that
$$\begin{aligned} M_{i,j} &= \sum_{\ell=1}^{n} \alpha_\ell^{i-1} \cdot \alpha_\ell^{j-1} \\ &= (V^T V)_{i,j} \end{aligned}$$

It follows that
$$\det M = (\det V)^2 = \prod_{1 \le i < j \le n} (\alpha_j - \alpha_i)^2$$

which is nonzero since the α_i are all distinct (because the field-extension was separable). □

Now we can prove our main result regarding integral extensions:

LEMMA 6.5.18. *Suppose that A is integrally closed domain whose field of fractions is F. Let $F \subset H$ be a separable extension of fields of degree n, and let B be the integral closure of A in H. Then there exists a basis $\{v_1,\ldots,v_n\}$ for H over F such that*
$$B \subseteq \{v^1,\ldots,v^n\} \cdot A$$
If A is noetherian, this implies that B is a finitely generated module over A.

REMARK. Roughly speaking, this says that a finite extension of fields induces a finite extension of integrally closed rings.

PROOF. Let $\{u_1,\ldots,u_n\}$ be a basis for H over F. Each of the u_i satisfies an algebraic equation
$$a_n u_i^n + \cdots + a_0 = 0$$
and multiplying by a_n^{n-1} gives us a monic polynomial in $(a_n u_i)$ so it is integral over A. It follows that $a_n u_i \in B$ and — without loss of generality — we may assume that the basis elements $u_i \in B$.

This does not prove the result: we have only shown that every element of H can be expressed in terms of B and F.

Let $\{v^1,\ldots,v^n\}$ be the dual basis defined by the trace form, via proposition 6.5.15 on the previous page. This exists because the trace form is nondegenerate, by lemma 6.5.17 on the preceding page.

If $x \in B$, let

(6.5.4) $$x = \sum_{i=1}^{n} c_i v^i$$

, where the $c_i \in F$. Note that $x \cdot u_i \in B$ since x and each of the u_i are in B. We claim that
$$b_{H/F}(x, u_i) = T_{H/F}(x \cdot u_i) \in A$$
This is because $x \cdot u_i$ satisfies a monic polynomial with coefficients in A — and $T_{H/F}(x \cdot u_i)$ is the negative of the coefficient of X^{n-1} (see definition 7.3.2 on page 273 and the remark following it). We use the properties

of the dual basis to conclude

$$T_{H/F}(x \cdot u_i) = T_{H/F}\left(\left(\sum_{j=1}^{n} c_j v^j\right) \cdot u_i\right)$$
$$= \sum_{j=1}^{n} c_j \cdot T_{H/F}(v^j \cdot u_i)$$
$$= c_i$$

So, in equation 6.5.4 on the preceding page, the c_i were elements of A all along and the conclusion follows. □

EXERCISES.

1. Let $R \subset T$ be an inclusion of rings and let $\bar{R}$ be its integral closure in T. Show, for any multiplicative set S, that $S^{-1}\bar{R}$ is the integral closure of $S^{-1}R$ in $S^{-1}T$.

2. Suppose R is an integral domain with field of fractions F and H is a finite extension of F. If $x \in H$ show that there exists an element $w \in R$ such that $r \cdot x$ is integral over R.

3. Let $R \subset T$ be an inclusion of rings with the property that T is a finitely-generated module over R. Now let $T \subset F$ where F is a field. Show that the integral closure of T in F is the same as the integral closure of R in F.

CHAPTER 7

Fields

"Cantor illustrated the concept of infinity for his students by telling them that there was once a man who had a hotel with an infinite number of rooms, and the hotel was fully occupied. Then one more guest arrived. So the owner moved the guest in room number 1 into room number 2; the guest in room number 2 into number 3; the guest in 3 into room 4, and so on. In that way room number 1 became vacant for the new guest.

What delights me about this story is that everyone involved, the guests and the owner, accept it as perfectly natural to carry out an infinite number of operations so that one guest can have peace and quiet in a room of his own. That is a great tribute to solitude."

— Smilla Qaavigaaq Jaspersen, in the novel *Smilla's Sense of Snow*, by Peter Høeg (see [57]).

7.1. Definitions

Field theory has a distinctly different "flavor" from ring theory due to the fact that the only ideal a field has is the *zero-ideal*. This means that homomorphisms of fields are always injective. Whereas we speak of homomorphisms of rings, we usually speak of *extensions* of fields and often denoted

$$\begin{array}{c} \Omega \\ | \\ | \\ F \end{array}$$

We begin by recalling definition 5.1.2 on page 108:

DEFINITION 7.1.1. A *field* is a commutative integral domain whose nonzero elements have multiplicative inverses. If F is a field, the set of nonzero elements is denoted $F^\times$ and is an *abelian group* (under multiplication).

If we define $m \cdot 1$ for $m \in \mathbb{Z}$, $m > 0$ as the sum of m copies of 1, then the smallest positive integral value of m such that $m \cdot 1 = 0$ is called the *characteristic* of the field. If $m \cdot 1 \neq 0$ for all values of m, the field is said to be of *characteristic* 0. The subfield generated by elements of the form $m \cdot 1$ is called the *prime field* of F.

REMARK. It follows that *every* field contains a subfield isomorphic to $\mathbb{Z}_p$ for some prime p, or $\mathbb{Q}$.

DEFINITION 7.1.2. *If $F \subset G$ is an extension of fields, then G is a vector space over F (see definition 6.2.1 on page 165). The dimension of G as a vector space (see proposition 6.2.8 on page 167) is called the degree of the extension, denoted $[G:F]$.*

REMARK. For instance, $\mathbb{R} \subset \mathbb{C}$ is an extension and $\{1, i\}$ is a basis for $\mathbb{C}$ as a vector-space over $\mathbb{R}$.

PROPOSITION 7.1.3. *If it is not 0, the characteristic of a field must be a prime number.*

PROOF. Suppose $0 < m$ is the characteristic of a field, $\mathbb{F}$, and $m = a \cdot b \in \mathbb{Z}$. Then $a, b \neq 0 \in \mathbb{F}$ and $a \cdot b = 0 \in \mathbb{F}$, which contradicts the fact that a field is an integral domain. □

DEFINITION 7.1.4. *If $\mathbb{F}$ is a field, an algebra over $\mathbb{F}$ is a vector space V over $\mathbb{F}$ that has a multiplication-operation*

$$V \times V \to V$$

The identity element $1 \in F$ defines an inclusion

$$\begin{aligned} \mathbb{F} &\to V \\ x &\mapsto x \cdot 1 \end{aligned}$$

REMARK. For instance the polynomials rings $\mathbb{F}[X_1, \ldots, X_n]$ are algebras over $\mathbb{F}$.

The complex numbers $\mathbb{C}$ are an algebra over $\mathbb{R}$.

Strictly speaking, algebras over fields are not required to be commutative or even associative. For instance, the quaternions and Cayley numbers are algebras over $\mathbb{R}$ — see section 9 on page 323.

An immediate consequence of definition 7.1.2 is:

PROPOSITION 7.1.5. *If F is a finite field, the number of elements in F must be p^n, where p is some prime number and n is a positive integer.*

PROOF. The characteristic of F must be some prime, p, by proposition 7.1.3. It follows that $\mathbb{Z}_p \subset F$ so F is a vector space over $\mathbb{Z}_p$. If $n = [F:\mathbb{Z}_p]$, is the dimension of that vector space, then F has p^n elements. □

Examples of fields are easy to find:
  ▷ The familiar examples: $\mathbb{Q}$, $\mathbb{R}$, and $\mathbb{C}$. They are fields of characteristic 0.
  ▷ If R is any integral domain and $S = R \setminus \{0\}$, then S is a multiplicative set in the sense of definition 6.4.1 on page 248 and $S^{-1}R$ is a field. This is called the field of fractions of R.
  ▷ If $\mathbb{F}$ is a field, the set of *rational functions* with coefficients in $\mathbb{F}$, denoted $\mathbb{F}(X)$, (with round rather than square brackets) is a field. This is the field of fractions of the polynomial ring, $\mathbb{F}[X]$.
  ▷ If p is a prime, $\mathbb{Z}_p$ is a field of characteristic p.

The following innocuous-looking result solved a great mystery of ancient Greece:

PROPOSITION 7.1.6. *Let $E \subset F$ and $F \subset G$ be finite extensions of fields. Then*
$$[G:E] = [G:F] \cdot [F:E]$$

PROOF. Let $\{x_1, \ldots, x_n\} \in G$ be a basis for it over F and let $\{y_1, \ldots, y_m\} \in F$ be a basis for it over E. So every element of G can be expressed as a linear combination of the x_i

(7.1.1) $$g = \sum_{i=1}^{n} f_i x_i$$

with the $f_i \in F$. Each of the f_i is given by

(7.1.2) $$f_i = \sum_{j=1}^{m} e_{i,j} y_j$$

which means that
$$g = \sum_{\substack{i=1,\ldots,n \\ j=1,\ldots,m}} e_{i,j} x_i y_j$$

which shows that the $n \cdot m$ elements $\{x_i \cdot y_j\}$ span G over F. To see that they are linearly independent, set $g = 0$ in equation 7.1.1 on page 263. The linear independence of the x_i implies that $f_i = 0$, $i = 1, \ldots, n$. These, and the linear independence of the y_j imply that $e_{i,j} = 0$ for $i = 1, \ldots, n$ and $j = 1, \ldots, m$ which proves the result. □

DEFINITION 7.1.7. If $F \subset G$ is an inclusion of fields, then $\alpha \in G$ is said to be *algebraic* over F if it is a root of a polynomial with coefficients in F. If $\alpha \in G$ is not algebraic, it is said to be *transcendental*.

The notation, $F(\alpha) \subset G$, represents the field of rational functions of α.

REMARK. For instance, if we think of
$$\mathbb{Q} \subset \mathbb{C}$$
then $\sqrt{2}$ is algebraic over $\mathbb{Q}$, but e is not (this is not obvious!).

In comparing $F(\alpha)$ with the ring $F[\alpha]$, it is not hard to see that:
(1) $F(\alpha)$ is the smallest subfield of G containing F and α.
(2) $F(\alpha)$ is the field of fractions of $F[\alpha]$.

PROPOSITION 7.1.8. *Let $F \subset G$ be an inclusion of fields and let $f: F[X] \to G$ be the unique homomorphism that sends X to $\alpha \in G$. Then α is algebraic over F if and only if $\ker f \neq 0$, in which case $\ker f = (p(X))$ and $p(X)$ is called the minimal polynomial of α.*

The minimal polynomial is always irreducible.

REMARK. The minimal polynomial is the lowest-degree polynomial such that $f(\alpha) = 0$. If $g(X)$ is any polynomial with the property that $g(\alpha) = 0$, then $f(X)|g(X)$. See example 7.3.1 on page 272 for techniques for computing it.

This result implies that $\alpha \in G$ is transcendental if and only if the homomorphism f is injective.

The numbers π and e are well-known to be transcendental — see [43].

PROOF. The kernel of f is just the polynomials that vanish when evaluated at α. This kernel is a principal ideal because $F[X]$ is a principal ideal domain — see corollary 5.3.10 on page 119.

If $f(X) = p(X) \cdot q(X)$ then
$$p(\alpha)q(\alpha) = 0$$
implies that $p(\alpha) = 0$ or $q(\alpha) = 0$. If p and q are of lower degree than f, it would contradict the minimality of $f(X)$. □

Consider
$$\mathbb{Q} \subset \mathbb{C}$$
and form the extension field
$$\mathbb{Q}(\sqrt{2})$$
which is the field of all possible rational functions
$$\frac{\sum_{i=1}^{m} p_i (\sqrt{2})^i}{\sum_{j=1}^{n} q_j (\sqrt{2})^j}$$
where the $p_i, q_j \in \mathbb{Q}$ — or the smallest subfield of $\mathbb{C}$ containing $\mathbb{Q}$ and $\sqrt{2}$.

Upon reflection, it becomes clear that we can always have $n, m \leq 1$ since $(\sqrt{2})^2 \in \mathbb{Q}$, so every element of $\mathbb{Q}(\sqrt{2})$ is really of the form
$$\frac{a + b\sqrt{2}}{c + d\sqrt{2}}$$
with $a, b, c, d \in \mathbb{Q}$.

We can even clear out the denominator because
$$\frac{a + b\sqrt{2}}{c + d\sqrt{2}} = \frac{a + b\sqrt{2}}{c + d\sqrt{2}} \cdot \frac{c - d\sqrt{d}}{c - d\sqrt{2}} = \frac{ac - 2bd + \sqrt{2}(bc - ad)}{c^2 - 2d^2}$$
$$= \frac{ac - 2bd}{c^2 - 2d^2} + \frac{bc - ad}{c^2 - 2d^2}\sqrt{2}$$

We have just proved that
$$\mathbb{Q}(\sqrt{2}) = \mathbb{Q}[\sqrt{2}]$$
This is no accident — it is true for *all* finite algebraic extensions:

LEMMA 7.1.9. *Let $F \subset G$ be an extension of fields and suppose $\alpha \in G$ is algebraic over F. Then*
$$F(\alpha) = F[\alpha]$$
and $[F(\alpha):F]$ is equal to the degree of the minimal polynomial of α.

REMARK. Note: algebraic extensions can be *infinite* (see exercise 3 on page 287).

PROOF. If α is algebraic over F, it has a minimal polynomial $p(X) \in F[X]$ (see definition 7.1.8 on the previous page) which is irreducible so the ideal $(p(X))$ is prime. Since the ring $F[X]$ is a principal ideal domain (see corollary 5.3.9 on page 119) the ideal $(p(X))$ is also *maximal* and
$$F[\alpha] = F[X]/(p(X))$$
is a field (see proposition 5.3.2 on page 117), so it is equal to $F(\alpha)$. □

EXAMPLE 7.1.10. Since $2^{1/3}$ is algebraic over $\mathbb{Q}$, we have $\mathbb{Q}[2^{1/3}] = \mathbb{Q}(2^{1/3})$, a field. The quantity $2^{1/3}$ is a root of the polynomial $X^3 - 2$, and Eisenstein's Criterion (see theorem 5.6.8 on page 149) shows that this is irreducible over $\mathbb{Q}$ (using $p = 2$), so it is the *minimal polynomial* of $2^{1/3}$.

We can solve famous unsolved problems from ancient Greece:

EXAMPLE 7.1.11. Speaking of $2^{1/3}$, one famous problem the ancient Greek geometers puzzled over is that of *doubling the cube* — using straight-edge and compass constructions. In other words, they wanted to construct $\sqrt[3]{2}$ via their geometric techniques. It can be shown that ancient Greek compass-and-straightedge techniques can construct
   (1) all integers
   (2) the square root of any number previously constructed (by drawing a suitable circle).
   (3) the sum, difference, product and quotient of any two numbers previously constructed.
Consequently, the numbers they constructed all lay in fields of the form
$$(7.1.3) \qquad F_n = \mathbb{Q}(\sqrt{\alpha_1})(\sqrt{\alpha_2}) \cdots (\sqrt{\alpha_n})$$
where each α_i is contained in the field to the left of it. Since the minimal polynomial of $\sqrt{\alpha_{i+1}}$ is $X^2 - \alpha_{i+1} \in F_i[X]$, lemma 7.1.9 on the preceding page implies that $[F_{i+1}:F_i] = 2$ and proposition 7.1.6 on page 263 implies that $[F_n:\mathbb{Q}] = 2^n$. But $[\mathbb{Q}(\sqrt[3]{2}):\mathbb{Q}] = 3$ and $3 \nmid 2^n$ for any n, so $\sqrt[3]{2} \notin F_n$ for any n.

So the problem of constructing $\sqrt[3]{2}$ is literally unsolvable by ancient Greek techniques.

EXERCISES.

1. Suppose $F \subset H$ is a finite extension of fields and $\alpha \in H$. If n is the degree of the minimum polynomial of α, show that $n \mid [H:F]$.

2. Show that the polynomial $X^3 + 3X + 3 \in \mathbb{Q}[X]$ is irreducible so that $H = \mathbb{Q}[X]/(X^3 + 3X + 3)$ is a field. If a generic element of H is written as $a + bX + cX^2$, compute the product
$$(a + bX + cX^2)(d + eX + fX^2)$$

7.2. Algebraic extensions of fields

DEFINITION 7.2.1. An extension of fields, $E \subset F$, is said to be *algebraic* if every element $x \in F$ is algebraic over E. If an extension is not algebraic, it is *transcendental*.

REMARK. For instance, $\mathbb{Q} \subset \mathbb{Q}(\sqrt{2})$ is algebraic and $\mathbb{Q} \subset \mathbb{R}$ and $\mathbb{F} \subset \mathbb{F}(X)$ are transcendental extensions, where $\mathbb{F}$ is any field.

PROPOSITION 7.2.2. *If $E \subset F$ is a finite field extension, then it is algebraic.*

PROOF. Suppose $[F:E] = n$ and let $x \in F$. Then the powers
$$\{1, x, x^2, \ldots, x^n\}$$
must be linearly dependent over E so we get a nontrivial algebraic equation
$$a_1 + a_1 x + \cdots + a_n x^n = 0$$
with $a_i \in E$. □

Extensions containing roots of polynomials always exist:

COROLLARY 7.2.3. *Let F be a field and let $f(X) \in F[X]$ be a polynomial. Then there exists an extension $F \subset \Omega$ such that Ω contains a root of f.*

PROOF. Factor f as
$$f(X) = p_1(X)^{\alpha_1} \cdots p_k(X)^{\alpha_k}$$
where the $p_i(X)$ are irreducible. This can be done (and is even unique) by corollary 5.3.7 on page 118. As in the proof of 7.1.9 on page 264, the quotient
$$E = F[X]/(p_1(X))$$
is a field containing F.

The image, α, of X under the quotient-mapping
$$F[X] \to F[X]/(p_1(X)) = E$$
has the property that $p_1(\alpha) = f(\alpha) = 0$. □

COROLLARY 7.2.4. *Let F be a field and let $f(X) \in F[X]$ be a polynomial. Then there exists an extension $F \subset \Omega$ such that*
$$f(X) = \prod_{k=1}^{\deg(f)} (X - \alpha_k) \in \Omega[X]$$

REMARK. This extension, Ω, is called a *splitting field* for $f(X)$. We can write
$$\Omega = F[\alpha_1, \ldots, \alpha_d]$$
where $d = \deg f$.

The solution to exercise 1 on page 287 shows that these splitting fields are *unique* up to isomorphism.

PROOF. This follows by an inductive application of corollary 7.2.3. We construct a field Ω_1 that contains a root, α, of $f(X)$. If $f(X)$ splits into linear factors in Ω_1, we are done. Otherwise, factor $f(X)$ as
$$f(X) = (X - \alpha)^k \cdot g(X) \in \Omega_1[X]$$
where $g(X)$ is relatively prime to $X - \alpha$, and construct an extension Ω_2 of Ω_1 that contains a root of $g(X)$. Eventually this process terminates with a field Ω that contains all of the roots of $f(X)$. □

The following statement seems clear, but it should be said:

COROLLARY 7.2.5. *If F is a field, a polynomial $p(X) \in F[X]$ of degree n has at most n roots in F.*

EXAMPLE 7.2.6. Consider the polynomial $X^3 - 1$ over $\mathbb{Q}$. It already factors as
$$X^3 - 1 = (X-1)(X^2 + X + 1)$$
and the second factor is irreducible by example 5.6.9 on page 150, so we construct the extension-field $\mathbb{Q}[X]/(X^2 + X + 1)$ and define α to be in the image of X under the projection
$$\mathbb{Q}[X] \to \mathbb{Q}[X]/(X^2 + X + 1) = G$$
$$X \mapsto \alpha$$
so that $\alpha \in G$ satisfies the identity $\alpha^2 + \alpha + 1 = 0$. Since α is a root of $X^2 + X + 1$, we have
$$(X - \alpha) \,|\, X^2 + X + 1$$
In fact the quotient is

$$
\begin{array}{r}
X + \alpha + 1 \\
X - \alpha \,\big)\, \overline{X^2 + X + 1 } \\
X^2 - \alpha X \\
\hline
(1+\alpha)X + 1 \\
(1+\alpha)X - \alpha(1+\alpha) \\
\hline
1 + \alpha + \alpha^2
\end{array}
$$

so we get the splitting in G
$$X^3 - 1 = (X - 1)(X - \alpha)(X + 1 + \alpha)$$

Since a polynomial splits into linear factors in its splitting field, one might expect the greatest common divisor of two polynomials to depend on the field in which one computes it. It is interesting that this does *not* happen:

PROPOSITION 7.2.7. *Let $F \subset \Omega$ be an inclusion of fields and let $f(X), g(X) \in F[X]$ be polynomials. Then*
$$g_F(X) = \gcd(f(X), g(X)) \in F[X]$$
is also their greatest common divisor in $\Omega[X]$.

REMARK. Since Ω could be the splitting field of $f(X)$ and $g(X)$, the greatest common divisor of these polynomials (up to units) is
$$\prod_{i=1}^{n}(X - \alpha_i)$$
where the α_i are *all* of the roots that $f(X)$ and $g(X)$ have in common. Somehow this product always defines an element of $F[X]$ (even though the α_i are *not* in F).

PROOF. Let us pass to a field $K \supset \Omega$ that is a splitting field for f and g. Suppose $f(X)$ and $g(X)$ have the following common roots in K:

$$\alpha_1, \ldots, \alpha_n$$

Then $g_F(X)$ also splits into linear factors and

(7.2.1) $$g_F(X) = \prod_{k=1}^{t}(X - \alpha_{j_k})$$

where $\{\alpha_{j_1}, \ldots, \alpha_{j_t}\}$ is, possibly, a subset of $\{\alpha_1, \ldots, \alpha_n\}$ such that the product in equation 7.2.1 lies in $F[X]$. If this product lies in $F[X]$, it is also in $\Omega[X]$, so the greatest common divisor calculated in this larger field will have these factors, at *least*. We conclude that

$$g_F(X) | g_\Omega(X)$$

where $g_\Omega(X)$ is the greatest common divisor calculated in $\Omega[X]$.

On the other hand, the Euclidean algorithm (proposition 5.3.5 on page 118) implies that there exist $a(X), b(X) \in F[X]$ such that

$$a(X) \cdot f(X) + b(X) \cdot g(X) = g_F(X)$$

so $g_\Omega(X) | g_F(X)$. □

There is an interesting result regarding repeated roots of a polynomial:

LEMMA 7.2.8. *Let F be a field and let $f(X) \in F[X]$ be a polynomial with a splitting field Ω. Then $f(X)$ has a repeated root in Ω if and only if*

$$f(X), f'(X)$$

have a common root. This occurs if and only if $\mathrm{Res}(f, f', X) = 0$ *(in the notation of definition 6.2.41 on page 188) which happens if and only if*

$$\gcd(f(X), f'(X)) \neq 1$$

REMARK. This is interesting because the criteria, $\mathrm{Res}(f, f', X) = 0$ or $\gcd(f(X), f'(X)) \neq 1$, make no direct reference to Ω.

Note that, in characteristic $p \neq 0$, the derivative of X^p is 0.

PROOF. The first statement follows by the chain-rule and the product-formula for derivatives. If

$$f(X) = (X - \alpha)^k g(X)$$

with $\alpha > 1$, then

$$f'(X) = k(X - \alpha)^{k-1} g(X) + (X - \alpha)^k g'(X)$$

which will share a root with $f(X)$ regardless of whether the characteristic of F divides k (for instance, the characteristic of F (and, therefore, Ω) might be p and k might be a multiple of p).

The second statement follows from proposition 6.2.42 on page 189 and the statement about the greatest common divisor follows from proposition 7.2.7 on the preceding page. □

This result tells us something important about irreducible polynomials:

LEMMA 7.2.9. *Let $f(X) \in F[X]$ be an irreducible polynomial of degree n with splitting field Ω, and suppose that F is of characteristic 0. Then, in the factorization*

$$f(X) = \prod_{i=1}^{n}(X - \alpha_i) \in \Omega[X]$$

the α_i are all distinct.

REMARK. This argument *fails* if the characteristic of F is $p \neq 0$. In this case, we can have an irreducible polynomial, $f(X^p)$, that has repeated roots.

PROOF. Since the characteristic of F is 0, and $f(X)$ is not constant, $f'(X) \neq 0$.

If $f(X)$ had a repeated factor, we would have

$$\gcd(f(X), f'(X)) = p(X) \neq 1$$

with $\deg p(X) < \deg f(X)$ since $\deg f'(X) < \deg f(X)$, which would contradict the irreducibility of $f(X)$. □

In characteristic p, it is possible to say *exactly* what may prevent an irreducible polynomial from having distinct roots:

PROPOSITION 7.2.10. *Let F be a field of characteristic p and suppose $f(X) \in F[X]$ is an irreducible nonconstant polynomial with repeated roots. Then there exists an irreducible polynomial $g(X) \in F[X]$ whose roots are all distinct such that*

$$f(X) = g(X^{p^t})$$

for some integer $t > 0$.

PROOF. If $f(X)$ has repeated roots, then it has roots in common with $f'(X)$. If $f'(X) \neq 0$, the greatest common divisor of $f(X)$ and $f'(X)$ would be a lower-degree polynomial that divides $f(X)$ — contradicting its irreducibility. It follows that $f'(X) = 0$, which is only possible if all of the exponents in $f(X)$ are multiples of p (since p is a prime and the coefficients of f are relatively prime to p). In this case,

$$g_1(X) = f(X^{1/p})$$

is a well-defined polynomial that is *still* irreducible: any nontrivial factorization of $g_1(X)$ implies one for $f(X)$. If $g_1'(X) = 0$, repeat this process. Since each iteration lowers degree by a factor of p, after a finite number of steps we arrive at an irreducible polynomial

$$g_t(X) = g(X) = f(X^{1/p^t})$$

with $g'(X) \neq 0$. □

DEFINITION 7.2.11. Let F be a field and $f(X) \in F[X]$ be a polynomial. Then $f(X)$ will be called *separable* if it factors into a product of *distinct* linear factors in a splitting field.

If $F \subset \Omega$ is an inclusion of fields and $\alpha \in \Omega$ is algebraic over F, then α will be called a *separable element* if its minimal polynomial is separable.

The field Ω will be called a *separable extension of F* if every element of Ω is separable over F.

REMARK. "Separable" = roots are *separated*. The whole question of separability is moot unless the fields in question have characteristic $p \neq 0$.

DEFINITION 7.2.12. A field F is said to be *perfect* if every finite extension of F is separable.

REMARK. This is equivalent to saying that all irreducible polynomial have distinct roots. Most of the fields that we have dealt with have been perfect. Perfect fields include:

  ▷ Any field of characteristic 0 (lemma 7.2.9 on the previous page).
  ▷ Finite fields (theorem 7.6.7 on page 289).
  ▷ Algebraically closed fields (definition 7.5.1 on page 283).

It is interesting that finite algebraic extensions of fields can always be generated by a single element:

THEOREM 7.2.13. *Let $F \subset H$ be an extension of infinite fields and suppose $\alpha_1, \ldots, \alpha_n \in H$ are algebraic over F. In addition, suppose $\alpha_2, \ldots, \alpha_n$ are separable over F.*

Then there exists an element $\beta \in H$ such that

$$F[\alpha_1, \ldots, \alpha_n] = F[\beta]$$

REMARK. The element β is called a *primitive element* and this result is often called the *primitive element theorem*.

PROOF. We will prove it for $n = 2$ — a simple induction proves the general case. We will show that $F[\alpha, \beta] = F[\gamma]$

Let α, β have minimal polynomials $f(X), g(X) \in F[X]$, respectively and let $\Omega \supset H$ be a splitting field for $f(X)$ and $g(X)$. Then $f(X)$ and $g(X)$ have roots

$$\alpha = \alpha_1, \ldots, \alpha_n$$
$$\beta = \beta_1, \ldots, \beta_m$$

respectively, with the β_i all distinct. For $j \neq 1$, the equation

$$\alpha_i + X\beta_j = \alpha_1 + X\beta_1 = \alpha + X\beta$$

has exactly one solution

$$x_{i,j} = \frac{\alpha_i - \alpha}{\beta - \beta_j}$$

If we choose $c \in F$ different from any of these elements (using the fact that F is infinite), we get

$$\alpha_i + c\beta_j \neq \alpha + c\beta$$

unless $i = j = 1$. We claim $\gamma = \alpha + c\beta$ will satisfy the hypotheses of this theorem.

The polynomials $g(X)$ and $h(X) = f(\gamma - cX) \in F[\gamma][X]$ have a β as a common root:

$$\begin{aligned} g(\beta) &= 0 \\ h(\beta) &= f(\gamma - c\beta) \\ &= f(\alpha) \\ &= 0 \end{aligned}$$

By the choice of c above, they will *only* have β as a common root because $\gamma - c\beta_j \neq \alpha_i$ for any $i \neq 1$ or $j \neq 1$. It follows that

$$\gcd(g(X), h(X)) = X - \beta$$

Proposition 7.2.7 on page 267 implies that the greatest common divisor has its coefficients in the field in which the polynomials have theirs, so

$$\beta \in F[\gamma]$$

On the other hand, we also have $\alpha = \gamma - c\beta \in F[\gamma]$ so

$$F[\alpha, \beta] = F[\gamma]$$

□

EXERCISES.

1. If F is a field, show that

$$\begin{aligned} F(Y) &\to F(X) \\ Y &\mapsto X^2 \end{aligned}$$

makes $F(X)$ algebraic extension of $F(X)$ of degree 2.

2. If $F = \mathbb{Q}(2^{1/3})$, express

$$\frac{1}{2^{2/3} - 2^{1/3} + 1}$$

as a polynomial in $2^{1/3}$ (see lemma 7.1.9 on page 264).

3. Find a primitive element for the field $\mathbb{Q}[\sqrt{2}, \sqrt{3}]$ over $\mathbb{Q}$ and find its minimal polynomial.

4. Find the splitting field of $X^3 - 2$ over $\mathbb{Q}$.

7.3. Computing Minimal polynomials

The Elimination Property of Gröbner bases is useful for algebraic computations:

EXAMPLE 7.3.1. Suppose we have a field $F = \mathbb{Q}[\sqrt{2}, \sqrt{3}]$ (see lemma 7.1.9 on page 264) and want to know the minimal polynomial (see 7.1.8 on page 263) of $\alpha = \sqrt{2} + \sqrt{3}$. We regard F as a quotient

$$F = \mathbb{Q}[X, Y]/(X^2 - 2, Y^2 - 3)$$

Now form the ideal $\mathfrak{s} = (X^2 - 2, Y^2 - 3, A - X - Y) \subset \mathbb{Q}[X, Y, A]$ and *eliminate* X and Y by taking a Gröbner basis using lexicographic ordering with

$$X \succ Y \succ A$$

The result is

$$(1 - 10 A^2 + A^4, -11 A + A^3 + 2Y, 9A - A^3 + 2X)$$

and we claim that the minimal polynomial of α is

$$\alpha^4 - 10\alpha^2 + 1 = 0$$

It is a polynomial that α satisfies and generates $\mathfrak{s} \cap k[A]$, which is a principal ideal domain (see corollary 5.3.10 on page 119), so any other polynomial that α satisfies must be a multiple of it. Since the degree of this minimal polynomial is the same as $[F:\mathbb{Q}] = 4$ it follows that $F = \mathbb{Q}[\alpha]$. Indeed, the second and third terms of the Gröbner basis imply that

$$\sqrt{2} = \frac{\alpha^3 - 9\alpha}{2}$$

$$\sqrt{3} = -\frac{\alpha^3 - 11\alpha}{2}$$

so $\mathbb{Q}[\alpha] = \mathbb{Q}[\sqrt{2}, \sqrt{3}]$. This is an example of the Primitive Element Theorem (7.2.13 on page 270).

Here's a second example:

$F = \mathbb{Q}[2^{1/3}]$ and we want the minimal polynomial of

$$\alpha = \frac{1 + 2^{1/3}}{1 - 2^{1/3}}$$

We create the ideal $\mathfrak{b} = (X^3 - 2, (1-X)A - 1 - X)$ (the second term is a polynomial that α satisfies) and take a Gröbner basis to get

$$\mathfrak{b} = (3 + 3A + 9A^2 + A^3, 1 - 8A - A^2 + 4X)$$

so the minimal polynomial of α is

$$\alpha^3 + 9\alpha^2 + 3\alpha + 3 = 0$$

The second term of the Gröbner basis implies that

$$2^{1/3} = \frac{\alpha^2 + 8\alpha - 1}{4}$$

so that $\mathbb{Q}[2^{1/3}] = \mathbb{Q}[\alpha]$.

EXERCISES.

1. If $F = \mathbb{Q}[2^{1/2}, 2^{1/3}]$ find the minimum polynomial of $\alpha = 2^{1/2} + 2^{1/3}$.

7.3.1. Norm and trace. The *norm* of a field element is an important concept that we have seen before in example 5.3.13 on page 119.

DEFINITION 7.3.2. If $F \subset H$ is a finite extension of fields, then H is a finite-dimensional vector space over F with basis $\{x_1, \ldots, x_n\}$ where $n = [H:F]$. If $\alpha \in H$, then
  (1) $m_\alpha = \alpha \cdot *: H \to H$, i.e. the matrix of the linear transformation of H (as a vector-space over F) defined by multiplication by α, and with respect to the basis $\{x_1, \ldots, x_n\}$,
  (2) $\chi_\alpha(X) = \det(X \cdot I - m_\alpha) \in F[X]$ is the characteristic polynomial of m_α (see definition 6.2.51 on page 195) and called the *characteristic polynomial of α*,
  (3) $N_{H/F}(\alpha) = \det m_\alpha \in F$ is the determinant of m_α, and is called the *norm* of α.
  (4) $T_{H/F}(\alpha) = \text{Tr}(m_\alpha) \in F$ is the trace of the matrix m_α (see definition 6.2.53 on page 196), and is called the *trace* of α.

REMARK. See theorem 8.8.1 on page 318 for an alternate formulation of these terms.

The terms are closely related
$$(7.3.1) \qquad \chi_\alpha(0) = (-1)^n N_{H/F}(\alpha)$$
where $n = [H:F]$. To see this, just plug $X = 0$ into $I \cdot X - m_\alpha$ and take the determinant. If the characteristic polynomial is of degree n, the trace is $-a_{n-1}$, where a_{n-1} is the coefficient of X^{n-1}.

For instance, suppose $F = \mathbb{Q}$ and $H = \mathbb{Q}[\sqrt{2}]$ with basis $\{1, \sqrt{2}\}$. Then the effect of $\alpha = a + b\sqrt{2}$ on this basis is
$$\begin{aligned} \alpha \cdot 1 &= a + b\sqrt{2} \\ \alpha \cdot \sqrt{2} &= 2b + a\sqrt{2} \end{aligned}$$
so the matrix m_α is
$$m_\alpha = \begin{bmatrix} a & 2b \\ b & a \end{bmatrix}$$
with characteristic polynomial
$$\chi_\alpha(X) = X^2 - 2aX + a^2 - 2b^2$$
and norm
$$N_{H/F}(\alpha) = a^2 - 2b^2$$
and trace
$$T_{H/F}(\alpha) = 2a$$
The basic properties of matrices imply that

LEMMA 7.3.3. *Under the assumptions of definition 7.3.2 on the preceding page, we have*

(1) *the characteristic polynomial, norm, and trace of an element do not depend on the basis used to compute them,*
(2) $N_{H/F}(1) = 1$
(3) *for all* $\alpha, \beta \in H$, $N_{H/F}(\alpha \cdot \beta) = N_{H/F}(\alpha) \cdot N_{H/F}(\beta)$
(4) *for all* $\alpha, \beta \in H$, $T_{H/F}(\alpha + \beta) = T_{H/F}(\alpha) + T_{H/F}(\beta)$
(5) *for all* $\alpha \in H$, $N_{H/F}(\alpha) = 0$ *if and only if* $\alpha = 0$

PROOF. See exercises 17 on page 201 and 18 on page 201.

In a splitting field for $\chi_\alpha(X)$, the characteristic polynomial satisfies

$$\chi_\alpha(X) = \prod_{j=1}^{n}(X - \lambda_j) = X^n + c_{n-1}X^{n-1} + \cdots + c_0$$

where the λ_j are the eigenvalues of m_α, which do not depend on the basis. The determinant is equal to $(-1)^n c_0$, so it is also independent of basis. The same is true for the trace, since it is equal to $-c_{n-1}$.

The second statement follows from the fact that m_1 is the identity matrix.

The third statement follows from the basic properties of determinants: the composite of α and β, as linear transformations, is $m_\alpha \cdot m_\beta = m_{\alpha \cdot \beta}$, and

$$\det(m_\alpha \cdot m_\beta) = \det(m_\alpha) \cdot \det(m_\beta)$$

The fourth statement follows from the fact that

$$m_{\alpha+\beta} = m_\alpha + m_\beta$$

And the fifth follows from the third and the fact that any nonzero element $\alpha \in H$ has a multiplicative inverse α^{-1} so that

$$N_{H/F}(\alpha) \cdot N_{H/F}(\alpha^{-1}) = 1$$

□

We can also say something about the characteristic polynomial of an element

PROPOSITION 7.3.4. *Under the assumptions of definition 7.3.2 on the previous page,*

$$m_*: H \to \mathrm{Mat}(F, n)$$

is a homomorphism into the ring of $n \times n$ matrices with coefficients in F. It follows that an element $\alpha \in H$ satisfies its characteristic polynomial, i.e.

$$\chi_\alpha(\alpha) = 0$$

PROOF. We have already seen that $m_\alpha \cdot m_\beta = m_{\alpha \cdot \beta}$ and it is not hard to see that $m_{\alpha+\beta} = m_\alpha + m_\beta$, which proves the first statement. The second follows from the first and the Cayley-Hamilton theorem (see 6.2.57 on page 199), which states that $\chi_\alpha(m_\alpha) = 0$. □

Clearly, if $F \subset H$ is an extension of fields and $\alpha \in F$, $N_{H/F}(\alpha) = \alpha^{[H:F]}$.
Here's another example of norms of field extensions:
Let $F = \mathbb{Q}$ and let $H = \mathbb{Q}[\gamma]$ where γ is a root of the polynomial
$$p(X) = X^3 + X^2 + X + 2$$
Eisenstein's Criterion (see theorem 5.6.8 on page 149) shows that $p(X)$ is irreducible over $\mathbb{Q}$ so we can construct $\mathbb{Q}[\gamma]$ as the quotient
$$\mathbb{Q}[X]/(p(X))$$
and γ is the element that X maps to under the projection to the quotient. Our basis for $\mathbb{Q}[\gamma]$ is $\{1, \gamma, \gamma^2\}$, where $\gamma^3 = -2 - \gamma - \gamma^2$, and
$$\gamma^4 = -2\gamma - \gamma^2 - \gamma^3 = 2 - \gamma$$
A general element of this field is
$$\alpha = a + b\gamma + c\gamma^2$$
and the effect of this element on the basis is
$$\begin{aligned}
\alpha \cdot 1 &= a + b\gamma + c\gamma^2 \\
\alpha \cdot \gamma &= a\gamma + b\gamma^2 + c\gamma^3 \\
&= -2c + (a-c)\gamma + (b-c)\gamma^2 \\
\alpha \cdot \gamma^2 &= a\gamma^2 + b\gamma^3 + c\gamma^4 \\
&= 2(c-b) - (c+b)\gamma + (a-b)\gamma^2
\end{aligned}$$
so we get the matrix
$$m_\alpha = \begin{bmatrix} a & -2c & 2(c-b) \\ b & a-c & -(c+b) \\ c & b-c & a-b \end{bmatrix}$$
with determinant
$$N_{H/F}(\alpha) = a^3 - a^2 b - ca^2$$
$$+ 5acb - 3ac^2 + ab^2 + 2cb^2 - 2b^3 - 2bc^2 + 4c^3$$
and characteristic polynomial
$$\chi_\alpha(X) =$$
$$X^3 - (3a - b - c)X^2 - \left(3c^2 - 5cb - b^2 - 3a^2 + 2ca + 2ab\right)X$$
$$- a^3 + a^2 b + ca^2 - 5acb + 3ac^2 - ab^2 - 2cb^2 + 2b^3 + 2bc^2 - 4c^3$$
and trace
$$T_{H/F}(\alpha) = 3a - b - c$$
Although an element of a field *satisfies* its characteristic polynomial, this does not mean the characteristic polynomial *is* its minimal polynomial. In fact:

LEMMA 7.3.5. *Let $F \subset H$ be a finite field-extension and let $\alpha \in H$ have minimal polynomial $p(X) \in F[X]$. Then*
$$\chi_\alpha(X) = p(X)^{[H:F[\alpha]]}$$

PROOF. Let $\{x_i\}$ be a basis for H over $F[\alpha]$ and let $\{y_j\}$ be a basis for $F[\alpha]$ over F. Then $\{x_i y_j\}$ is a basis for H over F (see proposition 7.1.6 on page 263). The effect of α on this basis is to act on the $\{y_j\}$ and leave the $\{x_i\}$ fixed. This means

$$m_\alpha = \begin{bmatrix} A & 0 & \cdots & 0 \\ 0 & A & \ddots & \vdots \\ \vdots & \ddots & \ddots & 0 \\ 0 & \cdots & 0 & A \end{bmatrix}$$

where $A = m_\alpha$ computed *in* $F[\alpha]$, and this block-matrix has $[H: f[\alpha]]$ rows and columns.

In $F[\alpha]$, the characteristic polynomial is a polynomial that α satisfies, hence is contained in the principal ideal $(p(X)) \subset F[X]$ and is of the same degree as $p(X)$ so it is a multiple of $p(X)$ by a unit $u \in F$. Since both polynomials are monic, we must have $u = 1$.

The conclusion follows from the properties of a determinant of a block matrix. □

EXERCISES.

2. If $H = \mathbb{Q}[2^{1/3}]$ compute the norm and characteristic polynomial of a general element.

3. If $H = \mathbb{Q}[\sqrt{2}, \sqrt{3}]$ compute the norm and characteristic polynomial of a general element.

7.4. Primitive roots of unity

7.4.1. Cyclotomic Polynomials . We begin this section with a discussion of the *cyclotomic*[1] *polynomials*. These are minimal polynomials of primitive n^{th} roots of unity.

DEFINITION 7.4.1. Let $n \geq 1$ be an integer and define the nth *cyclotomic polynomial*,

$$\Phi_1(X) = X - 1$$

if $n = 1$ and

$$\Phi_n(X) = \prod_{\substack{\gcd(d,n)=1 \\ 1 \leq d < n}} \left(X - e^{2\pi i d/n} \right)$$

---

[1] From Greek for "circle dividing."

if $n > 1$.

Compare this to the proof of lemma 4.3.5 on page 42.

Given the cyclic group $\mathbb{Z}_n$, and an integer d such that $d \mid n$, let Φ_d be the set of generators of the cyclic group of order d generated by n/d. Since every element of $\mathbb{Z}_n$ generates some cyclic subgroup, we get

$$\tag{7.4.1} \mathbb{Z}_n = \bigsqcup_{d \mid n} \Phi_d$$

as *sets*. This is the essence of the proof of lemma 4.3.5 on page 42 (which merely counts the sets).

We conclude that

THEOREM 7.4.2. *If $n > 0$ is an integer, then*

$$\tag{7.4.2} X^n - 1 = \prod_{d \mid n} \Phi_d(X)$$

It follows that the $\Phi_n(X)$ are monic polynomials with integer coefficients.

PROOF. Since the roots of $X^n - 1$ are all distinct (see lemma 7.2.8 on page 268), it follows that, in $\mathbb{C}[X]$

$$X^n - 1 = \prod_{k=0}^{n-1} \left(X - e^{2\pi i k/n} \right)$$
$$= \prod_{k \in \mathbb{Z}_n} \left(X - e^{2\pi i k/n} \right)$$
$$= \prod_{d \mid n} \left(\prod_{k \in \Phi_d} \left(X - e^{2\pi i k/n} \right) \right)$$

by equation 7.4.1. To prove equation 7.4.2 note that Φ_d is the set of generators of the cyclic subgroup of $\mathbb{Z}_n$ generated by n/d, i.e.,

$$\left\{ \frac{n}{d}, \ldots, \frac{n \cdot j}{d}, \ldots \right\}$$

where $1 \le j \le d - 1$ and $\gcd(j, d) = 1$. It follows that

$$\prod_{k \in \Phi_d} \left(X - e^{2\pi i k/n} \right) = \prod_{\substack{\gcd(j,d)=1 \\ 1 \le j < d}} \left(X - e^{2\pi i (jn/d)/n} \right)$$
$$= \prod_{\gcd(j,d)=1} \left(X - e^{2\pi i j/d} \right)$$
$$= \Phi_d(X)$$

The final statement follows by induction. We know $\Phi_1(X) = X - 1$ is monic with integer coefficients. The conclusion follows from

(1) The product of monic polynomials with integer coefficients is one

(2) the division algorithm for polynomials and

$$\Phi_n(X) = \frac{X^n - 1}{\prod_{\substack{d \mid n \\ d < n}} \Phi_d(X)}$$

□

Here are the first few such polynomials:

$\Phi_1(X) = X - 1$
$\Phi_2(X) = X + 1$
$\Phi_3(X) = X^2 + X + 1$
$\Phi_4(X) = X^2 + 1$
$\Phi_5(X) = X^4 + X^3 + X^2 + X + 1$
$\Phi_6(X) = X^2 - X + 1$

If p is a *prime*

$$\Phi_p(X) = \frac{X^p - 1}{X - 1} = 1 + X + \cdots + X^{p-1}$$

LEMMA 7.4.3. *Let $f(X) \in \mathbb{Z}[X]$ be a polynomial such that $f(X) \mid X^n - 1$ and let k be an integer such that $\gcd(k, n) = 1$. If ζ is a root of $f(X)$, then so is ζ^k.*

PROOF. We will show that, if p is a prime with $p \nmid n$, then ζ^p is a root of $f(X)$, which will imply the conclusion (since k can be factored into a product of powers of such primes).

We prove this by contradiction. Suppose $f(\zeta^p) \neq 0$. Then

$$f(X) = (X - \zeta_1) \cdots (X - \zeta_k)$$

where the ζ_i are roots of unity and, therefore, powers of $\omega = e^{2\pi i/n}$ and ζ^p is *not* included in this list. It follows that $f(\zeta^p)$ is a product of *differences* of powers of ω — and, therefore divides the *discriminant*, Δ, of $X^n - 1$ (see definition 5.5.20 on page 144).

Exercise 10 on page 145 shows that $\Delta = \pm n^n$. On the other hand

$$f(X^p) \equiv f(X)^p \pmod{p}$$

see lemma 7.6.1 on page 287 and proposition 3.3.2 on page 22 (coupled with the fact that $\phi(p) = p - 1$, so $a^p \equiv a \pmod{p}$), which implies that

$$f(\zeta^p) \equiv f(\zeta)^p \equiv 0 \pmod{p}$$

This implies that $p \mid f(\zeta^p)$ which divides $\Delta = \pm n^n$ in $\mathbb{Z}[\omega]$.

Since p and Δ are in $\mathbb{Z}$, it implies that $p \mid \Delta$ in $\mathbb{Z}[\omega]$ which also applies in $\mathbb{Z}$ (see exercise 1 on the next page). This contradicts the assumption $p \nmid n$.
□

THEOREM 7.4.4. *If $n > 0$ is an integer, then $\Phi_n(X)$ is irreducible over $\mathbb{Q}$.*

REMARK. This implies the claim made at the beginning of this section: that $\Phi_n(X)$ is the *minimal polynomial* of a primitive n^{th} root of unity.

PROOF. If ζ is a root of $\Phi_n(X)$ and $\Phi_n(X) \mid X^n - 1$, lemma 7.4.3 on the facing page implies that ζ^k must also be a root of $\Phi_n(X)$ for all $k \in \mathbb{Z}_n^\times$. Since $\omega = e^{2\pi i/n}$ is a root of $\Phi_n(X)$, ω^k must *also* be a root of $\Phi_n(X)$ for all $0 < k < n$ with $\gcd(k, n) = 1$. But these are precisely the roots of $\Phi_n(X)$ — and no others. Since $\Phi_n(X)$ is *minimal* in this sense, it must be irreducible (any proper factor would contain an incomplete set of roots). □

EXERCISES.

1. Fill in a detail in the proof of lemma 7.4.3 on the preceding page. If $\alpha, \beta \in \mathbb{Z} \subset \mathbb{Z}[\gamma]$ (where γ is a primitive n^{th} root of unity) and $\alpha \mid \beta$ in $\mathbb{Z}[\gamma]$, then $\alpha \mid \beta$ in $\mathbb{Z}$.

2. If ω is a primitive n^{th} root of unity, what is $[\mathbb{Q}(\omega):\mathbb{Q}]$?

7.4.2. Compass and straightedge constructions revisited. We'll revisit problems from ancient Greece begun in Example 7.1.11 on page 265 regarding compass and straightedge constructions. Recall that we showed that one could only construct elements of fields described in equation 7.1.3 on page 265, i.e., fields F with

$$[F:\mathbb{Q}] = 2^k$$

for some integer $k \geq 0$. This implied that doubling the cube was impossible by compass and straightedge constructions since it would entail constructing $\mathbb{Q}[\sqrt[3]{2}]$, with $[\mathbb{Q}[\sqrt[3]{2}]:\mathbb{Q}] = 3$ and $3 \nmid 2^k$ for any k.

Here, we'll be concerned with constructing *angles*. Bisecting an angle is a well-known construction with a compass and straightedge but the problem of *trisecting* an angle was a famous unsolved problem of ancient Greek mathematics.

Actually, we will concern ourselves with the more general questions of *constructing* angles. It is clearly easy to construct 90° and 60° angles (for the latter, just construct an equilateral triangle).

What does it mean to construct an angle $360°/n = 2\pi/n$? If we construct a line making this angle with the x-axis and intersect it with the unit circle, we have constructed the point $e^{2\pi i/n} \in \mathbb{C}$, i.e. we have constructed a primitive n^{th} root of unity and the field

$$\mathbb{Q}[e^{2\pi i/n}]$$

The results of the previous section (and the solution of exercise 2) imply that

$$\left[\mathbb{Q}[e^{2\pi i/n}]:\mathbb{Q}\right] = \phi(n)$$

so that the angle $360°/n = 2\pi/n$ is constructible if and only if $\phi(n) = 2^k$ for some k.

Recall equation 3.3.9 on page 25: If
$$n = p_1^{k_1} \cdots p_t^{k_t}$$
where the p_i are primes, then
$$\phi(n) = p_1^{k_1-1}(p_1 - 1) \cdots p_t^{k_t-1}(p_t - 1)$$
If this is to divide 2^k, we must have
  (1) if $p_i \neq 2$, then $k_i = 1$.
  (2) if $p_i \neq 2$, then $p_i - 1 = 2^j$ for some j.
We conclude that $2\pi/n$ is constructible if and only if
$$n = 2^k \cdot p_1 \cdots p_m$$
where p_i is a prime with $p_i - 1 = 2^{j_i}$ for all i. This leads to the interesting question of Fermat Numbers and Fermat Primes.

DEFINITION 7.4.5. If $n > 0$ is an integer, then n is a *Fermat number* if $n = 2^k + 1$ for some positive integer, k. It is a *Fermat Prime* if it is also a prime number.

PROPOSITION 7.4.6. *If n is a Fermat prime, then $n = 2^{2^k} + 1$ for some positive integer k.*

REMARK. Fermat conjectured that *all* numbers of this form were primes. in [38], Euler showed that $641 \mid F_5$ so it is *not* prime (he also showed how to construct a regular 17-gon, thereby constructing the angle $2\pi/17$).

PROOF. In the expression $2^k + 1$, suppose k has an odd factor b, so $k = a \cdot b$. Then
$$2^k + 1 = 2^{ab} + 1 = (2^a + 1)\left(2^{a(b-1)} - 2^{a(b-2)} + \cdots + 1\right)$$
□

The first few Fermat Primes are
  (1) $F_0 = 3$
  (2) $F_1 = 5$
  (3) $F_2 = 17$
  (4) $F_3 = 257$
  (5) $F_4 = 65537$
It is conjectured that no higher Fermat Primes exist.

We conclude that

THEOREM 7.4.7. *The angle $2\pi/n$ is constructible by straightedge and compass if and only if*
$$n = 2^k \cdot p_1 \cdots p_t$$
where the p_i are distinct Fermat primes.

This also resolves the question of *trisecting* an angle: if we can trisect *any* angle, we can trisect (easily-constructed) $60°$, i.e. construct the angle $20° = 2\pi/18$. But $18 = 2 \cdot 3^2$ so $\phi(18) = 6$, which is not a power of 2.

7.4.3. Primes in arithmetic progression. In this section, we will explore number-theoretic implications of the properties of cyclotomic polynomials. Dirichlet's theorem on primes in arithmetic progressions extends Euclid's theorem on the existence of an infinite number of primes.

> Johann Peter Gustav Lejeune Dirichlet (1805 – 1859) was a German mathematician who made deep contributions to number theory (creating the field of analytic number theory), and to the theory of Fourier series and other topics in mathematical analysis; he is credited with being one of the first mathematicians to give the modern formal definition of a function.

THEOREM 7.4.8 (Dirichlet's Theorem). *If $n, m > 1$ are integers with $\gcd(n, m) = 1$, there exist an infinite number of primes, p, such that*

$$p \equiv m \pmod{n}$$

In fact, Dirichlet also proved that the *density* of primes are equal for all $m \in \mathbb{Z}_n^\times$, i.e. that their statistical densities are all the same: $1/\phi(n)$ — see [28]. Proof of this involves analytic number theory and is beyond the scope of this book.

We will prove a limited special case (see [45] and [80]):

THEOREM 7.4.9. *If $n > 1$ is an integer, there exist an infinite number of primes, p, such that*

$$p \equiv 1 \pmod{n}$$

REMARK. Actually, we will prove a slightly stronger result — that there are an infinite number of primes in a kind of "geometric" progression, i.e., we get an infinite sequence of primes $\{p_1, \dots\}$ such that for any $j > 1$ $p_j \equiv 1 \pmod{p_i}$ for all $i < j$, as well as $p_j \equiv 1 \pmod{n}$.

We begin with

LEMMA 7.4.10. *Let k and n be positive integers, and let p be a prime. If $p \nmid n$ and $p \mid \Phi_n(k)$ then $p \equiv 1 \pmod{n}$.*

PROOF. Recall equation 7.4.2 on page 277:

$$X^n - 1 = \prod_{d \mid n} \Phi_d(X)$$

Since $p \mid \Phi_n(k)$, we know that $p \mid k^n - 1$ so that $k^n \equiv 1 \pmod{p}$, and $k \in \mathbb{Z}_p^\times$. If d is the order of k in $\mathbb{Z}_p^\times$, then $d \mid n$. Suppose $n = d \cdot m$. We will show that $m = 1$.

Let $\bar{x}$ denote the image of x in $\mathbb{Z}_p$ and $\overline{f(X)}$ denote the image of $f(X) \in \mathbb{Z}[X]$ in $\mathbb{Z}_p[X]$.

Suppose $k \neq 1$. Since d is the order of $\bar{k}$ in $\mathbb{Z}_p^\times$, we know that $\bar{k}^d = 1$ so that $p \mid k^d - 1$. We get

(7.4.3) $$X^d - 1 = \prod_{e \mid d} \Phi_e(X)$$

so that

(7.4.4)
$$X^n - 1 = \Phi_n(X) \prod_{\substack{j \mid n \\ 1 < j < n}} \Phi_j(X)$$
$$= \Phi_n(X) h(X) \prod_{e \mid d} \Phi_e(X)$$
$$= \Phi_n(X) h(X) (X^d - 1)$$

where $h(X) \in \mathbb{Z}[X]$ represents the factors in equation 7.4.4 not also in equation 7.4.3 on the preceding page.

Since $p \mid \Phi_n(k)$, so $\overline{\Phi_n(k)} = 0$, then $\overline{(X-k)} \mid \overline{\Phi_n(X)}$. Since $p \mid k^d - 1$, we see that $\overline{X^d - 1}$ has $\bar{k}$ as a zero, so that $\overline{X - k} \mid \overline{X^n - 1}$. It follows that $\overline{(X-k)}^2 \mid \overline{(X^n - 1)}$ and
$$\overline{(X^n - 1)} = \overline{(X - k)^2 r(X)}$$

If we take formal derivatives of both sides, we get
$$\bar{n} X^{n-1} = \overline{2(X-k) r(X)} + \overline{(X-k)^2 r'(X)}$$

Since $p \nmid n$ and $p \nmid k$ we get $\overline{nk^{n-1}} \neq 0$. On the other hand, the right side of this equation vanishes if we set $X = k$. This is a contradiction based on the assumption that $m > 1$.

It follows that the order of $\bar{k}$ in $\mathbb{Z}_p^\times$ is n. Since the order of $\mathbb{Z}_p^\times$ is $p - 1$, we conclude that $n \mid p - 1$ and $p \equiv 1 \pmod{n}$. □

Now we are in a position to prove theorem 7.4.9 on the previous page.

CLAIM. If m is any positive integer

(7.4.5)
$$\gcd(\Phi_m(km), m) = 1$$

for all $k \geq 1$.

PROOF. Suppose q is any prime that divides both $\Phi_m(km)$ and m. Then $\overline{km} \in \mathbb{Z}_q[X]$ is a zero of $\overline{\Phi_m(X)}$, so $\overline{X - km} \mid \overline{\Phi_m(X)}$. Since q also divides m we see that $\overline{km} = 0 \in \mathbb{Z}_q$ so that $\overline{X} \mid \overline{\Phi_m(X)}$ or $X \mid \Phi_m(X)$ and equation 7.4.2 on page 277 implies that $X \mid X^m - 1$ which is a contradiction. □

Now, note that if $m > 1$ then
$$|\Phi_m(km)| > 1$$

for all sufficiently large integers k. If $|\Phi_m(km)| \leq 1$ for all k, then one of the polynomials $\Phi_m(mX)$, $\Phi_m(mX) - 1$, or $\Phi_m(mX) + 1$ has infinitely many zeroes, so is the zero-polynomial, which is a contradiction.

So, choose k_1 large enough for $|\Phi_n(k_1 n)| > 1$. Then some prime $p_1 \mid \Phi_n(k_1 n)$ and equation 7.4.5 implies that $p_1 \nmid n$. Lemma 7.4.10 on the preceding page implies that $p_1 \equiv 1 \pmod{n}$. Next, choose k_2 large enough that $|\Phi_{np_1}(k_1 p_1 n)| > 1$ and let $p_2 \mid \Phi_{np_1}(k_1 p_1 n)$. Then $p_2 \nmid p_1 n$ and we conclude $p_2 \equiv 1 \pmod{np_1}$ which implies $p_2 \equiv 1 \pmod{n}$. In this fashion, we get an infinite sequence of primes $p_1, p_2, \ldots$ that are all congruent to 1 modulo n. This proves theorem 7.4.9 on the previous page.

7.5. Algebraically closed fields

These fields play important parts in many areas of algebra and algebraic geometry.

DEFINITION 7.5.1. A field Ω is said to be *algebraically closed* if any polynomial $p(X) \in \Omega[X]$ can be factored into linear terms

$$p(X) = f_0 \cdot \prod_{k=1}^{\deg p}(X - \alpha_k) \in \Omega[X]$$

with $f_0 \in \Omega$.

REMARK. This is equivalent to saying that $p(x)$ has $\deg p$ roots in Ω.

EXAMPLE. The Fundamental Theorem of Algebra (see theorem 8.9.1 on page 321) implies that the field $\mathbb{C}$ is algebraically closed.

DEFINITION 7.5.2. Let $F \subset \Omega$ be an extension of fields. Then Ω is defined to be an algebraic closure of F if
   (1) Ω is algebraically closed.
   (2) given any extension $F \subset G$ with G algebraically closed, Ω is isomorphic to a subfield, Ω' of G that contains F.

REMARK. The field of complex numbers is clearly the algebraic closure of $\mathbb{R}$.

If they exist, algebraic closures are essentially unique:

THEOREM 7.5.3. *Let F be a field and let Ω_1 and Ω_2 be algebraic closures of F. Then there exists an isomorphism*

$$f: \Omega_1 \to \Omega_2$$

such that $f|F = 1: F \to F$.

PROOF. Define a pair (E, τ) to consist of a subfield $E \subset \Omega_1$ such that $F \subset E$ and a monomorphism $\tau: E \to \Omega_2$, such that $\tau|F = 1$. At least one such pair exists because we can simply define $E = F \hookrightarrow \Omega_2$.

Define $(E_1, \tau_1) \prec (E_2, \tau_2)$ if $E_1 \subset E_2$ and $\tau_2|E_1 = \tau_1$. Then every chain

$$(E_1, \tau_1) \prec (E_2, \tau_2) \prec \cdots$$

has a maximal element, (E, τ): Simply define

$$E = \bigcup_i E_i$$

and define $\tau|E_i = \tau_i$. It follows, by Zorn's lemma (see lemma 14.2.12 on page 465) that we can find a maximal element among all of the (E, τ). Call this $(\bar{E}, \bar{\tau})$. We claim that $\bar{E} = \Omega_1$. If not, we could find a nontrivial algebraic extension $\bar{E}[\alpha]$ with minimal polynomial $p(x)$ and extend $\bar{\tau}$ to a map

$$\begin{aligned} g: \bar{E}[\alpha] &\to \Omega_2 \\ \alpha &\mapsto \beta \end{aligned}$$

where $\beta \in \Omega_2$ is a root of $\bar{\tau}(p(x))$. This is a contradiction. We also claim that $\bar{\tau}(\Omega_1) = \Omega_2$ since its image will be an algebraically closed subfield of Ω_2. □

It turns out that *every* field has an algebraic closure. We will fix a field F and explicitly construct its algebraic closure using a construction due to Artin (see [69]).

We need a lemma first:

LEMMA 7.5.4. *Let F be a field and let $f_1(X), \ldots, f_k(X) \in F[X]$ be polynomials. Then there exists an extension $F \subset \Omega$ such that Ω contains a root of each of the $f_i(X)$.*

PROOF. This follows from corollary 7.2.3 on page 266 and induction. □

DEFINITION 7.5.5. Let S denote the set of all monic, irreducible polynomials in $F[x]$ — this is infinite (just mimic Corollary 3.1.10 on page 16).

Form the polynomial ring $F[\{S_f\}]$ with an indeterminate, S_f, for each $f \in S$ and form the ideal $\mathfrak{M} = (\{f(S_f)\})$ — generated by indeterminates representing monic irreducible polynomials *plugged into those very polynomials.*

PROPOSITION 7.5.6. *The ideal, $\mathfrak{M} \subset F[\{S_f\}]$, defined in 7.5.5 is proper.*

PROOF. We will show that $1 \notin \mathfrak{M}$. Let

$$x = \sum_{k=1}^{n} a_k \cdot f_k(S_{f_k}) \in \mathfrak{M}$$

be some element, where $f_k \in S$. We will set $x = 1$ and get a contradiction.

Let Ω denote an extension of F containing one root, α_k, of each of the n polynomials $f_k(S_{f_k})$. Now define a homomorphism

$$\mathbb{F}[\{S_f\}] \to \Omega$$

(7.5.1) $\quad\quad\quad\quad\quad\quad\quad\quad S_{f_k} \mapsto \alpha_k$

(7.5.2) $\quad\quad\quad\quad\quad\quad\quad\quad S_{f'} \mapsto 0$

for $k = 1, \ldots, n$, where $f' \notin \{f_1, \ldots, f_n\}$. This is clearly possible since the S_{f_k} are all indeterminates. The equation $x = 1$ maps to $0 = 1$, a contradiction. □

REMARK. This argument is delicate: The existence of the mapping in equation 7.5.1 requires a *separate* indeterminate for each monic irreducible polynomial.

THEOREM 7.5.7. *An algebraic closure, Ω, exists for F. If the cardinality of F is infinite, then the cardinality of Ω is equal to that of F.*

REMARK. The Fundamental Theorem of Algebra (theorem 8.9.1 on page 321) implies that the algebraic closure of $\mathbb{R}$ is $\mathbb{C}$.

PROOF. Let $\mathfrak{M} \subset F[\{S_f\}]$ be as in proposition 7.5.6. Since $\mathfrak{M}$ is proper, proposition 5.2.11 on page 115 implies that it is contained in some maximal ideal $\mathfrak{M}'$. Define

$$\Omega_1 = F[\{S_f\}]/\mathfrak{M}'$$

This will be a field, by lemma 5.3.2 on page 117. This field will contain roots of all monic irreducible polynomials in $F[X]$. If it is algebraically closed, we are done. Otherwise, continue this construction to form a field Ω_2 containing Ω_1 and all roots of monic irreducible polynomials in $\Omega_1[X]$.

We obtain a (possibly infinite) chain of fields

$$F \subset \Omega_1 \subset \Omega_2 \subset \cdots$$

If any of the Ω_k are algebraically closed, then

$$\Omega_n = \Omega_k$$

for all $n > k$ since the only monic irreducible polynomials in Ω_k will be linear ones.

Define

$$\Omega = \bigcup_{i=1}^{\infty} \Omega_i$$

We claim that this is algebraically closed. Any polynomial $f(X) \in \Omega[X]$ is actually contained in $\Omega_k[X]$ for some value of k, and its roots will be contained in Ω_{k+1}.

The statement about cardinalities follows from the corresponding property of each of the $\Omega_i[\{S_f\}]$. □

EXAMPLE 7.5.8. The Fundamental Theorem of Algebra (8.9.1 on page 321) implies that $\mathbb{C}$ is algebraically closed. It follows that it is the algebraic closure of $\mathbb{R}$ and that the *only* nontrivial algebraic extension of $\mathbb{R}$ is $\mathbb{C}$.

When we deal with the rationals, things become much more complicated:

EXAMPLE 7.5.9. The algebraic closure of $\mathbb{Q}$ is called the *algebraic numbers* and written $\bar{\mathbb{Q}}$. It cannot equal $\mathbb{C}$ because it is *countable*, by theorem 7.5.7 on the preceding page. The structure of $\bar{\mathbb{Q}}$ is extremely complex and not well understood.

The uniqueness of algebraic closures have some interesting consequences:

DEFINITION 7.5.10. Let F be a field and let $\alpha \in \bar{F}$ be an element of the algebraic closure of F. Then the minimal polynomial $f(X)$ of α splits into linear factors

$$f(X) = \prod_{i=1}^{\deg f} (X - \alpha_i)$$

with $\alpha_1 = \alpha$. The $\{\alpha_i\}$ are called the *conjugates* of α.

REMARK. The conjugates of α are uniquely determined by α because $\bar{F}$ is uniquely determined up to an isomorphism.

The characteristic polynomial of α in $F[\alpha]$ is the minimal polynomial (it is of the same degree and α satisfies it) and the discussion following 7.3.4 on page 274 shows that the conjugates of α are just the eigenvalues of the matrix m_α in definition 7.3.2 on page 273.

For instance, if $z = a + bi \in \mathbb{C}$, then the minimal polynomial of z is its characteristic polynomial over $\mathbb{R}$ (see 7.3.2 on page 273), namely

$$X^2 - 2aX + a^2 + b^2$$

and the other root of this polynomial is $a - bi$, the usual complex conjugate.

The conjugates of an algebraic element are related to its norm:

LEMMA 7.5.11. *Let $F \subset H$ be a finite extension of fields and let $\alpha \in H$. Then*

$$N_{H/F}(\alpha) = \left(\prod_{j=1}^{m} \alpha_j\right)^{[H:F[\alpha]]}$$

$$T_{H/F}(\alpha) = [H:F[\alpha]] \cdot \sum_{j=1}^{m} \alpha_j$$

where the $\{\alpha_j\}$ run over the conjugates of α (with $\alpha = \alpha_1$).

PROOF. Let the minimal polynomial of α be $p(X) \in F[X]$, of degree m. Then, in an algebraic closure of F

$$p(X) = X^m + c_{n-1}X^{m-1} + \cdots + c_0 = \prod_{j=1}^{m}(X - \alpha_j)$$

from which it follows that

$$c_0 = (-1)^m \prod_{j=1}^{m} \alpha_j$$

The conclusion follows from lemma 7.3.5 on page 275 and equation 7.3.1 on page 273. The statement about the trace follows from the fact that the trace of a matrix is the sum of its eigenvalues. $\square$

Here is another interesting property of conjugates of an algebraic element:

LEMMA 7.5.12. *If F is a field and $\alpha \in \bar{F}$ is an element of the algebraic closure of F, then there exists isomorphisms of fields*

$$\begin{aligned} F[\alpha] &\to F[\alpha'] \\ f &\mapsto f \text{ for all } f \in F \\ \alpha &\mapsto \alpha' \end{aligned}$$

where α' is any conjugate of α.

REMARK. To make this more precise: regard $F[\alpha]$ as a vector-space with F-basis $\{1, \alpha, \alpha^2, \ldots, \alpha^{n-1}\}$ (if the minimal polynomial of α is of degree n). Then map vector-spaces $F[\alpha] \to F[\alpha']$ via the change of basis

$$\alpha^j \mapsto (\alpha')^j$$

for $j = 0, \ldots, n-1$. This lemma says that this defines a field-isomorphism.

This elementary result is the basis of a deep field of mathematics called Galois Theory — see chapter 8 on page 297.

PROOF. Each conjugate of α satisfies the same minimal polynomial, $p(X) \in F[X]$, as α and we have
$$F[\alpha] = F[X]/(p(X)) = F[\alpha']$$
□

EXERCISES.

1. If F is a field and $f(X) \in F[X]$ is a polynomial, show that any two splitting fields, G_1, G_2 of $f(X)$ are isomorphic via isomorphism
$$g: G_1 \to G_2$$
whose restriction to $F \subset G_1$ is the identity map. Consequently, we can speak of *the* splitting field of $f(X)$.

2. Compute the conjugates of an element $\gamma = a + b\, 2^{1/3} \in \mathbb{Q}[2^{1/3}]$.

3. Can an algebraic extension be of *infinite degree*?

4. Are the conclusions of lemma 7.1.9 on page 264 true for the extension $\mathbb{Q} \subset \bar{\mathbb{Q}}$?

7.6. Finite fields

Finite fields can be completely classified — one of those rare areas of mathematics that have an exhaustive solution.

We begin with a lemma:

LEMMA 7.6.1. *Let F be a ring or field of characteristic p. If $\alpha, \beta \in F$ then*
$$(\alpha + \beta)^p = \alpha^p + \beta^p$$
A simple induction shows that
$$(\alpha + \beta)^{p^k} = \alpha^{p^k} + \beta^{p^k}$$

PROOF. This follows from the binomial theorem
$$(\alpha + \beta)^p = \sum_{ik=1}^{p} \frac{p!}{(p-k)! \cdot k!} \alpha^k \beta^{p-k}$$
so all terms except the first and the last have a factor of p in the numerator that is not canceled by any factor in the denominator. □

We know, from proposition 7.1.3 on page 262 that the characteristic of a finite field is a prime p and proposition 7.1.5 on page 262 implies that the size of a finite field is p^k for some $k > 0$.

We first show that finite fields of order p^k exist for all primes p and all integers $k \geq 1$:

LEMMA 7.6.2. *Let $g_k(X) = X^{p^k} - X \in \mathbb{Z}_p[X]$. Then the roots of $g_k(X)$ in the algebraic closure, $\bar{\mathbb{Z}}_p$, of $\mathbb{Z}_p$ form a field of order p^k.*

PROOF. First note that $g'_k(X) = -1 \in \mathbb{Z}_p[X]$ so it has no repeated roots, by lemma 7.2.8 on page 268 — meaning it has p^k roots in $\bar{\mathbb{Z}}_p$. Note that
(1) $0, 1 \in \bar{\mathbb{Z}}_p$ are in the set of roots, and if α and β are two such roots then:
(2) $\alpha^{p^k} = \alpha$ and $\beta^{p^k} = \beta$, so
 (a) $(\alpha \cdot \beta)^{p^k} = \alpha^{p^k} \cdot \beta^{p^k} = \alpha \cdot \beta$, so $\alpha \cdot \beta$ also satisfies $g_k(\alpha \cdot \beta) = 0$ and the set of roots is closed under multiplication,
 (b) $(\alpha + \beta)^{p^k} = \alpha^{p^k} + \beta^{p^k}$ by lemma 7.6.1 on the previous page, so $g_k(\alpha + \beta) = 0$ and the set of roots is closed under addition.
(3) multiplying all nonzero roots by a fixed root, $\alpha \neq 0$, merely permutes them because
$$\alpha \cdot \beta_1 = \alpha \cdot \beta_2 \implies \beta_1 = \beta_2$$
because $\bar{\mathbb{Z}}_p$ is an integral domain. So there exists a root γ such that $\alpha \cdot \gamma = 1$.

It follows that the set of p^k roots of $g_k(X)$ constitute a field. □

Now that we know fields of order p^k exist, we prove that they are unique:

LEMMA 7.6.3. *Let F be any field of order p^k and let $\alpha \in F$ be any element. Then*
$$\alpha^{p^k} - \alpha = 0$$
It follows that F is isomorphic to the field of order p^k constructed in lemma 7.6.2.

PROOF. This is just Lagrange's theorem, applied to the multiplicative group, F^*, of F. In other words, take the product of all nonzero elements of F
$$\delta = \prod_{i=1}^{p^k-1} \alpha_i$$
and multiply each element by α to get
$$\alpha^{p^k-1} \cdot \delta = \prod_{i=1}^{p^k-1} \alpha \cdot \alpha_i$$
Since multiplication by α simply permutes the elements of F^*, we get
$$\alpha^{p^k-1} \cdot \delta = \delta$$
or $\alpha^{p^k-1} = 1$. □

DEFINITION 7.6.4. *The unique field of order p^k is denoted $\mathbb{F}_{p^k}$.*

REMARK. So $\mathbb{F}_p = \mathbb{Z}_p$. The notation $\mathrm{GF}(p^k)$ ("Galois Field") is sometimes used for $\mathbb{F}_{p^k}$ in honor of Galois who is responsible for all of the material in this section.

DEFINITION 7.6.5. The *Frobenius map*, $\mathcal{F}_p\colon \mathbb{F}_{p^k} \to \mathbb{F}_{p^k}$ is defined to send $\alpha \in \mathbb{F}_{p^k}$ to $\alpha^p \in \mathbb{F}_{p^k}$.

PROPOSITION 7.6.6. *The Frobenius map is an automorphism of finite fields.*

PROOF. By the definition, it clearly preserves multiplication.

If $\alpha, \beta \in \mathbb{F}_{p^k}$ note that $\mathcal{F}_p^k(\alpha) = \alpha^{p^k} = \alpha$ because α is a root of $X^{p^k} - X \in \mathbb{Z}_p[X]$ in the algebraic closure of $\mathbb{Z}_p$, so $\mathcal{F}_p(\alpha) = \mathcal{F}_p(\beta)$ implies that $\mathcal{F}_p^k(\alpha) = \mathcal{F}_p^k(\beta) = \alpha = \beta$. It follows that $\mathcal{F}_p$ is a injective. In addition,

$$\mathcal{F}_p(\alpha + \beta) = (\alpha + \beta)^p = \alpha^p + \beta^p = \mathcal{F}_p(\alpha) + \mathcal{F}_p(\beta)$$

by lemma 7.6.1 on page 287, so it also preserves addition.

Since $\mathbb{F}_{p^k}$ is *finite*, $\mathcal{F}_p$ must be $1-1$. □

Note that $\mathbb{F}_{p^k} \subset \mathbb{F}_{p^\ell}$ if and only if $k|\ell$, since $\mathbb{F}_{p^\ell}$ must be a vector-space over $\mathbb{F}_{p^k}$ and both are vector-spaces over $\mathbb{F}_p = \mathbb{Z}_p$. With this in mind, we can explicitly describe the algebraic closure of *all* finite fields of characteristic p:

THEOREM 7.6.7. *Let p be a prime number. Then the algebraic closure of all finite fields of characteristic p is*

$$\bar{\mathbb{F}}_p = \bigcup_{k=1}^{\infty} \mathbb{F}_{p^{k!}}$$

The Frobenius map

$$\mathcal{F}_p\colon \bar{\mathbb{F}}_p \to \bar{\mathbb{F}}_p$$

is an automorphism and the finite field $\mathbb{F}_{p^\ell} \subset \mathbb{F}_{p^{\ell!}} \subset \bar{\mathbb{F}}_p$ is the set of elements of $\bar{\mathbb{F}}_p$ fixed by $\mathcal{F}_p^\ell$ (i.e. elements $x \in \bar{\mathbb{F}}_p$ such that $\mathcal{F}_p^\ell(x) = x$).

REMARK. Note that $\bar{\mathbb{F}}_p$ is an infinite field since it contains subfields of order p^k for all k. This implies:

All algebraically closed fields are infinite.

PROOF. If $f(X) \in \mathbb{F}_{p^k}[X]$ is a polynomial, it splits into linear factors in some finite extension, G, of $\mathbb{F}_{p^k}$, by corollary 7.2.4 on page 266. It follows that G is a finite field that contains $\mathbb{F}_{p^k}$ — i.e. $\mathbb{F}_{p^\ell}$ for some ℓ that is a multiple of k. Consequently $f(X)$ splits into linear factors in $\bar{\mathbb{F}}_p$. It follows that $\bar{\mathbb{F}}_p$ is algebraically closed and it is the smallest field containing all of the $\mathbb{F}_{p^k}$, so it must be the algebraic closure of all of them.

Since the Frobenius map is an automorphism of all of the $\mathbb{F}_{p^k}$ it must be an automorphism of $\bar{\mathbb{F}}_p$. The statement that $\mathcal{F}_p^\ell(x) = x$ implies that x is a root of $X^{p^\ell} - X = 0$ so the final statement follows from lemma 7.6.2 on the preceding page. □

LEMMA 7.6.8. *If G is a finite group of order n with the property that the equation $x^k = 1$ has at most k solutions, then G is cyclic.*

REMARK. In a *non-cyclic* group, the equations $x^k = 1$ can have *more* than k solutions. For instance, in the group $\mathbb{Z}_3 \oplus \mathbb{Z}_3$, the equation $3x = 0$ (written additively) has 9 solutions.

PROOF. If $d|n$ and an element, x, of order d exists, then it generates a cyclic subgroup $(x) = \{1, x, \ldots, x^{d-1}\}$ — which has $\phi(d)$ *distinct* generators. The hypothesis implies that all solutions to the equation $x^d = 1$ are elements of (x). It follows that all elements of *order d* are *generators* of (x) and that there are $\phi(d)$ of them. For each $d|n$ the set of elements of order d is either

▷ *empty*, if there are *no* elements of order d,
▷ *nonempty* with $\phi(d)$ members.

Equation 4.3.2 on page 42 implies that the number of elements of G is $< n$ unless elements of order d exist for *all* $d|n$ — including n itself. An element of order n generates G and implies it is cyclic. □

THEOREM 7.6.9. *If $\mathbb{F}_{p^n}$ is a finite field, its multiplicative group, $\mathbb{F}_{p^n}^\times$, is cyclic of order $p^n - 1$.*

PROOF. If $x \in \mathbb{F}_{p^n}^\times$, the solution to exercise 1 on page 121 implies that the equation $x^k = 1$ has, at most k solutions for all integers $k > 0$. The conclusion follows immediately from lemma 7.6.8 on the preceding page. □

Among other things, this implies the *Primitive Element Theorem* (see theorem 7.2.13 on page 270) for finite fields.

The minimum polynomial of a generator of $\mathbb{F}_{p^n}^\times$ over $\mathbb{Z}_p$ is called a *primitive polynomial* and such polynomials are heavily used in cryptography (see [87]).

EXERCISES.

1. Let $p = x^6 + x^2 + 1$ and $q = x^3 + x + 1$ be two elements of the finite field $\mathbb{F}_{2^8} = \mathbb{Z}_2[x]/(x^8 + x^4 + x^3 + x + 1)$. Compute $p \cdot q$.

2. Let $f = x^3 + x + 1$ be an elements of the finite field $\mathbb{F}_{2^8} = \mathbb{Z}_2[x]/(x^8 + x^4 + x^3 + x + 1)$. Compute f^{-1}. Hint: use the extended Euclidean algorithm to find polynomials $R(x)$ and $S(x)$ such that

$$f \cdot R(x) + (x^8 + x^4 + x^3 + x + 1) \cdot S(x) = 1$$

Such polynomials exist because $x^8 + x^4 + x^3 + x + 1$ is *irreducible* over $\mathbb{Z}_2$.

7.7. Transcendental extensions

We will characterize transcendental extensions of fields and show that they have transcendence bases similar to the way vector spaces have bases (see table 7.7.1 on the following page). A great deal of this material originated in the work of the German mathematician, Ernst Steinitz (1871–1928) in his seminal paper, [**102**].

DEFINITION 7.7.1. Consider an inclusion of fields $F \subset \Omega$. Elements $\alpha_1, \ldots, \alpha_m \in \Omega$ will be called *algebraically independent* over F if the natural map

$$F[X_1, \ldots X_m] \to \Omega$$
$$X_i \mapsto \alpha_i$$

is *injective*. If they aren't independent, they are said to be algebraically *dependent* over F.

REMARK. In other words, the $\alpha_1, \ldots, \alpha_m \in \Omega$ are algebraically dependent if there exists a polynomial f with coefficients in F such that

$$f(\alpha_1, \ldots, \alpha_n) = 0$$

in Ω.

They are algebraically *independent* if any equation of the form

$$\sum c_{i_1, \ldots, i_m} \alpha_1^{i_1} \cdots \alpha_m^{i_m} = 0$$

implies that all of the $\{c_{i_1, \ldots, i_m}\}$ vanish. Note the similarity between this condition and the definition of linear independence in linear algebra. As we will see, this is not a coincidence, and the theory of transcendence bases is similar to that of bases of vector spaces.

EXAMPLE.
(1) A single element $\alpha \in \Omega$ is algebraically independent if it is transcendental over F.
(2) The numbers π and e are probably algebraically independent over $\mathbb{Q}$ but this has not been proved.
(3) An infinite set $\{\alpha_i\}$ is independent over F if and only if every finite subset is independent.
(4) If $\alpha_1, \ldots, \alpha_n$ are algebraically independent over F, then

$$F[X_1, \ldots, X_n] \to F[\alpha_1, \ldots, \alpha_n]$$
$$f(X_1, \ldots, X_n) \mapsto f(\alpha_1, \ldots, \alpha_n)$$

is injective, hence an isomorphism. This isomorphism extends to the fields of fractions. In this case, $F(\alpha_1, \ldots, \alpha_n)$ is called a *pure transcendental extension* of F.
(5) The Lindemann–Weierstrass theorem (see [**43**] and [**12**]) proves that if $\alpha_1, \ldots, \alpha_n$ are algebraic numbers that are linearly independent over $\mathbb{Q}$, then $e^{\alpha_1}, \ldots, e^{\alpha_n}$ are algebraically independent over $\mathbb{Q}$.

We can characterize algebraic elements of a field extension:

LEMMA 7.7.2. *Let* $f \subset \Omega$ *be an extension of fields with* $\gamma \in \Omega$ *and let* $A \subset \Omega$ *be some set of elements. The following conditions are equivalent:*

(1) γ *is algebraic over* $F(A)$.
(2) *There exist* $\beta_1, \ldots, \beta_t \in F(A)$ *such that* $\gamma^t + \beta_1 \gamma^{t-1} + \cdots + \beta_t = 0$.
(3) *There exist* $\beta_0, \ldots, \beta_t \in F[A]$ *such that* $\beta_0 \gamma^t + \beta_1 \gamma^{t-1} + \cdots + \beta_t = 0$.
(4) *There exists an* $f(X_1, \ldots, X_m, Y) \in F[X_1, \ldots, X_m, Y]$ *and* $\alpha_1 \cdots, \alpha_m \in A$ *such that* $f(\alpha_1, \ldots, \alpha_m, Y) \neq 0$ *but* $f(\alpha_1, \ldots, \alpha_m, \gamma) = 0$.

PROOF. Clearly statement 1 $\implies$ statement 2 $\implies$ statement 3 $\implies$ statement 1 — so those statements are equivalent.

Statement 4 $\implies$ statement 3: Write $f(X_1, \ldots, X_m, Y)$ as a polynomial in Y with coefficients in $\Gamma[X_1, \ldots, X_m]$, so

$$f(X_1, \ldots, X_m, Y) = \sum f_i(X_1, \ldots, X_m) Y^i$$

Then statement 3 holds with $\beta_i = f_i(\alpha_1, \ldots, \alpha_m)$.

Statement 3 $\implies$ statement 4: The β_i in statement 3 can be expressed as polynomials in a finite number of elements $\alpha_1, \ldots, \alpha_m \in A$

$$\beta_i = f_i(\alpha_1, \ldots, \alpha_m)$$

and we can use the polynomial

$$f(X_1, \ldots, X_m, Y) = \sum f_i(X_1, \ldots, X_m) Y^i$$

in statement 4. $\square$

When γ satisfies the conditions in the lemma, it is said to be *algebraically dependent on A over F*.

Table 7.7.1 illustrates the many similarities the theory of transcendence bases has with linear algebra.

Linear algebra	Transcendence
linearly independent	algebraically independent
$A \subset \text{Span}(B)$	A algebraically dependent on B
basis	transcendence basis
dimension	transcendence degree

TABLE 7.7.1. Analogy with linear algebra

Continuing our analogy with linear algebra, we have the following result, which shows that we can swap out basis elements:

LEMMA 7.7.3. EXCHANGE LEMMA: *Let* $\{\alpha_1, \ldots, \alpha_t\}$ *be a subset of* Ω. *If* $\beta \in \Omega$ *is algebraically dependent on* $\{\alpha_1, \ldots, \alpha_t\}$ *but not on* $\{\alpha_1, \ldots, \alpha_{t-1}\}$, *then* α_t *is algebraically dependent on* $\{\alpha_1, \ldots, \alpha_{t-1}, \beta\}$.

REMARK. Compare this with the argument in proposition 6.2.8 on page 167.

PROOF. Since β is algebraically dependent on $\{\alpha_1, \ldots, \alpha_t\}$, there exists a polynomial $f(X_1, \ldots, X_i, Y)$ with coefficients in F such that
$$f(\alpha_1, \ldots, \alpha_t, Y) \neq 0 \quad f(\alpha_1 \ldots, \alpha_t, \beta) = 0$$
Write f as a polynomial in X_t:
$$f(X_1, \ldots, X_t, Y) = \sum z_i(X_1, \ldots, X_{t-1}, Y) X_t^i$$

Because $f(\alpha_1, \ldots, \alpha_t, Y) \neq 0$ at least one of the z_i, say $z_{i_0}(\alpha_1, \ldots, \alpha_{t-1}, Y)$ is not the zero polynomial.

Because β is not algebraically dependent on $\{\alpha_1, \ldots, \alpha_{t-1}\}$, it follows that $z_{i_0}(\alpha_1, \ldots, \alpha_{t-1}, \beta) \neq 0$. Therefore $f(\alpha_1, \ldots, \alpha_{t-1}, X_t, \beta) \neq 0$.

But, because $f(\alpha_1, \ldots, \alpha_{t-1}, \alpha_t, \beta) = 0$, it follows that α_t is algebraically dependent on $\{\alpha_1, \ldots, \alpha_{t-1}, \beta\}$. □

LEMMA 7.7.4. *If C is algebraically dependent on B and B is algebraically dependent on A, then C is algebraically dependent on A.*

PROOF. If γ is algebraic over a field E that is algebraic over F, then γ is algebraic over F. Apply this with $E = F(A \cup B)$ and $F = F(A)$. □

Now we are ready to prove the main result

THEOREM 7.7.5. *Let $F \subset \Omega$ be an extension of fields, let $A = \{\alpha_1, \ldots, \alpha_t\}$ and $B = \{\beta_1, \ldots, \beta_m\}$ be two subsets of Ω, and suppose*

(1) A is algebraically independent over F.
(2) A is algebraically dependent on B over F.

Then $t \leq m$.

PROOF. Let ℓ be the number of elements A and B have in common. If this is t, the conclusion follows, so assume it is $< t$.

Write
$$B = \{\alpha_1, \ldots, \alpha_\ell, \beta_{\ell+1}, \ldots, \beta_m\}$$

Since $\alpha_{\ell+1}$ is algebraically dependent on B, but not on $\{\alpha_1, \ldots, \alpha_t\}$, there will be a β_j with $\ell + 1 \leq j \leq m$ such that $\alpha_{\ell+1}$ is algebraically dependent on $\{\alpha_1, \ldots, \alpha_\ell, \beta_{\ell+1}, \ldots, \beta_j\}$ but not on $\{\alpha_1, \ldots, \alpha_\ell, \beta_{\ell+1}, \ldots, \beta_{j-1}\}$.

The Exchange lemma 7.7.3 on the preceding page shows that β_j is algebraically dependent on
$$B_1 = B \cup \{\alpha_{\ell+1}\} \setminus \{\beta_j\}$$
So B is algebraically dependent on B_1 and A is algebraically dependent on B_1. Now we have $\ell + 1$ elements in common between A and B_1.

If $\ell + 1 < t$ repeat this process, using the Exchange property to swap elements of A for elements of B. We will eventually get $\ell = t$, and $t \leq m$. □

THEOREM 7.7.6. *Let $F \subset \Omega$ be an inclusion of fields. Then there exists a (possibly infinite) set of elements $\{\alpha_1, \ldots, \alpha_k\} \in \Omega$ such that the set $\{\alpha_1, \ldots, \alpha_k\}$ is algebraically independent over F, and Ω is an algebraic extension of $F(\alpha_1, \ldots, \alpha_k)$*

The number k is uniquely determined by Ω and is called the transcendence degree of Ω over F.

PROOF. All chains
$$A_1 \subset A_2 \subset \cdots$$
of sets of algebraically independent elements have an upper bound, namely their union. Zorn's lemma (14.2.12 on page 465) implies that there exists a maximal set of algebraically independent elements. If this set is finite and
$$\{\alpha_1,\ldots,\alpha_s\}$$
and
$$\{\beta_1,\ldots,\beta_t\}$$
are two maximal algebraically independent sets, theorem 7.7.5 on the previous page implies that $s \leq t$ and $t \leq s$ so $s = t$. □

EXAMPLE. The Lindemann–Weierstrass theorem (see [43] and [12]) proves that if $\alpha_1,\ldots,\alpha_n$ are algebraic numbers that are linearly independent over $\mathbb{Q}$, then $\mathbb{Q}(e^{\alpha_1},\ldots,e^{\alpha_n})$ has transcendence degree n over $\mathbb{Q}$.

DEFINITION 7.7.7. A *transcendence basis* for Ω over F is an algebraically independent set A, such that Ω is algebraic over $F(A)$.

If there is a *finite set* $A \subset \Omega$ such that Ω is algebraic over $F(A)$, then Ω has a *finite* transcendence basis over F. Furthermore, *every* transcendence basis of Ω over F is finite and has the same number of elements.

EXAMPLE 7.7.8. Let $p_1,\ldots,p_m$ be the elementary symmetric polynomials in $X_1,\ldots X_m$.

CLAIM. The field $F(X_1,\ldots,X_m)$ is algebraic over $F(p_1,\ldots,p_m)$.

Consider a polynomial $f(X_1,\ldots,X_n) \in F(X_1,\ldots,X_m)$. Theorem 5.5.18 on page 142 shows that the product
$$\prod_{\sigma \in S_n} (T - f(X_{\sigma(1)},\ldots,X_{\sigma(n)}))$$
over all permutations of the variables, is a polynomial with coefficients in $F(p_1,\ldots,p_m)$.

It follows that the set $\{p_1,\ldots,p_m\}$ must contain a transcendence basis for $F(X_1,\ldots,X_m)$ over F.

Since the size of a transcendence basis is unique, the $\{p_1,\ldots,p_m\}$ must be a transcendence basis and $F(X_1,\ldots,X_m)$ must be an algebraic extension of $F(p_1,\ldots,p_m)$.

Here's an example from complex analysis:

EXAMPLE 7.7.9. Let Ω be the field of meromorphic functions on a compact complex manifold.

The only meromorphic functions on the Riemann sphere are the rational functions in z. It follows that Ω is a pure transcendental extension of $\mathbb{C}$ of transcendence degree 1.

EXERCISES.

1. Use the Lindemann–Weierstrass theorem to prove that π is transcendental.

2. Show that the extension

has an uncountable degree of transcendence.

CHAPTER 8

Galois Theory

"Galois at seventeen was making discoveries of epochal significance in the theory of equations, discoveries whose consequences are not yet exhausted after more than a century."
— E. T. Bell.

8.1. Before Galois

Galois Theory involves studying when one can solve polynomial equations

(8.1.1) $$a_n X^n + a_{n-1} X^{n-1} + \cdots + a_1 X + a_0 = 0$$

via *radicals* — a term that will be made clear.

The first case to be solved was with $n = 2$

$$X^2 + bX + c = 0$$

with roots given by the Quadratic Formula — familiar to all students

$$X = \frac{-b \pm \sqrt{b^2 - 4a}}{2}$$

Note that $\sqrt{b^2 - 4a}$ is a *radical* — a quantity easily computed with algebraic properties that are easy to understand (i.e., its *square* is $b^2 - 4a$).

The Quadratic Formula dates to 830 A. D. or, depending on one's point of view, millennia earlier[1].

The solutions of this problem for $n = 3$ and $n = 4$ were some of the great mathematical triumphs of the Italian Renaissance. In renaissance Italy, there were public mathematical contests, "bills of mathematical challenge" (*cartelli di matematica disfida*). They were public written or oral contests modeled after knightly duels, with juries, notaries, and witnesses. The winner often received a cash prize, but the real prize was fame and the promise of paying students.

Like magicians, mathematicians often closely guarded their secret methods.

In 1539, Niccolò Tartaglia revealed his solution to the cubic equation

$$X^3 + aX^2 + bX + c = 0$$

---

[1]Purely geometric solutions were known to the ancient Greeks and, possibly, the ancient Egyptians. The process of *completing the square* dates to around 830 A. D. in a treatise by Al Khwarizmi.

to Gerolamo Cardano, a Milanese physician and mathematician — whom he swore to secrecy. This involved a complex series of transformations Tartaglia detailed in a poem[2]:

(1) First, substitute $X = y - \frac{a}{3}$ to get

$$y^3 + y\left(b - \frac{a^2}{3}\right) + \frac{2a^3}{27} - \frac{ba}{3} + c = 0$$

where the *quadratic* term has been eliminated. Incidentally, this is a general operation that kills off the $n - 1^{st}$ term in equations like 8.1.1 on the previous page.

(2) Now substitute

$$y = w - \frac{1}{3w}\left(b - \frac{a^2}{3}\right)$$

to get

$$w^3 + \frac{1}{w^3}\left(\frac{a^2b^2}{27} - \frac{ba^4}{81} - \frac{b^3}{27} + \frac{a^6}{729}\right) + \frac{2a^3}{27} - \frac{ab}{3} + c = 0$$

If we multiply this by w^3, we get a *quadratic* equation in w^3 which we can solve in the usual way.

(3) Then we take the cube root of w^3 and do all the *inverse substitutions* to get x. To get all three roots of the original cubic equation, we must take *all* three cube roots of w^3, namely

$$w, w \cdot \left(-\frac{1}{2} + i\frac{\sqrt{3}}{2}\right), w \cdot \left(-\frac{1}{2} - i\frac{\sqrt{3}}{2}\right)$$

Again, we have a solution involving *radicals:* square roots and cube roots.

> Niccolò Fontana Tartaglia (1499/1500 – 1557) was a mathematician, architect, surveyor, and bookkeeper in the then-Republic of Venice (now part of Italy). Tartaglia was the first to apply mathematics to computing the trajectories of cannonballs, known as ballistics, in his *Nova Scientia,* "A New Science."
> He had a tragic life. As a child, he was one of the few survivors of the massacre of the population of Brescia by French troops in the War of the League of Cambrai. His wounds made speech difficult or impossible for him, prompting the nickname Tartaglia ("stammerer").

When Cardano broke his oath of secrecy by publishing the formula in his book *Ars Magna,* a decade-long rivalry between him and Tartaglia ensued. To this day, the method outlined above is called Cardano's Formula — even though Cardano credited it to Tartaglia in his book.

This solution of the cubic equation is notable because it *requires* the use of complex numbers — even in cases where the roots are all *real* (in such cases, the imaginary parts cancel out at the end).

---

[2] Modeled on Dante's *Divine Comedy.*

> Gerolamo Cardano (1501 – 1576) was an Italian mathematician, physician, biologist, physicist, chemist, astrologer, astronomer, philosopher, writer, and gambler. He was one of the most influential mathematicians of the Renaissance. He wrote more than 200 works on science.

The case $n = 4$ was solved by Lodovico Ferrari in 1540 and published in 1545. We present Descartes's treatment of the solution:

Given a general quartic equation

$$X^4 + aX^3 + bX^2 + cX + d = 0$$

we first set $X = x - \frac{a}{4}$ to eliminate the X^3-term:

$$x^4 + qx^2 + rx + s = 0$$

Now we write

(8.1.2) $$x^4 + qx^2 + rx + s = (x^2 + kx + \ell)(x^2 - kx + m)$$

— this is possible because there is no x^3-term in the product.
Equating equal powers of x in equation 8.1.2 gives

$$\ell + m - k^2 = q$$
$$k(m - \ell) = r$$
$$\ell m = s$$

The first two equations imply that

$$2m = k^2 + q + 2r/k$$
$$2\ell = k^2 + q - 2r/k$$

If we plug these into $\ell m = s$, we get

$$k^6 + 2qk^4 + (q^2 - 4s)k^2 - r^2 = 0$$

Since this is *cubic* in k^2, we can use the method for solving cubic equations and determine k, m, and ℓ. At this point the solution of quadratic equations gives the four roots of our original problem — all of which are algebraic expressions involving *radicals*.

If this solution is written down as an equation, it takes up more than five pages.

The reader might wonder what alternatives exist to solving equations by radicals. There are purely *numeric* methods for finding the roots of a polynomial to an arbitrary accuracy. The problem is that the roots' *structural* algebraic properties are impossible to determine.

Expressions involving radicals are even our preferred way of *writing down* irrational algebraic numbers.

Although many mathematicians studied the cases where $n \geq 5$, they made no significant progress. The problem would remain open for more than 250 years — until Abel, Ruffini, and Galois.

8.2. Galois

We will explore the life-work of Évariste Galois, much of which he wrote down days before he died.

> Évariste Galois (1811 – 1832) was a French mathematician who, while still in his teens, was able to determine a necessary and sufficient condition for a polynomial to be solvable by radicals. His work laid the foundations for Galois theory, group theory, and two major branches of abstract algebra. He died at age 20 from wounds suffered in a duel.

Given a field extension

$$\begin{array}{c} E \\ | \\ F \end{array}$$

Galois theory studies the group of automorphisms of E that leave F fixed. This group of automorphisms has powerful mathematical properties, classifying all of the subfields of F, for instance.

We begin with a definition

DEFINITION 8.2.1. The *Galois group* of an extension

$$\begin{array}{c} E \\ | \\ F \end{array}$$

denoted $\text{Gal}(E/F)$ is defined by

$$\text{Gal}(E/F) = \{x \in \text{Aut}(E) | x(f) = f, \text{ for all } f \in F\}$$

EXAMPLE 8.2.2. For instance, consider the extension

$$\begin{array}{c} \mathbb{C} \\ | \\ \mathbb{R} \end{array}$$

Complex conjugation is known to be a automorphism of $\mathbb{C}$ that fixes $\mathbb{R}$ that generates the Galois group, so

$$\text{Gal}(\mathbb{C}/\mathbb{R}) = \mathbb{Z}_2$$

In fact, elements of the Galois group of a field-extension can be regarded as *generalizations* of complex conjugation.

8.3. Isomorphisms of fields

PROPOSITION 8.3.1. *Let $g: F_1 \to F_2$ be an isomorphism of fields, with induced isomorphism of polynomial rings $\bar{\sigma}: F_1[X] \to F_2[X]$. Let $f(X) \in F_1[X]$ be an irreducible polynomial and $\bar{f}(X) = \bar{\sigma}(f(X)) \in F_2[X]$. If α_1 is a root of $f(X)$ and α_2 is a root of $\bar{f}(X)$, there exists a unique isomorphism $G: F_1(\alpha_1) \to F_2(\alpha_2)$ extending g:*

$$\begin{array}{ccc} F_1(\alpha_1) & \xrightarrow{G} & F_2(\alpha_2) \\ | & & | \\ F_1 & \xrightarrow{g} & F_2 \end{array}$$

REMARK. Suppose $F_1 = F_2 = \mathbb{Q}$ and g is the identity map. Since $X^2 - 2$ is irreducible over $\mathbb{Q}$ (see theorem 5.6.8 on page 149) with roots $\pm\sqrt{2}$, this result implies that the map

$$\begin{aligned} \mathbb{Q}(\sqrt{2}) &\to \mathbb{Q}(\sqrt{2}) \\ \sqrt{2} &\mapsto -\sqrt{2} \end{aligned}$$

defines an isomorphism of fields. It is true that $\mathrm{Gal}(\mathbb{Q}(\sqrt{2})/\mathbb{Q}) = \mathbb{Z}_2$, generated by this automorphism.

PROOF. The isomorphism $\bar{g}\colon F_1[X] \to F_2[X]$ carries the ideal $(f(X))$ isomorphically to the ideal $(\bar{f}(X))$ so it induces an isomorphism

$$\frac{F_1[X]}{(f(X))} \to \frac{F_2[X]}{(\bar{f}(X))}$$

We define G to be the composite

$$F_1(\alpha_1) \overset{\cong}{\to} \frac{F_1[X]}{(f(X))} \to \frac{F_2[X]}{(\bar{f}(X))} \overset{\cong}{\to} F_2[\alpha_2]$$

This is unique because every element of $F_1(\alpha_1) = F_1[\alpha_1]$ is a polynomial in α_1 with coefficients in F_1. Its coefficients are mapped via g, and its image under G is uniquely determined by the way α_1 is mapped. □

PROPOSITION 8.3.2. *Let $g\colon F_1 \to F_2$ be an isomorphism of fields, with induced isomorphism of polynomial rings $\bar{g}\colon F_1[X] \to F_2[X]$. Let $f(X) \in F_1[X]$ be a polynomial and $\bar{f}(X) = \bar{g}(f(X)) \in F_2[X]$. In addition, let E_1 be a splitting field for $f(X)$ and E_2 be a splitting field for $\bar{f}(X)$. Then*

(1) *there exists an isomorphism $G\colon E_1 \to E_2$ extending g,*
(2) *if $f(X)$ is separable (see definition 7.2.11 on page 269), there exist $[E_1\colon F_1]$ distinct extensions G.*

REMARK. Lemma 7.2.9 on page 269 implies that *all* polynomials are separable if F has characteristic zero.

PROOF. We prove these statements by induction on $[E_1\colon F_1]$.

If $[E_1\colon F_1] = 1$, then $f(X)$ already splits into linear factors in $F_1[X]$ and it follows that $\bar{f}(X)$ also does, so that $E_2 = F_2$.

If $[E_1\colon F_1] > 1$, let $p(X)$ be an irreducible factor of $f(X)$ of degree ≥ 2, and let $\bar{p}(X) = \bar{g}(p(X))$. If α_1 is a root of $p(X)$ and α_2 is a root of $\bar{p}(X)$, proposition 8.3.1 on the facing page implies that there exists an isomorphism $\phi\colon F_1(\alpha_1) \to F_2(\alpha_2)$ extending g. Since E_1 is also a splitting field for $f(X)$ over $F_1(\alpha_1)$,

$$[E_1\colon F] = [E_1\colon F_1(\alpha_1)] \cdot [F(\alpha_1)\colon F]$$

and $[F(\alpha_1)\colon F] = \deg p(X) \geq 2$, induction implies that there exists an isomorphism G extending ϕ.

We prove the second statement by induction too. If $[E_1\colon F_1] = 1$, then $E_1 = F_1$ and there is clearly only one extension of $g\colon G = g$. So we assume $[E_1\colon F_1] > 1$. Let $p(X)$ be an irreducible factor of $f(X)$. If $\deg p(X) = 1$, then $p(X)$ contributes nothing new to E_1 and we may replace $f(X)$ by $f(X)/p(X)$ and continue.

If $\deg p(X) = d > 1$, let α be a root of $p(X)$ and let $\bar{p}(X) = \bar{g}(p(X)) \in F_2[X]$. In this case,
$$[E_1:F_1] = [E_1:F_1(\alpha)] \cdot [F_1(\alpha):F_1]$$
and proposition 8.3.1 on page 300 implies that there are d *distinct* isomorphisms (this is where we use separability of $f(X)$)
$$\beta: F_1(\alpha) \to F_2(\bar{\alpha})$$
extending g, where $\bar{\alpha}$ is a root of $\bar{p}(X)$. Since $[E_1:F_1(\alpha)] = [E_1:F_1]/d < [E_1:F_1]$, the induction hypothesis implies that each of these β's is covered by $[E_1:F_1]/d$ distinct isomorphisms $G: E_1 \to E_2$, giving a total of $d \cdot ([E_1:F_1]/d) = [E_1:F_1]$ isomorphisms covering g. □

We will compute some splitting fields and Galois groups:

EXAMPLE 8.3.3. If $\omega = e^{2\pi i/3}$, we know that $\omega^3 - 1 = 0$. The polynomial $X^3 - 1 \in \mathbb{Q}[X]$ is not irreducible. It factors as
$$X^3 - 1 = (X-1)(X^2 + X + 1)$$
and the factor $X^2 + X + 1$ is irreducible over $\mathbb{Q}$ (if not, its factors would be *linear* and imply that the cube roots of 1 are *rational*) — also see section 7.4.1 on page 276. It follows that

$$\mathbb{Q}(\omega)$$
$$|$$
$$\mathbb{Q}$$

is of degree 2, with $\mathbb{Q}$-basis $\{1, \omega\}$. The identity $\omega^2 + \omega + 1 = 0$ above, implies that $\omega^2 = -1 - \omega$.

The Galois group is $\mathbb{Z}_2$, with a generator that swaps ω with $\omega^2 = \omega^{-1}$. With respect to the $\mathbb{Q}$-basis given, it is

$$\begin{bmatrix} 1 & -1 \\ 0 & -1 \end{bmatrix}$$

Now we consider the polynomial $X^3 - 2 \in \mathbb{Q}[X]$. Its roots are $2^{1/3}, 2^{1/3}\omega, 2^{1/3}\omega^2 = -2^{1/3} - 2^{1/3}\omega$. Since
$$\omega = (2^{1/3})^2 2^{1/3} \omega / 2$$
we conclude that its splitting field is $\mathbb{Q}[\omega, 2^{1/3}]$. The Galois group of $\mathbb{Q}[\omega, 2^{1/3}]$ over $\mathbb{Q}$ is generated by α, given by
$$\omega \mapsto -1 - \omega$$
$$-1 - \omega \mapsto \omega$$
$$2^{1/3} \mapsto 2^{1/3}$$

and β, given by

$$\omega \mapsto \omega$$
$$2^{1/3} \mapsto 2^{1/3}\omega$$
$$2^{2/3} \mapsto -2^{2/3} - 2^{2/3}\omega$$
$$2^{1/3}\omega \mapsto -2^{1/3} - 2^{1/3}\omega$$
$$2^{2/3}\omega \mapsto 2^{2/3}$$

EXERCISES.

1. Suppose F is a field with an extension $H = F(\alpha_1, \ldots, \alpha_k)$ and suppose $f: H \to H$ is an automorphism such that
   a. $f(\beta) = \beta$ for all $\beta \in F$
   b. $f(\alpha_i) = \alpha_i$ for $i = 1, \ldots, n$

Show that f is the identity map. It follows that, if $f, g: H \to H$ are two automorphisms that fix F and map the α_i in the same way, then $f = g$.

8.4. Roots of Unity

Since we are exploring solutions of polynomials via radicals, we are interested in extensions of the form $F(\alpha^{1/n})$. Since we are also interested in *splitting fields*, we also need to understand extensions of the form $\mathbb{Q}(\omega)$ where $\omega = e^{2\pi i/n}$.

DEFINITION 8.4.1. If ω is an nth root of unity — i.e., if $\omega^n = 1$ — then ω is *primitive* if $\omega^i \neq 1$ for all $0 < i < n$.

REMARK. For example, $e^{2\pi i/n} \in \mathbb{C}$ is a primitive n^{th} root of unity, as is $e^{2\pi i k/n} \in \mathbb{C}$ for $k \in \mathbb{Z}_n^\times$, while $e^{2\pi i(n/2)/n}$ is not (assuming n is even).

We consider the Galois groups of simple field extensions in two cases:

THEOREM 8.4.2. *If F is a field and $E = F(\omega)$, with ω a primitive n^{th} root of unity, then $\text{Gal}(E/F)$ is isomorphic to a subgroup of $\mathbb{Z}_n^\times$, hence is finite abelian.*

PROOF. Exercise 1 implies that an automorphism

$$f: E \to E$$

that fixes F is completely determined by its effect on ω. Let $\sigma_i(\omega) = \omega^i$. For this to define an automorphism of $\langle \omega \rangle$, we must have $\gcd(i, n) = 1$ so that $i \in \mathbb{Z}_n^\times$. We define a map

$$g: \text{Gal}(E/F) \to \mathbb{Z}_n^\times$$

that sends σ_i to $i \in \mathbb{Z}_n^\times$. This is a homomorphism because
$$(\sigma_i \circ \sigma_j)(\omega) = \sigma_i(\omega^j) = \omega^{ij} = \sigma_{ij}(\omega)$$

□

If a field *already* has roots of unity and we form the splitting field of a simple polynomial, we get

THEOREM 8.4.3. *If F contains a primitive n^{th} root of unity, $f(X) = X^n - a \in F[X]$, and H is the splitting field of $f(X)$, then there exists an injective homomorphism*
$$\alpha: \mathrm{Gal}(H/F) \to \mathbb{Z}_n$$

REMARK. In both cases, the Galois group is *finite abelian* — hence a direct sum of suitable cyclic groups.

PROOF. If ω is a primitive n^{th} root of unity and β is a root of $f(X)$, then all the roots are
$$\{\beta, \beta\omega, \ldots, \beta\omega^{n-1}\}$$
If $\sigma \in \mathrm{Gal}(H/F)$, then $\sigma(\beta) = \beta\omega^i$, and this defines a map $\alpha(\sigma) = i \in \mathbb{Z}_n$. If $\gamma \in \mathrm{Gal}(H/F)$, then $\gamma(\omega) = \omega$ since $\omega \in F$. If $\gamma_1(\beta) = \beta\omega^{i_1}$ and $\gamma_2(\beta) = \beta\omega^{i_2}$, then
$$\gamma_1\gamma_2(\beta) = \gamma_1(\beta\omega^{i_2}) = \beta\omega^{i_1}\omega^{i_2} = \beta\omega^{i_1+i_2}$$
which implies that
$$\alpha(\gamma_1\gamma_2) = i_1 + i_2$$
so that α is a homomorphism. That this is an *injection* follows from exercise 1 on the preceding page. □

COROLLARY 8.4.4. *If p is a prime, F is a field with a primitive p^{th} root of unity, and $f(X) = X^p - a \in F[X]$ has splitting field E, then*
▷ $\mathrm{Gal}(E/F) = 1$ *or*
▷ $\mathrm{Gal}(E/F) = \mathbb{Z}_p$

PROOF. In both cases, theorem 8.4.3 implies that $\mathrm{Gal}(E/F) \subset \mathbb{Z}_p$. The only subgroups of $\mathbb{Z}_p$ are $\{1\}$ and $\mathbb{Z}_p$ (see proposition 4.3.2 on page 41). □

8.5. Group characters

We will study properties of isomorphisms and *automorphisms* of fields.

DEFINITION 8.5.1. *If G is a group and F is a field, a character from G to F is a homomorphism of groups*
$$\sigma: G \to F^\times$$

REMARK. We could have defined c as a homomorphism $\sigma: G \to F$ but, if $c(g) = 0$ for any $g \in G$, then $\sigma(h) = 0$ for *all* $h \in G$, so σ would not be very interesting.

This is a group-representation of degree-1 and defines a simple application of group-representation theory. See chapter 11 on page 387 for more on this subject.

DEFINITION 8.5.2. A set of characters $\sigma_1,\ldots,\sigma_k\colon G \to F^\times$ are said to be *dependent* if there exist $\alpha_1,\ldots,\alpha_k \in F$ not all equal to 0, such that

$$\alpha_1\sigma_1(g) + \cdots + \alpha_k\sigma_k(g) = 0$$

for all $g \in G$. Otherwise, they are said to be *independent*.

REMARK. Compare this to the notion of linear independence for vectors (see 6.2.2 on page 165).

Clearly, if $\sigma_i = \sigma_j$ for any $i \neq j$, then the set is dependent because we can write $\alpha_j = -\alpha_i$ with all of the other α's equal to 0. It is quite remarkable that is the *only* way a set of characters can be dependent:

PROPOSITION 8.5.3. *If G is a group, F is a field and*

$$\sigma_1,\ldots,\sigma_k$$

are distinct characters from G to F then they are independent.

REMARK. Two distinct vectors can easily be dependent, but the same is not true of distinct characters.

PROOF. Clearly, a single character must be independent, since it must be nonzero. Suppose we have proved that all sets of $< n$ distinct characters are independent and we have a dependency relation

$$\alpha_1\sigma_1(g) + \cdots + \alpha_n\sigma_n(g) = 0$$

for all $g \in G$. Note that we can assume that all of the α's are nonzero since the vanishing of any of them would imply a dependency relation of a set of $< n$ distinct characters. Multiplying by α_n^{-1} gives us

(8.5.1) $$\beta_1\sigma_1(g) + \cdots + \sigma_n(g) = 0$$

where $\beta_i = \alpha_n^{-1} \cdot \alpha_i$. Since $\sigma_1 \neq \sigma_n$, there exists an element $h \in G$ such that $\sigma_1(h) \neq \sigma_n(h)$. Replace g in equation 8.5.1 by $h \cdot g$ to get

$$\beta_1\sigma_1(h \cdot g) + \cdots + \sigma_n(h \cdot g) = 0$$
$$\beta_1\sigma_1(h)\sigma_1(g) + \cdots + \sigma_n(h)\sigma_n(g) = 0$$

and multiplying by $\sigma_n(h)^{-1}$ gives

(8.5.2) $$\beta_1\sigma_n(h)^{-1} \cdot \sigma_1(h)\sigma_1(g) + \cdots + \sigma_n(g) = 0$$

Since $\sigma_1(h) \neq \sigma_n(h)$, it follows that $\sigma_n(h)^{-1} \cdot \sigma_1(h) \neq 1$ and $\beta_1\sigma_n(h)^{-1} \cdot \sigma_1(h) \neq \beta_1$. If we subtract equation 8.5.2 from equation 8.5.1 we get

$$(\beta_1\sigma_n(h)^{-1} \cdot \sigma_1(h) - \beta_1)\sigma_1(g) + \cdots + 0 = 0$$

where we know that $\beta_1\sigma_n(h)^{-1} \cdot \sigma_1(h) - \beta_1 \neq 0$. This is a dependency relation with $< n$ terms and a contradiction. □

COROLLARY 8.5.4. *Every set of distinct automorphisms of a field is independent.*

PROOF. Automorphisms of a field are characters from the multiplicative group of the field into it. □

DEFINITION 8.5.5. If E is a field and $G = \text{Aut}(E)$ is the group of automorphisms. If $H \subset G$ is a subset, define
$$E^H = \{e \in E | h(e) = e \text{ for all } h \in H\}$$
called the *fixed field* of H.

REMARK. Note that we require H to fix elements of E^H *pointwise* rather than preserving the *set* E^H.

It is not hard to see that E^H is a subfield of E. It is also not hard to see that $H_1 \subset H_2$ implies that
$$E^{H_2} \subset E^{H_1}$$

Consider the following situation: $G = \{g_1, \ldots, g_n\}$ is a set of distinct automorphisms of a field, E, and E^G is the fixed subfield. Now let $x \in E \setminus E^G$. Since x is mapped non-trivially by G, and the set G is independent, *intuition* suggests that
$$\{g_1(x), \ldots, g_n(x)\}$$
will be *linearly independent* for a suitable x.

This is indeed the case:

LEMMA 8.5.6. *If E is a field and $G = \{g_1, \ldots, g_n\}$ is a set of distinct automorphisms of E, then*
$$[E : E^G] \geq n$$

PROOF. We prove it by contradiction. Suppose $[E : E^G] = k < n$ and let $\{e_1, \ldots, e_k\}$ be a basis for E over E^G. Consider the system of k linear equations over E in n unknowns
$$g_1(e_1)x_1 + \cdots + g_n(e_1)x_n = 0$$
$$\vdots$$
(8.5.3) $$g_1(e_k)x_1 + \cdots + g_n(e_k)x_n = 0$$

Since $k < n$, there exists a nonvanishing solution $(x_1, \ldots, x_n)$. For and $e \in E$, we have $e = \sum_{i=1}^{k} c_i e_i$. For $i = 1, \ldots, k$, multiply the ith row of equation 8.5.3 on page 306 by c_i to get
$$c_1 g_1(e_1)x_1 + \cdots + c_1 g_n(e_1)x_n = 0$$
$$\vdots$$
(8.5.4) $$c_k g_1(e_k)x_1 + \cdots + c_k g_n(e_k)x_n = 0$$
or
$$g_1(c_1 e_1)x_1 + \cdots + g_n(c_1 e_1)x_n = 0$$
$$\vdots$$
(8.5.5) $$g_1(c_k e_k)x_1 + \cdots + g_n(c_k e_k)x_n = 0$$
and add up all the rows of the result to get
$$g_1(e)x_1 + \cdots + g_n(e)x_n = 0$$
for an *arbitrary* $e \in E$. This contradicts the independence of the g_i. □

The alert reader may wonder whether strict inequality can occur, i.e. whether it is possible for $[E:E^G] > n$. The answer is yes for the following reason:

If $g_1, g_2 \in G$, then the composite, $g_1 \circ g_2$, will *also* fix E^G since each of the g_i fixes it. If $g_1 \circ g_2 \notin G$, let $G' = G \cup \{g_1 \circ g_2\}$ so that $|G'| = n+1$, then
$$E^G = E^{G'}$$
and the previous result shows that
$$[E:E^G] = [E:E^{G'}] \geq n+1$$

It turns out that this is the *only* way we can have $[E:E^G] > n$:

THEOREM 8.5.7. *If E is a field and $G = \{g_1, \ldots, g_n\}$ is a set of distinct automorphisms of E that form a group (i.e. composites and inverses of elements of G are in G), then*
$$[E:E^G] = n$$

PROOF. It will suffice to show that $[E:E^G] \leq n$, which we will prove by contradiction. Let $\{e_1, \ldots, e_{n+1}\}$ be linearly independent elements of E over E^G (i.e. linearly independent vectors with coefficients in E^G).

Now consider the system of n equations in $n+1$ unknowns:
$$(8.5.6) \qquad g_1(e_1)x_1 + \cdots + g_1(e_{n+1})x_{n+1} = 0$$
$$\vdots$$
$$g_1(e_1)x_1 + \cdots + g_1(e_{n+1})x_{n+1} = 0$$
where the $x_i \in E^G$. Since there are more unknowns than equations, the system has nontrivial solutions. Choose a solution with the least number of nonzero components and re-index the e_i so that the nonzero components come first. We get a solution like
$$(x_1, \ldots, x_k, 0, \ldots, 0)$$
where all of the x_i are nonzero. We will assume that $x_k = 1$ (multiplying the equations by x_k^{-1}, if necessary). We also note that $k > 1$ since $k = 1$ implies that $g_i(e_1) = 0$.

We also conclude that not all of the $x_i \in E^G$, since the row corresponding to $1 \in G$ (which exists since G is a group) would contradict the linear independence of the e_i. Again, by re-indexing the e_i, if necessary, assume that $x_1 \notin E^G$.

It follows that there exists a g_α such that $g_\alpha(x_1) \neq x_1$.

If
$$g_i(e_1)x_1 + \cdots + g_i(e_k) = 0$$
is an arbitrary row of the system in 8.5.6, act on it via g_α to get
$$(8.5.7) \qquad g_\alpha \cdot g_i(e_1) g_\alpha(x_1) + \cdots + g_\alpha \cdot g_i(e_k) = 0$$
Now note that there exists a $\beta \in \{1, \ldots n\}$, with $g_\alpha \cdot g_i = g_\beta \in G$, and that there is a βth row of the system in 8.5.6:
$$g_\beta(e_1)x_1 + \cdots + g_\beta(e_k) = 0$$

If we subtract equation 8.5.7 on the preceding page from this (using the fact that $g_\alpha \cdot g_i = g_\beta$, so that $g_\alpha \cdot g_i(e_k) - g_\beta(e_k) = 0$) we get

$$g_\beta(e_1)(g_\alpha(x_1) - x_1) + \cdots + g_\beta(e_{k-1})(x_{k-1} - g_\alpha(x_{k-1})) = 0$$

Since $\{g_1, \ldots, g_n\} = \{g_\alpha \cdot g_1, \ldots, g_\alpha \cdot g_n\}$, as *sets*, the result of performing this construction on *all* rows of equation 8.5.6 on the previous page, *reproduces* these rows (permuting them) — and the entire system.

It follows that there is a solution of the system 8.5.6 on the preceding page with $\leq k-1$ nonzero components, contracting the *minimality* of k. □

In many cases, the *fixed* fields of groups of automorphisms *determine* the automorphisms fixing them:

COROLLARY 8.5.8. *If E is a field and G_1 and G_2 are finite subgroups of* $\mathrm{Aut}(E)$ *with* $E^{G_1} = E^{G_2}$, *then* $G_1 = G_2$.

PROOF. Clearly, $G_1 = G_2$ implies that $E^{G_1} = E^{G_2}$. Conversely, assume $E^{G_1} = E^{G_2}$ and $g \in G_1$. Then g fixes E^{G_2}. If $g \notin G_2$ then E^{G_2} is fixed by $|G_2| + 1$ distinct automorphisms, namely element of G_2 and g. Lemma 8.5.6 on page 306 implies that

$$[E : E^{G_2}] \geq |G_2| + 1$$

while theorem 8.5.7 on the preceding page implies that

$$[E : E^{G_2}] = |G_2|$$

a contradiction. It follows that any $g \in G_1$ is also in G_2, and symmetry implies that every element of G_2 is also in G_1. □

8.6. Galois Extensions

We will be concerned with a particularly "well-behaved" class of field-extensions.

DEFINITION 8.6.1. Let

be an extension of fields with Galois group G (see definition 8.2.1 on page 300). The definition of Galois group implies that

$$F \subset E^G$$

If $F = E^G$, we will call the field extension a *Galois* extension or a *normal* extension.

REMARK. Both terms (Galois and normal) are used equally often, so the reader should be familiar with both.

It is possible to completely characterize when extensions are Galois:

LEMMA 8.6.2. *Let*

be a finite extension of fields. The following conditions are equivalent:

(1) If $G = \mathrm{Gal}(E/F)$, then $F = E^G$ — i.e. the extension is Galois.
(2) Every irreducible polynomial in $F[X]$ with a root in E is separable and has all of its roots in E.
(3) E is the splitting field of some separable polynomial in $F[X]$.

REMARK. A polynomial $f \in F[X]$ is separable if it factors into distinct linear factors in the algebraic closure.

PROOF. We prove 1 $\implies$ 2. If $p(X) \in F[X]$ is irreducible with a root $\alpha \in E$, let $\alpha_i = g_i(\alpha) \in E$ where $g_i \in G$ are all the distinct values that result when g_i runs over the elements of G. Let

$$f(X) = \prod_{i=1}^{n}(X - \alpha_i)$$

The coefficients of $f(X)$ are symmetric functions of the $\{\alpha_i\}$ (see equation 5.5.13 on page 142). Since elements of G permute the α_i, they *fix* the coefficients of $f(X)$ and statement 1, implies that these coefficients are in F. Since every root of $f(X)$ is also one of $p(X)$, it follows that they have nontrivial factors in common. Since $p(X)$ is irreducible, it follows that $p(X)|f(X)$. This implies that

(1) $p(X)$ is *separable* (its factors occur once), since $f(X)$ is (by construction)
(2) *all* roots of $p(X)$ lie in E.

Now we show that 2 $\implies$ 3. If $\alpha \in E \setminus F$, let $p_1(X)$ be its minimal polynomial (see definition 7.1.8 on page 263). This is irreducible, and statement 2 implies that all of its roots lie in E. If its roots in E are $\{\alpha_1 = \alpha, \ldots, \alpha_k\}$, let $F_1 = F[\alpha_1, \ldots, \alpha_k] \subset E$. If $F_1 = E$, we are done. If not, let $\beta \in E \setminus F_1$ with minimal polynomial $p_2(X)$ and continue the argument above. The process terminates in a finite number, t, of steps, and E is the splitting field of $p_1(X) \cdots p_t(X)$.

Suppose that E is the splitting field of a separable polynomial $f(X) \in F[X]$. The implication 3 $\implies$ 1 follows from proposition 8.3.2 on page 301, which implies that the identity map of F has precisely $[E:F]$ extensions to E. These are, of course, the automorphisms of E that fix F, or elements of $G = \mathrm{Gal}(E/F)$.

Now, note that (by theorem 8.5.7 on page 307 and proposition 7.1.6 on page 263):

$$[E:F] = |G| = [E:E^G] \cdot [E^G:F]$$

so that $[E^G:F] = 1$ and $E^G = F$. □

Here's a very simple example:

EXAMPLE 8.6.3. The polynomial $X^2 - 2 \in \mathbb{Q}[X]$ splits into

$$X^2 - 2 = (X - \sqrt{2})(X + \sqrt{2})$$

in $\mathbb{Q}[\sqrt{2}] = \mathbb{Q}(\sqrt{2})$ and the Galois group of the extension

$$\begin{array}{c}\mathbb{Q}(\sqrt{2}) \\ | \\ \mathbb{Q}\end{array}$$

is $\mathbb{Z}_2$, where the nontrivial operation is the map
$$\sqrt{2} \mapsto -\sqrt{2}$$

A more subtle example is:

EXAMPLE 8.6.4. Suppose k is a field and consider the symmetric group, S_n, acting on $E = k[X_1, \ldots, X_n]$ by permuting the X_i. These actions define automorphisms of E and we know that $E^{S_n} = F$ (see section 5.5.6 on page 142) where
$$F = k[\sigma_1, \ldots, \sigma_n]$$
and the σ_i are the elementary symmetric polynomials on the X_j — see equations 5.5.13 on page 142. On the face of it, we don't know whether the corresponding extension of fields of fractions

$$\begin{array}{c} E \\ | \\ F \end{array}$$

is Galois — we don't know whether $\text{Gal}(E/F) = S_n$. Although it must be a subgroup of some symmetric group, there might, conceivably, exist *more* automorphisms of E over F.

This extension is, indeed, Galois because E is the *splitting field* of the polynomial in $F[T]$:
$$T^n - \sigma_1 T^{n-1} + \sigma_2 T^{n-1} + \cdots + (-1)^n \sigma_n = \prod_{i=1}^{n}(T - X_i)$$
over F. Since any permutation of the X_i fixes F we conclude that $\text{Gal}(E/F) = S_n$.

COROLLARY 8.6.5. *If*

$$\begin{array}{c} E \\ | \\ F \end{array}$$

is a Galois extension and $F \subset B \subset E$ is an intermediate subfield of E, then

$$\begin{array}{c} E \\ | \\ B \end{array}$$

is a Galois extension and $\text{Gal}(E/B) \subset \text{Gal}(E/F)$ is the subgroup of automorphisms that fix B (pointwise[3]) as well as F.

REMARK. It is not hard to see that the subgroup, $\text{Gal}(E/B)$, *uniquely* determines B, since $B = E^{\text{Gal}(E/B)}$.

PROOF. The hypothesis implies that there exists a separable polynomial $f(X) \in F[X]$ such that E is its splitting field. It is not hard to see that $f(X) \in B[X]$ as well, with splitting field E, so lemma 8.6.2 on page 308 implies the conclusion. The statement about Galois groups is clear.

---

[3]I.e., the automorphisms fix *each element* of B.

COROLLARY 8.6.6. *Let*

$$E \\ | \\ F$$

be a Galois extension. Then there exists a finite number of subfields of E containing F.

PROOF. This is because $\text{Gal}(E/F)$ has a finite number of subgroups. □

Given two intermediate fields we can define a relation between them:

DEFINITION 8.6.7. *Let E be a field with subfields F, B_1, and B_2 with $F \subset B_1 \cap B_2$. If there exists an element $g \in \text{Gal}(E/F)$ such that $g(B_1) = B_2$, we say that B_1 and B_2 are* conjugates.

PROPOSITION 8.6.8. *Let*

$$E \\ | \\ F$$

be a Galois extension with conjugate subfields B_1, and B_2, and suppose $g(B_1) = B_2$ for $g \in \text{Gal}(E/F)$. Then $\text{Gal}(E/B_2) = \text{Gal}(E/B_1)^g$ (see definition 4.4.5 on page 44).

PROOF. If $x \in \text{Gal}(E/B_1)$, then xg^{-1} maps B_2 to B_1 and gxg^{-1} is an automorphism of E that fixes B_2 — i.e., an element of $\text{Gal}(E/B_2)$. Conversely, a similar argument shows that $g^{-1}yg \in \text{Gal}(E/B_1)$ if $y \in \text{Gal}(E/B_2)$ and that these operations define a 1-1 correspondence between the two Galois groups. □

THEOREM 8.6.9. *If*

$$E \\ | \\ B \\ | \\ F$$

is a Galois extension with intermediate field B, the following conditions are equivalent:

(1) *B is equal to its conjugates (i.e., it has no nontrivial conjugates).*
(2) *If $x \in \text{Gal}(E/F)$, then $x|B \in \text{Gal}(B/F)$.*
(3) *The extension*

$$B \\ | \\ F$$

is Galois.

In all of these cases, restriction defines a surjective homomorphism

$$\text{Gal}(E/F) \to \text{Gal}(B/F)$$

with kernel $\text{Gal}(E/B)$ so that

$$\text{Gal}(E/B) \triangleleft \text{Gal}(E/F)$$

and
$$\text{Gal}(B/F) \cong \frac{\text{Gal}(E/F)}{\text{Gal}(E/B)}$$

PROOF. 1 $\implies$ 2 is clear.

Assuming 2, we note that restriction defines a group homomorphism

(8.6.1) $\qquad f\colon \text{Gal}(E/F) \to \text{Gal}(B/F)$

Since E is Galois over B, lemma 8.6.2 on page 308 implies that E is the splitting field of some polynomial in $B[X]$, and proposition 8.3.2 on page 301 implies that every automorphism of B extends to one (in fact, to $[E:B]$ of them) of E.

This means the homomorphism f in 8.6.1 is *surjective*. It follows that
$$B^{\text{Gal}(B/F)} = B^{\text{Gal}(E/F)} \subset E^{\text{Gal}(E/F)} = F$$

The definition of Galois group implies that
$$F \subset B^{\text{Gal}(B/F)}$$
always, so we conclude that $B^{\text{Gal}(B/F)} = F$ and the conclusion follows from definition 8.6.1 on page 308.

Now we show that 3 $\implies$ 1. Since B is a Galois extension of F, lemma 8.6.2 on page 308 implies that
$$B = F(\alpha_1, \ldots, \alpha_n)$$
where the α_i are all the roots of a polynomial, $f(X)$, in $F[X]$. If $g \in \text{Gal}(E/F)$ and $\alpha \in B$ is any root of $f(X)$, we claim that $g(\alpha)$ is also a root of $f(X)$: if $f(X)$ is
$$X^n + b_{n-1}X^{n-1} + \cdots + b_0$$
with $b_i \in F$, then
$$\alpha^k + b_{k-1}\alpha^{k-1} + \cdots + b_0 = 0$$
and, if we apply g to it, we get
$$g(\alpha)^k + b_{k-1}g(\alpha)^{k-1} + \cdots + b_0 = 0$$
since $g(b_i) = b_i$, so $g(\alpha)$ is a root of $f(X)$. It follows that g *permutes* the α_i (since they are *all* the roots of $f(X)$) via a permutation $\sigma \in S_n$ and
$$g(B) = F(g(\alpha_1), \ldots, g(\alpha_n)) = F(\alpha_{\sigma(1)}, \ldots, \alpha_{\sigma(n)}) = B$$

As for the final statements,
$$\text{Gal}(E/B) \triangleleft \text{Gal}(E/F)$$
follows from proposition 8.6.8 on the preceding page, which implies that $\text{Gal}(E/B)^g = \text{Gal}(E/B)$ for all $g \in \text{Gal}(E/F)$.

Restriction induces a surjective map
$$\text{Gal}(E/F) \to \text{Gal}(B/F)$$
and the kernel is precisely the elements that fix B, namely $\text{Gal}(E/B)$. The final statement follows from proposition 4.4.8 on page 45. $\square$

We will be particularly interested in a certain class of field-extensions:

8.6. GALOIS EXTENSIONS

DEFINITION 8.6.10. If $m \geq 2$ is an integer, a field extension

$$\begin{array}{c} E \\ | \\ F \end{array}$$

is a *pure extension of type m* if $E = F(\alpha)$, where $\alpha^m \in F$. A tower of extensions

$$F = E_0 \subset E_1 \subset \cdots \subset E_k$$

is a *radical tower* if $E_i \subset E_{i+1}$ is pure for all i. In this case, E_k is called a *radical extension* of F.

EXERCISES.

1. If $F \subset E$ is a radical extension, show that there exists a tower of pure extensions

$$F \subset E_1 \subset \cdots \subset E_k = E$$

such that each extension $E_i \subset E_{i+1}$ is pure of type p_i, where p_i is a *prime*. Hint: Look at the solution to exercise 2 on page 87.

2. If

$$\begin{array}{c} E \\ | \\ F \end{array}$$

is a finite field extension, show that there exists a field extension

$$\begin{array}{c} G \\ | \\ E \end{array}$$

that is the splitting field of a polynomial $f(X) \in F[X]$.

3. Suppose we are given an extension

$$\begin{array}{c} E \\ | \\ F \end{array}$$

If B and C are subfields of E, their *compositum* $B \vee C$ is the intersection of all the subfields of E containing B and C. Show that if $\alpha_1, \ldots, \alpha_n \in E$, then

$$F(\alpha_1) \vee \cdots \vee F(\alpha_n) = F(\alpha_1, \ldots, \alpha_n)$$

4. Show that the splitting field constructed in exercise 2 is given by

$$K = E_1 \vee \cdots \vee E_n$$

where each E_i is isomorphic to E via an isomorphism that fixes F.

5. In exercise 2, if

$$\begin{array}{c} E \\ | \\ F \end{array}$$

is a radical extension, then so is

$$\begin{array}{c} G \\ | \\ F \end{array}$$

8.7. Solvability by radicals

In this section, we can give a group-theoretic necessary condition for a polynomial to be solvable by radicals.

DEFINITION 8.7.1. If F is a field, $f(X) \in F[X]$ is a polynomial, then $f(X)$ is said to be *solvable by radicals* if there exists a radical extension

$$\begin{array}{c} E \\ | \\ F \end{array}$$

such that E contains a splitting field of $f(X)$.

REMARK. This simply means that one can write all of the roots of $f(X)$ as algebraic expressions involving radicals.

LEMMA 8.7.2. *Let F be a field of characteristic 0, let $f(X) \in F[X]$ be solvable by radicals, and let E be a splitting field of $f(X)$. Then*

(1) there is a radical tower

$$F \subset R_1 \subset \cdots \subset R_N$$

with $E \subset R_N$ and R_N the splitting field of some polynomial over F, and with each $R_i \subset R_{i+1}$ a pure extension of type p_i where p_i is prime.

(2) If F contains the p_i^{th} roots of unity for all i, then the Galois group $\mathrm{Gal}(E/F)$ is solvable.

PROOF. Exercise 2 on the previous page implies that there is an extension

$$E \subset S$$

where S is the splitting field of a polynomial $F(X) \in F[X]$, and exercise 5 on the preceding page implies that

$$F \subset S = R_N$$

is a radical extension. Exercise 1 on the previous page implies that each stage of this extension is pure of type p_j, where p_j is a prime. Since F contains the p_i^{th} roots of unity, and exercise 8.4.3 on page 304 implies that each extension

(8.7.1) $$R_i \subset R_{i+1}$$

is the splitting field of a polynomial. Lemma 8.6.2 on page 308 implies that extension 8.7.1 is Galois.

Let

$$G_i = \mathrm{Gal}(R_N/R_i)$$

We get a subnormal series
$$G_N = 1 \subset G_{N-1} \subset \cdots \subset G_0$$
and theorem 8.6.9 on page 311 implies that
$$G_i/G_{i-1}$$
is a cyclic group, so that G_N is solvable. Since $E \subset G_N$, $\mathrm{Gal}(E/F)$ is a quotient of $\mathrm{Gal}(G_N/F)$ (by theorem 8.6.9 on page 311) hence *also* solvable (by exercise 5 on page 87). □

We can eliminate the hypotheses involving roots of unity:

THEOREM 8.7.3. *If F is a field of characteristic 0 and $f(X) \in F[X]$ that is solvable by radicals and the splitting field of f is H, then $\mathrm{Gal}(H/F)$ is a solvable group.*

PROOF. The hypothesis implies that
$$F \subset R_1 \subset \cdots \subset R_N$$
with each $R_i \subset R_{i+1}$ a pure extension of type a prime p_i and with $H \subset R_N$, and R_N is the splitting field of a polynomial $g(X) \in F[X]$. If t is the least common multiple of the p_i and ω is a primitive t^{th} root of unity, then $R_N(\omega)$ is the splitting field of $(X^t - 1)g(X)$.

We get a new tower
$$(8.7.2) \qquad F \subset F(\omega) \subset R_1(\omega) \subset \cdots \subset R_N(\omega)$$

Theorem 8.6.9 on page 311
$$\mathrm{Gal}(R_N(\omega)/F(\omega)) \triangleleft \mathrm{Gal}(R_N(\omega)/F)$$
and
$$\mathrm{Gal}(F(\omega)/F) \cong \frac{\mathrm{Gal}(R_N(\omega)/F)}{\mathrm{Gal}(R_N(\omega)/F(\omega))}$$
which is an abelian group, by theorem 8.4.2 on page 303.

Theorem 8.4.3 on page 304 implies that
$$\frac{\mathrm{Gal}(R_{i+1}(\omega)/F)}{\mathrm{Gal}(R_i(\omega)/F)} \cong \mathrm{Gal}(R_{i+1}(\omega)/R_i(\omega))$$
are cyclic groups for all i. Theorem 8.4.2 on page 303 implies that $\mathrm{Gal}(F(\omega)/F)$ is *finite abelian,* hence a direct sum of cyclic groups.

Exercise 2 on page 87 implies that $\mathrm{Gal}(R_N(\omega)/F)$ is solvable. Since $\mathrm{Gal}(H/F)$ is a quotient of $\mathrm{Gal}(R_N(\omega)/F)$, it must *also* be solvable. □

Now we consider whether there exists a "formula" for the roots of a polynomial.

Recall the field extension
$$k(X_1, \ldots, X_n)$$
$$|$$
$$k(\sigma_1, \ldots, \sigma_n)$$

discussed in Example 8.6.4 on page 310. If this extension is *solvable by radicals*, an expression for the roots of

$$f(T) = T^n - \sigma_1 T^{n-1} + \sigma_2 T^{n-1} + \cdots + (-1)^n \sigma_n = \prod_{i=1}^{n}(T - X_i)$$

involving radicals of coefficients (i.e., the σ_i) is a *formula* for them since one can plug arbitrary values into the indeterminates.

THEOREM 8.7.4 (Abel-Ruffini). *If $n \geq 5$, there exists no general formula (involving radicals) for the roots of a polynomial of degree n.*

REMARK. This is also known as Abel's Impossibility Theorem. Ruffini gave an incomplete proof in 1799 and Niels Hendrik Abel proved it in 1824.

PROOF. If such a formula exists, S_n must be a solvable group. If it is, then exercise 5 on page 87 implies that the subgroup

$$A_n \triangleleft S_n$$

is also solvable. Theorem 4.5.15 on page 55 implies that this is not so. □

Niels Henrik Abel (1802 – 1829) was a Norwegian mathematician who made contributions to a variety of fields. His most famous single result is the first complete proof of the impossibility of solving the general quintic equation in radicals. This question had been unresolved for 250 years. He also did research in elliptic functions, discovering Abelian functions. Abel made his discoveries while living in poverty and died at the age of 26.

Paolo Ruffini (1765 – 1822) was an Italian mathematician and philosopher. He is most well-known for his attempts to prove that polynomials of degree five are not solvable by radicals. He was a professor of mathematics at the University of Modena and a medical doctor who conducted research on typhus.

Galois Theory carries the Abel-Ruffini Theorem several steps further, as the following example shows

EXAMPLE 8.7.5. The polynomial

(8.7.3) $$f(X) = X^5 - 80X + 5 \in \mathbb{Q}[X]$$

is irreducible by Eisenstein's Criterion (theorem 5.6.8 on page 149) with the prime 5. If H is the splitting field of $f(X)$, we claim that $\text{Gal}(H/\mathbb{Q}) = S_5$. The plot

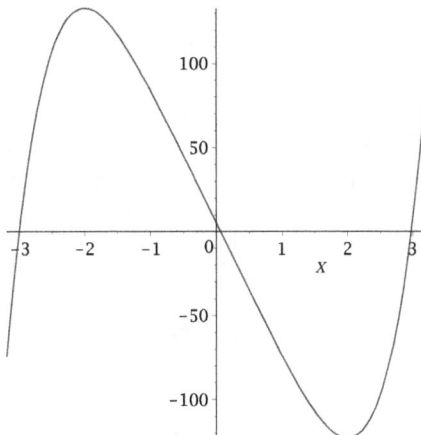

shows that it has three real roots and a pair of complex conjugate roots, say $\{\alpha, \bar{\alpha}, \beta_1, \beta_2, \beta_3\}$. Let $\tau \in \text{Gal}(H/\mathbb{Q})$ be complex conjugation. Since $f(X)$ is of degree 5, we have $5 \mid [H:\mathbb{Q}]$ and $5 \mid |\text{Gal}(H/\mathbb{Q})|$.

Sylow's First Theorem (4.8.1 on page 81), implies that it has a nontrivial Sylow 5-group, which has an element of order 5. The only elements of order 5 in S_5 are 5-cycles $(i_1, \ldots, i_5)$.

By relabeling elements, if necessary, we can assume $(1,2,3,4,5) \in \text{Gal}(H/\mathbb{Q}) \subset S_5$, and exercise 6 on page 58 implies that $\text{Gal}(H/\mathbb{Q}) = S_5$.

This is an odd and even disturbing result: not only is there no *general formula* for the roots of $f(X)$; one cannot even *write down* these roots in *radical notation* (if one magically found out what they were). In other words, *no* expression of the form

$$\sqrt[3]{1 + \sqrt[4]{2 + \sqrt{10}}}$$

however complicated — will equal these roots.

This leads to deep questions on how one can do concrete *calculations* with algebraic numbers. In general, the most one can say is that

"This number is a root of a certain irreducible polynomial."

This isn't saying much — for instance, these roots can have wildly different properties (three of the roots of equation 8.7.3 on the preceding page are real, two are complex).

Radical notation allows one to compute numerical values with arbitrary accuracy and allows one to deduce algebraic properties of numbers.

8.8. Galois's Great Theorem

In this section we will prove a converse to theorem 8.7.3 on page 315: if a polynomial's splitting field has a solvable Galois group, the polynomial is solvable by radicals.

We first need a result involving the *characteristic polynomial* of field elements (first mentioned in definition 7.3.2 on page 273.

THEOREM 8.8.1. *Let*

(8.8.1)
$$\begin{array}{c} E \\ | \\ F \end{array}$$

be a finite Galois extension with Galois group $G = \operatorname{Gal}(E/F)$. *If* $\alpha \in E$, *then*

$$\chi_\alpha(X) = \prod_{\sigma \in G}(X - \sigma(\alpha))$$

In particular

$$N_{E/F}(\alpha) = \prod_{\sigma \in G} \sigma(\alpha)$$

and

$$T_{E/F}(\alpha) = \sum_{\sigma \in G} \sigma(\alpha)$$

PROOF. Let $p(X)$ be the *minimal polynomial* (see definition 7.1.8 on page 263) of α over F, and let $d = [E\colon F(\alpha)]$.

Lemma 7.3.5 on page 275 implies that

$$\chi_\alpha(X) = p(X)^d$$

and lemma 8.6.2 on page 308 implies that *all* of the roots of $p(X)$ lie in E. Furthermore, proposition 8.3.1 on page 300 implies that G maps a *fixed* root of $p(X)$ to all of the others, so we get

$$p(X) = \prod_{\sigma_i \in G}(X - \sigma_i(\alpha))$$

where the $\sigma_i \in G$ are the elements that send α to *distinct* images — there are $\deg p(X)$ of them.

A given $\sigma \in G$ has $\sigma(\alpha) = \sigma_i(\alpha)$ if and only if $\sigma \in \sigma_i H$ where $H = \operatorname{Gal}((E/F(\alpha))$, and $|H| = d$. It follows that

$$\prod_{\sigma \in G}(X - \sigma(\alpha)) = \left(\prod_{\sigma_i \in G}(X - \sigma_i(\alpha))\right)^d = p(X)^d = \chi_\alpha(X)$$

□

The properties of the norm are well-known — see lemma 7.3.3 on page 274. In the setting of equation 8.8.1, the norm defines a homomorphism

$$N_{E/F}\colon E^\times \to F^\times$$

The following result says something about the *kernel* of this homomorphism:

THEOREM 8.8.2 (Hilbert's Theorem 90). *Let*

(8.8.2)
$$\begin{array}{c} E \\ | \\ F \end{array}$$

be a finite Galois extension with Galois group $G = \text{Gal}(E/F)$ that is cyclic of order n with generator $\sigma \in G$. If $\alpha \in E^\times$, then $N_{E/F}(\alpha) = 1$ if and only if there exists $\beta \in E^\times$ such that
$$\alpha = \beta \cdot \sigma(\beta)^{-1}$$

REMARK. This first appeared in a treatise on algebraic number theory by David Hilbert (available in English translation in [53]).

See theorem 13.3.3 on page 456 for a *cohomological* statement of this theorem.

PROOF. First, we verify that
$$\begin{aligned}
N_{E/F}(\beta \cdot \sigma(\beta)^{-1}) &= N_{E/F}(\beta) \cdot N_{E/F}(\sigma(\beta)^{-1}) \\
&= N_{E/F}(\beta) \cdot N_{E/F}(\sigma(\beta))^{-1} \\
&= N_{E/F}(\beta) \cdot N_{E/F}(\beta)^{-1} \\
&= 1
\end{aligned}$$

Conversely, suppose $N_{E/F}(\alpha) = 1$.

Define
$$\begin{aligned}
\delta_0 &= \alpha \\
\delta_1 &= \alpha \sigma(\alpha) \\
&\vdots \\
\delta_{i+1} &= \alpha \sigma(\delta_i)
\end{aligned}$$

so that
$$\delta_{n-1} = \alpha \sigma(\alpha) \sigma^2(\alpha) \cdots \sigma^{n-1}(\alpha)$$
which, by theorem 8.8.1 on the preceding page, is equal to $N_{E/F}(\alpha) = 1$.

Since the characters $\{1, \sigma, \sigma^2, \ldots, \sigma^{n-1}\}$ are independent (see definition 8.5.2 on page 305 and proposition 8.5.3 on page 305), there exists $\gamma \in E$ such that
$$\beta = \delta_0 \gamma + \delta_1 \sigma(\gamma) + \cdots + \delta_{n-2} \sigma^{n-2}(\gamma) + \sigma^{n-1}(\gamma) \neq 0$$
since $\delta_{n-1} = 1$. If we act on this via σ, we get
$$\begin{aligned}
\sigma(\beta) &= \sigma(\delta_0)\sigma(\gamma) + \cdots + \sigma(\delta_{n-2})\sigma^{n-1}(\gamma) + \sigma^n(\gamma) \\
&= \alpha^{-1}\left(\delta_1 \sigma(\gamma) + \cdots + \delta_{n-2}\sigma^{n-2}(\gamma) + \delta_{n-1}\sigma^{n-1}(\gamma)\right) + \gamma \\
&= \alpha^{-1}\left(\delta_1 \sigma(\gamma) + \cdots + \delta_{n-2}\sigma^{n-2}(\gamma) + \delta_{n-1}\sigma^{n-1}(\gamma)\right) + \alpha^{-1}\delta_0\gamma \\
&= \alpha^{-1}\beta
\end{aligned}$$

□

We apply this to

COROLLARY 8.8.3. *If*

$$\begin{array}{c} E \\ | \\ F \end{array}$$

is a Galois extension of degree p, a prime, and F has a primitive p^{th} root of unity, then $E = F(\beta)$, where $\beta^p \in F$.

PROOF. If ω is a primitive p^{th} root of unity, $\omega \in F$ and $N_{E/F}(\omega) = \omega^p = 1$.

Since $\text{Gal}(E/F) \neq \{1\}$, corollary 8.4.4 on page 304 implies that $\text{Gal}(E/F) \cong \mathbb{Z}_p$ — let σ be a generator. Then theorem 8.8.2 on page 318 implies that there exists $\beta \in E$ such that
$$\omega = \beta \cdot \sigma(\beta)^{-1}$$
or $\sigma(\beta) = \beta \cdot \omega^{-1}$. This implies that $\sigma(\beta^p) = \beta^p \cdot \omega^{-p} = \beta^p$. It follows that $\beta^p \in F$. The fact that $\omega \neq 1$ implies that $\beta \notin F$.

Since $[E:F] = p$, a prime, there are no intermediate fields between E and F, so $E = F(\beta)$. □

LEMMA 8.8.4. *Let E be the splitting field of a polynomial $f(X) \in F[X]$, and let $\bar{F}$ be an extension of F. If $\bar{E}$ is the splitting field of $f(X)$ in $\bar{F}$ then restriction defines an injective homomorphism*
$$\text{Gal}(\bar{E}/\bar{F}) \to \text{Gal}(E/F)$$

PROOF. Let $E = F(\alpha_1, \ldots \alpha_k)$ and $\bar{E} = \bar{F}((\alpha_1, \ldots \alpha_k)$. Any automorphism of $\bar{E}$ that fixes $\bar{F}$ also fixes F, and so is determined by its effect on the α_i (see exercise 1 on page 303). It follows that the homomorphism
$$\text{Gal}(\bar{E}/\bar{F}) \to \text{Gal}(E/F)$$
defined by $\sigma \mapsto \sigma|E$ is injective. □

We come to the main object of this section:

THEOREM 8.8.5 (Galois). *Let F be a field of characteristic 0 and let*

$$\begin{array}{c} E \\ | \\ F \end{array}$$

be a Galois extension. Then $\text{Gal}(E/F)$ is a solvable group if and only if there exists a radical extension R of F with $E \subset R$.

A polynomial $f(X) \in F[X]$ has a solvable Galois group if and only if it is solvable by radicals.

PROOF. Theorem 8.7.3 on page 315 gives the "if" part of the statement.

Suppose $G = \text{Gal}(E/F)$ is solvable. We will proceed by induction on $[E:F]$. Since G is solvable, it contains a normal subgroup $H \triangleleft G$ such that $[G:H] = p$, a prime (see exercise 1 on page 87).

Since G is a solvable group, $H = \text{Gal}(E/E^H) \subset G$ is a solvable group and induction implies that there exists a tower of subfields

(8.8.3) $$E^H \subset R_1 \subset \cdots \subset R_N$$

where each term is pure of type some prime.

Consider $E^H \subset E$. Since $H \triangleleft G$, E^H is a Galois extension of F. If we assume F contains a primitive p^{th} root of unity, then corollary 8.8.3 on the previous page implies that $F \subset E^H$ is pure of type p and we can splice it onto equation 8.8.3 on the facing page to get the required tower of field-extensions.

If F does *not* contain a primitive p^{th} root of unity, let $\bar{F} = F(\omega)$, where ω is a primitive p^{th} root of unity, and let $\bar{E} = E(\omega)$. Note that $F \subset \bar{E}$ is a Galois extension because, if E is the splitting field of $f(X)$, the $\bar{E}$ is the splitting field of $f(X)(X^p - 1)$. It also follows that $\bar{F} \subset \bar{E}$ is a Galois extension and lemma 8.8.4 on the preceding page implies that

$$\text{Gal}(\bar{E}/\bar{F}) \subset \text{Gal}(E/F)$$

and is, therefore, a solvable group. Induction implies the existence of a tower of pure extensions

(8.8.4) $$\bar{F} \subset \bar{R}_1 \subset \cdots \subset \bar{R}_N$$

with $E \subset \bar{E} \subset \bar{R}_N$. Since $F \subset \bar{F} = F(\omega)$ is a pure extension, we can splice this onto equation 8.8.4 to get the required result. $\square$

EXERCISES.

1. If G is a finite p-group, show that G is solvable.

8.9. The fundamental theorem of algebra

As an application of Galois theory, we will prove

THEOREM 8.9.1 (Fundamental Theorem of Algebra). *If $p(X) \in \mathbb{C}[X]$ is a nonconstant polynomial, then*

$$p(X) = 0$$

has a solution in the complex numbers.

REMARK. In other words, the field, $\mathbb{C}$, is algebraically closed. This also implies that the algebraic closure of $\mathbb{R}$ is $\mathbb{C}$, and that the only proper algebraic extension of $\mathbb{R}$ is $\mathbb{C}$.

This is usually proved using complex analysis. We will need a *little* bit of analysis

FACT 8.9.2 (Intermediate value property). *If $f(X) \in \mathbb{R}[X]$ and $f(x_1) < 0$ and $f(x_2) > 0$, there exists a value x with $x \in (x_1, x_2)$ and $f(x) = 0$.*

COROLLARY 8.9.3. *If $\alpha \in \mathbb{R}$ with $\alpha > 0$, then α has a square root.*

PROOF. Consider the polynomial $X^2 - \alpha$ and set $x_1 = 0$ and $x_2 = \alpha + 1$. $\square$

COROLLARY 8.9.4. *If $\alpha \in \mathbb{C}$, then α has a square root.*

PROOF. If $x = a + bi$, let $r = \sqrt{a^2 + b^2}$ (which exists by corollary 8.9.3 on the previous page) and write

$$\alpha = re^{i\theta}$$

The required square root is $\sqrt{r} \cdot e^{i\theta/2}$. □

COROLLARY 8.9.5. *The field, $\mathbb{C}$, has no extensions of degree 2.*

PROOF. If $F = \mathbb{C}[\alpha]$ is a degree-2 extension, the minimal polynomial of α is quadratic, which means its roots exist in $\mathbb{C}$ — by corollary 8.9.4 on the preceding page. □

LEMMA 8.9.6. *Every polynomial $f(X) \in \mathbb{R}[X]$ of odd degree has a real root.*

PROOF. This is a variation on the proof of corollary 8.9.3 on the previous page. If $f(X) = X^n + a_{n-1}X^{n-1} + \cdots + a_0$, where n is odd, set $t = 1 + \sum_{i=0}^{n-1} |a_i|$. We claim that $f(t) > 0$ and $f(-t) < 0$. Since $|a_i| \leq t - 1$ for all i, we get

$$\begin{aligned} |a_{n-1}t^{n-1} + \cdots + a_0| &\leq (t-1)(t^{n-1} + \cdots + 1) \\ &= t^n - 1 < t^n \end{aligned}$$

It follows that $f(t) > 0$ because the term t^n dominates all the others. The same reasoning shows that $f(-t) < 0$. □

COROLLARY 8.9.7. *There are no field extensions of $\mathbb{R}$ of odd degree > 1.*

PROOF. Let $\mathbb{R} \subset E$ be an extension. If $\alpha \in E$, then the minimal polynomial of α must have even degree, by lemma 8.9.6. It follows that $[\mathbb{R}(\alpha):\mathbb{R}]$ is even which means that $[E:\mathbb{R}] = [E:\mathbb{R}(\alpha)][\mathbb{R}(\alpha):\mathbb{R}]$ is also even. □

Now we are ready to prove theorem 8.9.1 on the previous page.

We claim that it suffices to prove every polynomial in $\mathbb{R}[X]$ has a complex root. If $f(X) \in \mathbb{C}[X]$, then $f(X) \cdot \overline{f(X)} \in \mathbb{R}[X]$, and has complex roots if and only if $f(X)$ does.

If $f(X) \in \mathbb{R}[X]$ is irreducible, let E be a splitting field of $f(X)(X^2 + 1)$ containing $\mathbb{C}$. Then $\mathbb{R} \subset E$ is a Galois extension (see lemma 8.6.2 on page 308). Let $G = \text{Gal}(E/\mathbb{R})$ and let $|G| = 2^n m$ where m is odd.

Theorem 4.8.1 on page 81 implies that G has a subgroup, H, of order 2^n. Let $F = E^H$ be the fixed field.

Corollary 8.6.5 on page 310 implies that $[E:F] = 2^n$ with Galois group H. It follows that $[E:\mathbb{R}] = [E:F] \cdot [F:\mathbb{R}] = 2^n m$, so that $[F:\mathbb{R}] = m$.

Corollary 8.9.7 implies that $m = 1$, so that $|G| = 2^n$. Let $S \subset G$ be the subgroup corresponding to $\text{Gal}(E/\mathbb{C})$. If $2^n > 1$, it follows that $n \geq 1$ so that $|S| > 1$. This is also a 2-group, so it is solvable by exercise 1 on the preceding page. It follows that it has a subgroup, T, of index 2, and E^T is an extension of $\mathbb{C}$ of degree 2 — which contradicts corollary 8.9.5.

We conclude that $E = \mathbb{C}$.

CHAPTER 9

Division Algebras over $\mathbb{R}$

"The next grand extensions of mathematical physics will, in all likelihood, be furnished by quaternions."
— Peter Guthrie Tait, *Note on a Quaternion Transformation*, 1863

9.1. The Cayley-Dickson Construction

Algebras are vector-spaces with a multiplication defined on them. As such they are *more* and *less* general than rings:

▷ *more* general in that they may fail to be associative.
▷ *less* general in that they are always vector-spaces over a field.

DEFINITION 9.1.1. An *algebra* over a field, $\mathbb{F}$, is a vector space, V, over $\mathbb{F}$ that is equipped with a form of *multiplication*:

$$\mu: V \times V \to V$$

that is *distributive* with respect to addition and subtraction
(1) $\mu(x+y,z) = \mu(x,z) + \mu(y,z)$
(2) $\mu(x,y+z) = \mu(x,y) + \mu(x,z)$
for all $x,y,z \in V$, and is also well-behaved with respect to *scalar multiplication*: $\mu(kx,y) = \mu(x,ky) = k\mu(x,y)$ for $k \in \mathbb{F}$.

An algebra is called a *division algebra* if

▷ it contains an *identity element*, $1 \in V$ such that $\mu(1,x) = x = \mu(x,1)$ for all $x \in V$.
▷ Given any nonzero $x, y \in V$, there exists *unique* elements $q_1, q_2 \in V$ such that $y = \mu(q_1, x) = \mu(x, q_2)$.

An *involution* over an algebra is a homomorphism (with respect to '+')

$$\iota: V \to V$$

with $\iota^2 = 1: V \to V$.

REMARK. Division algebras have left and right quotients which may differ (if they are non-associative). For *associative* algebras, the condition of being a division algebra is equivalent to elements having multiplicative inverses.

There are several inequivalent definitions of division algebras in the literature: one defines them as algebras without zero-divisors, and another defines them as algebras for which nonzero elements have multiplicative inverses. If an algebra is finite-dimensional, some of these become equivalent — see exercise 5 on the following page.

In this chapter, we will several explore division algebras over $\mathbb{R}$ — in fact, *all* of them.

Given an algebra with an involution, the Cayley-Dickson Construction creates one of twice the dimension that also has an involution:

DEFINITION 9.1.2 (Cayley-Dickson Construction). Let A be an algebra with an involution given by $\iota(x) = x^*$ for all $x \in A$. Then we can impose an algebra structure on $A \oplus A$ — all ordered pairs of elements (a, b) with $a, b \in A$ with addition defined elementwise — and multiplication defined by

$$(9.1.1) \qquad (a, b)(c, d) = (ac - d^*b, da + bc^*)$$

This has an involution defined by

$$(9.1.2) \qquad (a, b)^* = (a^*, -b)$$

If we apply this construction to the *real numbers* (as a one-dimensional algebra over $\mathbb{R}$) and with the trivial involution (i.e., the identity map), we get a *two-dimensional* algebra over the real numbers — the *complex* numbers (see definition 2.1.1 on page 6).

EXERCISES.

1. If A is an algebra over $\mathbb{R}$ with involution $*: A \to A$, let $B = A \oplus A$ be the Cayley-Dickson construction on A. Show that
   a. if $xx^* \in \mathbb{R}$ for all $x \in A$, then $yy^* \in \mathbb{R}$ for all $y \in B$, and y has a *right* inverse, i.e. there exists y^{-1} such that $yy^{-1} = 1$.
   b. if $x^*x \in \mathbb{R}$ for all $x \in A$, then $y^*y \in \mathbb{R}$ for all $y \in B$, and y has a *left* inverse.
   c. if $xx^* = x^*x$ for all $x \in A$, then $yy^* = y^*y$ for all $y \in B$
   d. if $xx^* \geq 0 \in \mathbb{R}$ for all $x \in A$ and $xx^* = 0 \implies x = 0$, then the same is true for all $y \in B$.

2. Let A be an algebra with an involution, $*: A \to A$, over $\mathbb{R}$ and B be the result of the Cayley-Dickson construction on A. Show that, if

$$(9.1.3) \qquad (xy)^* = y^*x^*$$

for all $x, y \in A$, then

$$(uv)^* = v^*u^*$$

for all $u, v \in B$.

3. How is it possible for an algebra to *simultaneously* have:
   ▷ multiplicative inverses for all nonzero elements
   ▷ zero-divisors?

4. Show that an algebra with the properties described in exercise 3 *cannot* be a division algebra.

5. If A is a finite-dimensional algebra over $\mathbb{R}$ with no zero-divisors, show that every nonzero element has a *multiplicative inverse*.

6. If A is an n-dimensional algebra over $\mathbb{R}$, we can represent each element, $x \in A$, by an $n \times n$ matrix $m_x\colon \mathbb{R}^n \to \mathbb{R}^n$ that represents the action of x by left-multiplication. If A is associative, show that $m_{x \cdot y} = m_x m_y$ (matrix-product) — so matrices faithfully represent the multiplication on A.

If A is *not* associative, what is the relationship between $m_{x \cdot y}$ and $m_x m_y$?

9.2. Quaternions

For years, Hamilton tried to find an analogue to complex numbers for points in *three-dimensional* space. One can easily add and subtract points in $\mathbb{R}^3$— but multiplication and division remained elusive.

In a 1865 letter from Hamilton to his son Archibald, he wrote (see [60]):

> ...Every morning in the early part of the above-cited month[1], on my coming down to breakfast, your (then) little brother William Edwin, and yourself, used to ask me: "Well, Papa, can you multiply triplets?" Whereto I was always obliged to reply, with a sad shake of the head: "No, I can only add and subtract them."...

The answer came to him in October 16, 1843 while he walked along the Royal Canal with his wife to a meeting of the Royal Irish Academy: It dawned on him that he needed to go to *four dimensions*. He discovered the formula

(9.2.1) $$i^2 = j^2 = k^2 = i \cdot j \cdot k = -1$$

and — in a famous act of vandalism — carved it into the stone of Brougham Bridge as he paused on it. A plaque commemorates this inspiration today.

What he discovered was:

DEFINITION 9.2.1. The *quaternions*, denoted $\mathbb{H}$, is an algebra over $\mathbb{R}$ whose underlying vector-space is $\mathbb{R}^4$. An element is written

$$x = a + b \cdot i + c \cdot j + d \cdot k \in \mathbb{R}^4$$

where i, j, k are called the *quaternion units*, and multiplication is defined by the identities

(9.2.2) $$i^2 = j^2 = k^2 = -1$$

(9.2.3) $$i \cdot j = k, \quad j \cdot k = i, \quad k \cdot i = j$$

(9.2.4) $$u \cdot v = -v \cdot u$$

where $u, v = i, j, k$, the quaternion units. The quantity $a \in \mathbb{R}$ is called the *scalar part* of x and $b \cdot i + c \cdot j + d \cdot k \in \mathbb{R}^3$ is called the *imaginary part*, or the *vector-part*.

---

[1]October, 1843.

If $a \in \mathbb{R}$ and $\mathbf{v} = \begin{bmatrix} v_1 \\ v_2 \\ v_3 \end{bmatrix} \in \mathbb{R}^3$, the *vector notation* $(a, \mathbf{v})$ represents the quaternion

$$a + \mathbf{v} = a + v_1 i + v_2 j + v_3 k \in \mathbb{H}$$

REMARK. Note that, unlike complex multiplication, quaternion-multiplication is *not* commutative — as shown in equation 9.2.4 on the previous page. It is left as an exercise to the reader to derive equations 9.2.3 on the preceding page and 9.2.4 on the previous page from

$$i^2 = j^2 = k^2 = i \cdot j \cdot k = -1$$

We can formulate quaternion multiplication in terms of the *vector form*:

PROPOSITION 9.2.2. *If $a + \mathbf{v}, b + \mathbf{w} \in \mathbb{H}$ are two quaternions in their vector form, then (see definitions 6.2.62 on page 208 and 6.2.71 on page 211)*

(9.2.5) $\qquad (a + \mathbf{v})(b + \mathbf{w}) = ab - \mathbf{v} \bullet \mathbf{w} + a \cdot \mathbf{w} + b \cdot \mathbf{v} + \mathbf{v} \times \mathbf{w}$

PROOF. We will prove the claim in the case where $a = b = 0$. The general case will follow by the distributive laws: statements 1 and 2 in definition 9.1.1 on page 323.

$$\begin{aligned}
(0 + v)(0 + w) &= vw_1 i^2 + v_2 w_2 j^2 + v_3 w_3 k^2 + v_1 w_2 ij + v_1 w_3 ik \\
&\quad + v_2 w_1 ji + v_2 w_3 jk \\
&\quad + v_3 w_1 ki + v_3 w_2 kj \\
&= -\mathbf{v} \bullet \mathbf{w} + v_1 w_2 k - v_1 w_3 j \\
&\quad - v_2 w_1 k + v_2 w_3 i \\
&\quad + v_3 w_1 j - v_3 w_2 i \\
&= -\mathbf{v} \bullet \mathbf{w} + i(v_2 w_3 - v_3 w_2) \\
&\quad + j(v_3 w_1 - v_1 w_3) + k(v_1 w_2 - v_2 w_1) \\
&= -\mathbf{v} \bullet \mathbf{w} + \mathbf{v} \times \mathbf{w}
\end{aligned}$$

— see 6.2.71 on page 211. □

DEFINITION 9.2.3. Let $x = a + b \cdot i + c \cdot j + d \cdot k \in \mathbb{H}$ be an element. Define its *conjugate*, denoted x^*, by

$$x^* = a - b \cdot i - c \cdot j - d \cdot k$$

REMARK. Compare this with *complex* conjugation.

Conjugation is well-behaved with respect to multiplication:

PROPOSITION 9.2.4. *If $x, y \in \mathbb{H}$, with $x = a + b \cdot i + c \cdot j + d \cdot k$, then*
(1) $x \cdot x^* = x^* x = a^2 + b^2 + c^2 + d^2 \in \mathbb{R}$
(2) $(x \cdot y)^* = y^* \cdot x^*$ *(note the order-reversal!).*

PROOF. If $x = a + \mathbf{v}$, with $\mathbf{v} = \begin{bmatrix} b \\ c \\ d \end{bmatrix} \in \mathbb{R}^3$, then $x^* = a - \mathbf{v}$ and equation 9.2.5 on the facing page implies that

$x \cdot x^* = a^2 + \mathbf{v} \bullet \mathbf{v} + a\mathbf{v} - a\mathbf{v} + \mathbf{v} \times \mathbf{v}$ (see proposition 6.2.72 on page 212)
$= a^2 + \|\mathbf{v}\|^2$
$= a^2 + b^2 + c^2 + d^2$

If $x = a + \mathbf{v}, y = b + \mathbf{w}$, then

$$(xy)^* = ab - \mathbf{v} \bullet \mathbf{w} - a \cdot \mathbf{w} - b \cdot \mathbf{v} - \mathbf{v} \times \mathbf{w}$$

and $x^* = a - \mathbf{v}, y = b - \mathbf{w}$ so

$$y^*x^* = ab - \mathbf{v} \bullet \mathbf{w} - b \cdot \mathbf{v} - a \cdot \mathbf{w} + \mathbf{w} \times \mathbf{v}$$
$$= ab - \mathbf{v} \bullet \mathbf{w} - a \cdot \mathbf{w} - b \cdot \mathbf{v} - \mathbf{v} \times \mathbf{w}$$

□

DEFINITION 9.2.5. If $x \in \mathbb{H}$, then the *norm* of x, denoted $\|x\| \in \mathbb{R}$, is defined by

$$\|x\| = \sqrt{x \cdot x^*}$$

We immediately have

PROPOSITION 9.2.6. *If $x, y \in \mathbb{H}$, then $\|x\| = 0$ if and only if $x = 0$ and $\|x \cdot y\| = \|x\| \cdot \|y\|$.*

PROOF. The first statement follows immediately from proposition 9.2.4 on the preceding page. The second follows from

$$\|x \cdot y\|^2 = x \cdot y \cdot (x \cdot y)^*$$
$$= x \cdot y \cdot y^* \cdot x^*$$
$$= \|y\|^2 x \cdot x^*$$
$$= \|y\|^2 \|x\|^2$$

□

The discovery of quaternions was regarded as a major breakthrough in the mid 1800's. Many physical laws, including Maxwell's equations for electromagnetic fields were stated in terms of quaternions. This initial enthusiasm faded when people realized:

(1) most applications do not need the *multiplicative* structure of quaternions,
(2) unlike vectors, quaternions are limited to three or four dimensions[2].

---

[2]Depending on one's point of view.

The ebbs and flows of history are fascinating and ironic: What Hamilton and his colleagues regarded as his crowning achievement — quaternions — faded in importance, while his "lesser" discoveries like Hamiltonian energy-functions turned out to be vital to the development of *quantum mechanics*.

In the late 20th century, quaternions found new applications with the advent of computer graphics: *unit* quaternions express *rotations* in $\mathbb{R}^3$ more efficiently than any other known representation. They are widely used in computer games and virtual reality systems[3].

In quantum mechanics, the state of a system is given by a wave function whose values lie in $\mathbb{C}$ — see [48]. In Dirac's *relativistic* wave-equations, the wave function takes its values in $\mathbb{H}$ — see [27] (a reprint of Dirac's original book). Those equations apply to electrons and other spin-1/2 particles, and predicted the existence of positrons (or anti-electrons) long before the particles were observed.

Now we will focus on the relationship between quaternions and rotations in $\mathbb{R}^3$.

Proposition 9.2.6 on the preceding page immediately implies that:

PROPOSITION 9.2.7. *The set of quaternions, $x \in \mathbb{H}$ with $\|x\| = 1$ forms a group called S^3.*

REMARK. The quaternion group, Q, defined on page 106, consists of $\{\pm 1, \pm i, \pm j, \pm k\} \subset S^3$.

Since $\|x\| = 1$, it follows that $\|x\|^2 = x \cdot x^* = 1$ so that $x^{-1} = x^*$.

If $x = a + bi + cj + dk$, the condition $\|x\| = 1$ implies that
$$a^2 + b^2 + c^2 + d^2 = 1$$
which is the equation of a 3-dimensional sphere in $\mathbb{R}^4$. This is the reason for the name S^3.

We need another version of the Euler formula:

THEOREM 9.2.8 (Quaternionic Euler Formula). *Let $\mathbf{u} \in \mathbb{R}^3$ be a unit-vector and let $\alpha \in \mathbb{R}$ be a scalar. Then*
$$e^{\alpha \cdot \mathbf{u}} = \cos \alpha + \sin \alpha \cdot \mathbf{u}$$

PROOF. By equation 9.2.5 on page 326,

(9.2.6) $$(\alpha \cdot \mathbf{u})^2 = -\alpha^2$$

since $\mathbf{u} \times \mathbf{u} = 0$ (see proposition 6.2.72 on page 212) and $\mathbf{u} \bullet \mathbf{u} = 1$. It follows that
$$(\alpha \cdot \mathbf{u})^{2k} = (-1)^k \alpha^{2k}$$
$$(\alpha \cdot \mathbf{u})^{2k+1} = (-1)^k \alpha^{2k+1} \cdot \mathbf{u}$$

The conclusion follows by plugging these equations into the power-series for e^y — equation 2.1.3 on page 7. □

---

[3]Most such systems hide the quaternions they use. In the OpenSimulator and Second Life virtual-reality systems, rotations are *explicitly* called quaternions (somewhat inaccurately).

9.2. QUATERNIONS

Now we will explore the relation between quaternions and rotations in $\mathbb{R}^3$. If we think of $\mathbb{R}^3$ as the subspace of imaginary quaternions, we run into trouble:
$$\mathbf{v} \cdot (\cos \alpha + \sin \alpha \cdot \mathbf{u}) = -\sin \alpha \cdot \mathbf{v} \bullet \mathbf{u} + \cos \alpha \cdot \mathbf{v} + \sin \alpha \cdot \mathbf{v} \times \mathbf{u}$$
which is no longer in $\mathbb{R}^3$ since it has a *scalar component*, namely $-\sin \alpha \cdot \mathbf{v} \bullet \mathbf{u}$.

We next try *conjugating* $\mathbf{v}$ by $e^{\alpha \cdot \mathbf{u}}$
$$e^{\alpha \cdot \mathbf{u}} \cdot \mathbf{v} \cdot e^{-\alpha \cdot \mathbf{u}} = \cos^2 \alpha \cdot \mathbf{v}$$
$$+ \sin \alpha \cos \alpha \, (\mathbf{u}\mathbf{v} - \mathbf{v}\mathbf{u}) - \sin^2 \alpha \mathbf{u}\mathbf{v}\mathbf{u}$$
To simplify this, we note that
(9.2.7) $$\mathbf{u}\mathbf{v} - \mathbf{v}\mathbf{u} = 2\mathbf{u} \times \mathbf{v}$$
(see equation 9.2.5 on page 326) and
$$\mathbf{u}\mathbf{v}\mathbf{u} = \mathbf{u}\,(-\mathbf{v} \bullet \mathbf{u} + \mathbf{v} \times \mathbf{u})$$
$$= -(\mathbf{v} \bullet \mathbf{u})\,\mathbf{u} - \mathbf{u} \bullet (\mathbf{v} \times \mathbf{u}) + \mathbf{u} \times (\mathbf{v} \times \mathbf{u})$$
$$= -(\mathbf{v} \bullet \mathbf{u})\,\mathbf{u} + \mathbf{v}(\mathbf{u} \cdot \mathbf{u}) - \mathbf{u}(\mathbf{u} \cdot \mathbf{v}) \quad \text{(see exercise 25 on page 214)}$$
$$= -2\,(\mathbf{v} \bullet \mathbf{u})\,\mathbf{u} + \mathbf{v}$$
It follows that
$$e^{\alpha \cdot \mathbf{u}} \cdot \mathbf{v} \cdot e^{-\alpha \cdot \mathbf{u}} = \cos^2 \alpha \cdot \mathbf{v} + 2 \sin \alpha \cos \alpha \cdot \mathbf{u} \times \mathbf{v}$$
$$- \sin^2 \alpha \,(-2\,(\mathbf{v} \bullet \mathbf{u})\,\mathbf{u} + \mathbf{v})$$
$$= \left(\cos^2 \alpha - \sin^2 \alpha\right) \cdot \mathbf{v} + 2 \sin \alpha \cos \alpha \cdot \mathbf{u} \times \mathbf{v}$$
$$+ 2 \sin^2 \alpha \cdot (\mathbf{v} \bullet \mathbf{u})\,\mathbf{u}$$
$$= \cos(2\alpha) \cdot \mathbf{v} + \sin(2\alpha) \cdot \mathbf{u} \times \mathbf{v} + (1 - \cos(2\alpha)) \cdot (\mathbf{v} \bullet \mathbf{u})\,\mathbf{u}$$
$$= (\mathbf{v} \bullet \mathbf{u})\,\mathbf{u} + \cos(2\alpha)\,(\mathbf{v} - (\mathbf{v} \bullet \mathbf{u})) + \sin(2\alpha) \cdot \mathbf{u} \times \mathbf{v}$$
The last equation is *identical* to equation 6.2.43 on page 214 so we conclude:
> The conjugation $e^{\alpha \cdot \mathbf{u}} \cdot \mathbf{v} \cdot e^{-\alpha \cdot \mathbf{u}}$ is the result of rotating $\mathbf{v}$ on the axis defined by $\mathbf{u}$ by an angle of 2α.

and we get our result

THEOREM 9.2.9. *If* $\mathbf{v}, \mathbf{u} \in \mathbb{R}^3 \subset \mathbb{H}$ *are vectors with* $\|\mathbf{u}\| = 1$, *then the result of rotating* $\mathbf{v}$ *by an angle* θ *around the axis defined by* $\mathbf{u}$ *is the conjugation in* $\mathbb{H}$
$$\mathbf{r} = e^{(\theta/2) \cdot \mathbf{u}} \cdot \mathbf{v} \cdot e^{-(\theta/2) \cdot \mathbf{u}} \in \mathbb{R}^3 \subset \mathbb{H}$$

REMARK. There are several advantages of using quaternions over 3×3 orthogonal matrices (see corollary 6.2.69 on page 210) in computer graphics systems:
 (1) Quaternions have *four* data-values rather than *nine*.
 (2) Multiplying unit quaternions requires fewer operations than multiplying matrices. Computer graphics systems often require many composites of rotations.

(3) *Continuously varying rotations* are easy to implement in quaternions:
$$\frac{a(t)+b(t)i+c(t)j+d(t)k}{\|a(t)+b(t)i+c(t)j+d(t)k\|}$$
These are widely used in computer games, virtual reality systems, and avionic control systems. This is relatively difficult to do with other ways of defining rotations (like orthogonal matrices or rotations defined in terms of unit vectors and angles — using equation 6.2.43 on page 214).

(4) Quaternions do not have the *gimbal* problem — which often happens when one defines rotations in terms of angles. Certain values of these angles cause singularities: For instance, at latitudes of $\pm 90°$ (the Earth's north and south poles), all *longitudes* are the *same*. If you vary longitude and latitude *continuously*, the process *locks up* whenever the latitude reaches $\pm 90°$.

(5) It is easy to correct for round-off errors in working with quaternions. In performing computations with unit quaternions, round-off errors often give quaternions whose length is not quite 1. One can simply divide these by their norms and produce valid rotations. The same is *not* true of orthogonal matrices. Round off error can give matrices that are only *approximately* orthogonal. There is no simple way to find the actual orthogonal matrix "closest" to one of these.

Ferdinand Georg Frobenius (1849 – 1917) was a German mathematician, best known for his contributions to the theory of elliptic functions, differential equations, number theory, and group theory.

Next, we will discuss a remarkable result[4]:

THEOREM 9.2.10 (Frobenius's Theorem). *The only associative division algebras over $\mathbb{R}$ are $\mathbb{R}$, $\mathbb{C}$, and $\mathbb{H}$.*

REMARK. There are *several* famous theorems called "Frobenius's Theorem," including one in differential topology.

PROOF. Let A be an associative division algebra over $\mathbb{R}$. This is a vector space over $\mathbb{R}$, so suppose its basis is $b = \{v_1 = 1, \ldots, v_n\}$ — here $1 \in A$ generates a one-dimensional sub-algebra $\mathbb{R} \subset A$.

If $n = 1$, then $A = \mathbb{R}$ and we're *done*. Consequently, we'll assume $n > 1$.

For any $d \in A \setminus \mathbb{R}$, let $\mathbb{R}\langle d \rangle$ denote the two-dimensional vector space over $\mathbb{R}$ spanned by 1 and d.

CLAIM 9.2.11. $\mathbb{R}\langle d \rangle$ is the maximal commutative subset of all elements of A that commute with d. Furthermore, it is isomorphic to $\mathbb{C}$.

---

[4]The proof given here is due to Palais in [88]. It is shorter than the original.

PROOF. Let $F \subset A$ be the subspace of maximal dimension of elements which includes $\mathbb{R}\langle d \rangle$ and is commutative. If $x \in A$ commutes with everything in F, then $F + \mathbb{R} \cdot x$ is commutative, so must equal F. It follows that $x \in F$ — so everything that commutes with F is *in* F. If $x \neq 0 \in F$ then x^{-1} commutes with everything in F because $xy = yx \implies x^{-1}y = yx^{-1}$. It follows that $x^{-1} \in F$ so that F is a *field*.

Since $[F:\mathbb{R}]$ is *finite* (it is $\leq n$), proposition 7.2.2 on page 266 implies that the field-extension $\mathbb{R} \subset F$ is *algebraic*. The Fundamental Theorem of Algebra (8.9.1 on page 321) and example 7.5.8 on page 285 imply that $F \cong \mathbb{C}$. □

Since F is isomorphic to $\mathbb{C}$, we can select an element $i \in F$ such that $i^2 = -1$.

It follows that A is also a vector space over $\mathbb{C}$ (and that n *had* to be an even number), with the action of $\mathbb{C}$ on A being given by left-multiplication. If $n = 2$, $A = \mathbb{C}$ and we are *done*.

If $n > 2$, regard *right-multiplication* by $i \in F$ as defining a linear transformation $T: A \to A$:
$$Tx = x \cdot i \in A$$

Since $T^2 = -I$, its eigenvalues are $\pm i$ with corresponding *eigenspaces* $A_\pm$.

(1) If $x \in A_+$, then $Tx = ix$ so that $ix = xi$.
(2) If $x \in A_-$, then $Tx = -ix$, so that $ix = -xi$.
(3) $A = A_+ + A_-$. This follows from the fact that, for all $x \in A$,
$$x = \frac{1}{2}(x - ixi) + \frac{1}{2}(x + ixi)$$
where $x - ixi \in A_+$ and $x + ixi \in A_-$.
(4) $A = A_+ \oplus A_-$. This follows from the preceding statement and the fact that: $A_+ \cap A_- = 0$.
(5) $A_+ = \mathbb{C}$ and $x, y \in A_-$ implies that $xy \in A_+$ — because of statement 2 above and

(9.2.8) $\quad (xy)i = x(yi) = -x(iy) = -(xi)y = (ix)y = i(xy)$

(6) $\dim_\mathbb{C} A_- = 1$. If $x \in A_-$ is *any* nonzero element, statement 5 implies that multiplication by x defines an *isomorphism*[5] $x\cdot: A_- \to A_+$, so they must have the *same* dimension over $\mathbb{C}$. We claim that $x^2 \in \mathbb{R}$ and $x^2 < 0$. Claim 9.2.11 on the preceding page implies that $\mathbb{R}\langle x \rangle$ is a field isomorphic to $\mathbb{C}$, so it contains x^2. In addition, statement 5 implies that $x^2 \in A_+ = \mathbb{C}$ so it is in $\mathbb{C} \cap \mathbb{R}\langle x \rangle$. Since $x \notin \mathbb{C}$, $\mathbb{C} \cap \mathbb{R}\langle x \rangle = \mathbb{R}$. If $x^2 > 0$, it would have square roots $\pm\sqrt{x^2} \in \mathbb{R}$, and a total of *three* square roots in the field $\mathbb{R}\langle x \rangle$, which contradicts corollary 7.2.5 on page 267.
(7) A suitable multiple of $x \in A_-$, called j, has the property that $j^2 = -1$. If $k = ij$, then j and k form a basis for A_- and the set
$$\{1, i, j, k\}$$

---

[5]Because x has an *inverse* — since A is a *division* algebra.

forms a basis for A over $\mathbb{R}$. It is easy to verify that these satisfy the identities for the quaternion units of $\mathbb{H}$ — so we are done.

$\square$

EXERCISES.

1. If $\alpha, \beta \in \mathbb{R}$ and $\mathbf{u}_1, \mathbf{u}_2 \in \mathbb{R}^3 \subset \mathbb{H}$, is the equation $e^{\alpha \cdot \mathbf{u}_1} \cdot e^{\beta \cdot \mathbf{u}_2} = e^{\alpha \cdot \mathbf{u}_1 + \beta \cdot \mathbf{u}_2}$ valid?

2. Where did the proof of theorem 9.2.10 on page 330 use the fact that A was *associative*?

3. Show that the quaternions are the result of applying the Cayley-Dickson construction (definition 9.1.1 on page 324) to the complex numbers, where the involution on $\mathbb{C}$ is complex-conjugation.

4. Given a unit-quaternion
$$x = a + b \cdot i + c \cdot j + d \cdot k$$
with
$$x^{-1} = a - b \cdot i - c \cdot j - d \cdot k$$
Compute the orthogonal matrix for the rotation defined by $y \mapsto xyx^{-1}$. Recall that
$$a^2 + b^2 + c^2 + d^2 = 1$$

5. Show that the quaternionic conjugate can be expressed as an algebraic equation (in sharp contrast to the complex conjugate):
$$q^* = -\frac{1}{2}(q + i \cdot q \cdot i + j \cdot q \cdot j + k \cdot q \cdot k)$$

9.3. Octonions and beyond

Hamilton's success in constructing quaternions inspired mathematicians to research possibilities of higher-dimension division algebras over $\mathbb{R}$. Frobenius's Theorem implies that these cannot possibly be associative.

In 1843, John T. Graves, a friend of Hamilton's, discovered an eight-dimensional division algebra over $\mathbb{R}$ that he called octaves (now called octonions). Cayley independently discovered this algebra and published slightly earlier — causing many to call them Cayley Numbers.

DEFINITION 9.3.1. The *octonions*, $\mathbb{O}$, consists of $\mathbb{R}$-linear combinations of the 8 *octonion units*
$$\{1 = e_0, \ldots, e_7\}$$
where multiplication is distributive over addition and octonion units multiply via the table

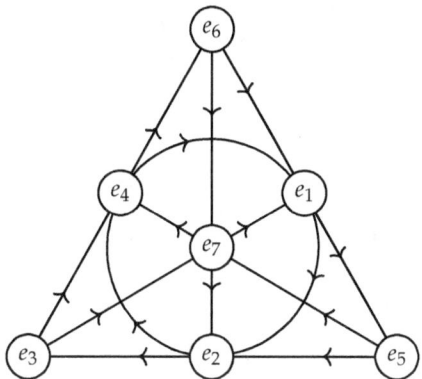

FIGURE 9.3.1.   Fano Diagram

×	e_1	e_2	e_3	e_4	e_5	e_6	e_7
e_1	-1	e_4	e_7	$-e_2$	e_6	$-e_5$	$-e_3$
e_2	$-e_4$	-1	e_5	e_1	$-e_3$	e_7	$-e_6$
e_3	$-e_7$	$-e_5$	-1	e_6	e_2	$-e_4$	e_1
e_4	e_2	$-e_1$	$-e_6$	-1	e_7	e_3	$-e_5$
e_5	$-e_6$	e_3	$-e_2$	$-e_7$	-1	e_1	e_4
e_6	e_5	$-e_7$	e_4	$-e_3$	$-e_1$	-1	e_2
e_7	e_3	e_6	$-e_1$	e_5	$-e_4$	$-e_2$	-1

REMARK. As with quaternions, we have $e_i^2 = -1$ and $e_i e_j = -e_j e_i$ for $i \neq j$. Unlike quaternions

$$(e_3 e_4)e_1 = e_5 \neq e_3(e_4 e_1) = -e_5$$

so the octonions are not associative.

One way to represent this multiplication table in compact form involves the Fano Diagram in figure 9.3.1.

Every pair (e_i, e_j) lies on one of the lines in this diagram (where we regard the circle in the center as a "line"). The third e-element on the line is the product — with the sign determined by whether one must traverse the line in the direction of the arrows (giving a positive result) or in the opposite direction. The lines are also interpreted as cycles, with ends wrapping around to the beginnings.

The bottom line of figure 9.3.1 implies that $e_5 e_2 = e_3$, and $e_5 e_3 = -e_2$. In the second case, we interpret the end of the line as wrapping around to the beginning — and we must traverse it in a direction *opposite* to the arrows.

Since octonions are non-associative, we consider "flavors of associativity."

DEFINITION 9.3.2. An algebra, A, is defined to be
  (1) *power-associative* if any single element generates an associative subalgebra,
  (2) *alternative* if any pair of elements generate an associative subalgebra.

REMARK. Associative algebras clearly satisfy both of these conditions.

Power-associative algebras are ones where expressions like x^4 are well-defined — otherwise, one would have to provide brackets specifying the order of the products, like $((xx)(xx))$.

THEOREM 9.3.3 (Artin's Theorem). *An algebra, A, is alternative if and only if*
$$(aa)b = a(ab)$$
$$(ba)a = b(aa)$$
for all $a, b \in A$.

PROOF. If A is alternative, the two equations given in the statement will be true.

Conversely, assume these two equations are true. Since a and b are *arbitrary*, we conclude that $(aa)a = a(aa)$ and $(bb)a = b(ba)$.

Linearity implies that
$$(a-b)((a-b)a) = a((a-b)a) - b((a-b)a)$$
$$= a(aa) - a(ba) - b(aa) + b(ba)$$

The hypothesis implies that this is equal to
$$((a-b)(a-b))a = (aa)a - (ab)a - b(aa) + (bb)a$$

We compare the two to get
$$a(aa) - a(ba) - b(aa) + b(ba) = (aa)a - (ab)a - b(aa) + (bb)a$$

and cancel $b(aa)$ to get
$$a(aa) - a(ba) + b(ba) = (aa)a - (ab)a + (bb)a$$

subtract off $a(aa) = (aa)a$ and $b(ba) = (bb)a$ to get
$$-a(ba) = -(ab)a$$

so $a(ba) = (ab)a$. This implies that multiplication of *all possible* combinations of a's and b's is associative. We conclude that the algebra generated by a and b is associative. □

THEOREM 9.3.4. *Let A be an associative algebra with involution $*: A \to A$, and let $B = A \oplus A$ be the Cayley-Dickson construction applied to A. If*
$$aa^* = a^*a \text{ and } a + a^*$$
commute with all elements of A, and $(ab)^* = b^*a^*$ for all elements $a, b \in A$, then B is an alternative algebra.

PROOF. We verify the hypotheses of theorem 9.3.3. If $x = (a, b)$ and $y = (c, d)$, the
$$xx = (aa - b^*b, ba + ba^*) = (aa - b^*b, b(a + a^*))$$

and

(9.3.1)
$$(xx)y = ((aa - b^*b)c - d^*b(a + a^*), daa - db^*b + b(a + a^*)c^*)$$
$$= (aac - b^*bc - d^*b(a + a^*), daa - db^*b + b(a + a^*)c^*)$$

Since
$$xy = (ac - d^*b, da + bc^*)$$

we get
$$x(xy) = (a(ac - d^*b) - (da + bc^*)^*b, (da + bc^*)a + b(ac - d^*b)^*)$$
$$= (a(ac - d^*b) - (cb^* + a^*d^*)b, (da + bc^*)a + b(c^*a^* - b^*d))$$
$$= (aac - ad^*b - cb^*b - a^*d^*b, daa + bc^*a + bc^*a^* - bb^*d)$$

(9.3.2)
$$= (aac - (a + a^*)d^*b - cb^*b, daa + bc^*(a + a^*) - bb^*d)$$

Now we simply compare the right sides of equations 9.3.1 and 9.3.2, using the fact that $bb^* = b^*b$ and $a + a^*$ commute with all elements of A to see that they are equal.

Another tedious computation verifies the *second* condition in theorem 9.3.3 on the facing page. □

As with quaternions, we define

DEFINITION 9.3.5. If $x = a_0 + a_1 e_1 + \cdots + a_7 e_7 \in \mathbb{O}$ with the $a_i \in \mathbb{R}$, then its *conjugate*, x^* is defined by

$$x^* = a_0 - a_1 e_1 - \cdots - a_7 e_7$$

LEMMA 9.3.6. *The result of performing the Cayley-Dickson construction (definition 9.1.2 on page 324) on the quaternions is isomorphic to the octonions. We conclude that* $\mathbb{O}$ *is an alternative algebra.*

PROOF. It suffices to prove the first statement for the *units*, following the correspondence

$$1 \leftrightarrow (1, 0)$$
$$e_1 \leftrightarrow (i, 0)$$
$$e_7 \leftrightarrow (-j, 0)$$
$$e_3 \leftrightarrow (k, 0)$$
$$e_6 \leftrightarrow (0, 1)$$
$$e_5 \leftrightarrow (0, -i)$$
$$e_2 \leftrightarrow (0, j)$$
$$e_4 \leftrightarrow (0, -k)$$

Then, it is just a matter of checking that the Cayley-Dickson product matches the table in definition 9.3.1 on page 332.

Exercises 1 on page 324 and 2 on page 324 imply that $xx^* \in \mathbb{R}$ for all $x \in \mathbb{O}$ and $(xy)^* = y^* x^*$ for all $x, y \in \mathbb{O}$. In addition, the *definition* of the involution implies that $x + x^* \in \mathbb{R}$ for all $x \in \mathbb{O}$. The conclusion follows from theorem 9.3.4 on the preceding page. □

As with the complex numbers and quaternions,

DEFINITION 9.3.7. If $x \in \mathbb{O}$, define $\|x\| = \sqrt{xx^*}$, the *norm* of x. If $\|x\| = 0$, then $x = 0$.

REMARK. Exercise 1 on page 324 implies that $xx^* > 0$ for all $x \neq 0$.

LEMMA 9.3.8. *If* $x, y \in \mathbb{O}$ *then there exists an associative subalgebra* $S_{x,y} \subset \mathbb{O}$ *with* $x, y, x^*, y^* \in S_{x,y}$.

PROOF. Write

$$\Re(x) = \frac{x + x^*}{2} \in \mathbb{R}$$
$$\Im(x) = \frac{x - x^*}{2}$$

The subalgebra of $\mathbb{O}$ generated by $\Im(x)$ and $\Im(y)$ is associative because $\mathbb{O}$ is *alternative*. Define $S_{x,y}$ to be this subalgebra *plus* $\mathbb{R}$. Since the elements of $\mathbb{R}$ commute and associate with everything in $\mathbb{O}$, we conclude that $S_{x,y}$ is associative.

Since $\Re(x) \pm \Im(x), \Re(y) \pm \Im(y)$, are in $S_{x,y}$, it follows that $x, y, x^*, y^* \in S_{x,y}$. □

This norm is well-behaved in the sense that:

PROPOSITION 9.3.9. *If* $x, y \in \mathbb{O}$, *then* $\|x \cdot y\| = \|x\| \cdot \|y\|$.

REMARK. John T. Graves originally proved this using brute-force computation from the table in definition 9.3.1 on page 332.

This has an interesting number-theoretic implication: if we set all the coefficients in x and y to *integers*, it implies that a product of two sums of eight perfect squares is *also* a sum of eight perfect squares.

PROOF. If we try to mimic the proof of proposition 9.2.6 on page 327, we wind up with
$$\|x \cdot y\|^2 = (x \cdot y) \cdot (x \cdot y)^*$$
$$= (x \cdot y) \cdot (y^* \cdot x^*)$$

Lemma 9.3.8 on the previous page implies that we can continue the proof of proposition 9.2.6 on page 327 to get
$$(x \cdot y) \cdot (y^* \cdot x^*) = x \cdot (y \cdot y^*) \cdot x^*$$
$$= \|y\|^2 x \cdot x^*$$
$$= \|y\|^2 \|x\|^2$$

□

COROLLARY 9.3.10. *The octonions have no zero-divisors.*

PROOF. If $x, y \in \mathbb{O}$ are both nonzero, $\|x\| \neq 0$ and $\|y\| \neq 0$, so $x \cdot y$ has a *nonzero* norm and *cannot* be 0. □

We finally have:

THEOREM 9.3.11. *The octonions constitute a division algebra.*

PROOF. Given nonzero $x, y \in \mathbb{O}$, we have
$$y^{-1} = \frac{y^*}{yy^*}$$
and $x = (x \cdot y^{-1})y = x(y^{-1}y)$ since x, y, y^* lie in an associative subalgebra (lemma 9.3.8 on the previous page). □

Octonions are not as well understood as quaternions and complex numbers. They have applications to string-theory — see [11].

It turns out that $\mathbb{R}$, $\mathbb{C}$, $\mathbb{H}$, and $\mathbb{O}$ are *all possible* division-algebras over $\mathbb{R}$: Using advanced algebraic topology, Hopf proved that the dimension of a finite-dimensional division algebra over $\mathbb{R}$ must be a power of 2 — see [54].

In 1958, Bott and Milnor extended this result by proving that division algebras over $\mathbb{R}$ must have dimension 1, 2, 4, or 8 — see the later account in [16]. The book [5] contains the shortest proof the author has ever seen. Hurwitz proved (among other things) that an 8-dimensional division algebra over $\mathbb{R}$ is isomorphic to $\mathbb{O}$.

If we apply the Cayley-Dickson Construction to the octonions, we get a sixteen-dimensional algebra over $\mathbb{R}$ called the *sedenions*. Nonzero sedenions have multiplicative inverses but the algebra also has zero-divisors, so it is not a division algebra (see exercise 4 on page 324).

EXERCISES.

1. In equation 9.2.6 on page 328, **u** is a purely imaginary unit quaternion. What happens if it isn't a *unit* quaternion?

2. Prove an octonionic Euler Formula similar to theorem 2.1.2 on page 7.

Hint: Use the Cayley-Dickson construction (definition 9.1.2 on page 324).

Is this a special case of an Euler Formula for algebras of dimension 2^n that result from applying the Cayley-Dickson construction n-times to $\mathbb{R}$?

CHAPTER 10

A taste of category theory

"The language of categories is affectionately known as 'abstract nonsense,' so named by Norman Steenrod. This term is essentially accurate and not necessarily derogatory: categories refer to 'nonsense' in the sense that they are all about the 'structure,' and not about the 'meaning,' of what they represent."
— Paolo Aluffi, *Algebra:* Chapter 0

10.1. Introduction

Category theory is a field as general as set theory that can be applied to many areas of mathematics. It is concerned with the patterns of mappings between mathematical structures and the types of conclusions one can draw from them.

Eilenberg and MacLane developed it with applications to algebraic topology in mind, see [31]. Today, it has applications to many other fields, including computer science — [90]

Once derided as "general nonsense," it has gained acceptance over time. Readers who want more than the "drive-by" offered here are invited to look at MacLane's classic, [71].

Here is an example of the kind of reasoning that category theory uses:

Suppose you want to define the *product* of two mathematical objects, A and B. One way to proceed is to say that $A \times B$ has the following *universal property*:

(1) There exist maps from $A \times B$ to A and B (projections to the factors).
(2) Given *any* maps $f: Z \to A$ and $g: Z \to B$, there is a *unique* map
$$f \times g: Z \to A \times B$$
compatible with the maps from Z to A, B.

This is more succinctly stated with *commutative diagrams*. In a diagram like

(10.1.1)
$$\begin{array}{ccc} U & \xrightarrow{r} & V \\ {\scriptstyle t}\downarrow & & \downarrow{\scriptstyle s} \\ W & \xrightarrow{b} & X \end{array}$$

the arrows represent maps. We will say this diagram commutes if, whenever one can reach a node along different paths, the composite maps one encounters are equal. For instance, the statement that diagram 10.1.1 commutes is equivalent to saying $s \circ r = b \circ t$.

DEFINITION 10.1.1. We can define $A \times B$ by saying that,

(1) it has projection-maps $p_1: A \times B \to A$ and $p_2: A \times B \to B$
(2) whenever we have a diagram with solid arrows

(10.1.2)
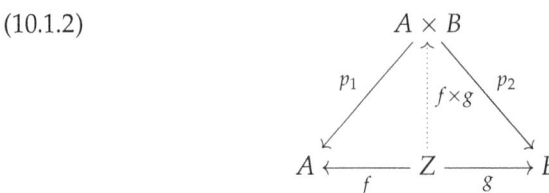

where Z is an arbitrary "object" that maps to A and B — the dotted of arrow *exists*, is *unique*, and makes the whole diagram commute.

In other words, we define $A \times B$ by a general structural property that does not use the inner workings or A or B.

DEFINITION 10.1.2. A *category*, $\mathscr{C}$, is a collection of *objects* and *morphisms*, which are maps between objects. These must satisfy the conditions:

(1) Given objects $x, y \in \mathscr{C}$, $\hom_\mathscr{C}(x, y)$ denotes the morphisms from x to y. This may be an empty set.
(2) Given objects $x, y, z \in \mathscr{C}$ and morphisms $f: x \to y$ and $g: y \to z$, the composition $g \circ f: x \to z$ is defined. In other words a dotted arrow exists in the diagram

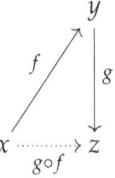

making it commute.
(3) Given objects $x, y, z, w \in \mathscr{C}$ and morphisms $f: x \to y, g: y \to z$, $h: z \to w$, composition is associative, i.e., $h \circ (g \circ f) = (h \circ g) \circ f: x \to w$. This can be represented by a commutative diagram:

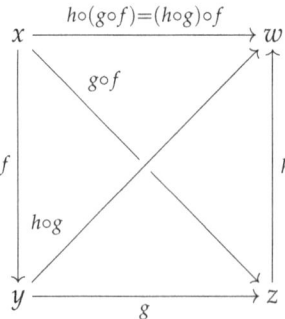

(4) Every object, $x \in \mathscr{C}$, has an identity map $1_x: x \to x$ such that, for any $f: x \to y$, $f \circ 1_x = 1_y \circ f = f: x \to y$. This is equivalent to

saying that the diagram

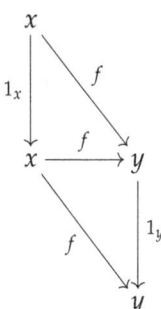

commutes.

After defining something so general, it is necessary to give many examples:
(1) The category, $\mathscr{V}$, of vector-spaces and linear transformations (when defining a category, one must specify the morphisms as well as the objects). Given two vector spaces, V and W, the set of morphisms, $\hom_{\mathscr{V}}(V,W)$, is *also* a vector-space.
(2) The category, $\mathscr{D}$, of vector-spaces where the only morphisms are *identity maps* from vector spaces to themselves. Categories in which the only morphisms are identity maps are called *discrete*. Discrete categories are essentially sets of objects.
(3) The category, $\mathscr{R}$, of rings where the morphisms are ring-homomorphisms.
(4) The category, $\mathscr{N}$, whose objects are positive integers and where the morphisms
$$m \to n$$
are all possible $n \times m$ matrices of real numbers. Composition of morphisms is just matrix-multiplication. This is an example of a category in which the morphisms aren't maps.
(5) The category, $\mathscr{S}$, of sets with the morphisms functions mapping one set to another.
(6) The category, $\mathscr{T}$, of topological spaces where the morphisms are continuous maps.
(7) The category, R-mod of modules over a ring, R. If $M,N \in R$-mod, the set of morphisms, $\hom_{R\text{-mod}}(M,N)$ is usually written $\hom_R(M,N)$.

DEFINITION 10.1.3. A category, $\mathscr{C}$, is called *concrete* if
(1) its objects are sets (possibly with additional structure)
(2) morphisms that are equal as set-mappings are equal in $\mathscr{C}$.

REMARK. All of the examples given above except $\mathscr{N}$ are concrete.

We will also need the dual concept of *coproduct*. A coproduct of a set of objects is essentially their union. So why not just call it the union? Well the categorical definition below is essentially the same as that of the product, except that all of the *arrows* in definition 10.1.1 on page 339 *are reversed* (hence, hardcore category-theorists insist that it is the *co*product):

DEFINITION 10.1.4. An object Z in a category is a *coproduct* of A and B if
(1) there exist maps $i_1: A \to Z$, $i_2: B \to Z$, and
(2) *Any* maps $f_1: A \to W$ $f_2: B \to W$ induce a *unique* map $g: Z \to W$ making the diagram

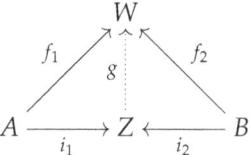

commute. If this is true, we write
$$Z = A \coprod B$$

REMARK. Note that the symbol for a coproduct is an inverted product-symbol — which looks vaguely like a union-symbol. This is appropriate since coproducts have the structural properties of a union. As before, the universal property of coproducts imply that if they exist, they are unique.

Products *map to* their factors, and coproducts have *maps from* their factors. In some cases, products and coproducts are the same.

Coproducts are special cases of something called a push-out:

DEFINITION 10.1.5. Given a diagram

$$\begin{array}{ccc} A & \xrightarrow{i_1} & B \\ {\scriptstyle i_2}\downarrow & & \\ C & & \end{array}$$

an object, Z, is called a *push-out* if it fits into a diagram

$$\begin{array}{ccc} A & \xrightarrow{i_1} & B \\ {\scriptstyle i_2}\downarrow & & \downarrow{\scriptstyle p_1} \\ C & \xrightarrow{p_2} & Z \end{array}$$

with the following *universal property*:

Whenever there exist morphisms v_1 and v_2 making the diagram

$$\begin{array}{ccc} A & \xrightarrow{i_1} & B \\ {\scriptstyle i_2}\downarrow & & \downarrow{\scriptstyle v_1} \\ C & \xrightarrow{v_2} & Q \end{array}$$

commute, there exists a *unique morphism* $j\colon Z \to Q$ making the diagram

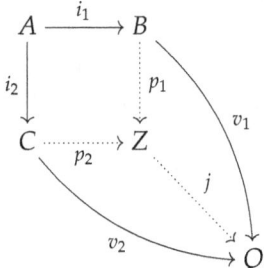

commute.

We also have the dual concept of pull-back — essentially reverse all arrows:

DEFINITION 10.1.6. Given a diagram

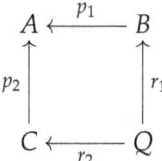

an object, Z, is called a *pull-back* if it fits into a diagram

$$\begin{array}{ccc} A & \xleftarrow{p_1} & B \\ {\scriptstyle p_2}\uparrow & & \uparrow{\scriptstyle v_1} \\ C & \xleftarrow{v_2} & Z \end{array}$$

with the following *universal property*:

Whenever there exist morphisms r_1 and r_2 making the diagram

$$\begin{array}{ccc} A & \xleftarrow{p_1} & B \\ {\scriptstyle p_2}\uparrow & & \uparrow{\scriptstyle r_1} \\ C & \xleftarrow{r_2} & Q \end{array}$$

commute, there exists a *unique morphism* $h\colon Q \to Z$ making the diagram

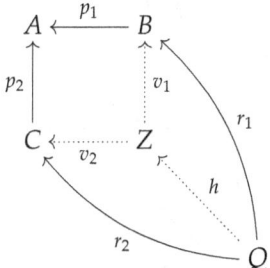

commute.

EXAMPLE 10.1.7. Products, coproducts, push-outs, and pull-backs depend strongly on the category (many categories do not even *have* these constructions):
(1) In the category of sets, the *union* is a coproduct. The Cartesian product is the product, so coproducts and products are very different.
(2) In the category of modules over a ring, the direct sum is the coproduct as well as the product.
(3) In the category of groups, the *free product* is the coproduct.

Category theory expresses the familiar concepts of monomorphism and epimorphism in "arrow-theoretic" terms:

DEFINITION 10.1.8. A morphism $f\colon A \to B$ between objects of a category is:
(1) a *monomorphism* if, for any other object C and any two morphisms $g_1, g_2\colon C \to A$
$$f \circ g_1 = f \circ g_2 \implies g_1 = g_2$$
(2) an *epimorphism* if, for any other object C and any two morphisms $g_1, g_2\colon B \to C$
$$g_1 \circ f = g_1 \circ f \implies g_1 = g_2$$

EXERCISES.

1. If A and B are objects of a category $\mathscr{C}$, show that, for any object $W \in \mathscr{C}$
$$\hom_{\mathscr{C}}(W, A \times B) = \hom_{\mathscr{C}}(W, A) \times \hom_{\mathscr{C}}(W, B)$$

2. Prove the statement above that in the category of modules over a ring, the product and coproduct of two modules V and W is $V \oplus W$. Is the same thing true for infinite products and coproducts?

3. In a category $\mathscr{C}$, if
$$f\colon A \to B$$
is a monomorphism, show that
$$\hom_{\mathscr{C}}(C, A) \xrightarrow{\hom_{\mathscr{C}}(1, f)} \hom_{\mathscr{C}}(C, B)$$
is a monomorphism in the category of sets.

4. If $\mathscr{A}b$ is the category of abelian groups, show that a map is a monomorphism if and only if it is *injective* (in the usual sense) and is an epimorphism if and only if it is *surjective*.

5. In the category of modules over a commutative ring, show that push-outs exist, i.e. that the push-out of

$$\begin{array}{ccc} M & \xrightarrow{i_1} & N \\ {\scriptstyle i_2}\downarrow & & \\ W & & \end{array}$$

is

$$\begin{array}{ccc} M & \xrightarrow{i_1} & N \\ {\scriptstyle i_2}\downarrow & & \downarrow{\scriptstyle p_1} \\ W & \xrightarrow{p_2} & Z \end{array}$$

where

$$Z = \frac{N \oplus W}{i(M)}$$

and

$$i = (i_1, -i_2): M \to N \oplus W$$

6. In the category of modules over a commutative ring, show that pull-backs exist, i.e. that the pull-back of

$$\begin{array}{ccc} M & \xleftarrow{p_1} & N \\ {\scriptstyle p_2}\uparrow & & \\ W & & \end{array}$$

is

$$\begin{array}{ccc} M & \xleftarrow{p_1} & N \\ {\scriptstyle p_2}\uparrow & & \uparrow{\scriptstyle v_1} \\ W & \xleftarrow{v_2} & Z \end{array}$$

where

$$Z = \{(n, w) \in N \oplus W \,|\, p_1(n) = p_2(w) \in M\}$$

10.2. Functors

A *functor* from one category to another is a kind of function of objects *and* morphisms.

DEFINITION 10.2.1. Let $\mathscr{C}$ and $\mathscr{D}$ be categories. A *functor*

$$f: \mathscr{C} \to \mathscr{D}$$

is a function from the objects of $\mathscr{C}$ to those of $\mathscr{D}$ — i.e., if $x \in \mathscr{C}$ then $f(x) \in \mathscr{D}$ with the following additional property:

If $h: x \to y$ is a morphism in $\mathscr{C}$, then f defines, either

▷ a morphism $f(h): f(x) \to f(y)$ in $\mathscr{D}$ — in which case f is called a *covariant functor* or just a *functor*, or
▷ a morphism $f(h): f(y) \to f(x)$ in $\mathscr{D}$ — in which case f is called a *contravariant functor*.

In addition $f(1_x) = 1_{f(x)}$ and $f(j \circ h) = f(j) \circ f(h)$, if f is covariant or $f(j \circ h) = f(h) \circ f(j)$ is f is contravariant.

REMARK. Functors play an extremely important part in algebraic topology, particularly contravariant ones.

Here are some examples:

(1) a functor $f: \mathscr{S} \to \mathscr{S}$ from the category of sets to itself. If $x \in \mathscr{S}$, $f(x) = 2^x$, the power-set or set of all subsets of x. If $d: x \to y$ is a set-mapping and $z \subset x$ is a subset, then $d|z: z \to y$ is a set-mapping whose images is a subset of y. It follows that d induces a natural map

$$2^d: 2^x \to 2^y$$

and this is what we define $f(d)$ to be.

(2) We can define $f: \mathscr{V} \to \mathscr{V}$ to send a real vector-space, $x \in \mathscr{V}$ to its dual, x^* — the vector-space of all linear transformations $\eta: x \to \mathbb{R}$.

If $m: x \to y$ is a linear transformation, and $\mu: y \to \mathbb{R}$ is an element of y^*, the composite $\eta \circ m: x \to \mathbb{R}$ is an element of x^*. We get a natural map $m^*: y^* \to x^*$ and we set $f(m) = m^*$. It follows that f is a *contravariant* functor.

(3) If $\mathscr{F}$ is the category of finite dimensional vector spaces, it is well-known that $x^{**} = x \in \mathscr{F}$, so the functor f defined in statement 2 above actually is a contravariant *isomorphism of categories*

$$f: \mathscr{F} \to \mathscr{F}$$

(4) If $\mathscr{G}$ is the category of groups and $\mathscr{R}$ is that of commutative rings, we can define a functor

$$g_n: \mathscr{R} \to \mathscr{G}$$

that sends a commutative ring $r \in \mathscr{R}$ to $GL_n(r)$, the group of $n \times n$ matrices whose determinant is a unit of r. Since homomorphisms of rings send units to units, it follows that any homomorphism of rings

$$h: r \to s$$

induces a natural homomorphism of groups $g_n(h): GL_n(r) \to GL_n(s)$.

We can classify functors in various ways:

DEFINITION 10.2.2. A functor $f: \mathscr{C} \to \mathscr{D}$ is:

(1) an *isomorphism* if it is a bijection of objects and morphisms,

(2) *full* if it is "surjective on morphisms" — i.e., *every* morphism $g: f(c_1) \to f(c_2) \in \mathscr{D}$ is of the form $f(t)$ where t is a morphism $t: c_1 \to c_2$ (if f is covariant). In the contravariant case, reverse the arrows in $\mathscr{C}$ or $\mathscr{D}$ (but not both).

(3) *faithful* if it is "injective on morphisms" — i.e., given morphisms $m_1, m_2: c_1 \to c_2$ $f(m_1) = f(m_2)$ always implies that $m_1 = m_2$.

For instance, concrete categories are commonly defined as categories that have a faithful functor to the category of sets.

Isomorphism of categories is too stringent a condition in practice. Equivalence of categories is slightly weaker but very useful. To define it, we need:

DEFINITION 10.2.3. Suppose $\mathscr{C}$ is a category and $f: \mathscr{C} \to \mathscr{C}$ is a functor such that $f(x)$ is isomorphic to x for all $x \in \mathscr{C}$. A *natural isomorphism*

$$j_x: x \to f(x)$$

is an isomorphism defined for all objects $x \in \mathscr{C}$ with the property that, for any morphism $g: x \to y$ the diagram

$$\begin{array}{ccc} x & \xrightarrow{j_x} & f(x) \\ {\scriptstyle g}\downarrow & & \downarrow{\scriptstyle f(g)} \\ y & \xrightarrow[j_y]{} & f(y) \end{array}$$

commutes.

REMARK. The thing that makes an isomorphism *natural* is that it is defined for all objects and in a way compatible with all maps between them. Prior to the introduction of category theory, it was common to call certain maps natural without giving any precise definition.

This is a special case of a *natural transformation* of functors:

DEFINITION 10.2.4. If $\mathscr{C}, \mathscr{D}$ are categories and $f, g: \mathscr{C} \to \mathscr{D}$ are functors, a *natural transformation*

$$t: f \to g$$

is a morphism

$$t(x): f(x) \to g(x)$$

defined for all $x \in \mathscr{C}$, such that, for any morphism $m: x \to y$ the diagram

$$\begin{array}{ccc} f(x) & \xrightarrow{t(x)} & g(x) \\ {\scriptstyle f(m)}\downarrow & & \downarrow{\scriptstyle g(m)} \\ f(y) & \xrightarrow[t(y)]{} & g(y) \end{array}$$

commutes.

REMARK. In the notation of definition 10.2.3 on the previous page, a natural *isomorphism* is a natural transformation from the identity functor to f.

It is possible to form a category out of all of the functors between two categories. Natural transformations are the *morphisms* in this "category of functors."

Here's an example of a natural isomorphism:

EXAMPLE 10.2.5. If V is a vector-space, there is a morphism of vector-spaces
$$V \to V^{**}$$
that sends $v \in V$ to the linear function, $t \in V^{**}$, on V^* with $t(r) = r(v)$ for $r \in V^*$. It clearly commutes with all maps of vector-spaces. This is well-known to be an isomorphism if V is finite-dimensional.

And here is one of a natural transformation:

EXAMPLE 10.2.6. If V is a vector-space, define
$$\begin{aligned} f(V) &= V \oplus V \\ g(V) &= V \end{aligned}$$

Now, for every vector-space, V, define
$$t(V): f(V) \to g(V)$$
to be the homomorphism that sends $(v_1, v_2) \in V \oplus V$ to $v_1 + v_2 \in V$. This is easily verified to be a natural transformation.

In considering when two categories are "equivalent," it turns out that requiring them to be isomorphic is usually too restrictive. Instead, we require them to be equivalent in the following sense:

DEFINITION 10.2.7. Given categories $\mathscr{C}$ and $\mathscr{D}$, a pair of functors
$$\begin{aligned} f: \mathscr{C} &\to \mathscr{D} \\ g: \mathscr{D} &\to \mathscr{C} \end{aligned}$$
define an *equivalence of categories* if there exist natural isomorphisms
$$j_x: x \to g \circ f(x)$$
for all $x \in \mathscr{C}$ and
$$k_y: y \to f \circ g(x)$$
for all $y \in \mathscr{D}$.

EXERCISES.

1. Let f be the functor defined in statement 2 on page 346 above and suppose we have a morphism
$$m\colon V_1 \to V_2$$
between vector spaces that is represented by a matrix, A. Describe the matrix-representation of
$$f(m)\colon V_2^* \to V_1^*$$

2. If $\mathscr{F}$ is the category of finite-dimensional vector-spaces, show that the functor f defined in statement 2 on page 346 above is an equivalence of categories
$$f\colon \mathscr{F} \to \mathscr{F}$$
Why isn't it an isomorphism?

3. Find a functor from the category of sets to itself that does *not* always send injective maps to injective maps.

10.3. Adjoint functors

Adjoint functors are ones that *complement* each other in a certain sense. They occur naturally in many settings — Daniel Kan was the first to recognize these patterns (see [59]) and develop a general concept.

As often happens in category theory, the definition is very cryptic without several examples:

DEFINITION 10.3.1. Given two categories, $\mathscr{A}$ and $\mathscr{B}$, functors
$$f\colon \mathscr{A} \to \mathscr{B}$$
$$g\colon \mathscr{B} \to \mathscr{A}$$
are said to be *adjoint* if there exists a natural isomorphism
(10.3.1) $$\hom_{\mathscr{A}}(x, g(y)) = \hom_{\mathscr{B}}(f(x), y)$$
for all $x \in \mathscr{A}$ and $y \in \mathscr{B}$. In this situation, f is called a *left-adjoint* to g and g is called a *right-adjoint* to f. The collection, $(f, g, \mathscr{A}, \mathscr{B})$ is called an *adjunction*.

REMARK. Note that (with rare exceptions) f and g are not *inverses* of each other.

Our terminology was taken from Hilbert space theory: U_1 and U_2 are adjoint *operators* if
$$\langle U_1 x, y \rangle = \langle x, U_2 y \rangle$$
in the Hilbert space, where $\langle *, * \rangle$ is the inner product[1]. Kan was inspired by this equation's similarity (in appearance, not function!) to equation 10.3.1 on page 349 to name his constructs "adjoints". Hilbert-space adjoints are not adjoints in *our* sense except in certain odd settings (see [10]).

---

[1] In *finite dimensions*, U_1 and U_2 are *matrices*, and U_2 is the conjugate-transpose of U_2.

Here is an example of a common pattern — where one of the functors forgets extra structure an object has and regards it as something more primitive (these are called *forgetful functors*):

EXAMPLE 10.3.2. Let $\mathscr{V}_k$ be the category of vector-spaces over a field, k, and let $\mathscr{S}$ be the category of sets. The functor
$$g\colon \mathscr{V}_k \to \mathscr{S}$$
simply maps a vector space onto the set of its elements — it *forgets* the extra structure a vector-space has. The functor
$$f\colon \mathscr{S} \to \mathscr{V}_k$$
maps a set $x \in \mathscr{S}$ to
$$f(x) = \bigoplus_{y \in x} k \cdot x$$
— the vector-space with basis x. Any set-mapping $t\colon x \to f(V)$ extends *uniquely* to a vector-space homomorphism
$$f(t)\colon f(x) \to V$$
since a homomorphism of vector-spaces is determined by its effect on basis-elements. On the other hand, any homomorphism of vector-spaces *is a unique* map of their nonzero elements (regarded as *sets*) so we get a natural equality
$$\hom_{\mathscr{S}}(x, g(y)) = \hom_{\mathscr{V}_k}(f(x), y)$$
for all $y \in \mathscr{V}_k$ and $x \in \mathscr{S}$.

Here's another example of adjoint functors where forgetful functors are *not* involved:

EXAMPLE 10.3.3. Suppose $\mathscr{C}$ is some category and assume that the categorical product (defined in 10.1.1 on page 339) in $\mathscr{C}$ exists. Strictly speaking, it is a functor
$$\prod\colon \mathscr{C} \times \mathscr{C} \to \mathscr{C}$$
where $\mathscr{C} \times \mathscr{C}$ is the category of pairs (x, y) for $x, y \in \mathscr{C}$ and morphisms are defined in a similar way. Now consider the diagonal functor

(10.3.2) $$\Delta\colon \mathscr{C} \to \mathscr{C} \times \mathscr{C}$$

that sends every $x \in \mathscr{C}$ to $(x, x) \in \mathscr{C} \times \mathscr{C}$. The definition of product, and diagram 10.1.2 on page 340 implies that every pair of morphisms
$$\begin{array}{ccc} x & \to & y \\ x & \to & z \end{array}$$
induces a unique morphism $x \to y \prod z$. Such pairs of morphisms are really morphisms
$$\Delta x \to (y, z) \in \mathscr{C} \times \mathscr{C}$$
so we get an equivalence
$$\hom_{\mathscr{C} \times \mathscr{C}}(\Delta x, (y, z)) = \hom_{\mathscr{C}}(x, y \prod z)$$
which implies that $\prod$ is a right-adjoint to Δ. In this case, the adjunction involves a functor of two variables.

Exercises.

1. Example 10.3.3 on the facing page shows that the diagonal functor
$$\Delta: \mathscr{C} \to \mathscr{C} \times \mathscr{C}$$
in equation 10.3.2 on the preceding page is a left-adjoint to the product functor
$$\prod: \mathscr{C} \times \mathscr{C} \to \mathscr{C}$$
Show that it is a right-adjoint to the coproduct functor (definition 10.1.4 on page 342), showing that a functor can be a left-adjoint to one functor and a right-adjoint to another.

10.4. Limits

Limits in category theory are universal constructions somewhat like the union construction in the introduction. We will look at something similar but more complex:

DEFINITION 10.4.1. Suppose $\mathscr{C}$ is a category and I is a partially ordered set of indices. Suppose $\{X_\alpha\}$, for $\alpha \in I$, is a sequence of objects of $\mathscr{C}$. Whenever $\alpha \leq \beta$ suppose there is a morphism
$$f_{\alpha,\beta}: X_\alpha \to X_\beta$$
and whenever $\alpha \leq \beta \leq \gamma$ the diagram

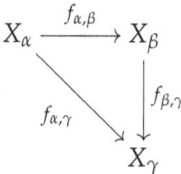

commutes. Then the *direct limit*, $\varinjlim X_\alpha$ has

(1) morphisms $\phi_\alpha: X_\alpha \to \varinjlim X_\alpha$ that make the diagrams

(10.4.1)
$$\begin{array}{ccc} X_\alpha & \xrightarrow{f_{\alpha,\beta}} & X_\beta \\ & \searrow{\phi_\alpha} & \downarrow{\phi_\beta} \\ & & \varinjlim X_\alpha \end{array}$$

commute for all $\alpha, \beta \in I$ with $\alpha \leq \beta$.

(2) the *universal property* that whenever there is an object $Z \in \mathscr{C}$ and morphisms $h_\alpha \colon X_\alpha \to Z$ for all $\alpha \in I$ that make the diagrams

$$X_\alpha \xrightarrow{f_{\alpha,\beta}} X_\beta$$
$$h_\alpha \searrow \quad \downarrow h_\beta$$
$$Z$$

commute for all $\alpha \leq \beta$, then there exists a *unique morphism* $u \colon \varinjlim X_\alpha \to Z$ that makes the diagrams

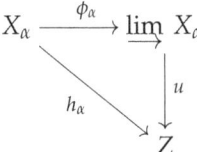

commute for all $\alpha \in I$.

REMARK. To roughly summarize: whenever the X's map to some object, Z, in a way compatible with the $f_{\alpha,\beta}$'s, the direct limit *also* maps to Z.

Some authors require I to be a *directed set*, i.e., for any $\alpha, \beta \in I$ there exists some $\gamma \in I$ with $\alpha \leq \gamma$ and $\beta \leq \gamma$.

Suppose we have two objects K_1 and K_2 that satisfy all of the conditions listed above. Then statement 2 above implies the existence of *unique* maps

$$K_1 \xrightarrow{f} K_2$$
$$K_2 \xrightarrow{g} K_1$$

The composites

$$K_1 \xrightarrow{g \circ f} K_1$$
$$K_2 \xrightarrow{f \circ g} K_2$$

are also unique maps satisfying all of the conditions in definition 10.4.1 on the preceding page. But the respective *identity maps* satisfy these conditions, so we must have

$$g \circ f = 1_{K_1}$$
$$f \circ g = 1_{K_2}$$

Note that the word "unique" is crucial to this discussion. Also note that we have not promised that direct limits exist — only that *if* they exist, they are unique (up to isomorphism). Whether they exist depends on the category.

Hardcore category-theorists prefer the term "filtered colimit" for direct limit. It is also sometimes called the inductive limit. The term "direct limit" seems to be favored by algebraists.

In the case where $\mathscr{C}$ is a concrete category (see definition 10.1.3 on page 341) we can explicitly construct the direct limit.

PROPOSITION 10.4.2. *Let $\mathscr{C}$ be a concrete category (see definition 10.1.3 on page 341) and assume the notation of definition 10.4.1 on page 351. Then*

(10.4.2) $$\varinjlim X_\alpha = \coprod_{\alpha \in I} X_\alpha / \sim$$

the coproduct (see definition 10.1.4 on page 342) or union modulo an equivalence relation, $\sim$, defined by

$$x_\alpha \sim x_\beta$$

for $x_\alpha \in X_\alpha$, $x_\beta \in X_\beta$ if and only if there exists a $\gamma \in I$ with $\alpha \leq \gamma$ and $\beta \leq \gamma$ and

$$f_{\alpha,\gamma}(x_\alpha) = f_{\beta,\gamma}(x_\beta)$$

The maps $\phi_\alpha \colon X_\alpha \to \varinjlim X_\alpha$ are the composites

(10.4.3) $$X_\alpha \to \coprod_{\alpha \in I} X_\alpha \to \coprod_{\alpha \in I} X_\alpha / \sim$$

REMARK. So the maps $f_{\alpha,\beta}$ "glue together" the pieces, X_α, in the union. Elements of the X_α are equivalent if they eventually get glued together.

Concrete categories include the category of rings, vector spaces, and sets. Coproducts of *sets* are just their union. Coproducts of *vector spaces* are their direct sum (see exercise 2 on page 344). Coproducts of *rings* are more complicated and so is the corresponding definition of direct limit.

PROOF. Checking the commutativity of diagrams 10.4.1 on page 351 is straightforward.

If we have morphisms $h_\alpha \colon X_\alpha \to Z$ for all $\alpha \in I$, the disjoint union also maps to Z:

$$\coprod h_\alpha \colon \coprod_\alpha X_\alpha \to Z$$

and in a *unique* way compatible with the inclusions $X_\alpha \hookrightarrow \coprod_\alpha X_\alpha$. The commutativity of the diagrams

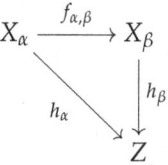

implies that *equivalent* elements under $\sim$ will map to the *same* element of Z via $\coprod h_\alpha$, so that we get a well-defined map

$$u = \coprod h_\alpha / \sim \colon \coprod_\alpha X_\alpha / \sim \to Z$$

that is *unique* (because $\coprod h_\alpha$ was). Since our construction has the same universal property as the direct limit, it must be isomorphic to it (in a unique way). □

The reader may still find the concept of direct limit hard to grasp. We claim that direct limits are a kind of "generalized union," something implied the following:

PROPOSITION 10.4.3. *Assuming the notation of definition 10.4.1 on page 351 and that $\mathscr{C}$ is a concrete category, we have*

$$\tag{10.4.4} \varinjlim X_\alpha = \bigcup_{\alpha \in I} \phi_\alpha(X_\alpha)$$

If all of the maps $f_{\alpha,\beta}$ are injective, then the maps $\phi_\alpha \colon X_\alpha \to \varinjlim X_\alpha$ are also injective.

REMARK. If the $f_{\alpha,\beta}$ are injective, the direct limit is *literally* a union of the X_α.

If they are *not* injective, and $\mathscr{C}$ is a category of groups, rings, or vector-spaces, the direct limit essentially divides out by the kernels of the $f_{\alpha,\beta}$ — "forcing them" to be injective — and *then* takes the union.

PROOF. Equation 10.4.4 follows immediately from equations 10.4.2 and 10.4.3 on the preceding page.

If all of the $f_{\alpha,\beta}$ are injective, then the only way two elements $x_1, x_2 \in X_\alpha$ can become equivalent is for $x_1 = x_2 \in X_\alpha$. □

EXAMPLE 10.4.4. Suppose I is the set of positive integers and $i \leq j$ is $i|j$. Let

$$R_n = \mathbb{Z}\left[\frac{1}{n}\right]$$

Then

$$f_{n,m} \colon R_n \to R_m$$

when $n|m$, is defined to send

$$\frac{1}{n} \mapsto \frac{k}{m}$$

where $k = m/n$. We claim that $\varinjlim R_n = \mathbb{Q}$. The maps $f_{n,m}$ are all injective and each $R_n \subset \mathbb{Q}$ so

$$\varinjlim R_n = \bigcup_{n=1}^{\infty} R_n \subset \mathbb{Q}$$

Since every possible denominator occurs in some R_n this inclusion must actually be an equality.

DEFINITION 10.4.5. In the notation of definition 10.4.1 on page 351, a subset $I' \subset I$ is said to be *cofinal*, if for every $\alpha \in I$, there exists a $\beta \in I'$ such that $\alpha \leq \beta$.

REMARK. Cofinal subsets are important because they *determine* colimits and limits:

PROPOSITION 10.4.6. *In the notation of definition 10.4.1 on page 351, if $I' \subset I$ is a cofinal subset then*

$$\varinjlim X_\beta = \varinjlim X_\alpha$$

where α runs over I and β runs over I'.

REMARK. This is significant because direct limits are sometimes easier to compute with cofinal subsets.

PROOF. This follows immediately from the universal properties: since all X_β map to $\varinjlim X_\alpha$, we get a unique map

$$\varinjlim X_\beta \to \varinjlim X_\alpha$$

Since every $\alpha \in I$ is $\leq \beta$ for some $\beta(\alpha) \in I'$, we get unique maps from all of the $X_\alpha \to X_{\beta(\alpha)}$ inducing a unique map to

$$\varinjlim X_\alpha \to \varinjlim X_\beta$$

□

Recall the concept of rings of fractions in definition 6.4.3 on page 249. We can define this in terms of a universal property:

PROPOSITION 10.4.7. *In the category, $\mathscr{R}$, of commutative rings the pair $(S^{-1}A, \iota)$ has the universal property: every element of S maps to a unit in $S^{-1}A$, and any other homomorphism $f: A \to B$ with this property factors uniquely through ι:*

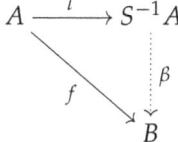

PROOF. If β exists

$$s\frac{a}{s} = a \implies \beta(s)\beta\left(\frac{a}{s}\right) = \beta(a) = f(a)$$

so just *define*

$$\beta\left(\frac{a}{s}\right) = f(a)f(s)^{-1}$$

Now

$$\frac{a}{s} = \frac{b}{t} \implies z(at - bs) = 0 \text{ for some } z \in S$$

and this implies

$$f(a)f(t) - f(b)f(s) = 0$$

since $f(z)$ is a unit. □

Modules of fractions also have a universal property:

PROPOSITION 10.4.8. *If M is a module over a ring A with multiplicative set $S \subset A$ and N is a module over $S^{-1}A$ then N is also a module over A via the standard inclusion $\iota: A \to S^{-1}A$. Any homomorphism*

$$f: M \to N$$

over A extends uniquely to a homomorphism of $S^{-1}A$ modules

$$\bar{f}: S^{-1}M \to N$$

that makes the diagram

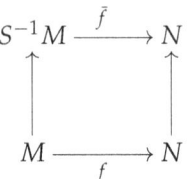

where the vertical maps are homomorphisms of modules covering the map $\iota\colon A \to S^{-1}A$.

PROOF. Left as an exercise to the reader. □

One bonus of this approach is the following (this is very similar to example 10.4.4 on page 354):

COROLLARY 10.4.9. *Suppose A is a commutative ring with a multiplicative set $S \subset A$. Define an order on the elements of S via:*

$s_1 \leq s_2$ if there exists an element $x \in R$, such that $s_2 = x \cdot s_1$. Define maps $f_{s,t}\colon A_s \to A_t$ for $s, t \in S$ with $t = x \cdot s$ by

$$\frac{a}{s} \mapsto \frac{a \cdot x}{t}$$

Then

$$S^{-1}A = \varinjlim A_s$$

REMARK. Recall the notation A_h in definition 6.4.5 on page 250.

The proof below almost seems like "cheating" — we ignore algebraic subtleties and give an "arrow-theoretic" argument. This was one of the early complaints against category theory (and [**31**]).

The philosophy of category theory is that if one can prove something *merely* by analyzing patterns of mappings, one should do so.

PROOF. The ring of fractions, $S^{-1}A$, and the direct limit, $\varinjlim A_s$, have the same universal property. □

If we *reverse* all of the arrows that occur in the diagrams of definition 10.4.1 on page 351, we get another important construction — the *inverse limit*:

DEFINITION 10.4.10. *Suppose $\mathscr{C}$ is a category and I is a partially ordered set of indices. Suppose $\{X_\alpha\}$, for $\alpha \in I$, is a sequence of objects of $\mathscr{C}$. Whenever $\alpha \leq \beta$ suppose there is a morphism*

$$f_{\alpha,\beta}\colon X_\alpha \leftarrow X_\beta$$

and whenever $\alpha \leq \beta \leq \gamma$ the diagram

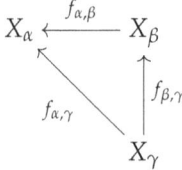

commutes. Then the inverse limit, $\varprojlim X_\alpha$ has

(1) morphisms $\pi_\alpha\colon X_\alpha \leftarrow \varprojlim X_\alpha$ that make the diagrams

(10.4.5)
$$\begin{array}{ccc} X_\alpha & \xleftarrow{f_{\alpha,\beta}} & X_\beta \\ {}_{\pi_\alpha}\nwarrow & & \uparrow {}_{\pi_\beta} \\ & \varprojlim X_\alpha & \end{array}$$

commute for all $\alpha, \beta \in I$ with $\alpha \leq \beta$.

(2) the *universal property* that whenever there is an object $Z \in \mathscr{C}$ and morphisms $h_\alpha\colon X_\alpha \leftarrow Z$ for all $\alpha \in I$ that make the diagrams

$$\begin{array}{ccc} X_\alpha & \xleftarrow{f_{\alpha,\beta}} & X_\beta \\ {}_{h_\alpha}\nwarrow & & \uparrow {}_{h_\beta} \\ & Z & \end{array}$$

commute for all $\alpha \leq \beta$, then there exists a *unique morphism* $u\colon \varprojlim X_\alpha \leftarrow Z$ that makes the diagrams

$$\begin{array}{ccc} X_\alpha & \xleftarrow{\pi_\alpha} & \varprojlim X_\alpha \\ {}_{h_\alpha}\nwarrow & & \uparrow {}_{u} \\ & Z & \end{array}$$

commute for all $\alpha \in I$.

REMARK. So anything that maps to all of the X's in a way compatible with the maps $f_{\alpha,\beta}$ also maps to the inverse limit.

Since the inverse limit has a universal property, it is unique up to isomorphism (if it exists at all!). Hardcore category-theorists prefer the term "limit" for the inverse limit.

As with the direct limit, we have an explicit construction of the inverse limit in categories of groups, rings, and vector-spaces:

PROPOSITION 10.4.11. *Let $\mathscr{C}$ be a category of groups, rings, or vector-spaces and assume the notation of definition 10.4.10 on the preceding page. Then*

$$\varprojlim X_\alpha \subset \prod_{\alpha \in I} X_\alpha$$

is the subset of (possibly infinite) sequences

$$(\ldots, x_\alpha, \ldots)$$

where $x_\alpha \in X_\alpha$ for all $\alpha \in I$, and with the property that, whenever $\alpha \leq \beta$, $f_{\alpha,\beta}(x_\beta) = x_\alpha$.

The maps $\pi_\beta\colon X_\beta \leftarrow \varprojlim X_\alpha$ are the composites

$$\varprojlim X_\alpha \hookrightarrow \prod_{\alpha \in I} X_\alpha \to X_\beta$$

where $\prod_{\alpha \in I} X_\alpha \to X_\beta$ is just projection to a factor.

REMARK. Whereas the direct limit *glues together* the X's via the $f_{\alpha,\beta}$, the inverse limit *selects* infinite sequences compatible with the $f_{\alpha,\beta}$. If $\mathscr{C}$ is a category of groups, rings, or vector-spaces, then $f_{\alpha,\beta}$ will preserve this structure and the inverse limit will also have it.

PROOF. We only have to verify that this construction has the same universal property as the inverse limit. If $Z \in \mathscr{C}$ and has maps $h_\beta: Z \to X_\beta$ for all $\beta \in I$, then we get a *unique* map
$$\prod h_\alpha: Z \to \prod_{\alpha \in I} X_\alpha$$
—see the definition of product in diagram 10.1.2 on page 340 and extend it to an arbitrary number of factors. The commutativity of the diagrams

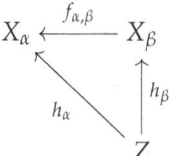

implies that the image of $\prod h_\alpha$ will *actually* lie within $\varprojlim X_\alpha \subset \prod X_\alpha$. This verifies the universal property. □

As we noted earlier, direct limits are "generalized unions". Under some circumstances, inverse limits are like "generalized intersections:"

PROPOSITION 10.4.12. *Under the assumptions of proposition 10.4.11, suppose the $f_{\alpha,\beta}: X_\beta \to X_\alpha$ are injective for all $\alpha, \beta \in I$. Then so is*
$$\pi_\beta: \varprojlim X_\alpha \to X_\beta$$
for all $\beta \in I$.

If there exists $X \in \mathscr{C}$ such that
$$X_\alpha \subset X$$
for all $\alpha \in I$ and $\alpha \leq \beta$ if and only if $X_\beta \subset X_\alpha$ where $f_{\alpha,\beta}: X_\beta \to X_\alpha$ is the inclusion, then
$$\varprojlim X_\alpha = \bigcap_{\alpha \in I} X_\alpha$$

PROOF. If all of the $f_{\alpha,\beta}$ are injective, then a sequence
$$(\ldots, x_\alpha, \ldots)$$
is *uniquely* determined by its αth member: the βth element, x_β, to the right of x_α will be $f_{\alpha,\beta}^{-1}(x_\alpha)$, and this is unique. It follows that the projection map
$$\varprojlim X_\beta \to X_\alpha$$
is injective. Its image is the set of all $x_\alpha \in X_\alpha$ of the form $f_{\alpha,\beta}(x_\beta)$ for *all* β with $\alpha \leq \beta$.

Moving on to the second statement, we have proved that
$$\varprojlim X_\alpha \subset \bigcap_{\alpha \in I} X_\alpha$$

Equality follows from both objects having the same universal property. □

From the proof of proposition 10.4.12 on the preceding page, it is clear that $x_\beta = f_{\alpha,\beta}^{-1}(x_\alpha)$ (if it exists). If $f_{\alpha,\beta}$ is *not* injective, *all* elements of $f_{\alpha,\beta}^{-1}(x_\alpha)$ give rise to new sequences from the β-position on. For instance:

PROPOSITION 10.4.13. *Under the assumptions of proposition 10.4.11, suppose that the set of indices, I, is the disjoint union of $\{I_j\}$ — i.e. no $\alpha \in I_j$ is comparable with any $\beta \in I_k$ with $j \neq k$. Then*

$$\varprojlim_{\alpha \in I} X_\alpha = \prod_j \varprojlim_{\beta \in I_j} X_\beta$$

PROOF. Since the I_j are disjoint they have no influence over each other — all sequences from I_j are paired with all sequences from I_k, $j \neq k$. $\square$

It follows that $\varprojlim X_\alpha$ can be very large indeed:

EXAMPLE 10.4.14. Let I be positive integers ordered in the usual way, let $p \in \mathbb{Z}$ be a prime, and let $X_n = \mathbb{Z}_{p^n}$ for all n. The maps $f_{n,m}: \mathbb{Z}_{p^m} \to \mathbb{Z}_{p^n}$ are reduction modulo p^n (where $n \leq m$).

Then

$$\mathbb{Z}_{(p)} = \varprojlim X_n$$

is called the *p-adic integers* and its field of fractions is called the *p-adic numbers*, $\mathbb{Q}_{(p)}$. Reduction modulo p^n (for all n) defines an injection

$$\mathbb{Z} \hookrightarrow \mathbb{Z}_{(p)}$$

and, like $\mathbb{R}$, $\mathbb{Z}_{(p)}$ is *uncountable* for all p. These rings were first described by Hensel in 1897 (see [52]), with a definition *wildly* different from ours. Hensel showed that one could define infinite series in $\mathbb{Q}_{(p)}$ like that for e^x with many number-theoretic applications.

Technically, elements of $\mathbb{Z}_{(p)}$ are "infinite series"

$$n_0 + n_1 \cdot p + n_2 \cdot p^2 + \cdots$$

such that $0 \leq n_i < p$ for all i. The image $\mathbb{Z} \subset \mathbb{Z}_{(p)}$ consists of the "series" that terminate after a finite number of terms. Two such "series" are equal if all corresponding n_i's are equal. Define a metric on $\mathbb{Z}$ via the p-adic valuation defined in 15.1.3 on page 468

$$d(m_1, m_2) = \left(\frac{1}{2}\right)^{v_p(m_1 - m_2)}$$

so $v_p(m_1 - m_2)$ is the highest power of p such that

$$p^k | (m_1 - m_2)$$

Then $\mathbb{Z}_{(p)}$ is the *completion* of $\mathbb{Z}$ in this metric, and two elements, P, Q of $\mathbb{Z}_{(p)}$ are equal if and only if

$$\lim_{i \to \infty} d(P_i, Q_i) = 0$$

where P_i and Q_i are, respectively, the ith partial sums of P and Q.

> Kurt Wilhelm Sebastian Hensel (1861 – 1941) was a German mathematician born in Königsberg. He is known for work in number theory, including the introduction of p-adic numbers.

Here's another interesting example:

EXAMPLE 10.4.15. Let A be a ring and let $\mathfrak{m} = (X) \subset A[X]$ be an ideal. Then
$$A[[X]] = \varprojlim A[X]/\mathfrak{m}^n$$

On the other hand, if there is a top index in I, the inverse limit is well-behaved:

PROPOSITION 10.4.16. *Under the assumptions of proposition 10.4.11 on page 357, suppose there exists $\gamma \in I$ such that $\alpha \leq \gamma$ for all $\alpha \in I$. Then*
$$\varprojlim X_\alpha = X_\gamma$$

PROOF. We could do an algebraic analysis of this statement, but it is easier to "cheat," so our proof is: they both have the same universal property. □

EXERCISES.

1. In the category of groups, free groups can be defined using a universal property. What is it?

2. Let $\mathscr{C}$ be a category and let $\mathscr{C}_\infty$ be the category of infinite sequences
$$\cdots \to x_2 \to x_1$$
of morphisms of objects of $\mathscr{C}$. Then
$$\varprojlim *: \mathscr{C}_\infty \to \mathscr{C}$$
is a functor. Show that this is an adjoint of the functor
$$\begin{aligned} \Delta_\infty: \mathscr{C} &\to \mathscr{C}_\infty \\ x &\mapsto \cdots \xrightarrow{1} x \xrightarrow{1} x \end{aligned}$$

3. Suppose $\{X_\alpha\}$, for $\alpha \in I$, is a sequence of objects of a concrete category, $\mathscr{C}$. Whenever $\alpha \leq \beta$ suppose there is a morphism
$$f_{\alpha,\beta}: X_\alpha \to X_\beta$$
and whenever $\alpha \leq \beta \leq \gamma$ the diagram

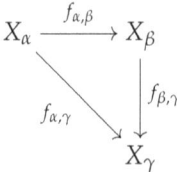

commutes. If $x, y \in X_\alpha$ map to the same element of $\varinjlim X_\alpha$, show that there exists a $\beta \geq \alpha$ such that $f_{\alpha,\beta}(x) = f_{\alpha,\beta}(y)$.

4. Show that Prüfer group (see example 4.1.10 on page 38) is given by the direct limit
$$\mathbb{Z}/p^\infty = \varinjlim \mathbb{Z}/p^n\mathbb{Z}$$
where $\mathbb{Z}/p^n\mathbb{Z} \hookrightarrow \mathbb{Z}/p^{n+1}\mathbb{Z}$ multiplies elements by p. Contrast this with example 10.4.14 on page 359.

10.5. Abelian categories

An abelian category is essentially one in which morphisms of objects have kernels and cokernels. The standard example is the category of modules over a commutative ring. The official definition is:

DEFINITION 10.5.1. A category $\mathscr{A}$ is *abelian* if:
(1) it has *products* and *coproducts* of all pairs of objects,
(2) it has a *zero object* (which behaves like an identity for products and coproducts),
(3) all morphisms have a kernel and cokernel:

   (a) if $A \xrightarrow{f} B$ is a morphism, there exists a monomorphism $K \xrightarrow{m} A$ such that $f \circ m = 0$, and if $C \xrightarrow{g} A$ is any morphism with $f \circ g = 0$, there exists a *unique* morphism $v: C \to K$ such that

   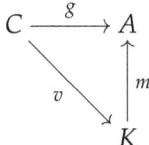

   commutes.

   (b) if $A \xrightarrow{f} B$ is a morphism, there exists an epimorphism $B \xrightarrow{e} E$ such that $e \circ f = 0$, and if $g: B \to D$ is any morphism with $g \circ f = 0$, then there exists a *unique* morphism $v: E \to D$ such that

   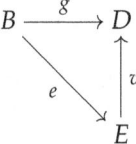

(4) the set of morphisms between two objects, $\hom_\mathscr{A}(A,B)$, has the structure of an abelian group for which composition is distributive over sums.

If $F: \mathscr{A} \to \mathscr{A}'$ is a functor between abelian categories, F is said to be *additive* if, whenever we have morphisms $g_1, g_2: M \to N$ in $\mathscr{A}$,
$$F(g_1 + g_2) = F(g_1) + F(g_2): F(M) \to F(N)$$

REMARK. Since kernels and cokernels are defined by universal properties, they are unique up to isomorphism. Also note that the kernel and cokernel defined *here* are "arrow-theoretic" versions of the more familiar algebraic concepts — i.e., morphisms.

Examples of *additive* functors include $M \otimes_{\mathbb{Z}} *$ and $\hom_{\mathscr{A}}(M, *)$. The functor $F\colon \mathscr{A}b \to \mathscr{A}b$ that sends an abelian group G to $G \otimes_{\mathbb{Z}} G$ is an example of a functor that is *not* additive (it is "quadratic").

The concept of a projective module is well-defined for an abelian category

DEFINITION 10.5.2. If $A, B \in \mathscr{A}$ are objects of an abelian category with $f\colon A \to B$ an epimorphism (see definition 10.1.8 on page 344), an object P is *projective* if, for any morphism $g\colon P \to B$, there exists a morphism $\ell\colon P \to A$ that fits into a commutative diagram

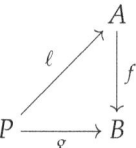

The category $\mathscr{A}$ will be said to have *enough projectives* if, for any object A there exists a projective object P and an epimorphism $P \to A$.

REMARK. For instance, the category of modules over a ring *always* has enough projectives because every module is the surjective image of a *free* module.

If we reverse all of the arrows in 10.5.2, we get a definition of *injective* objects:

DEFINITION 10.5.3. If $A, B \in \mathscr{A}$ are objects of an abelian category with $f\colon B \to A$ a monomorphism (see definition 10.1.8 on page 344), an object I is *injective* if, any morphism $g\colon B \to I$, there exists a morphism $e\colon A \to I$ that fits into a commutative diagram

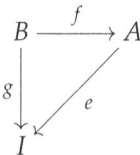

The category $\mathscr{A}$ will be said to have *enough injectives* if, for any object A there exists an injective object I and a monomorphism $A \to I$.

REMARK. Homomorphisms into injective objects *extend* to other objects containing them.

The categorical property of a product in definition 10.1.1 on page 339 implies that arbitrary *products* of injective objects are injective.

EXAMPLE 10.5.4. Theorem 4.6.20 on page 70 shows that injective objects in the category of abelian groups are precisely the *divisible* abelian groups, like $\mathbb{Q}$, $\mathbb{R}$, or $\mathbb{Q}/\mathbb{Z}$.

Over the category of modules, we have a criterion for injectivity:

PROPOSITION 10.5.5 (Baer's Criterion). *If R is a commutative ring, an R-module, I is injective if and only if every homomorphism $\mathfrak{J} \to I$ from an ideal $\mathfrak{J} \subset R$ extends to a homomorphism $R \to I$.*

REMARK. In other words, in the category of modules over a ring, injectivity only has to be verified for *ideals* of the ring.

PROOF. The only-if part follows from the definition of injective modules.

Conversely, suppose $A \subset B$ are R-modules and $f: A \to I$ is a homomorphism and we consider extensions to submodules B' with

$$A \subset B' \subseteq B$$

These extensions are partially ordered by inclusion. Zorn's lemma (14.2.12 on page 465) implies that there is a maximal one, B' say. If $B' \neq B$, we will get a contradiction. If $b \in B \setminus B'$ then $\mathfrak{J}(b) = \{r \in R | r \cdot b \in B'\}$ is an ideal of R and $\mathfrak{J}(b) \xrightarrow{\cdot b} B' \xrightarrow{f} I$ defines a homomorphism into I. The hypotheses imply that this extends to all of R, so $b \in B'$. □

Since all abelian groups are modules over $\mathbb{Z}$ and all ideals of $\mathbb{Z}$ are of the form (m) for $m \in \mathbb{Z}$, Baer's Criterion implies that

PROPOSITION 10.5.6. *An abelian group, G, is injective if and only if it is divisible (see definition 4.6.17 on page 69).*

Since quotients of divisible groups are divisible (see proposition 4.6.18 on page 69), we conclude:

PROPOSITION 10.5.7. *Any quotient of an injective object in $\mathscr{A}b$ is injective in $\mathscr{A}b$.*

EXAMPLE 10.5.8. In the category of abelian groups, $\mathscr{A}b$, $\mathbb{Q}$ and $\mathbb{Q}/\mathbb{Z}$ are injective.

This allow us to conclude that the category of abelian groups has enough injectives (see [8]):

PROPOSITION 10.5.9. *If A is an abelian group and*

$$A = \frac{F}{K}$$

where F is a free abelian group, then $F \otimes_\mathbb{Z} \mathbb{Q}$ is injective and

$$A \hookrightarrow \frac{F \otimes_\mathbb{Z} \mathbb{Q}}{K}$$

is an injective abelian group containing A so the category of abelian groups, $\mathscr{A}b$, has enough injectives.

It is interesting that this result immediately extends to the category of modules over an arbitrary ring (see [30]):

PROPOSITION 10.5.10. *If R is a ring and M is an R-module and I is an injective abelian group, then*

$$I(R) = \hom_{\mathscr{A}b}(R, I)$$

is an injective R-module — with R acting on the first factor via

$$(r' \cdot \varphi)(r) = \varphi(r' \cdot r)$$

for $\varphi \in \hom_{\mathscr{A}b}(R, I)$. In addition, there exists an injective R-module N and an inclusion

$$M \hookrightarrow N$$

so that the category of modules over R, $\mathscr{M}_R$, has enough injectives.

PROOF. Suppose $A \subset B$ is an inclusion of R-modules and $g\colon A \to \hom_{\mathscr{A}b}(R, I)$ is a homomorphism. We will show that this extends to B. Define a natural map

$$\begin{aligned} \iota\colon I(R) &\to I \\ f &\mapsto f(1) \end{aligned}$$

The composite $\iota \circ g\colon A \to I$, regarded as a map of abelian groups, extends to $\bar{g}\colon B \to I$ and we define

$$\begin{aligned} G\colon B &\to \hom_{\mathscr{A}b}(R, I) \\ b &\mapsto (r \mapsto \bar{g}(r \cdot b)) \end{aligned}$$

— a homomorphism of R-modules and the desired extension.

To prove the second statement, note the existence of a monomorphism

$$\begin{aligned} f\colon M &\to \hom_{\mathscr{A}b}(R, M) \\ m &\mapsto (r \mapsto r \cdot m) \end{aligned}$$

of R-modules. If we "forget" the module structure of M and regard it only as an abelian group, there exists an injective abelian group and a morphism

$$g\colon M \to I$$

The composite

$$M \xrightarrow{f} \hom_{\mathscr{A}b}(R, M) \xrightarrow{\hom_{\mathscr{A}b}(1,g)} \hom_{\mathscr{A}b}(R, I)$$

is a monomorphism (see exercise 4 on the facing page). □

EXERCISES.

1. If M and N are modules over a ring, show that there is a natural isomorphism
$$M \oplus N \cong N \oplus M$$

2. If A is an abelian group, show that $\hom(A, \mathbb{Q}/\mathbb{Z}) = 0$ if and only if $A = 0$.

3. Show that if we have a monomorphism
$$f: A \to B$$
where A is injective, there exists a map
$$g: B \to A$$
such that $g \circ f = 1: A \to A$.

4. If $\mathscr{A}$ is an abelian category and
$$0 \to A \xrightarrow{r} B \xrightarrow{s} C \to 0$$
is an exact sequence — i.e., $r = \ker s$ and $s = \operatorname{coker} r$, show that
$$0 \to \hom_{\mathscr{A}}(D, A) \xrightarrow{\hom_{\mathscr{A}}(1,r)} \hom_{\mathscr{A}}(D, B) \xrightarrow{\hom_{\mathscr{A}}(1,s)} \hom_{\mathscr{A}}(D, C)$$
is exact.

10.6. Direct sums and Tensor products

The standard example of an abelian category is $\mathscr{M}_R$ — the category of modules over a commutative ring, R.

There are various operations with modules that are most easily understood in terms of category theory. The simplest is the direct sum

DEFINITION 10.6.1. If $M_1, M_2 \in \mathscr{M}_R$, the *direct sum*, $M_1 \oplus M_2 \in R\text{-mod}$, is the module of pairs
$$(m_1, m_2) \in M_1 \oplus M_2$$
with R acting via
$$r \cdot (m_1, m_2) = (r \cdot m_1, r \cdot m_2)$$
for all $r \in R$, $m_1 \in M_1$ and $m_2 \in M_2$.

REMARK. This is just a straightforward generalization of the concept of direct sum of vector-spaces — and the direct sum is a product and coproduct in the category of R-modules.

For instance, the free module R^n is a direct sum
$$R^n = \underbrace{R \oplus \cdots \oplus R}_{n \text{ factors}}$$

The direct sum is a functor of two variables:

PROPOSITION 10.6.2. *If $f_1: M_1 \to N_1$ and $f_2: M_2 \to N_2$ are morphisms in R-mod, then there is an induced morphism*
$$\begin{aligned} f_1 \oplus f_2: M_1 \oplus M_2 &\to N_1 \oplus N_2 \\ (m_1, m_2) &\mapsto (f_1(m_1), f_2(m_2)) \end{aligned}$$
for all $m_1 \in M_1$ and $m_2 \in M_2$. In addition, $\ker(f_1 \oplus f_2) = \ker f_1 \oplus \ker f_2$.

PROOF. The only thing that needs to be proved is the statement about the kernels. Clearly $\ker f_1 \oplus \ker f_2 \subset \ker(f_1 \oplus f_2)$. If (m_1, m_2) maps to 0 in $N_1 \oplus N_2$, we must have $f_1(m_1) = f_2(m_2) = 0$ so this proves $\ker(f_1 \oplus f_2) \subset \ker f_1 \oplus \ker f_2$. □

The following concept also originated with linear algebra, but is more complex than the direct sum. It is another functor of two variables:

DEFINITION 10.6.3. If $M_1, M_2 \in R\text{-mod}$, then define the *tensor product* of M_1 and M_2 over R
$$M_1 \otimes_R M_2$$
to be the R-module that is a quotient of the free R-module generated by symbols $\{m_1 \otimes m_2\}$ with $m_1 \in M_1, m_2 \in M_2$ subject to the identities

(1) $(r \cdot m_1) \otimes m_2 = m_1 \otimes (r \cdot m_2) = r \cdot (m_1 \otimes m_2)$ (defines the R-action on $M_1 \otimes_R M_2$) for all $r \in R, m_1 \in M_1, m_2 \in M_2$,
(2) $(m_1 + m_1') \otimes m_2 = m_1 \otimes m_2 + m_1' \otimes m_2$ for all $m_1, m_1' \in M_1, m_2 \in M_2$,
(3) $m_1 \otimes (m_2 + m_2') = m_1 \otimes m_2 + m_1 \otimes m_2'$ for all $m_1 \in M_1, m_2, m_2' \in M_2$.

REMARK. Rule 1 implies that $0 \otimes m_2 = m_1 \otimes 0 = 0$. Here is another way to define the tensor product:

Form the free abelian group $\mathbb{Z}[M_1 \times M_2]$. Its elements are formal linear combinations of symbols $[m \times n]$ for all $m \in M_1$ and $n \in M_2$. Then $M_1 \otimes_R M_2 = \mathbb{Z}[M_1 \times M_2]/W$, where $W \subset \mathbb{Z}[M_1 \times M_2]$ is the subgroup generated by

(1) $[r \cdot m_1 \times m_2] - [m_1 \times (r \cdot m_2)], [e \cdot m_1 \times m_2] - e \cdot [m_1 \times m_2]$, for all $e \in \mathbb{Z}, r \in R, m_1 \in M_1$, and $m_2 \in M_2$
(2) $[(m_1 + m_1') \times m_2] - [m_1 \times m_2] - [m_1' \times m_2]$, for all $m_1, m_1' \in M_1, m_2 \in M_2$,
(3) $[m_1 \times (m_2 + m_2')] - [m_1 \otimes m_2] - [m_1 \otimes m_2']$ for all $m_1 \in M_1, m_2, m_2' \in M_2$.

The R-module structure is defined by setting $r \cdot [m_1 \times m_2] = [r \cdot m_1 \times m_2]$ for all $r \in R, m_1 \in M_1, m_2 \in M_2$

EXAMPLE 10.6.4. If $M \in R\text{-mod}$, then
$$R \otimes_R M \xrightarrow{\cong} M$$
$$r \otimes m \mapsto r \cdot m$$

Clearly this map is surjective. If $r \otimes m$ is in the kernel, then $r \cdot m = 0 \in M$. In this case, rule 1 in definition 10.6.3 implies that
$$r \cdot 1 \otimes m \sim 1 \otimes r \cdot m = 0$$

In the category-theoretic sense, $M \otimes_R N$ is neither a product nor a coproduct. It does have a universal property, though:

PROPOSITION 10.6.5. *Let M, N, and T be modules over a commutative ring, R, and let*
$$f \colon M \times N \to T$$
be a mapping with the property that

(1) $f|m \times N \to T$ is an R-module-homomorphism for any $m \in M$
(2) $f|M \times n \to T$ is an R-module homomorphism for any $n \in N$
(3) $f(r \cdot m, n) = f(m, r \cdot n)$ for all $m \in M, n \in N$, and $r \in R$

Then there exists a unique map
$$g: M \otimes_R N \to T$$
that makes the diagram

$$\begin{array}{ccc} M \times N & \xrightarrow{c} & M \otimes_R N \\ & \searrow f & \downarrow g \\ & & T \end{array}$$

commute, where $c(m, n) = m \otimes n$, *for all* $m \in M$ *and* $n \in N$.

REMARK. Here $M \times N$ is simply a Cartesian product of *sets*. A map satisfying statements 1 and 2 above is said to be *bilinear*.

The canonical map
$$c: M \times N \to M \otimes_R N$$
is not surjective in general since $M \otimes_R N$ consists of *formal linear combinations* of symbols $m \otimes n$. Elements of $M \otimes_R N$ in the image of c are called *decomposable tensors* or elementary tensors. The paper [49] gives criteria for elements of $M \otimes_R N$ to be decomposable when R is a field.

This result implies that an decomposable tensor $m \otimes n$ *vanishes* if and only if every bilinear map
$$F: M \times N \to T$$
sends $m \times n$ to 0.

PROOF. The *only* map $\mathbb{Z}[M \times N] \to T$ compatible with f is
$$\begin{array}{rcl} \mathbb{Z}[M \times N] & \to & T \\ m \times n & \mapsto & f(m, n) \end{array}$$
for all $m \in M$ and $n \in N$. The defining relations for $M \otimes_R N$ and the conditions on the map f imply that this gives a well-defined map
$$\begin{array}{rcl} M \otimes N & \to & T \\ m \otimes n & \mapsto & f(m, n) \end{array}$$
for all $m \in M$ and $n \in N$. Since any such map must lift to a map $\mathbb{Z}[M \times N] \to T$, this must be *unique*. □

Tensor-products are functors of two variables:

PROPOSITION 10.6.6. *Let* $f: V_1 \to V_2$ *and* $g: W_1 \to W_2$ *be homomorphisms of vector-spaces. Then there is a natural map*
$$\begin{array}{rcl} f \otimes g: V_1 \otimes_k W_1 & \to & V_2 \otimes_k W_2 \\ (f \otimes g)(v \otimes w) & = & f(v) \otimes g(w) \end{array}$$

REMARK. Exercise 5 on page 377 gives some idea of what the homomorphism $f \otimes g$ looks like.

Tensor products are distributive over direct sums, a property that allows us to do many computations:

PROPOSITION 10.6.7. *Let M, N, T be modules over the commutative ring R. Then there are standard isomorphisms*
$$M \otimes_R (N \oplus T) = M \otimes_R N \oplus M \otimes_R T$$
and
$$(M \oplus N) \otimes_R T = M \otimes_R T \oplus N \otimes_R T$$

PROOF. We will prove the first case: the second is similar. We could use a detailed algebraic argument, but it is easier to "cheat" and use the universal property of a tensor product.

We will show that, given any bilinear map $z \colon M \times (N \oplus T) \to Z$, where Z is an R-module, there exists a *unique* homomorphism $d \colon M \otimes_R N \oplus M \otimes_R T \to Z$ making the diagram

(10.6.1)
$$M \times (N \oplus T) \xrightarrow{b} M \otimes_R N \oplus M \otimes_R T$$
$$\downarrow d$$
$$z \searrow Z$$

commute. Here, b is a bilinear map taking the place of the c map in 10.6.5 on page 366. This will show that $M \otimes_R N \oplus M \otimes_R T$ has the same universal property as $M \otimes_R (N \oplus T)$ so it must be isomorphic to it.

We begin by constructing a bilinear map $b \colon M \times (N \oplus T) \to M \otimes_R N \oplus M \otimes_R T$ via $b(m, (n, t)) = (m \otimes n, m \otimes t)$ for all $m \in M$, $n \in N$, and $t \in T$. This is easily verified to be bilinear:

(1) for any fixed $m_0 \in M$, $\ell(n, t) = b(m_0, (n, t)) = (m_0 \otimes n, m_0 \otimes t)$ for all $n \in N$ and $t \in T$, defines an R-module homomorphism
$$\ell \colon N \oplus T \to M \otimes_R N \oplus M \otimes_R T$$
since the composites
$$N \to m_0 \otimes N \subset M \otimes_R N$$
$$T \to m_0 \otimes T \subset M \otimes_R T$$
are module-homomorphisms.

(2) a similar argument shows that for any fixed $n_0 \in N$ and $t_0 \in T$, the map $\ell(m) = b(m, (n_0, t_0))$, for all $m \in M$ defines a module homomorphism
$$\ell \colon M \to M \otimes_R N \oplus M \otimes_R T$$

Now, suppose we have a bilinear map
$$z \colon M \times (N \oplus T) \to Z$$
We will show that there exists a unique map
$$d \colon M \otimes_R N \oplus M \otimes_R T \to Z$$
that makes diagram 10.6.1 commute.

We define d on the direct summands of $M \otimes_R N \oplus M \otimes_R T$:

(1) $d_1 \colon M \otimes_R N \to Z$ *must send* $m \otimes n$ to $z(m,(n,0)) \in Z$ for all $m \in M$ and $n \in N$ so we define $d_1(m \otimes n) = z(m,(n,0))$. The bilinearity of z implies that $d_1 | M \otimes n_0 \colon M \to Z$ is a module homomorphism for any fixed $n_0 \in N$ and $d_1 | m_0 \otimes N \colon N \to Z$ is also a module homomorphism. It follows that d_1 is a module homomorphism.
(2) We define $d_2 \colon M \otimes_R T \to Z$ by $d_2(m \otimes t) = z(m,(0,t))$. This is the only definition compatible with z and an argument like that used above shows that it is a module-homomorphism.

We set
$$d = d_1 + d_2 \colon M \otimes_R N \oplus M \otimes_R T \to Z$$
This is a module-homomorphism that makes diagram 10.6.1 on the preceding page commute. It is *unique* because it is uniquely determined on the two summands. □

COROLLARY 10.6.8. *If $M \in R$-mod, R, then*
$$M \otimes_R R^n = \underbrace{M \oplus \cdots \oplus M}_{n \text{ times}}$$
and
$$R^n \otimes_R R^m = R^{n \cdot m}$$

PROOF. This follows from example 10.6.4 on page 366, proposition 10.6.7 on the preceding page and induction on n. □

If R is an algebra over another ring S, we can define the structure of an R-module on $A \otimes_S B$ by $f \cdot (a \otimes b) = (r \cdot a \otimes r \cdot b)$, for $r \in R$. We can also define an R-action on groups of homomorphisms:

DEFINITION 10.6.9. *If M and N are R-modules, where R is an S-algebra, then*
$$\mathrm{Hom}_R(M,N)$$
denotes morphisms that are R-linear (i.e. morphisms of R-mod) and
$$\mathrm{Hom}_S(M,N)$$
are morphisms of S-mod, i.e. morphisms that are S-linear. Then we can equip $\mathrm{Hom}_S(M,N)$ with the structure of an R-module via the rule

If $f \in \mathrm{Hom}_S(M,N)$ is such that $f(m) = n$, then $(r \cdot f)(m) = f(r \cdot m)$.

We have important relations between $\hom_S$ and $\hom_R$:

PROPOSITION 10.6.10. *If $A, B, C \in R$-mod, where R is an S-algebra, then there exists a unique isomorphism*
$$s \colon \mathrm{Hom}_R(A, \mathrm{Hom}_S(B,C)) \to \mathrm{Hom}_R(A \otimes_S B, C)$$

REMARK. This is clearly natural with respect to all homomorphisms of A, B, or C.

PROOF. We define the map by $s(\varphi)(a \otimes b) = \varphi(a)(b)$. If $s(\varphi) = 0$, if is the 0-map for all b or a, so it vanishes in $\operatorname{Hom}_R(A, \operatorname{Hom}_S(B, C))$. It follows that s is injective. If $f \in \operatorname{Hom}_R(A \otimes_S B, C)$ then $f(a, *)$ for a fixed $a \in A$ defines a function $B \to C$ which is S-linear. This implies that s is surjective. □

Suppose $\mathfrak{a} \subset R$ is an ideal and M is a module over R. Then it is easy to see that $\mathfrak{a} \cdot M \subset M$ is a submodule and we have

PROPOSITION 10.6.11. *If $M \in R$-mod and $\mathfrak{a} \subset R$ is an ideal, there exists a natural isomorphism:*

$$q: M \otimes_R \left(\frac{R}{\mathfrak{a}}\right) \to \frac{M}{\mathfrak{a} \cdot M}$$
$$m \otimes r \mapsto R \cdot m \pmod{\mathfrak{a} \cdot M}$$

PROOF. It is not hard to see that q is surjective. Consider the composite

$$M = M \otimes_R R \xrightarrow{1 \otimes p} M \otimes_R \left(\frac{R}{\mathfrak{a}}\right)$$

where $p: R \to R/\mathfrak{a}$ is the projection. The surjectivity of q implies that $\ker 1 \otimes p \subset \mathfrak{a} \cdot M$. On the other hand, if $x \in \mathfrak{a}$, $x \cdot m \otimes 1 \sim m \otimes x \cdot 1 = 0 \in M \otimes_R (R/\mathfrak{a})$, by rule 1 in definition 10.6.3 on page 366. This shows that $\mathfrak{a} \cdot M \subset \ker 1 \otimes p$ and that q is also *injective*. □

We can use tensor-products to convert modules over one ring into modules over another:

PROPOSITION 10.6.12. *Let M be a module over a commutative ring R and let $f: R \to S$ be a homomorphism of rings. Then S is a module over R and*

$$M \otimes_R S$$

is a module over S with S-action given by

$$t \cdot m \otimes s = m \otimes st$$

for all $s, t \in S$ and $m \in M$.

REMARK. This operation is called a change of base. If $R \hookrightarrow S$ is an inclusion, it is called extension of scalars, the idea being that the action of R on M is "extended" to the larger ring, S.

Recall the concept of a module of fractions, defined in section 6.4 on page 248.

PROPOSITION 10.6.13. *If $M \in R$-mod and $S \subset R$ is a multiplicative set, then*

$$S^{-1}M \cong M \otimes_R (S^{-1}R)$$

is a module over $S^{-1}R$. If $\mathfrak{p} \subset R$ is a prime ideal and $S = R \setminus \mathfrak{p}$, then $S^{-1}R = R_\mathfrak{p}$ and

$$M_\mathfrak{p} \cong M \otimes_R R_\mathfrak{p}$$

and is a module over $R_\mathfrak{p}$.

REMARK. As defined in definition 6.4.3 on page 249, $S^{-1}M$ is a module over R. Proposition 10.6.12 on the preceding page shows that $S^{-1}M$ is *also* a module over $S^{-1}R$. If $S^{-1}R$ is the field of fractions, then $S^{-1}M$ is a *vector space* over that field.

PROOF. The map
$$f: M \otimes_R (S^{-1}R) \to S^{-1}M$$
is defined by $f(m \otimes s^{-1}r) = r \cdot m/s$ for all $m \in M$, $s \in S$, and $r \in R$. If $r_1/s_1 \equiv r_2/s_2 \in S^{-1}R$, then $u \cdot (s_2 r_1 - s_1 r_2) = 0$ for some $u \in S$, and
$$u \cdot (s_2 r_1 \cdot m - s_1 r_2 \cdot m) = u \cdot (s_2 r_1 - s_1 r_2) \cdot m = 0$$
so f is well-defined. The inverse map
$$g: S^{-1}M \to M \otimes_R (S^{-1}R)$$
is defined by $g(m/s) = m \otimes s^{-1}$. If $m_1/s_1 \equiv m_2/s_2 \in S^{-1}M$ then
$$u \cdot (s_2 \cdot m_1 - s_1 \cdot m_2) = 0$$
for some $u \in S$, or $u s_2 \cdot m_1 = u s_1 \cdot m_2$, so
$$u s_2 \cdot m_1 \otimes u^{-1} s_1^{-1} s_2^{-1} = u s_1 \cdot m_2 \otimes u^{-1} s_1^{-1} s_2^{-1}$$
By rule 1 of definition 10.6.3 on page 366, both sides of this equation are equal to
$$\begin{aligned} u s_2 \cdot m_1 \otimes u^{-1} s_1^{-1} s_2^{-1} &= m_1 \otimes s_1^{-1} \\ u s_1 \cdot m_2 \otimes u^{-1} s_1^{-1} s_2^{-1} &= m_2 \otimes s_2^{-1} \end{aligned}$$
It follows that $g(m_1/s_1) = g(m_2/s_2)$, so g is well-defined and clearly the inverse of f. □

It is easy to verify that tensor products preserve *surjectivity* of maps:

PROPOSITION 10.6.14. *If $M \in R$-mod and $f: N \to T$ is a surjective morphism in R-mod, then*
$$1 \otimes f: M \otimes_R N \to M \otimes_R T$$
is also surjective.

REMARK. If
$$0 \to A \to B \to C \to 0$$
is an exact sequence of modules and we take the tensor product of this with M, the resulting sequence is exact on the *right*
$$M \otimes_R A \to M \otimes_R B \to M \otimes_R C \to 0$$
and we say that the *functor* $M \otimes_R *$ is *right-exact*. This sequence might not be exact on the left — $M \otimes_R A \to M \otimes_R B$ might not be an inclusion. For instance if
$$f = \times 2: \mathbb{Z} \to \mathbb{Z}$$
and $M = \mathbb{Z}_2$, then
$$f \otimes 1 = 0: \mathbb{Z} \otimes_{\mathbb{Z}} \mathbb{Z}_2 = \mathbb{Z}_2 \to \mathbb{Z} \otimes_{\mathbb{Z}} \mathbb{Z}_2 = \mathbb{Z}_2$$

PROOF. If $\sum m_i \otimes t_i \in M \otimes_R T$ then it is the image of $\sum m_i \otimes n_i$, where $f(n_i) = t_i$. □

This leads to another consequence of Nakayama's Lemma:

COROLLARY 10.6.15. *Let R be a noetherian local ring with maximal ideal $\mathfrak{m}$, let M be a finitely generated R-module, and let*

$$f\colon M \to \frac{M}{\mathfrak{m} \cdot M} = M \otimes_R \left(\frac{R}{\mathfrak{m}}\right)$$

be the projection to the quotient. If $\{m_1, \ldots m_t\} \in M$ are elements with the property that $\{f(m_1), \ldots, f(m_t)\}$ generate $M/\mathfrak{m} \cdot M$, then $\{m_1, \ldots m_t\}$ generate M.

REMARK. Note that $R/\mathfrak{m}$ is a field so that $M/\mathfrak{m} \cdot M$ is a vector space.

PROOF. Let $M' \subset M$ be the submodule generated by $\{m_1, \ldots m_t\}$. Then M/M' is a finitely generated R module with the property that

$$\left(\frac{M}{M'}\right) \otimes_R \left(\frac{R}{\mathfrak{m}}\right) = 0$$

which implies that

$$\mathfrak{m} \cdot \left(\frac{M}{M'}\right) = \left(\frac{M}{M'}\right)$$

Corollary 6.3.35 on page 247 implies that $M/M' = 0$. □

We also get the interesting result

COROLLARY 10.6.16. *If R is a noetherian local ring with maximal ideal $\mathfrak{m}$, then finitely generated projective modules over R are free.*

PROOF. Let P be a projective module over R and let $p_1, \ldots, p_n \in P$ be a set of elements with the property that their image in

$$P \otimes_R \left(\frac{R}{\mathfrak{m}}\right) = \frac{P}{\mathfrak{m} \cdot P} = V$$

generate the vector-space V. Corollary 10.6.15 implies that $p_1, \ldots, p_n$ generate P. If R^n is a free module on generators $x_1, \ldots, x_n$, the homomorphism

$$f\colon R^n \to P$$

that sends x_i to p_i for $i = 1, \ldots, n$ is surjective. If $K = \ker f$, we get a short exact sequence

$$0 \to K \to R^n \xrightarrow{f} P \to 0$$

Since P is projective, this is split and we get an isomorphism

$$R^n \cong K \oplus P$$

(see exercise 16 on page 248). Now take the tensor product with $R/\mathfrak{m}$ to get

$$R^n \otimes_R \left(\frac{R}{\mathfrak{m}}\right) \cong P \otimes_R \left(\frac{R}{\mathfrak{m}}\right) \oplus K \otimes_R \left(\frac{R}{\mathfrak{m}}\right)$$

Since $R^n \otimes_R \left(\frac{R}{\mathfrak{m}}\right)$ and $P \otimes_R \left(\frac{R}{\mathfrak{m}}\right)$ are both n-dimensional vector-spaces over $R/\mathfrak{m}$, it follows that
$$K \otimes_R \left(\frac{R}{\mathfrak{m}}\right) = \frac{K}{\mathfrak{m} \cdot K} = 0$$
This implies that $K = \mathfrak{m} \cdot K$ and corollary 6.3.35 on page 247 implies that $K = 0$ and $P = R^n$. □

DEFINITION 10.6.17. A module $M \in \mathcal{M}_R$ will be called *flat* if the functor $M \otimes_R *$ preserves *injections* as well as surjections. In other words, M is flat if, whenever
$$N \to T$$
is an injective homomorphism of R-modules, so is
$$M \otimes_R N \to M \otimes T$$

REMARK. For instance, R is a flat module over itself, as example 10.6.4 on page 366 shows. In general, every free module, R^n, is flat over R, by proposition 10.6.7 on page 368.

The term *flat module* first appeared in Serre's paper, [98].

Flat modules are very useful because:

PROPOSITION 10.6.18. *Let R be a commutative ring and let A be a flat R-module. If*
$$\cdots \xrightarrow{f_{n+1}} M_{n+1} \xrightarrow{f_n} M_n \xrightarrow{f_{n-1}} M_{n-1} \to \cdots$$
is an exact sequence in $\mathcal{M}_R$, then so is
$$\cdots \xrightarrow{1 \otimes f_{n+1}} A \otimes_R M_{n+1} \xrightarrow{1 \otimes f_n} A \otimes_R M_n \xrightarrow{1 \otimes f_{n-1}} A \otimes_R M_{n-1} \to \cdots$$

PROOF. The exactness of the (long) sequence above is equivalent to saying that the short sequences
$$0 \to \text{im } f_n \to M_n \to \text{im } f_{n-1} \to 0$$
are exact (see definition 6.3.7 on page 224) for all n. Since tensor products preserve surjections (proposition 10.6.14 on page 371), we know that $\text{im}(1 \otimes f_n) = A \otimes_R (\text{im } f_n)$ for all n (and this is for *any* module, A, not just a flat one). The conclusion follows by the fact that flat modules preserve *injections* as well as surjections (definition 10.6.17). □

The following result describes a very important class of flat modules:

LEMMA 10.6.19. *Let R be a commutative ring and let S be a multiplicative set. Then $S^{-1}R$ is a flat module over R.*

PROOF. If $f: N \to T$ is an injective homomorphism of R-modules, we will show that
$$S^{-1}R \otimes_R N \to S^{-1}R \otimes_R T$$
is also injective. We replace these tensor products by modules of fractions, using proposition 10.6.13 on page 370, to get the equivalent map
$$S^{-1}N \to S^{-1}T$$

Extend this to
$$N \to S^{-1}N \to S^{-1}T$$
An element $n \in N$ maps to zero in $S^{-1}T$ if and only if $s \cdot n = 0$ for some $s \in S$ (see definition 6.4.3 on page 249). If this happens, n also maps to 0 in $S^{-1}N$ so the map is injective. $\square$

We know that free modules are flat by proposition 10.6.7 on page 368. It turns out that

PROPOSITION 10.6.20. *Projective modules are flat.*

REMARK. Projective modules over $\mathbb{Z}$ have elements that are *not* multiples of 2. On the other hand, lemma 10.6.19 on the previous page shows that $\mathbb{Q}$ is a flat module over $\mathbb{Z}$ that *cannot* be projective since *all* of its elements are divisible by 2.

PROOF. Let P be a projective module over a ring R and let Q be another (projective) module such that $P \oplus Q = R^n$. If
$$f: M \to N$$
is an injective homomorphism, we know that
$$f \otimes 1: M \otimes_R R^n \to N \otimes_R R^n$$
is also injective, and is equal to
$$f \otimes 1: M \otimes_R (P \oplus Q) \to N \otimes_R (P \oplus Q)$$
which is equal to
$$(f \otimes 1_P) \oplus (f \otimes 1_Q): M \otimes_R P \oplus M \otimes_R Q \to N \otimes_R P \oplus N \otimes_R Q$$
where 1_P and 1_Q are the identity maps of P and Q, respectively. Since $(f \otimes 1_P) \oplus (f \otimes 1_Q)$ is injective, proposition 10.6.2 on page 365 implies that $f \otimes 1_P$ and $f \otimes 1_Q$ must be injective too. $\square$

It is interesting to see what forming modules of fractions does to prime filtrations:

COROLLARY 10.6.21. *Let M be a module over a ring R and let $S \subset R$ be a multiplicative set. Let*
$$0 = M_0 \subset M_1 \subset \cdots \subset M_n = M$$
with
$$\frac{M_{i+1}}{M_i} \simeq \frac{R}{\mathfrak{p}_i}$$
for prime ideals $\mathfrak{p}_i \subset R$, be the prime filtration of M. Then the prime filtration of $S^{-1}R \otimes_R M$ is
$$0 = S^{-1}R \otimes_R M_{j_0} \subset S^{-1}R \otimes_R M_{j_1} \subset \cdots \subset S^{-1}R \otimes_R M_{j_t} = S^{-1}R \otimes_R M$$
where
$$\frac{S^{-1}R \otimes_R M_{j_{i+1}}}{S^{-1}R \otimes_R M_{j_i}} \simeq \frac{S^{-1}R}{\mathfrak{p}_{j_i} \cdot S^{-1}R}$$
where $\{\mathfrak{p}_{j_0}, \ldots, \mathfrak{p}_{j_t}\} \subseteq \{\mathfrak{p}_0, \ldots, \mathfrak{p}_n\}$ is the subset of prime ideals that do not contain any elements of S.

PROOF. Since $S^{-1}R$ is flat over R, every short exact sequence
$$0 \to \frac{R}{\mathfrak{p}_i} \to M_i \to M_{i+1} \to 0$$
gives rise to
$$0 \to S^{-1}R \otimes_R \left(\frac{R}{\mathfrak{p}_i}\right) \to S^{-1}R \otimes_R M_i \to S^{-1}R \otimes_R M_{i+1} \to 0$$
and the short exact sequence
$$0 \to S^{-1}R \otimes_R \mathfrak{p}_i \to S^{-1}R \otimes_R R \to S^{-1}R \otimes_R \left(\frac{R}{\mathfrak{p}_i}\right) \to 0$$
where $S^{-1}R \otimes_R \mathfrak{p}_i = \mathfrak{p}_i \cdot S^{-1}R$ and $S^{-1}R \otimes_R R = S^{-1}R$. If $\mathfrak{p}_i$ contains an element of S, $\mathfrak{p}_i \cdot S^{-1}R = S^{-1}R$ and the quotient
$$S^{-1}R \otimes_R \left(\frac{R}{\mathfrak{p}_i}\right)$$
will be the trivial ring. It follows that those primes do not participate in the prime filtration of $S^{-1}R \otimes_R M$. □

We conclude this section with a converse to lemma 10.6.19 on page 373:

LEMMA 10.6.22. *Let R be a noetherian ring and let A be a finitely-generated R-module. Then $A_\mathfrak{m}$ is a free $R_\mathfrak{m}$-module for all maximal ideals $\mathfrak{m} \subset R$ if and only if A is projective.*

REMARK. In other words, locally free modules are *projective*.

PROOF. Since A is finitely-generated, there exists a finitely-generated free module, F, and a surjective homomorphism
$$f: F \to A$$
inducing surjective homomorphisms
$$f_\mathfrak{m} = 1 \otimes f: F_\mathfrak{m} \to A_\mathfrak{m}$$
Since $A_\mathfrak{m}$ is free, there exist splitting maps
$$g_\mathfrak{m}: A_\mathfrak{m} \to F_\mathfrak{m}$$
with $f_\mathfrak{m} \circ g_\mathfrak{m} = 1: A_\mathfrak{m} \to A_\mathfrak{m}$ for all maximal ideals $\mathfrak{m} \subset R$. Since A is finitely-generated, there exists an element $s_\mathfrak{m} \in R \setminus \mathfrak{m}$ for each maximal ideal such that
$$s_\mathfrak{m} \cdot f_\mathfrak{m} \circ g_\mathfrak{m}(A) \subset A \subset A_\mathfrak{m}$$
i.e., $s_\mathfrak{m}$ "clears the denominators" of $f_\mathfrak{m} \circ g_\mathfrak{m}(A)$. Let $\mathfrak{G}$ denote the ideal generated by all of the $s_\mathfrak{m}$. Since R is noetherian, $\mathfrak{G}$ is generated by some finite set of the $s_\mathfrak{m}$
$$\mathfrak{G} = (s_{\mathfrak{m}_1}, \ldots, s_{\mathfrak{m}_t})$$
If $\mathfrak{G} \subsetneq R$, then it is contained in some maximal ideal, which contradicts the fact that it contains an element not in every maximal ideal. We conclude that $\mathfrak{G} = R$ and that there exist elements $\{x_1, \ldots, x_t\}$ such that
$$\sum_{i=1}^{t} x_i \cdot s_{\mathfrak{m}_i} = 1$$

If we define
$$g = \sum_{i=1}^{t} x_i \cdot f_{m_i} \circ g_{m_i} \colon A \to F$$
and $f \circ g = 1 \colon A \to A$, so A is a direct summand of F and is projective.

The *only if* part comes from corollary 10.6.16 on page 372. □

We conclude this section with a generalization of the *dual* of a module:

DEFINITION 10.6.23. If M is a module over a ring, R, define $M^* = \hom_R(M, R)$ — the dual of M. It is also a module over R (just let R act on it by multiplying the values of homomorphisms).

REMARK. Clearly, $R^* = R$ since a homomorphism $R \to R$ is determined by its effect on $1 \in R$. It is also not hard to see that the dual of a finitely generated free module is free of the same rank. If F is a free module, the isomorphism between F and F^* is not natural.

There *is* a natural isomorphism
$$F \to F^{**}$$
where we map $x \in F$ to the homomorphism $F^* \to R$ given by $f(x)$ for $f \in F^*$.

This and the way hom behaves with direct sums implies that:

COROLLARY 10.6.24. *Let P be a finitely generated projective module over a commutative ring R. Then P^* is also a finitely generated projective module and*
$$P = P^{**}$$

EXERCISES.

1. Suppose M is a module with submodules R and S, such that $M = R + S$ and $R \cap S = 0$. Show that
$$M \cong R \oplus S$$

2. Suppose M is a module over a ring, R, and $N \subset M$ is a submodule. If A is a flat module over R, show that $A \otimes_R N$ is a submodule of $A \otimes_R M$ and
$$\frac{A \otimes_R M}{A \otimes_R N} = A \otimes_R \left(\frac{M}{N}\right)$$

3. If M, N, T are modules over a ring, R, show that there are natural isomorphisms
$$\begin{aligned} M \otimes_R (N \otimes_R T) &\cong (M \otimes_R N) \otimes_R T \\ M \otimes_R N &\cong N \otimes_R M \end{aligned}$$

4. Let V be a vector space over a field, k, with basis $\{e_1, \ldots, e_n\}$ and let W be a vector space with basis $\{f_1, \ldots, f_n\}$. Show that
$$V \otimes_k W$$

is $n \cdot m$ dimensional, with basis

$$\{e_i \otimes f_j\}$$

$i = 1, \ldots, n$ and $j = 1, \ldots, m$.

Show that $\{e_i \otimes f_j\}$ are a basis for $V \otimes_k W$ even if they are infinite dimensional.

5. Suppose k is a field and $f \colon k^n \to k^m$ and $g \colon k^s \to k^t$ are given by $m \times n$ and $t \times s$ matrices A and B, respectively. What is the matrix representing $A \otimes B$?

6. If M and N are finite-dimensional vector-spaces over a field, k, show that there is a natural isomorphism

$$\hom_k(M, N) \cong M^* \otimes N$$

7. If $\mathfrak{a}, \mathfrak{b} \subset R$ are two ideals in a commutative ring, show that

$$\frac{R}{\mathfrak{a}} \otimes_R \frac{R}{\mathfrak{b}} = \frac{R}{\mathfrak{a} + \mathfrak{b}}$$

This implies that

$$\mathbb{Z}_n \otimes_\mathbb{Z} \mathbb{Z}_m = \mathbb{Z}_{\gcd(n,m)}$$

8. If R is a ring and M is a flat R-module. Show that

$$\mathfrak{a} \otimes_R M = \mathfrak{a} \cdot M$$

for all ideals $\mathfrak{a} \subset R$.

9. Show that tensor products commute with direct limits, i.e. if

$$M_0 \xrightarrow{f_0} \cdots \xrightarrow{f_{n-1}} M_n \xrightarrow{f_n} \cdots$$

is a direct system of modules over a ring R and N is an R-modules, show that

$$\left(\varinjlim M_j\right) \otimes_R N = \varinjlim (M_j \otimes_R N)$$

10. If $\{A_i, a_i\}$, $\{B_i, b_i\}$, and $\{C_i, c_i\}$ are three systems of homomorphisms of modules such that the diagram

$$\begin{array}{ccccccccc} 0 & \to & A_{i+1} & \xrightarrow{f_{i+1}} & B_{i+1} & \xrightarrow{g_{i+1}} & C_{i+1} & \to & 0 \\ & & \uparrow a_i & & \uparrow b_i & & \uparrow c_i & & \\ 0 & \to & A_i & \xrightarrow{f_i} & B_i & \xrightarrow{g_i} & C_i & \to & 0 \end{array}$$

commutes for all i and each row is exact, show that

$$0 \to \varinjlim A_i \xrightarrow{\varinjlim f_i} \varinjlim B_i \xrightarrow{\varinjlim g_i} \varinjlim C_i \to 0$$

is an exact sequence.

11. Suppose $f \colon R \to S$ is a homomorphism of rings with the property that S is a flat module over R. If $\alpha \in R$ is a non-zero-divisor, show that $f(\alpha) \in S$ is a non-zero-divisor.

12. Suppose $f\colon R \to S$ is a homomorphism of rings and M is a flat module over R. Show that $M \otimes_R S$ is a flat module over S.

13. Let $\mathbb{Z}$-mod be the category of modules over $\mathbb{Z}$ (otherwise known as abelian groups, $\mathscr{A}b$), the set, $\hom_{\mathbb{Z}\text{-mod}}(A, B)$, is naturally a module over $\mathbb{Z}$. For any $A, B, C \in \mathbb{Z}$-mod, show that there exists a natural isomorphism

$$\hom_{\mathbb{Z}\text{-mod}}(A \otimes_\mathbb{Z} B, C) \cong \hom_{\mathbb{Z}\text{-mod}}(A, \hom_{\mathbb{Z}\text{-mod}}(B, C))$$

so that the functors $* \otimes_\mathbb{Z} B$ and $\hom_{\mathbb{Z}\text{-mod}}(B, *)$ are adjoints.

14. Let $M, N \in R$-mod and let $S \subset R$ be a multiplicative set. Show that

$$S^{-1}R \otimes_R (M \otimes_R N) = (S^{-1}R \otimes_R M) \otimes_{S^{-1}R} (S^{-1}R \otimes_R N)$$

15. If M is a finitely generated projective module, show that M^* is also a finitely generated projective module.

10.7. Tensor Algebras and variants

In this section, we will discuss several algebras one can construct from modules over a ring. The most general case is the tensor algebra, with the symmetric and exterior algebras being quotients.

Historically, the first of these to appear were *exterior algebras,* described in [47] by Hermann Grassmann. Grassmann developed exterior algebras in the context of vector spaces — and many linear algebra constructs (like determinants) have elegant formulations in terms of exterior algebras[2].

Tensor algebras appeared later, in the context of category theory and are more general than exterior algebras.

DEFINITION 10.7.1. If R is a commutative ring and M is an R-module, define:

$$M^{\otimes n} = \underbrace{M \otimes_R \cdots \otimes_R M}_{n \text{ times}}$$

with $M^{\otimes 0} = R$ and $M^{\otimes 1} = M$. Given this definition, we define the *tensor algebra* over M:

$$T(M) = R \oplus M \oplus M^{\otimes 2} \oplus M^{\otimes 3} \oplus \cdots$$

This is a (noncommutative) algebra over R by defining

$$(m_1 \otimes \cdots \otimes m_s) \cdot (n_1 \otimes \cdots \otimes n_t) = m_1 \otimes \cdots \otimes m_s \otimes n_1 \otimes \cdots \otimes n_t$$

and extending this to all of $T(M)$ R-linearly.

REMARK. Tensor algebras are often called free algebras. Any module homomorphism

$$f\colon M \to N$$

induces a unique algebra-homomorphism

$$T(f)\colon T(M) \to T(N)$$

---

[2]Often called Grassmann algebras

Furthermore, if A is any algebra over R and $g\colon M \to A$ is a homomorphism of R-modules, there exists a unique homomorphism of R-algebras
$$T(M) \to A$$
whose restriction to $M = M^{\otimes 1}$ is g. If $\mathscr{A}_R$ is the category of R-algebras and $\mathscr{M}_R$ that of R-modules, let
$$F\colon \mathscr{A}_R \to \mathscr{M}_R$$
be the forgetful functor that maps an R-algebra to its underlying R-module (forgetting that we can multiply elements of this module), we get a natural isomorphism

(10.7.1) $$\hom_{\mathscr{A}_R}(T(M), A) \cong \hom_R(M, FA)$$

making $T(*)$ and F adjoints (compare with example 10.3.2 on page 350).

The tensor algebra is an example of a *graded ring* (see definition 15.3.1 on page 471) with
$$T(M)_n = M^{\otimes n}$$

Corollary 10.6.8 on page 369 immediately implies that

PROPOSITION 10.7.2. *If M is a free module of rank t (see example 6.3.2 on page 223) over a ring R, then $T_n(M)$ is a free module of rank t^n.*

We also have:

PROPOSITION 10.7.3. *If M is any module over a commutative ring, R, and $S \subset R$ is any multiplicative set, then*
$$T(S^{-1}R \otimes_R M) = S^{-1}R \otimes_R T(M)$$

PROOF. This follows immediately from the solution to exercise 14 on the preceding page. □

There are two important variants on tensor algebras that we need:

DEFINITION 10.7.4. Let M be a module over a commutative ring, R, and let $\mathfrak{s} \subset T(M)$ be the (two-sided) ideal generated by elements
$$x \otimes y - y \otimes x$$
for all $x, y \in M$. The quotient, $\mathcal{S}(M) = T(M)/\mathfrak{s}$, is called the *symmetric algebra* on M.

REMARK. This is clearly a commutative ring. Since $T(M)$ is not commutative, the ideal $\mathfrak{s}$ must be two-sided — it is the sum
$$\sum_{x,y \in M} T(M) \cdot (x \otimes y - y \otimes x) \cdot T(M)$$

Symmetric algebras also have a defining universal property:

PROPOSITION 10.7.5. *Let $\mathscr{C}_R$ denote the category of commutative algebras over a (commutative) ring R and let $\mathscr{M}_R$ denote the category of R-modules. There is a forgetful functor*
$$f\colon \mathscr{C}_R \to \mathscr{M}_R$$
that "forgets" the multiplication operation in an R-algebra (so it becomes a mere module). The symmetric algebra is an adjoint to f in the sense that
$$\hom_R(M, f(A)) = \hom_{\mathscr{C}_R}(\mathcal{S}M, A)$$

PROOF. We already know that
$$\hom_{\mathscr{A}_R}(T(M), A) \cong \hom_R(M, FA) \tag{10.7.2}$$
If A is a *commutative* algebra, then the map
$$T(M) \to A$$
factors through $\mathcal{S}M$:
$$T(M) \to \mathcal{S}M \to A$$
□

It is not hard to see that

PROPOSITION 10.7.6. *If M is a free module of rank t over a commutative ring, R, then*
$$\mathcal{S}(M) = R[X_1, \ldots, X_t]$$

PROOF. Suppose $\{e_1, \ldots, e_t\}$ is a free basis for M. It is straightforward to see that
$$e_{j_1}^{\otimes n_1} \otimes \cdots \otimes e_{j_\ell}^{\otimes n_\ell}$$
with $\sum n_i = n$ and $j_1 < \cdots < j_\ell$ is a free basis for $\mathcal{S}_n(M)$ — and these are in a 1-1 correspondence with monomials in the X_i of total degree n. □

The second variant of tensor algebras is called *exterior algebras* or Grassmann algebras in honor of Hermann Grassmann (since he first described them in [47]). For our purposes, they are more interesting than symmetric algebras and have more applications. Although Grassman originally defined them for vector-spaces over fields, this definition can easily be extended to modules over a commutative ring:

DEFINITION 10.7.7. If M is a module over a commutative ring, R, the *exterior algebra* over M is defined to be
$$\bigwedge M = T(M)/\mathfrak{a}$$
where $\mathfrak{a}$ is the two-sided ideal generated by elements $\{x \otimes x\}$ for all $x \in M$. This is a graded ring with
$$\bigwedge^n M = M^{\otimes n}/M^{\otimes n} \cap \mathfrak{a}$$
The product-operation is written $x \wedge y$ for $x, y \in \bigwedge M$.

REMARK. If $x, y \in M$, then
$$(x + y) \wedge (x + y) = 0$$
because of how the ideal $\mathfrak{a}$ is defined. The distributive laws implies that
$$\begin{aligned}(x + y) \wedge (x + y) &= x \wedge x + x \wedge y + y \wedge x + y \wedge y \\ &= x \wedge y + y \wedge x\end{aligned}$$
so $x \wedge y = -y \wedge x$ for elements of M. The level $\bigwedge^n M$ is generated by all expressions of the form
$$x_1 \wedge \cdots \wedge x_n$$
for $x_1, \ldots, x_n \in M$.

Exterior algebras have applications to fields as varied as differential geometry (see [101]), partial differential equations (see [19]) and physics (see [89]) — besides algebraic geometry. Grassmann's original definition was *axiomatic*, using axioms based on linearity, associativity, and anti-commutativity — and only for vector-spaces.

We have some direct-sum relations:

PROPOSITION 10.7.8. *If M and N are modules over R then*
(1) $T(M \oplus N) = T(M) \otimes_R T(N) \otimes_R T(M) \otimes_R \cdots$ — *as graded algebras (see definition 15.3.2 on page 472), i.e.,*
$$T(M \oplus N)_m = \bigoplus_{\sum_{j=1}^{\infty}(i_j + n_j) = m} T(M)_{i_1} \otimes_R T(N)_{n_1} \otimes_R \cdots$$
(2) $\mathcal{S}(M \oplus N) = \mathcal{S}(M) \otimes_R \mathcal{S}(N)$ — *as graded algebras, so*
$$\mathcal{S}(M \oplus N)_m = \bigoplus_{i+j=m} \mathcal{S}(M)_i \otimes_R \mathcal{S}(N)_j$$
(3) $\Lambda(M \oplus N) \cong \Lambda(M) \otimes_R \Lambda(N)$ — *as graded algebras, so*
$$\Lambda^m (M \oplus N) \cong \bigoplus_{i+j=m} \Lambda^i(M) \otimes_R \Lambda^j(N)$$

REMARK. Note that, in line 1 all but a finite number of the i_j, n_j must be 0.

PROOF. The first statement follows from the general properties of the tensor product.

The second statement follows from the first and the fact that the commutativity relations between $T(M)$ and $T(N)$ reduces the "infinite tensor product" to $T(M) \otimes_R T(N)$. Imposing the commutativity relations within $T(M)$ and $T(N)$ gives $\mathcal{S}(M) \otimes_R \mathcal{S}(N)$.

The third statement follows by a similar argument except that we may have to permute factors in an expression like $n_1 \wedge m_2 \wedge \ldots, \wedge m_i$ so that all of the m-factors occur to the left of the n-factors. This multiplies by ± 1, so we get an isomorphism. □

Here's an example of computations in an exterior algebra:

EXAMPLE. Let M be a free module over R on a free basis $\{e_1, e_2, e_3\}$ and let $v = 2e_1 + e_2 - e_3$ and $w = e_1 - 3e_2 + e_3$. Then
$$\begin{aligned} v \wedge w &= (2e_1 + e_2 - e_3) \wedge (e_1 - 3e_2 + e_3) \\ &= 2e_1 \wedge (e_1 - 3e_2 + e_3) + e_2 \wedge (e_1 - 3e_2 + e_3) \\ &\quad - e_3 \wedge (e_1 - 3e_2 + e_3) \\ &= 2e_1 \wedge e_1 - 2e_1 \wedge 3e_2 + 2e_1 \wedge e_3 \\ &\quad + e_2 \wedge e_1 - 3e_2 \wedge e_2 + e_2 \wedge e_3 \\ &\quad - e_3 \wedge e_1 + 3e_3 \wedge e_2 - e_3 \wedge e_3 \end{aligned}$$

Here, we have used the distributive rule several times. After applying the annihilation and linearity conditions, we get
$$\begin{aligned} v \wedge w &= -6e_1 \wedge e_2 + 2e_1 \wedge e_3 + e_2 \wedge e_1 + e_2 \wedge e_3 \\ &\quad - e_3 \wedge e_1 + 3e_3 \wedge e_2 \end{aligned}$$

And after "standardizing" by replacing any $e_j \wedge e_i$ by $-e_i \wedge e_j$ whenever $j > i$, we get
$$v \wedge w = -7e_1 \wedge e_2 + 3e_1 \wedge e_3 - 2e_2 \wedge e_3$$
Clearly, the set $\{e_1 \wedge e_2, e_1 \wedge e_3, e_2 \wedge e_3\}$ forms a free basis for $\Lambda^2 V$ (any relation between them would imply a relation between basis elements of $T_2(M)$).

In general, we have:

PROPOSITION 10.7.9. *Let M be a free module over R with free basis $\{e_1, \ldots, e_n\}$. Then $\Lambda^k M$ has a free basis consisting of symbols*
$$\{e_{i_1} \wedge \cdots \wedge e_{i_k}\}$$
for all sequences $1 \leq i_1 < i_2 < \cdots < i_k \leq n$. Consequently, the rank of $\Lambda^k V$ is $\binom{n}{k}$, and $\Lambda^k V = 0$ whenever $k > n$.

PROOF. By definition, $\Lambda^k V$ consists of all sequences $v_1 \wedge \cdots \wedge v_k$ and, using the linearity and distributivity properties, we can write these as linear combinations of all length-k sequences of basis elements
$$\{e_{j_1} \wedge \cdots \wedge e_{j_k}\}$$
The annihilation property implies that any such sequence with two *equal* indices will vanish. It also implies that we can arrange these indices in ascending order (multiplying terms by -1 if necessary). □

Proposition 10.7.3 on page 379 and the fact that $S^{-1}R$ is *flat* over R (see lemma 10.6.19 on page 373) imply that

PROPOSITION 10.7.10. *Let M be a module over a commutative ring, R, and let $S \subset R$ be a multiplicative set. Then*
$$\Lambda(S^{-1}R \otimes_R M) = S^{-1}R \otimes_R \Lambda M$$
$$\mathcal{S}(S^{-1}R \otimes_R M) = S^{-1}R \otimes_R \mathcal{S}(M)$$

PROOF. The fact that $S^{-1}R$ is flat over R implies that
$$S^{-1}R \otimes_R \left(\frac{T(M)}{\mathfrak{a}}\right) = \frac{S^{-1}R \otimes_R T(M)}{S^{-1}R \otimes_R \mathfrak{a}} = \frac{T(S^{-1}R \otimes_R M)}{\mathfrak{a}'}$$
where $\mathfrak{a}'$ is the form of $\mathfrak{a}$ in $T(S^{-1}R \otimes_R M)$. It follows that $\Lambda(S^{-1}R \otimes_R M) = S^{-1}R \otimes_R \Lambda M$. The proof for the symmetric algebra is entirely analogous. □

We will often be interested in certain elements of ΛM with an especially simple structure (particularly when we study Grassmannians):

DEFINITION 10.7.11. *If M is a module over a commutative ring, elements of $\Lambda^k M$ of the form*
$$m_1 \wedge \cdots \wedge m_k$$
for $m_i \in M$ will be said to be decomposable.

REMARK. An exterior algebra consists of formal linear combinations of decomposable elements. If $x \in \Lambda^k M$ is decomposable then

$$\begin{aligned} x \wedge x &= (m_1 \wedge \cdots \wedge m_k) \wedge (m_1 \wedge \cdots \wedge m_k) \\ &= 0 \end{aligned}$$

because of the annihilation condition. Suppose M is a free module on the free basis $\{e_1, e_2, e_3, e_4\}$ and

$$x = e_1 \wedge e_2 + e_3 \wedge e_4$$

Then

$$x \wedge x = 2e_1 \wedge e_2 \wedge e_3 \wedge e_4 \neq 0$$

so this x is *not* decomposable.

For the rest of this section, we will assume that R is a *field* so that modules over R are vector-spaces. The following result is key to understanding the geometric meaning of $\Lambda^k V$:

LEMMA 10.7.12. *Let $v_1, \ldots, v_k \in V$ be vectors in a vector space. Then, in $\Lambda^k V$,*

$$v_1 \wedge \cdots \wedge v_k = 0$$

if and only if the set $\{v_1, \ldots, v_k\}$ is linearly dependent.

PROOF. If they are linearly *independent*, they are part of a basis for V and proposition 10.7.9 on the preceding page implies that their wedge-product is part of a basis for $\Lambda^k V$, hence nonzero.

Suppose they are linearly dependent and, without loss of generality, suppose

$$v_1 = \sum_{j=2}^{k} a_j v_j$$

Then

$$\begin{aligned} v_1 \wedge \cdots \wedge v_k &= \sum_{j=2}^{k} a_j v_j \wedge v_2 \wedge \cdots \wedge v_k \\ &= 0 \end{aligned}$$

since each term in the sum on the right will have v_j equal to one of the vectors in $v_2 \wedge \cdots \wedge v_k$. □

COROLLARY 10.7.13. *Let $W \subset V$ be a k-dimensional subspace with basis $\{w_1, \ldots, w_k\}$. Then the element*

$$\bar{w} = w_1 \wedge \cdots \wedge w_k \in \Lambda^k V$$

determines W uniquely. In fact the kernel of the linear map

$$\bar{w} \wedge *: V \to \Lambda^{k+1} V$$

is precisely W.

REMARK. This gives a kind of geometric interpretation of a wedge-product like $w_1 \wedge \cdots \wedge w_k$: it represents a k-dimensional subspace of V, and $\Lambda^k V$ is "all formal linear combinations" of such subspaces.

In three dimensions, the *cross-product* is really a wedge-product in disguise, i.e. $v \times w$ is the wedge-product, $v \wedge w$, that represents the plane spanned by v and w. It "looks like" a vector because in $\mathbb{R}^3$ there is a 1-1 correspondence between planes and normal vectors to those planes. This is a special case of something called *Hodge duality*: if V is n-dimensional, a fixed element $\alpha \neq 0 \in \Lambda^n V$ defines an isomorphism

$$\Lambda^k V^* \to \Lambda^{n-k} V$$

where V^* is the dual of V (see 2 on page 346) — also n-dimensional.

PROOF. Lemma 10.7.12 on the previous page implies that, for any $v \in V$, $\bar{w} \wedge v = 0$ if and only if the set of vectors $\{w_1, \ldots, w_k, v\}$ is linearly dependent. Since the set $\{w_1, \ldots, w_k\}$ is linearly independent, it follows that $\bar{w} \wedge v = 0$ if and only if $v \in W$. □

We get a cool way to compute determinants:

LEMMA 10.7.14. *Suppose V is a vector space with basis $\{e_1, \ldots, e_n\}$ and A is an $n \times n$ matrix. If the columns of A are vectors $\{v_1, \ldots, v_n\}$ then*

$$v_1 \wedge \cdots \wedge v_n = \det A \cdot e_1 \wedge \cdots \wedge e_n$$

PROOF. We do induction on n. If $n = 1$, there is nothing to prove. Suppose the result is true for $(n-1) \times (n-1)$ matrices and $n-1$-dimensional vector spaces, and we are computing

$$v_1 \wedge \cdots \wedge v_n$$

Let $v = \sum_{i=1}^n a_i \cdot e_i$ and plug this into the formula. We get

$$v_1 \wedge \cdots \wedge v_n = \sum_{i=1}^n a_i \cdot e_i \wedge v_2 \wedge \cdots \wedge v_n$$

Consider the ith term of this, $a_i \cdot e_i \wedge v_2 \wedge \cdots \wedge v_n$. The vectors in $v_2 \wedge \cdots \wedge v_n$ will also be linear combinations of the e_j but the presence of e_i in the wedge product will annihilate all of their terms containing e_i, i.e.

$$a_i \cdot e_i \wedge v_2 \wedge \cdots \wedge v_n = a_i e_i \wedge v_2' \wedge \cdots \wedge v_n'$$

where $v_j' = v_j - $ (its ith component). In other words, v_j' will be a vector in an $(n-1)$-dimensional vector space that is the result of deleting e_i from V. By induction, we get

$$v_2' \wedge \cdots \wedge v_n' = M_{i,1}(A) \cdot e_1 \wedge \cdots \wedge e_{i-1} \wedge e_{i+1} \wedge \cdots \wedge e_n$$

where $M_{i,1}(A)$ is the $(i,1)^{\text{th}}$ *minor* — see definition 6.2.36 on page 184. We get

$$a_i \cdot e_i \wedge v_2 \wedge \cdots \wedge v_n = a_i M_{i,1}(A) e_i \wedge e_1 \wedge \cdots \wedge e_{i-1} \wedge e_{i+1} \wedge \cdots \wedge e_n$$

Shifting e_i into its proper place multiplies this by $(-1)^{i+1}$ so we get

$$a_i \cdot e_i \wedge v_2 \wedge \cdots \wedge v_n = (-1)^{i+1} a_i M_{i,1}(A) \cdot e_1 \wedge \cdots \wedge e_n$$

and
$$v_1 \wedge \cdots \wedge v_n = \left(\sum_{i=1}^n (-1)^{i+1} a_i M_{i,1}(A) \right) \cdot e_1 \wedge \cdots \wedge e_n$$
$$= \det A \cdot e_1 \wedge \cdots \wedge e_n$$
see proposition 6.2.37 on page 184. □

COROLLARY 10.7.15. *Let V be an n-dimensional vector space with k-dimensional subspace W, and suppose*
$$\{b_1, \ldots, b_k\}$$
is a basis for W. If
$$A \colon W \to W$$
is a change of basis, to a basis
$$\{c_1, \ldots, c_k\}$$
then
$$c_1 \wedge \cdots \wedge c_k = \det A \cdot b_1 \wedge \cdots \wedge b_k$$

PROOF. Extend the bases for W to bases for all of V, i.e.
$$\{b_1, \ldots, b_k, e_{k+1}, \ldots, e_n\}$$
and
$$\{c_1, \ldots, c_k, e_{k+1}, \ldots, e_n\}$$
The change of basis can be represented by an $n \times n$ matrix that is A extended by the identity matrix, i. e.,
$$A' = \begin{bmatrix} A & 0 \\ 0 & I \end{bmatrix}$$
Lemma 10.7.14 on the facing page implies that

$c_1 \wedge \cdots \wedge c_k \wedge e_{k+1} \wedge \cdots \wedge e_n$
$$= \det A' \cdot b_1 \wedge \cdots \wedge b_k \wedge e_{k+1} \wedge \cdots \wedge e_n$$
$$= \det A \cdot b_1 \wedge \cdots \wedge b_k \wedge e_{k+1} \wedge \cdots \wedge e_n$$
so
$$(c_1 \wedge \cdots \wedge c_k - \det A \cdot b_1 \wedge \cdots \wedge b_k) \wedge e_{k+1} \wedge \cdots \wedge e_n = 0$$
The conclusion follows from lemma 10.7.12 on page 383 since
$$x = c_1 \wedge \cdots \wedge c_k - \det A \cdot b_1 \wedge \cdots \wedge b_k$$
is not in the span of $z = e_{k+1} \wedge \cdots \wedge e_n$ so that $x \wedge z = 0$ implies $x = 0$. □

EXERCISES.

1. If
$$0 \to U \xrightarrow{f} V \xrightarrow{g} W \to 0$$
is an exact sequence of k-vector-spaces of dimensions, respectively, u, w, w, show that
$$\Lambda^v V \cong \Lambda^u U \otimes_k \Lambda^w W$$
and if the diagram
$$\begin{array}{ccccccccc}
0 & \longrightarrow & U_1 & \xrightarrow{f_1} & V_1 & \xrightarrow{g_1} & W_1 & \longrightarrow & 0 \\
& & {\scriptstyle a}\downarrow & & {\scriptstyle b}\downarrow & & {\scriptstyle c}\downarrow & & \\
0 & \longrightarrow & U_2 & \xrightarrow{f_2} & V_2 & \xrightarrow{g_2} & W_2 & \longrightarrow & 0
\end{array}$$
commutes and columns that are isomorphisms, then the diagram
$$\begin{array}{ccc}
\Lambda^v V_1 & \xrightarrow{\cong} & \Lambda^u U_1 \otimes_k \Lambda^w W_1 \\
{\scriptstyle \Lambda^w b}\downarrow & & \downarrow {\scriptstyle \Lambda^u a \otimes_k \Lambda^w c} \\
\Lambda^v V_2 & \xrightarrow{\cong} & \Lambda^u U_2 \otimes_k \Lambda^w W_2
\end{array}$$
also commutes (so it is natural with respect to isomorphisms of exact sequences).

2. If V is 3-dimensional with basis $\{e_1, e_2, e_3\}$, compute
$$(2e_1 + 3e_2 - e_3) \wedge (e_1 - e_2 + e_3)$$

3. Compute the determinant of
$$\begin{bmatrix} 0 & 0 & 2 & 0 \\ 1 & 0 & 0 & 1 \\ 0 & 3 & 0 & 0 \\ 2 & 0 & 0 & -1 \end{bmatrix}$$
using exterior products.

CHAPTER 11

Group Representations, a Drive-by

"Wigner's discovery about the electron permutation group was just the beginning. He and others found many similar applications and nowadays, group theoretical methods — especially those involving characters and representations, pervade all branches of quantum mechanics."
— George Whitelaw Mackey, *Proceedings of the American Philosophical Society.*

11.1. Introduction

Group-representations are a kind of dual to group-presentations in section 4.10.2 on page 89: a *presentation* maps a free group *onto* a group; *representation theory* maps the group *into or onto* something else. This "something else" is a group of linear transformations of a vector space.

We use knowledge of linear algebra to understand groups.

This is a vast subject that could fill several volumes thicker than the present book.

Here is the classical definition:

DEFINITION 11.1.1. Given a vector-space, V, over a field, k, a *representation* of a *group G over V* is a pair (ρ, V), where ρ is a homomorphism

$$\rho: G \to \text{GL}(V)$$

— see definition 6.2.58 on page 202.

Given two representations

$$\rho_1: G \to \text{GL}(V_1)$$
$$\rho_2: G \to \text{GL}(V_2)$$

a *homomorphism of representations* is a homomorphism of vector spaces

$$f: V_1 \to V_2$$

such that

$$\rho_2(g) = f^{-1} \circ \rho_1(g) \circ f$$

for all $g \in G$. This is an isomorphism if f is an isomorphism of vector spaces.

A representation, $\rho: G \to \text{GL}(V)$, is called *faithful* if the homomorphism, ρ, is injective. If $U \subset V$ is a sub-vector-space with the property that

$$\rho(g)(U) \subset U$$

then $\rho\colon G \to \mathrm{GL}(U)$ is called a *sub-representation* of G. A representation is *irreducible* or *simple* if it has no sub-representations (other than itself and the 0-map).

The vector-space, V, is called the *representation space*. The dimension of V is called the dimension of the representation.

We'll look at some examples of group-representations.

EXAMPLE. The trivial representation

$$\rho\colon G \to \mathrm{GL}(V)$$

that maps every element of G to the identity matrix.

EXAMPLE 11.1.2. Degree-1 representations are just group-homomorphisms

$$\rho\colon G \to \mathrm{GL}(k) = k^{\times}$$

They are just *group-characters* in the sense of definition 8.5.1 on page 304. In fact, the proofs in chapter 8 on page 297 can be regarded as simple applications of group-representation theory. It's important to note that the word 'character' is often used in group-representation theory in a *very* different sense than that definition. The *only* time the two meanings of 'character' coincide is with degree-1 representations.

If $G = \mathbb{Z}_3 = \{0,1,2\}$, we have the trivial representation, ρ_0, and

$$\rho_1\colon G \to \mathbb{C}^{\times}$$

with $\rho_1(1) = e^{2\pi i/3}$, so $\rho_1(2) = e^{4\pi i/3}$. We could also define $\rho_2(1) = e^{4\pi i/3}$, in which case, $\rho_2(2) = e^{8\pi i/3} = e^{2\pi i/3}$.

The following is simple but significant:

EXAMPLE 11.1.3. We also have a representation $\mathrm{sgn}\colon \mathbb{Z}_2 \to \mathbb{R}^{\times}$ defined by

$$0 \mapsto 1$$
$$1 \mapsto -1$$

— called the *sign-representation*.

REMARK. Given any group, G, with a homomorphism

$$f\colon G \to \mathbb{Z}_2$$

we get an induced sign representation of G:

$$\mathrm{sgn} \circ f\colon G \to \mathbb{R}^{\times}$$

Here's an example of a degree-2 representation:

EXAMPLE 11.1.4. Let $G = S_3$ and define
$$\rho(1) = \begin{bmatrix} 1 & 0 \\ 0 & 1 \end{bmatrix}$$
$$\rho((1,2)) = \begin{bmatrix} -\frac{1}{2} & \frac{\sqrt{3}}{2} \\ \frac{\sqrt{3}}{2} & \frac{1}{2} \end{bmatrix}$$
$$\rho((1,3)) = \begin{bmatrix} -\frac{1}{2} & -\frac{\sqrt{3}}{2} \\ -\frac{\sqrt{3}}{2} & \frac{1}{2} \end{bmatrix}$$
$$\rho((2,3)) = \begin{bmatrix} 1 & 0 \\ 0 & -1 \end{bmatrix}$$
$$\rho((1,2,3)) = \begin{bmatrix} -\frac{1}{2} & -\frac{\sqrt{3}}{2} \\ \frac{\sqrt{3}}{2} & -\frac{1}{2} \end{bmatrix}$$
$$\rho((1,3,2)) = \begin{bmatrix} -\frac{1}{2} & \frac{\sqrt{3}}{2} \\ -\frac{\sqrt{3}}{2} & -\frac{1}{2} \end{bmatrix}$$

This has a geometric interpretation: Let $\omega = e^{2\pi i/3} = -\frac{1}{2} + \frac{\sqrt{3}}{2}$ be a primitive cube root of 1. Then the three cube roots of 1 are $\{1, \omega, \omega^2\}$ and form a triangle in $\mathbb{C}$, which we identify with $\mathbb{R}^2$. If we number these three roots, 1, 2, and 3, respectively, S_3 permutes these points via the matrices given above.

We can also give representations for S_n:

EXAMPLE 11.1.5. Let V be a vector space over k with basis-elements $\{b_1, \ldots, b_n\}$. Then we can define a representation for S_n by having it permute the basis elements. This representation has two sub-representations, V_1 and T given by

$$V_1 = \left\{ \sum_{i=1}^{n} a_i b_i \,\middle|\, a_i \in k \text{ and } a_1 = a_2 = \cdots = a_n \right\}$$
$$T_n = \left\{ \sum_{i=1}^{n} a_i b_i \,\middle|\, a_i \in k \text{ and } \sum_{i=1}^{n} a_i = 0 \right\}$$

T_n is called the *standard representation* of S_n.

To give a more modern definition of group-representations, we need to recall the concept of a *group-ring* — see 5.1.6 on page 109, in our case the group-ring kG and modules (see section 6.3 on page 222). Since k is a field, these are also *algebras* as in definition 7.1.4 on page 262 and often called *group-algebras*.

DEFINITION 11.1.6. Let kG-mod and $\mathscr{V}_k$ be the categories of left kG-modules and vector-spaces over k, respectively. Define the forgetful functor (see example 10.3.2 on page 350) that maps a kG-module to its underlying vector-space over k:

$$\mathfrak{g} \colon kG\text{-mod} \to \mathscr{V}_k$$

Given a vector-space, V, over a field, k, a *representation* of a *group G over V* is a (left) kG-module, M, with $\mathfrak{g}(M) = V$. A *homomorphism of representations* is just a kG-module homomorphism

$$f: M_1 \to M_2$$

An isomorphism is a homomorphism that is an isomorphism of vector spaces. A representation, M, is *faithful* if it contains a submodule isomorphic to kG. An *irreducible* representation, M, has no sub-modules, other than M and the 0-submodule.

REMARK. Note that any module over kG is automatically a vector-space over k.

DEFINITION 11.1.7. If

$$\rho: G \to \mathrm{GL}(V)$$

is a representation of a group, G, define

$$V^G = \{x \in V | \rho(g)(x) = x \text{ for all } g \in G\}$$

the *G-stable subspace* of V.

REMARK. Compare this with definition 13.3.1 on page 455.

We immediately have:

LEMMA 11.1.8 (Schur's Lemma). *If $f: M_1 \to M_2$ is a homomorphism of irreducible representations of a group, G, over a field, k, then f is either an isomorphism or the 0-map. If follows that the ring, $\mathrm{Hom}_{kG}(M_1, M_1) = D$, is a division ring. If k is algebraically closed then $\mathrm{Hom}_{kG}(M_1, M_1) = k$, so every homomorphism $f: M_1 \to M_1$ is multiplication by a scalar in k.*

PROOF. The kernel of f is a submodule of M_1. Since M_1 is irreducible, this must either be the 0-module or all of M_1. If it is not all of M_1, then its image in M_2 is a submodule. Therefore its image must be *all* of M_2. This also implies that any homomorphism $f: M_1 \to M_1$ must be the zero-map or an *isomorphism*, which implies that $\mathrm{Hom}_{kG}(M_1, M_1)$ is a division-ring (multiplication is composition, and every nonzero element has an inverse).

If k is algebraically closed, f has an eigenvalue, λ, in k and

$$f - \lambda \cdot I: M_1 \to M_1$$

is a homomorphism that has a kernel. The first part of this lemma implies that $f - \lambda \cdot I = 0$, so $f = \lambda \cdot I$. □

Issai Schur, 1875 – 1941, was a Russian mathematician who worked in Germany for most of his life. He studied at the University of Berlin.
As a student of Ferdinand Georg Frobenius, he worked on group representations, but also in combinatorics and number theory and even theoretical physics. He is best known for his result on the existence of the Schur decomposition and for his work on group representations (lemma 11.1.8).

In analogy with group theory (definition 4.4.14 on page 47), we define

DEFINITION 11.1.9. Given a kG-module, M, the *socle* of M, denoted $\operatorname{Soc}(M)$ is the sum of its irreducible submodules, i.e.,

$$\operatorname{Soc}(M) = \sum N \subset M, \text{ where } N \text{ is irreducible}$$

Given group-representations, there are several operations we can perform with them. Essentially, they are the functors laid out in sections 10.6 on page 365 — all of which can be applied to group-representations. We focus on three of them:

DEFINITION 11.1.10. Given a field k, group, G, and two kG-modules M_1 and M_2 we can form
  (1) the *direct sum*, $M_1 \oplus M_2$, with kG module structure given in definition 10.6.1 on page 365, and
  (2) the *tensor product*, $M_1 \otimes_k M_2$, with kG module structure given in definition 10.6.3 on page 366 and

$$g \cdot (m_1 \otimes m_2) = (g \cdot m_1) \otimes (g \cdot m_2)$$

  for all $m_1 \in M_1, m_2 \in M_2$, and $g \in G$.
  (3) the Hom-functor, $\operatorname{Hom}_k(M_1, M_2)$ of linear transformations. Its kG module structure is defined by
      for all $f \in \hom(M_1, M_2)$ and $x \in M_1$, then $(g \cdot f)(x) = g \cdot f(g^{-1} \cdot x)$ for all $g \in G$

REMARK. An interesting special case of the hom functor is the *dual* of a representation: $M^* = \hom(M, k)$, where k is the trivial representation.

It is important to describe the effects of these constructions on representations as in definition 11.1.1 on page 387. Proofs will be left as an exercise to the reader (see the material in section 10.6 on page 365).

If (ρ_i, V_i), $i = 1, 2$ are representations of a group G and $g \in G$ is an arbitrary element with $\rho_1(g) = A$, an $n \times n$ and $\rho_2(g) = B$, an $m \times m$ matrix, then
  (1)

(11.1.1) $$(\rho_1 \oplus \rho_2)(g) = \begin{bmatrix} A & 0 \\ 0 & B \end{bmatrix}$$

  — the $(n+m) \times (n+m)$ block-matrix.
  (2) If

$$A = \begin{bmatrix} A_{1,1} & \cdots & A_{1,n} \\ \vdots & \ddots & \vdots \\ A_{n,1} & \cdots & A_{n,n} \end{bmatrix}$$

  then

(11.1.2) $$(\rho_1 \otimes \rho_2)(g) = \begin{bmatrix} A_{1,1}B & \cdots & A_{1,n}B \\ \vdots & \ddots & \vdots \\ A_{n,1}B & \cdots & A_{n,n}B \end{bmatrix}$$

  — the $nm \times nm$ Kronecker product. See exercise 5 on page 377.

(3)
$$\text{(11.1.3)} \quad \text{Hom}_k(\rho_1, \rho_2)(g) = \begin{bmatrix} \hat{A}_{1,1}B & \cdots & \hat{A}_{1,n}B \\ \vdots & \ddots & \vdots \\ \hat{A}_{n,1}B & \cdots & \hat{A}_{n,n}B \end{bmatrix}$$

— a Kronecker product, where $\hat{A} = (A^{-1})^t$, the transpose of the inverse.

DEFINITION 11.1.11. A representation (or module) is called *semisimple* if it is a direct sum of simple (i.e., irreducible) representations. A ring, R, is called left-semisimple if all left R-modules are semisimple.

EXERCISES.

1. Show that the definitions 11.1.1 on page 387 and 11.1.6 on page 389 are mathematically equivalent.

2. Suppose M is a simple (i.e., irreducible) module over a ring, R. Show that M is generated by a single element.

3. Show that the representations of S_3 in example 11.1.4 on page 388 and T_3 in example 11.1.5 on page 389 (done over $\mathbb{C}$) are isomorphic.

4. Show that the representations of S_3 in example 11.1.4 on page 388 and T_3 in example 11.1.5 on page 389 (done over $\mathbb{C}$) are irreducible.

5. If G is a finite group and (ρ, V) is a representation over $\mathbb{C}$, show that $\rho(g)$ is a diagonalizable matrix for every $g \in G$ and its eigenvalues have absolute value 1.

6. Prove equation 11.1.3.

7. Suppose G is a group, k is a field, and M_1 and M_2 are two kG-modules. Show that
$$\text{Hom}_k(M_1, M_2)^G = \text{Hom}_{kG}(M_1, M_2)$$
In other words, the stabilizer of the Hom-functor of two representations is the group of *homomorphisms* of the representations.

8. Let V be a finite-dimensional vector space with a subspace W and suppose
$$P: V \to W$$
is a projection onto W with the property that $P|W = 1: W \to W$. Show that $\text{Tr}(P) = \dim W$.

9. Suppose $\rho_1, \rho_2: G \to \text{GL}(V)$ are two faithful representations of a group G. If they are isomorphic, show that $\rho_1(G)$ and $\rho_2(G)$ are conjugate subgroups of $\text{GL}(V)$.

10. Suppose $\rho_1, \rho_2: G \to \mathrm{GL}(V)$ are two faithful representations of a group G. such that $\rho_1(G)$ and $\rho_2(G)$ are conjugate subgroups of $\mathrm{GL}(V)$. Show that there is an automorphism $\varphi: G \to G$ such that ρ_1 is isomorphic to $\rho_2 \circ \varphi$.

11. Suppose a module M has two irreducible submodules U_1, U_2 such that
$$M = U_1 + U_2$$
Show that
$$M \cong U_1 \oplus U_2$$

12. Show that a module, M, is semisimple if and only if every submodule $N \subset M$ has a *complement*, i.e. if there exists another module U such that
$$M \cong N \oplus U$$

13. If a module, M, is semisimple and $N \subset M$ is a submodule, show that N and M/N are semisimple.

14. If R is a ring that is semisimple as a left R-module, show that *all* left R-modules are semisimple.

11.2. Finite Groups

Representation-theory for finite groups is a vast subject, so we will barely scratch the surface!

EXAMPLE 11.2.1. Consider the dihedral group, D_{2n} with elements $\{1, f, \cdots, f^{n-1}, r, rf, \ldots, rf^{n-1}\}$, where f is rotation by $2\pi/n$ and r is reflection of the x-coordinate. We have representations over $\mathbb{R}$:
  ▷ the trivial representation that maps its $2n$ elements to 1.
  ▷ the sign representation that maps f^k to 1 and rf^k to -1, for $k = 1, \ldots, n-1$.
  ▷ and a two-dimensional representation, ρ, from its geometric description in example 4.5.16 on page 57:

$$\rho(f^k) = \begin{bmatrix} \cos \frac{2\pi k}{n} & -\sin \frac{2\pi k}{n} \\ \sin \frac{2\pi k}{n} & \cos \frac{2\pi k}{n} \end{bmatrix}$$
$$\rho(rf^k) = \begin{bmatrix} -\cos \frac{2\pi k}{n} & \sin \frac{2\pi k}{n} \\ \sin \frac{2\pi k}{n} & \cos \frac{2\pi k}{n} \end{bmatrix}$$

We begin with

THEOREM 11.2.2 (Maschke's Theorem). *Let G be a finite group and let k be a field with the property that $|G|$ is invertible in k. If $M \subset N$ is an inclusion of left kG-modules, then there exists a submodule $U \subset N$ with the property that*
$$N \cong M \oplus U$$

It follows that all representations of G over k are direct sums of irreducible representations.

REMARK. The condition on $|G|$ and the field, k means that k must be of characteristic 0 or, if it is of characteristic $p \neq 0$, that $p \nmid |G|$.

The submodule (or sub-representation) U is called a *complement* of M in N.

This result implies that, to classify *all* representations, it is only necessary to classify the *irreducible* ones.

PROOF. Note that $M \subset N$ is an inclusion of vector-spaces. Theorem 6.2.6 on page 166 implies that we can find a basis, $\{b_1, \ldots, b_m\}$ of M that extends to one, $\{b_1, \ldots, b_m, c_1, \ldots, c_k\}$, for N. It is easy to define a linear map of vector-spaces
$$f: N \to M$$
whose restriction to M is the identity map: just map the $\{b_i\}$ to themselves and map each $\{c_j\}$ to an arbitrary $\{b_i\}$ or 0. Now we define
$$\hat{f}(x) = \frac{1}{|G|} \sum_{g \in G} g^{-1} \cdot f(g \cdot x)$$
for all $x \in N$. First of all, $\hat{f}|M$ is still the identity map: if $x \in M$, we have $g \cdot x \in M$ because M is a sub-kG module and $f(g \cdot x) = g \cdot x$ so $g^{-1} g \cdot x = x$.

We claim that $\hat{f}$ is a kG-module homomorphism, i.e., we claim that, for any $h \in G$ $\hat{f}(h \cdot x) = h \cdot \hat{f}(x)$.

First, note that,
$$\hat{f}(x) = \frac{1}{|G|} \sum_{g \in G} g^{-1} \cdot f(g \cdot x) = \frac{1}{|G|} \sum_{g \in G} h^{-1} g^{-1} \cdot f(gh \cdot x)$$
for any fixed element $h \in G$. This is because, as g cycles through all the elements of G, so will gh. We have
$$\hat{f}(h \cdot x) = \frac{1}{|G|} \sum_{g \in G} g^{-1} \cdot f(gh \cdot x)$$
$$= h \frac{1}{|G|} \sum_{g \in G} h^{-1} g^{-1} \cdot f(gh \cdot x)$$
$$= h \cdot \hat{f}(x)$$

It follows that we have a kG-module homomorphism
$$\hat{f}: N \to M$$
with $\hat{f}|M = 1: M \to M$. If $U = \ker \hat{f}$, then $\hat{g} = 1 - \hat{f}: N \to U$ is a kG-module homomorphism with $\hat{g}|U = 1: U \to U$. It is not hard to see that
$$M \cap U = 0$$
and we get an isomorphism
$$(\hat{f}, \hat{g}) N \to M \oplus U$$
The conclusion follows from exercise 12 on the preceding page. □

> Heinrich Maschke (1853 – 1908) was a German mathematician whose most well-known accomplishment was Maschke's theorem.
> He earned his Ph.D. degree from the University of Göttingen in 1880. He came to the United States in 1891, and took up an Assistant Professor position at the University of Chicago in 1892

The condition that the order of the group be invertible is necessary:

EXAMPLE 11.2.3. Let $G = \mathbb{Z}_p$, generated by g with $g^p = 1$, and consider the two-dimensional representation on the vector-space, $\mathbb{Z}_p \oplus \mathbb{Z}_p$, given by
$$\rho(g) = \begin{bmatrix} 1 & 0 \\ x & 1 \end{bmatrix}$$
where $x \in G$ and not equal to the identity. This has an invariant subspace
$$\begin{bmatrix} 0 \\ 1 \end{bmatrix}$$
so it is *not* irreducible. On the other hand, it is not the direct sum of this invariant subspace with any other one-dimensional representation so it violates Maschke's Theorem.

Given the importance of irreducible representations, we should have a criterion for irreducibility:

PROPOSITION 11.2.4. *A representation, M, of a (not necessarily finite) group, G is irreducible if and only if, for every nonzero $v \in M$, the set of vectors*
$$S = \{g \cdot v | \forall g \in G\}$$
spans M.

PROOF. If $x \in G$, we claim that $x \cdot \text{Span}(S) = \text{Span}(S)$. This is because, if g runs over all the elements of G so does $x \cdot g$. It follows that $\text{Span}(S)$ is a sub-representation (or submodule). If $\text{Span}(S) \neq M$, then M has a proper sub-representation and is not irreducible. □

DEFINITION 11.2.5. If V is a vector-space, $\rho \colon G \to \text{GL}(V)$ is a representation, and $n \in \mathbb{Z}_{\geq 0}$, define
$$V^{\oplus n} = \begin{cases} 0 & \text{if } n = 0 \\ \oplus_{i=1}^n V & \text{otherwise} \end{cases}$$
or
$$\rho^{\oplus n} = \begin{cases} 0 & \text{if } n = 0 \\ \oplus_{i=1}^n \rho & \text{otherwise} \end{cases}$$

Maschke's Theorem shows that representations can be "factored":

PROPOSITION 11.2.6. *If G is a finite group, let $\{\iota_j\}$ be a list of its irreducible representations over $\mathbb{C}$. If $\rho \colon G \to \text{GL}(V)$ is a finite-dimensional representation of G over $\mathbb{C}$, there exists an isomorphism*

(11.2.1) $$\rho \to \iota_1^{\oplus m_1} \oplus \iota_2^{\oplus m_2} \oplus \cdots$$

for some integers $m_i \in \mathbb{Z}_{\geq 0}$.

PROOF. If ρ is irreducible, we're done. Otherwise, Maschke's Theorem (11.2.2 on page 393) implies that $\rho = \rho_1 \oplus \rho_2$. If we repeatedly apply this reasoning to the summands, we eventually get a direct sum of irreducible representations. Exercise 1 on page 364 implies that we can rearrange these summands to the form in equation 11.2.1 on the previous page. □

The reasoning used in Schur's Lemma (11.1.8 on page 390) implies that

LEMMA 11.2.7. *Let ξ be a representation of a group, let ι be an irreducible representation of the same group, and let*

$$f : \iota \to \xi$$

be a homomorphism. Then there are two possibilities:

  (1) *f is an isomorphism of ι with an imbedded sub-representation $\iota \subset \xi$, or*
  (2) *f is the zero-map.*

This implies that the factorization in proposition 11.2.6 on the previous page is *unique*:

COROLLARY 11.2.8. *If G is a finite group, let $\{\iota_j\}$ be a list of its irreducible representations over $\mathbb{C}$. The representations $\iota_1^{\oplus m_1} \oplus \iota_2^{\oplus m_2} \oplus \cdots$ and $\iota_1^{\oplus n_1} \oplus \iota_2^{\oplus n_2} \oplus \cdots$ are isomorphic if and only if $m_i = n_i$ for all i.*

PROOF. Let $p_i : \iota_1^{\oplus n_1} \oplus \iota_2^{\oplus n_2} \oplus \cdots \to \iota_i^{\oplus m_i}$ be the projection. Lemma 11.2.7 implies that, given any homomorphism

$$f : \iota_1^{\oplus m_1} \oplus \iota_2^{\oplus m_2} \oplus \cdots \to \iota_1^{\oplus n_1} \oplus \iota_2^{\oplus n_2} \oplus \cdots$$

the composite $p_i \circ f|_{\iota_j^{\oplus m_j}} = 0 : \iota_j^{\oplus m_j} \to \iota_i^{\oplus m_i}$ whenever $i \neq j$. This means that $f\left(\iota_j^{\oplus m_j}\right) \subset \iota_j^{\oplus n_j}$ for all j. This can be an isomorphism if and only if $m_j = n_j$. □

The following result will be useful in characterizing group-rings:

LEMMA 11.2.9. *If R is a ring,*

$$\text{Hom}_R(R, R) \cong R^{\text{op}}$$

where R^{op} is the opposite ring to R — it has the same elements as R but $r_1 \cdot r_2$ in R^{op} is equal to $r_2 \cdot r_1$ in R.

REMARK. Of course $(R^{\text{op}})^{\text{op}} = R$.

PROOF. A homomorphism $f \in \text{Hom}_R(R, R)$ is completely determined by where it sends $1 \in R$ — i.e., $f(x) = x \cdot f(1)$. We define

$$g : \text{Hom}_R(R, R) \to R^{\text{op}}$$
$$f \mapsto f(1)$$

We need R^{op} because

$$f_1 \circ f_2 \mapsto f_2(1) \cdot f_1(1)$$

It is straightforward to verify that g is an isomorphism. □

This leads to an interesting structural result:

THEOREM 11.2.10 (Artin-Wedderburn). *Suppose R is a finite-dimensional algebra over a field, k, with the property that*

$$R = \iota_1^{\oplus n_1} \oplus \cdots \oplus \iota_m^{\oplus n_m}$$

where the $\{\iota_j\}$ are pairwise non-isomorphic irreducible left R-modules (left-ideals of R). Then there is an isomorphism

$$R \cong M_{n_1}(D_1) \oplus \cdots \oplus M_{n_m}(D_m)$$

where the $\{D_j = \mathrm{Hom}_R(\iota_j, \iota_j)^{\mathrm{op}}\}$ are division-algebras and $M_n(D)$ is the ring of $n \times n$ matrices with entries in D — see example 5.1.5 on page 109.

If k is algebraically closed, all the D_i are equal to k, and

$$\dim_k \iota_j = n_j$$

PROOF. We will use lemma 11.2.9 on the facing page and compute $\mathrm{Hom}_R(R, R)^{\mathrm{op}} = R$. Note that, in any homomorphism

$$\iota_1^{\oplus n_1} \oplus \cdots \oplus \iota_m^{\oplus n_m} \to \iota_1^{\oplus n_1} \oplus \cdots \oplus \iota_m^{\oplus n_m}$$

the image of any summand $\iota_j^{\oplus n_j}$ will lie entirely in $\iota_j^{\oplus n_j}$, by Schur's Lemma (11.1.8 on page 390). Lemma 11.2.9 on the facing page implies that

$$R^{\mathrm{op}} = \mathrm{Hom}_R(R, R) = \mathrm{Hom}_R(\iota_1^{\oplus n_1}, \iota_1^{\oplus n_1}) \oplus \cdots \oplus \mathrm{Hom}_R(\iota_m^{\oplus n_m}, \iota_m^{\oplus n_m})$$

Furthermore

$$\mathrm{Hom}_R(\iota_j^{\oplus n_j}, \iota_j^{\oplus n_j})$$

is the ring of $n_j \times n_j$ matrices with entries in $\mathrm{Hom}_R(\iota_j, \iota_j) = D_j^{\mathrm{op}}$ by Schur's Lemma again. The $(j, i)^{\mathrm{th}}$ entry in this matrix will be a map from the ith summand in the domain to the jth summand in the range. We get

$$R^{\mathrm{op}} \cong M_{n_1}(D_1^{\mathrm{op}}) \oplus \cdots \oplus M_{n_k}(D_k^{\mathrm{op}})$$

The result follows by taking the opposite of everything in sight and noting that $M_{n_i}(D_i^{\mathrm{op}})^{\mathrm{op}} = M_{n_i}(D_i)$.

The final statement follows from

$$\dim_k M_{n_j}(k) = n_j^2 = n_j \cdot \dim_k \iota_j$$

$\square$

The reasoning used in solving exercise 14 on page 393 implies that:

COROLLARY 11.2.11. *Suppose R is a finite-dimensional semisimple algebra over a field, k, with the property that*

$$R = \iota_1^{\oplus n_1} \oplus \cdots \oplus \iota_m^{\oplus n_m}$$

where the $\{\iota_j\}$ are pairwise non-isomorphic irreducible left R-modules (left-ideals of R). Then the set $\{\iota_1, \ldots, \iota_m\}$ is a complete set of irreducible (simple) modules over R. If k is algebraically closed, we have $n_j = \dim_k \iota_j$ and

$$\dim_k R = n_1^2 + \cdots + n_m^2$$

REMARK. Maschke's Theorem (11.2.2 on page 393) implies that the hypotheses are satisfied if $R = kG$ for some finite group, G, and k is a field of characteristic that doesn't divide $|G|$. If k is also algebraically closed, the last formula implies that

$$|G| = n_1^2 + \cdots + n_m^2$$

which limits the number of possible irreducible representations and their dimensions.

PROOF. All R-modules are quotients of free R-modules (proposition 6.3.10 on page 225), which are direct sums of copies of R. It follows that all of the irreducible modules that can occur in *any* R-module must *already* be present in R itself. □

EXERCISES.

1. If k is a field and $M_n(k)$ is the ring of $n \times n$ matrices over k, show that
$$M_n(k) \cong M_n(k)^{\text{op}}$$

2. Show that the opposite of a matrix ring over a division-algebra is also a matrix-ring over a division algebra. Conclude that a left-semisimple algebra is also right-semisimple, so we may simply call them semisimple.

11.3. Characters

Now we will explore *invariants* of representations. These are easily-computed quantities that can characterize a representation up to isomorphism.

EXERCISE 11.3.1. Show that the standard representation of S_n in example 11.1.5 on page 389 is irreducible.

Given a matrix, A, representing a linear transformation

$$f: V \to V$$

note that the *trace* (see definition 6.2.53 on page 196), $\text{Tr}(A)$, is a well-defined invariant of f. In other words, it does not depend on the basis for V used to compute $\text{Tr}(A)$. This is because $\text{Tr}(P^{-1}AP) = \text{Tr}(A)$ for any invertible matrix P — see exercise 18 on page 201.

With this in mind, we define

DEFINITION 11.3.2. Given a vector-space, V, over a field, k, and a representation of a group G, (ρ, V), define the *character* of the representation to be a function

$$\zeta: G \to k$$

given by $\zeta(g) = \text{Tr}(\rho(g))$.

REMARK. The character of a representation is almost *never* a homomorphism (unless the representation is of degree 1)!

Isomorphic representations have the same characters (again, consider exercise 18 on page 201).

Recall the equivalence relation of conjugacy defined in proposition 4.7.11 on page 78. This results in a group, G, being subdivided into conjugacy classes as in equation 4.7.2 on page 78

$$G = \bigsqcup C_i$$

where all elements in each of the C_i are conjugate to each other.

DEFINITION 11.3.3. Let G be a finite group, let $S = \{C_1, \ldots, C_n\}$ be its conjugacy classes, and let k be a field. A *class function* of G is just a function

$$f: S \to k$$

Exercise 18 on page 201 immediately implies that

PROPOSITION 11.3.4. *Characters of group-representations are class functions.*

EXAMPLE 11.3.5. Consider the representation of S_3 in example 11.1.4 on page 388. Since conjugacy in S_n is determined by *cycle-structure* (see corollary 4.5.12 on page 54 and exercise 3 on page 58) we have three conjugacy classes in S_3

$$C_0 = \{1\}$$
$$C_1 = \{(1,2), (1,3), (2,3)\}$$
$$C_3 = \{(1,2,3), (1,3,2)\}$$

and the character of that representation is given by

$$\xi(C_0) = 2$$
$$\xi(C_1) = 0$$
$$\xi(C_2) = -1$$

DEFINITION 11.3.6. A finite group's character table has rows corresponding to irreducible representations and columns corresponding to conjugacy classes.

EXAMPLE 11.3.7. Here's a character table for S_3:

	$C_0 = \{1\}$	$C_1 = \{(1,2), (1,3), (2,3)\}$	$C_2 = \{(1,2,3), (1,3,2)\}$
triv	1	1	1
sgn	1	-1	1
std	2	0	-1

Here,
- triv is the one-dimensional *trivial representation* that sends all elements of S_3 to the identity,
- sgn is the one-dimensional *sign-representation*, and
- std is the two-dimensional *standard representation* in examples 11.1.4 on page 388 and 11.1.5 on page 389.

REMARK. It isn't obvious (but true) that these are all of the irreducible representations of S_3. We can make other observations that turn out not to be mere coincidences:
- $(\dim \text{triv})^2 + (\dim \text{sgn})^2 + (\dim \text{std})^2 = 1 + 1 + 4 = |S_3|$. Compare this to the Artin-Wedderburn Theorem, 11.2.10 on page 397.
- If ζ_1, ζ_2 are two *distinct* representations from the character table above, then
$$\sum_{g \in S_3} \zeta_1(g)\zeta_2(g) = 0$$

The reasoning used in Maschke's Theorem can be generalized

LEMMA 11.3.8. *Let $\rho: G \to \text{GL}(V)$ be a representation of a finite group and define*
$$T = \frac{1}{|G|} \sum_{g \in G} \rho(g)$$
Then
$$T: V \to V^G$$
is a projection onto the stable subspace of V (see definition 11.1.7 on page 390).

PROOF. If $x \in V$ is in the image of T then there exists $y \in V$ such that
$$x = \frac{1}{|G|} \sum_{g \in G} \rho(g)(y)$$
and
$$\rho(h)(x) = \rho(h) \frac{1}{|G|} \sum_{g \in G} \rho(g)(y)$$
$$= \frac{1}{|G|} \sum_{g \in G} \rho(h)\rho(g)(y)$$
$$= \frac{1}{|G|} \sum_{g \in G} \rho(hg)(y)$$
$$= x$$

since hg runs over all the elements of G. On the other hand, it is not hard to see that $x \in V^G$ implies that $T(x) = x$, so the conclusion follows. □

COROLLARY 11.3.9. *Let $\rho: G \to \text{GL}(V)$ be a representation of a finite group with character $\zeta: G \to \mathbb{C}$. Then*
$$\frac{1}{|G|} \sum_{g \in G} \zeta(g) = \dim V^G$$

PROOF. We use lemma 11.3.8 and the conclusion of exercise 8 on page 392 to conclude that
$$\frac{1}{|G|} \text{Tr}\left(\frac{1}{|G|} \sum_{g \in G} \rho(g)\right) = \dim V^G$$

But

$$\mathrm{Tr}\left(\frac{1}{|G|}\sum_{g\in G}\rho(g)\right) = \frac{1}{|G|}\sum_{g\in G}\mathrm{Tr}\,(\rho(g))$$

$$= \frac{1}{|G|}\sum_{g\in G}\xi(g)$$

□

For the remainder of this chapter, we'll assume that $k = \mathbb{C}$ and define

DEFINITION 11.3.10. If, ξ_1, ξ_2, are characters of representations of a finite group, G, over $\mathbb{C}$, we define the *inner product*

$$\langle \xi_1, \xi_2 \rangle = \frac{1}{|G|}\sum_{g\in G}\xi_1(g)\overline{\xi_2(g)}$$

(11.3.1)
$$= \frac{1}{|G|}\sum_{c\in S}|c|\cdot \xi_1(c)\overline{\xi_2(c)}$$

where S is the set of conjugacy-classes in G and $|s|$ is the number of elements in the conjugacy-class, s.

REMARK. Note that

$$\frac{|c|}{|G|} = \frac{1}{|Z_G(c)|}$$

where $|Z_G(c)|$ is the the number elements that commute with g (the centralizer of g) — see corollary 4.7.13 on page 79. We can rewrite equation 11.3.1 as

(11.3.2)
$$\langle \xi_1, \xi_2 \rangle = \sum_{c\in S}\frac{\xi_1(c)\overline{\xi_2(c)}}{|Z_G(c)|}$$

To understand the significance of this inner product, we need

PROPOSITION 11.3.11. *Let ρ_1, ρ_2 be two representations of a finite group G over $\mathbb{C}$ with characters ξ_1, ξ_2, respectively. Then the character of $\mathrm{Hom}_k(\rho_1, \rho_2)$ is*

$$\overline{\xi_1}\cdot \xi_2$$

PROOF. Equation 11.1.3 on page 392 implies that the matrix for $\mathrm{Hom}_k(\rho_1, \rho_2)(g)$ is the Kronecker product of the transpose of the inverse of the matrix for $\rho_1(g)$ with that of $\rho_2(g)$ for all $g \in G$. It follows that the *traces* satisfy

$$\mathrm{Tr}\,(\mathrm{Hom}(\rho_1, \rho_2)(g)) = \mathrm{Tr}\left(\rho_1(g)^{-1}\right)\cdot \xi_2(g)$$

since transposing leaves the trace unchanged. Exercise 214 on page 513 shows that $\rho_1(g)$ is diagonalizable and that its eigenvalues, $\{\lambda_1, \ldots, \lambda_k\}$ are on the complex unit circle. We have

$$\text{Tr}\left(\text{Hom}(\rho_1,\rho_2)(g)\right) = \left(\sum_{i=1}^{k}\lambda_i^{-1}\right)\cdot \xi_2(g)$$
$$= \left(\overline{\sum_{i=1}^{k}\lambda_i}\right)\cdot \xi_2(g)$$
$$= \overline{\left(\sum_{i=1}^{k}\lambda_i\right)}\cdot \xi_2(g)$$
$$= \overline{\xi_1}\cdot \xi_2$$

□

This immediately leads to an interpretation of the inner product:

COROLLARY 11.3.12. *Given two representations over* $\mathbb{C}$ *of a finite group,* G

$$\rho_1\colon G \to \text{GL}(V_1)$$
$$\rho_2\colon G \to \text{GL}(V_2)$$

with respective characters, ξ_1, ξ_2, *we have*

$$\langle \xi_1, \xi_2 \rangle = \dim \text{Hom}_{\mathbb{C}G}(V_1,V_2) = \dim \text{Hom}_{\mathbb{C}G}(\rho_1,\rho_2)$$

PROOF. The character of $\text{Hom}_{\mathbb{C}G}(\rho_1,\rho_2)$ is $\overline{\xi_1}\cdot\xi_2$ so corollary 11.3.9 on page 400 implies that

$$\langle \xi_1,\xi_2\rangle = \dim\overline{\text{Hom}_k(V_1,V_2)^G} = \dim\text{Hom}_k(V_1,V_2)^G$$

and exercise 7 on page 392 implies the conclusion. □

Schur's Lemma (11.1.8 on page 390) immediately implies that

COROLLARY 11.3.13. *If ξ_1 and ξ_2 are characters of distinct irreducible representations of G over* $\mathbb{C}$, *then*

$$\langle \xi_1,\xi_2\rangle = 0$$

and

$$\langle \xi_1,\xi_1\rangle = \langle \xi_2,\xi_2\rangle = 1$$

PROOF. The second statement follows from the fact that $\mathbb{C}$ is algebraically closed (see theorem 8.9.1 on page 321) so that a homomorphism from ξ_1 to itself is multiplication by a scalar, $c \in \mathbb{C}$ which implies that

$$\text{Hom}_{\mathbb{C}G}(\xi_1,\xi_1) = \mathbb{C}$$

□

REMARK. This has a number of fascinating consequences. For one thing, it shows that an irreducible representation is *completely determined* by its character: just form its inner product with characters of all the irreducible representations and pick the one that gives a nonzero result.

It also limits the *number* of possible irreducible representations: Since the *rows* of a character table must be orthogonal, they must number less than or equal to the number of *columns* — i.e., the number of conjugacy classes.

In fact we can say a great deal more:

THEOREM 11.3.14. *If G is a finite group and ρ_1, ρ_2 are two finite-dimensional representations over $\mathbb{C}$ with respective characters, ξ_1, ξ_2, then ρ_1 is isomorphic to ρ_2 if and only if $\xi_1 = \xi_2$.*
If

$$\rho_1 = \bigoplus_{j=1}^{n} \iota_j^{\oplus m_j}$$

then $\langle \xi_1, \xi_1 \rangle = \sum_{i=1}^{n} m_i^2$ and ρ_1 is simple if and only if

$$\langle \xi_1, \xi_1 \rangle = 1$$

PROOF. Clearly, $\rho_1 \cong \rho_2$ implies that $\xi_1 = \xi_2$.

Now suppose $\{\iota_k\}$ is a list of all irreducible representations of G over $\mathbb{C}$ with respective characters $\{\eta_k\}$. Corollary 11.2.8 on page 396 implies that there is a unique expression

$$\rho_1 = \bigoplus_{j=1}^{n} \iota_j^{\oplus m_j}$$

where $m_j \in \mathbb{Z}_{\geq 0}$ is the number of times ι_j occurs in the big direct sum. Clearly, the $\{m_j\}$ determine ρ_1 up to isomorphism.

The character of ρ_1 will satisfy

$$\xi_1 = \sum_{j=1}^{n} m_j \cdot \eta_j$$

so that corollary 11.3.13 on the preceding page implies that

$$m_j = \langle \xi_1, \eta_j \rangle$$

It follows that we can reconstruct ρ_1 completely from the numbers $\langle \xi_1, \eta_j \rangle$. It follows that if $\xi_1 = \xi_2$ the coefficients $\{m_k\}$ will be the same. □

EXAMPLE 11.3.15. Consider the permutation-representation, η, of S_3 over $\mathbb{C}$:

$$() \to \begin{bmatrix} 1 & 0 & 0 \\ 0 & 1 & 0 \\ 0 & 0 & 1 \end{bmatrix}$$

$$(1,2) \to \begin{bmatrix} 0 & 1 & 0 \\ 1 & 0 & 0 \\ 0 & 0 & 1 \end{bmatrix}$$

$$(1,3) \to \begin{bmatrix} 0 & 0 & 1 \\ 0 & 1 & 0 \\ 1 & 0 & 0 \end{bmatrix}$$

$$(2,3) \to \begin{bmatrix} 1 & 0 & 0 \\ 0 & 0 & 1 \\ 0 & 1 & 0 \end{bmatrix}$$

$$(1,2,3) \to \begin{bmatrix} 0 & 1 & 0 \\ 0 & 0 & 1 \\ 1 & 0 & 0 \end{bmatrix}$$

$$(1,3,2) \to \begin{bmatrix} 0 & 0 & 1 \\ 1 & 0 & 0 \\ 0 & 1 & 0 \end{bmatrix}$$

The corresponding character is:

$C_0 = \{1\}$	$C_1 = \{(1,2),(1,3),(2,3)\}$	$C_2 = \{(1,2,3),(1,3,2)\}$
3	1	0

and we compute (using the character table in example 11.3.7 on page 399):

$$\zeta \cdot \text{triv} = 1$$
$$\zeta \cdot \text{sign} = 0$$
$$\zeta \cdot \text{std} = 1$$

so $\eta = \text{triv} \oplus \text{std}$.

Consider the representation $\text{std} \otimes \text{std}$. It has a character table

$C_0 = \{1\}$	$C_1 = \{(1,2),(1,3),(2,3)\}$	$C_2 = \{(1,2,3),(1,3,2)\}$
4	0	1

and we compute

$$(\text{std} \otimes \text{std}) \cdot \text{triv} = 1$$
$$(\text{std} \otimes \text{std}) \cdot \text{sign} = 1$$
$$(\text{std} \otimes \text{std}) \cdot \text{std} = 1$$

so $\text{std} \otimes \text{std} = \text{triv} \oplus \text{sign} \oplus \text{std}$.

If we consider the regular representations of S_3 on $\mathbb{C}S_3$, we get a character

$C_0 = \{1\}$	$C_1 = \{(1,2),(1,3),(2,3)\}$	$C_2 = \{(1,2,3),(1,3,2)\}$
1	0	0

so $\mathbb{C}S_3 = \text{triv} \oplus \text{sign} \oplus \text{std} \oplus \text{std}$.

Our next project will be to determine the *number* of isomorphism classes of irreducible representations. We begin by extending the concept of center from group-theory (definition 4.4.9 on page 46) to ring-theory:

DEFINITION 11.3.16. If R is a ring, the *center* of R, denoted $Z(R)$ is the set of elements that commute with all elements of R, i.e.
$$Z(R) = \{r \in R | sr = rs, \text{for } \forall s \in R\}$$

We leave the proof of the following as an exercise to the reader:

PROPOSITION 11.3.17. *If G is a finite group and k is a field, then*
$$Z(kG) = \left\{r \in kG | g^{-1}rg = r, \text{for } \forall g \in G\right\}$$

COROLLARY 11.3.18. *Let G be a finite group, let k be a field, let $S = \{C_1, \ldots, C_n\}$ be the conjugacy classes of G, and let*
$$\sigma_i = \sum_{g \in C_i} g$$
Then $\sigma_i \in Z(kG)$ for $i = 1, \ldots n$. It follows that
$$\dim_k Z(kG) \geq n$$
— the number of conjugacy classes.

PROOF. Since all the elements of C_i are conjugate to each other, conjugating them by group-elements just permutes them and leaves the sum unchanged. Since distinct σ_i are linearly independent (for instance, $C_i \cap C_j = \emptyset$ for $i \neq j$), the conclusion follows. □

It's easy to characterize the center of a matrix-ring:

PROPOSITION 11.3.19. *Let k be a field and let $n > 0$ be an integer. Then*
$$Z(M_n(k)) = k \cdot I$$
— scalar multiples of the identity matrix.

PROOF. Let $E_{i,j}$ be the $n \times n$ matrix with 1 in position i,j and 0 elsewhere. If A is an arbitrary $n \times n$ matrix, then

$E_{i,j}A = $ matrix whose i^{th}-row is the jth row of A

$AE_{i,j} = $ matrix whose i^{th}-column is the jth column of A

The only way these can be equal is if every row and column of A has, at most, a single nonzero element — and these elements (in distinct rows and columns) are equal. □

Now we can prove our main result:

THEOREM 11.3.20. *If G is a finite group, then*
$$\dim_k Z(\mathbb{C}G) = \text{the number of distinct simple representations of } G$$
$$= \text{the number of conjugacy classes of } G$$

It follows that character-tables are always square. Furthermore, the $\{\sigma_i\}$ defined in corollary 11.3.18 are a basis for $Z(kG)$.

PROOF. The Artin-Wedderburn Theorem (11.2.10 on page 397) implies that
$$\mathbb{C}G \cong M_{n_1}(\mathbb{C}) \oplus \cdots \oplus M_{n_m}(\mathbb{C})$$
where each matrix-ring corresponds to a simple representation of G. Taking centers (and referring to proposition 11.3.19 on the preceding page) gives the first statement. Corollary 11.3.18 on the previous page implies that $\dim_k Z(\mathbb{C}G) \geq c$ where c is the number of conjugacy classes of G. Corollary 11.3.13 on page 402 and the fact that distinct characters are orthogonal implies that the number of distinct simple representations (which is equal to $\dim_k Z(\mathbb{C}G)$) is $\leq c$. These inequalities imply that the number of distinct simple representations is actually *equal* to c. □

Since character-tables are square, we conclude that

PROPOSITION 11.3.21. *The complex characters of a finite group form a basis for all complex-valued class functions of that group.*

We conclude by considering the standard representation of a finite group: $\mathbb{C}G$ as a left module over itself with character $\zeta_{\mathbb{C}G}$. Since the diagonal entries correspond to elements fixed by the action of a group-element, we get

$$\zeta_{\mathbb{C}G}(g) = \begin{cases} |G| & \text{if } g = 1 \\ 0 & \text{otherwise} \end{cases}$$

We have a decomposition

$$\mathbb{C}G = \bigoplus_{j=1}^{n} \iota_j^{\oplus m_j}$$

where the $\{\iota_j\}$ are all the simple complex representations. If ζ_j is the character of ι_j then we get a formula

$$\zeta_{\mathbb{C}G} = \sum_{j=1}^{n} m_j \zeta_j$$

We have proved

LEMMA 11.3.22. *If G is a finite group and $\{\zeta_1, \ldots, \zeta_n\}$ are characters of its simple representations over $\mathbb{C}$, $\{\iota_1, \ldots, \iota_n\}$, where $\dim_\mathbb{C} \iota_j = m_j = \zeta_j(1)$ then, for any $g \in G$*

$$\sum_{j=1}^{n} m_j \zeta_j(g) = \begin{cases} |G| & \text{if } g = 1 \\ 0 & \text{otherwise} \end{cases}$$

Just as rows of a character table are orthogonal, so are the columns (as in the remarks following example 11.3.7 on page 399):

THEOREM 11.3.23. *Let $\{\chi_i\}_{i=1..k}$ be characters of simple representations of a finite group G and let $\{1 = g_1, g_2, \ldots g_k\}$ be representatives of the conjugacy-classes of G. Then*

(11.3.3) $$\sum_{i=1}^{k} \chi_i(g_u)\overline{\chi_i(g_v)} = \begin{cases} 0 & \text{if } u \neq v \\ |Z_G(g_u)| & \text{otherwise} \end{cases}$$

Here, $Z_G(g_j)$ is the centralizer of g_j in G — see definition 4.7.12 on page 79.

PROOF. Let X be the $k \times k$ matrix for the character-table and let Z be the $k \times k$ diagonal matrix

$$Z = \begin{bmatrix} |Z_G(g_1)| & 0 & \cdots & 0 \\ 0 & |Z_G(g_2)| & \cdots & \vdots \\ \vdots & \vdots & \ddots & 0 \\ 0 & \cdots & 0 & |Z_G(g_k)| \end{bmatrix}$$

Equation 11.3.2 on page 401 is equivalent to

$$\bar{X} C^{-1} X^t = I$$

where $\bar{X}$ is the complex conjugate. Since all matrices here are invertible, we can write

$$\left(\bar{X} C^{-1}\right)^{-1} = X^t = C \bar{X}^{-1}$$

so that

$$X^t \bar{X} = C$$

which is equation 11.3.3 on the facing page. $\square$

Recall the concept of integral elements of a field developed in section 6.5 on page 251.

PROPOSITION 11.3.24. *If G is a finite group the center, $Z(\mathbb{Z}G)$, is integral over $\mathbb{Z}$. If $\{g_1 = 1, \ldots, g_k\}$ are representatives of the conjugacy-classes of G then $\hat{g}_i \in Z(\mathbb{Z}G)$ are integral over $\mathbb{Z}$, where $\hat{g}$ is the sum of all the elements conjugate to g. If $\lambda_1, \ldots, \lambda_k$ are algebraic integers over $\mathbb{Z}$, then*

$$\sum_{i=1}^{k} \lambda_i \hat{g}_i$$

is an algebraic integer over $\mathbb{Z}$. If $f: G \to \mathbb{C}$ is a class function then

$$z = \sum_{g \in G} f(g) \cdot g$$

is in $Z(\mathbb{C}G)$.

REMARK. In light of definition 6.5.1 on page 252, this means it's a root of a monic polynomial with integer coefficients. Proposition 6.5.5 on page 253 shows that algebraic integers form a ring.

PROOF. That elements of $Z(\mathbb{Z}G)$ are integral over $\mathbb{Z}$ follows from the fact that $\mathbb{Z}G$ is a finitely-generated $\mathbb{Z}$-module, and $Z(\mathbb{Z}G)$ is a submodule, so also finitely generated — see proposition 6.5.2 on page 252. The statement about the $\hat{g}_i$ follows from proposition 11.3.17 on page 405. The second statement follows from the fact that algebraic integers form a ring — again, see proposition 6.5.2 on page 252.

The statement about z follows from the fact that we can rewrite z as

$$\sum_{i=1}^{k} f(g_i) \cdot \hat{g}_i$$

since f is constant over elements of a conjugacy class. $\square$

PROPOSITION 11.3.25. *If ρ is a d-dimensional simple representation of a finite group, G, over $\mathbb{C}$ and $x \in Z(\mathbb{C}G)$ then*
$$\rho(x) = \lambda \cdot I$$
for some $\lambda \in \mathbb{C}$. in fact
$$\rho(x) = \frac{1}{d} \operatorname{Tr}(\rho(x)) \cdot I$$
If $x = \sum_{g \in G} \alpha_g g$, then
$$\rho(x) = \sum_{g \in G} \alpha_g \chi(g) \cdot I$$
where χ is the character of ρ.

PROOF. Since x commutes with every element of G, $\rho(x)$ commutes with all of $\rho(g)$ for any $g \in G$ — therefore the homomorphism
$$\rho(x) \colon \mathbb{C}^d \to \mathbb{C}^d$$
of vector-spaces is actually a homomorphism of $\mathbb{C}G$-modules. Schur's Lemma, 11.1.8 on page 390, implies that it is of the form $\lambda \cdot I$. The final conclusion follows from the fact that
$$\operatorname{Tr}(\lambda \cdot I) = d \cdot \lambda$$
□

We also conclude that

THEOREM 11.3.26. *The degrees of the simple complex representations of a finite group all divide the order of the group.*

PROOF. Let χ be the character of a degree-d simple complex representation, ρ. The final statement of proposition 11.3.24 on the preceding page implies that
$$x = \sum_{g \in G} \chi(g^{-1}) g$$
is in $Z(\mathbb{C}G)$. Proposition 11.3.25 implies that
$$\rho(x) = \frac{1}{d} \sum_{g \in G} \chi(g^{-1}) \chi(g) \cdot I = \frac{1}{d} \left(\sum_{g \in G} 1 \right) I = \frac{|G|}{d} \cdot I$$
Since x is integral over $\mathbb{Z}$ (by Proposition 11.3.24 on the preceding page) and $|G|/d \in \mathbb{Q}$, we conclude that $|G|/d \in \mathbb{Z}$ so $d \mid |G|$. □

EXERCISES.

1. If G is a finite group and $\rho\colon G \to \mathbb{C}^k$ is a representation, show that $\rho(g)$ is diagonalizable for any $g \in G$.

2. If G is a finite group and $\rho\colon G \to \mathbb{C}^k$ is a representation with character χ, and $g \in G$ is of order n, show that $\chi(g)$ is a sum of nth roots of unity (including roots whose order divides n).

3. If G is a finite group and $\rho\colon G \to \mathbb{C}^k$ is a representation with character χ, and $|\chi(1)| = |\chi(g)|$ for some element $g \in G$, show that $\rho(g)$ is multiplication by a scalar (root of unity)

4. If χ is the character of a complex representation of a finite group, G, show that $\chi(g^{-1}) = \overline{\chi(g)}$ for all $g \in G$.

5. Show that, if every element of a finite group G is conjugate to its inverse, every complex character is real-valued. Conversely, show that if every complex character of G is real-valued, then every element of G is conjugate to its inverse.

11.4. Examples

We have seen a character-table for S_3 in example 11.3.7 on page 399.

Now we will do S_4. It is well-known that conjugacy-class of an element of a symmetric group is determined by its *cycle-structure* — see exercise 3 on page 58. Our table will look like

	()	$(*,*)$	$(*,*)(*,*)$	$(*,*,*)$	$(*,*,*,*)$
size	1	6	3	8	6
triv	1	1	1	1	1
sgn	1	-1	1	1	-1

where the second row is the size of the respective conjugacy classes.

We consider the representation where S_4 acts on $\mathbb{C}^4$ by permuting the axes. The character is equal to the number of axes fixed by a given permutation, so we get

	()	$(*,*)$	$(*,*)(*,*)$	$(*,*,*)$	$(*,*,*,*)$
size	1	6	3	8	6
perm	4	2	0	1	0

This representation is not simple: it contains an invariant subspace isomorphic to the trivial representation. It is the representation that permutes the axes in the three-dimensional subspace of $\mathbb{C}^4$ with $\sum_{i=1}^{4} x_i = 0$. We subtract the trivial character from perm to get

	()	$(*,*)$	$(*,*)(*,*)$	$(*,*,*)$	$(*,*,*,*)$
size	1	6	3	8	6
std	3	1	-1	0	-1

The computation $\langle \text{std}, \text{std} \rangle = 1$ shows that it is simple (see theorem 11.3.14 on page 403). It is not hard to see that

$$\langle \text{std} \otimes \text{sgn}, \text{std} \otimes \text{sgn} \rangle = 1$$

so our character-table has 4 rows

	()	$(*,*)$	$(*,*)(*,*)$	$(*,*,*)$	$(*,*,*,*)$
size	1	6	3	8	6
triv	1	1	1	1	1
sgn	1	−1	1	1	−1
std	3	1	−1	0	−1
std ⊗ sgn	3	−1	−1	0	1

To get the fifth representation, note that the subgroup

$$K = \{1, (1,2)(3,4), (1,3)(2,4), (1,4)(2,3)\} \subset S_4$$

is normal[1]. The quotient is isomorphic to S_3 and the projection $p \colon S_4 \to S_3$ is given by

$$() \mapsto ()$$
$$(1,2) \mapsto (1,2)$$
$$(1,2)(3,4) \mapsto ()$$
$$(1,2,3,4) \mapsto (1,3)$$
$$(1,2,3) \mapsto (1,2,3)$$

Any representation $\rho \colon S_3 \to \mathrm{GL}(V)$ gives rise to a representation of S_4 by composition $p \circ \rho$. It follows that the representation in the bottom row of the table in example 11.3.7 on page 399 gives rise to one of S_4:

	()	$(*,*)$	$(*,*)(*,*)$	$(*,*,*)$	$(*,*,*,*)$
size	1	6	3	8	6
$\zeta_{p\circ\alpha}$	2	0	2	−1	0

A simple computation shows that

$$\langle \zeta_{p\circ\alpha}, \zeta_{p\circ\alpha} \rangle = 1$$

so it is irreducible (or simple). We could also have come to this conclusion using the fact that the representation of S_3 was simple.

This completes our character-table for S_4:

	()	$(*,*)$	$(*,*)(*,*)$	$(*,*,*)$	$(*,*,*,*)$
size	1	6	3	8	6
triv	1	1	1	1	1
sgn	1	−1	1	1	−1
std	3	1	−1	0	−1
std ⊗ sgn	3	−1	−1	0	1
$\zeta_{p\circ\alpha}$	2	0	2	−1	0

Based on the representations in example 11.2.1 on page 393, we get part of a character table over $\mathbb{R}$ for the dihedral group, D_{2n}:

---

[1] Conjugation leaves its cycle-structure intact.

	f^k	rf^k
triv	1	1
sgn	1	-1
ρ	$2\cos\frac{2\pi k}{n}$	0

11.5. Burnside's Theorem

We conclude our drive-by of representation theory with an application.

THEOREM 11.5.1 (Burnside's Theorem). *If G is a group of order $p^a q^b$, where p and q are distinct prime numbers and a and b are nonnegative integers, then G is solvable.*

REMARK. It follows that a finite simple group must have an order divisible by at least three distinct primes.

PROOF. We prove this by contradiction. Suppose $p^a q^b$ is the order of the smallest counterexample, G.

Claim 1: G is a simple group and $a > 0$.

If there existed a normal subgroup $H \triangleleft G$ then H and G/H would be smaller than G, hence solvable (since G is the smallest non-solvable group whose order is of the form $p^a q^b$). This implies that G is solvable too (see exercise 6 on page 87). Since G is simple, $Z(G) = \{1\}$.

If $a = 0$, then G is a finite q-group and is solvable by exercise 1 on page 321.

Claim 2: There is an element $g \in G$ that has q^d conjugates, for some $d > 0$.

The first Sylow Theorem (see theorem 4.8.1 on page 81) implies that there's a subgroup $H \subset G$ of order p^a. Since this is a p-group, its center is nontrivial (see theorem 4.7.14 on page 79). Pick a nontrivial element $g \in Z(H)$. This is not central in G because the center of G is trivial (it's a simple group). Regard G as acting on elements of G by conjugation, and recall the material in section 4.7 on page 73. Let $Z_G(g) \subset G$ be the stabilizer of g under this action (see definition 4.7.4 on page 74), i.e. the subgroup of all elements of G that *commute* with g.

Since $H \subset S_g$, $[G:S_g] | [G:H] = q^b$. It follows that

$$|\operatorname{Orbit}(g)| = [G:S_g] = \frac{|G|}{|S_g|} = q^d$$

(see proposition 4.7.7 on page 75 and corollary 4.7.13 on page 79) where the orbit of g is the number of distinct elements in the group-action, i.e. the number of distinct conjugates. The inverse, g^{-1}, also has this number of distinct conjugates.

Claim 3: There exists a nontrivial irreducible representation, ρ, of G such that its dimension m is not divisible by q and whose character, χ, has the property that $\chi(g) \neq 0$.

Recall that the dimension of the representation, m, is equal to $\chi(1)$.

Let $\{\chi_1, \ldots, \chi_k\}$ denote the irreducible characters of G, where χ_1 is the trivial character. Theorem 11.3.23 on page 406 implies that

$$\sum_{i=1}^{k} \chi_i(1)\overline{\chi_i(g)} = 0 = 1 + \sum_{i=2}^{k} \chi_i(1)\overline{\chi_i(g)} = 1 + \sum_{i=2}^{k} \chi_i(1)\chi_i(g^{-1})$$

If all of the algebraic integers $\chi_i(g^{-1})$ are divisible by q then

$$-\frac{1}{q} = \sum_{i=2}^{k} \frac{1}{q} \chi_i(1) \chi_i(g^{-1})$$

would be an algebraic integer, which it clearly isn't.

Claim 4: The complex number $q^d \chi(g)/m$ is an algebraic integer.

The number of conjugates of g is q^d. Let $\hat{g}$ be the sum of all these conjugates; we have $\hat{g} \in Z(\mathbb{C}G)$, by proposition 11.3.24 on page 407 and proposition 11.3.25 on page 408 implies that

$$\rho(\hat{g}) = \frac{1}{m} \sum_{h \sim g} \chi(h) \cdot I$$
$$= \frac{q^d \chi(g)}{m} \cdot I$$

(where $h \sim g$ means h is conjugate to g) and this is integral over $\mathbb{Z}$.

Claim 5: The complex number $\chi(g)/m$ is an algebraic integer.

This is due to the fact that $d \nmid m$, so we can find integers a and b such that

$$aq^d + bm = 1$$

(see lemma 3.1.5 on page 14) which gives

$$\frac{\chi(g)}{m} = a \frac{q^d \chi(g)}{m} + b\chi(g)$$

a sum of algebraic integers.

Claim 6: $|\chi(g)| = m$.

Set $\xi = \chi(g)/m$. Then the algebraic conjugates (see definition 3.1.5 on page 14) of ξ are all algebraic integers so their product $N(\xi)$ (see lemma 7.5.11 on page 286) is also an algebraic integer. Since it is $\pm$ the constant term of the minimal polynomial of ξ over $\mathbb{Q}$, it is rational, hence $0 \neq N(\xi) \in \mathbb{Z}$.

Now $\chi(g)$ is the sum of the m eigenvalues of $\rho(g)$, each of which is a root of unity. The triangle inequality implies that $|\chi(g)| \leq m$, and Galois theory implies that this is true of each of the algebraic conjugates of $\chi(g)$, so that all the algebraic conjugates of ξ have absolute value ≤ 1.

Consequently $0 \neq |N(\xi)| \leq 1$. Since it must be an integer, we have $|N(\xi)| = 1$, which proves the claim.

Consider the set

$$H = \{h \in G | |\chi(h)| = m\}$$

Exercise 3 on page 409 implies that $\rho(h)$ is multiplication by a scalar (root of unity). This implies that H is a subgroup of G. Furthermore, it's a *normal* subgroup since matrices $\lambda \cdot I$ *commute* with all other matrices. In fact, $H/\ker\rho$ is abelian since it is equal to the image of H under ρ in the ring of matrices — and diagonal matrices commute with each other.

Since H contains the element g, it is nontrivial. The simplicity of G then implies that $H = G$. Since ρ is not a trivial representation, the simplicity of G implies that $\ker\rho = 1$. This implies that $G = H/\ker\rho$ is *abelian*, which is a contradiction. □

CHAPTER 12

A little algebraic geometry

"Algebraic geometry seems to have acquired the reputation of being esoteric, exclusive, and very abstract, with adherents who are secretly plotting to take over all the rest of mathematics. In one respect this last point is accurate."
—David Mumford in [77].

12.1. Introduction

Algebraic geometry concerns itself with objects called *algebraic varieties*. These are essentially solution-sets of systems of algebraic equations.

Although restricting our attention to algebraic varieties might seem limiting, it has long been known that more general objects like compact smooth manifolds are diffeomorphic to real varieties — see [79][1] and [104]. The paper [1] even shows that many piecewise-linear manifolds, including ones with *no* smooth structure are homeomorphic to real varieties.

The reader interested in more than a *little bit* of algebraic geometry is invited to look at [99].

We begin with *algebraic sets*, whose geometric properties are completely characterized by a basic algebraic invariant called the *coordinate ring*. The main objects of study — algebraic varieties — are the result of gluing together multiple affine sets.

Throughout this discussion, k will denote a fixed *algebraically closed* field (see definition 7.5.1 on page 283). In classical algebraic geometry $k = \mathbb{C}$.

DEFINITION 12.1.1. An *n-dimensional affine space*, $\mathbb{A}^n = k^n$, regarded as a space in which geometric objects can be defined. An *algebraic set*, $\mathcal{V}(S)$, in k^n is the set of common zeros of some set S of polynomials in $k[X_1, \ldots, X_m]$:

$$\mathcal{V}(S) = \{(a_1, \ldots, a_n) \in \mathbb{A}^n | f(a_1, \ldots, a_n) = 0 \text{ for all } f(X_1, \ldots, X_n) \in S\}$$

REMARK. It is not hard to see that if the set of polynomials is larger, the set of common zeros will be smaller, i.e.,

$$S \subset S' \implies \mathcal{V}(S) \supset \mathcal{V}(S')$$

If $\mathfrak{a}$ is the *ideal* generated by the polynomials in S, we have $\mathcal{V}(\mathfrak{a}) = \mathcal{V}(S)$ so algebraic sets are described as $\mathcal{V}(\mathfrak{a})$ for some ideal $\mathfrak{a} \subseteq k[X_1, \ldots, X_m]$ (see definition 5.2.3 on page 111).

Recall that all ideals in $k[X_1, \ldots, X_n]$ are *finitely generated* by theorem 5.4.4 (the Hilbert Basis Theorem).

---

[1]Written by John Nash, the character of the film "A beautiful mind."

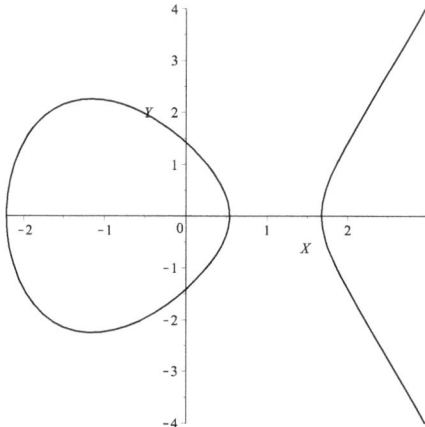

FIGURE 12.1.1.   An elliptic curve

EXAMPLE. For instance, we have

(1) If S is a system of homogeneous linear equations, then $\mathcal{V}(S)$ is a subspace of $\mathbb{A}^n$.
(2) If S consists of the single equation

$$Y^2 = X^3 + aX + b \text{ where } 4a^3 + 27b^2 \neq 0$$

then $\mathcal{V}(S)$ is an *elliptic curve*. The quantity, $4a^3 + 27b^2$ is the discriminant (see definition 5.5.20 on page 144) of the cubic polynomial $Y^2 = X^3 + aX + b$. Its non-vanishing guarantees that the polynomial has no repeated roots — see corollary 5.5.21 on page 144. Figure 12.1.1 shows the elliptic curve $Y^2 = X^3 - 2X + 1$. Elliptic curves over finite fields form the basis of an important cryptographic system.
(3) For the zero-ideal, $\mathcal{V}((0)) = \mathbb{A}^n$.
(4) $\mathcal{V}((1)) = \emptyset$,
(5) The algebraic subsets of $k = \mathbb{A}^1$ itself are finite sets of points since they are roots of polynomials.
(6) The *special linear group*, $SL(n,k) \subset \mathbb{A}^{n^2}$ — the group of $n \times n$ matrices with determinant 1. This is an algebraic set because the determinant is a polynomial of the matrix-elements — so that $SL(n,k)$ is the set of zeros of the polynomial, $\det(A) - 1$ for $A \in \mathbb{A}^{n^2}$. This is an example of an algebraic group, an algebraic set that is also a group under a multiplication-map that can be expressed as polynomial functions of the coordinates.
(7) If A is an $n \times m$ matrix whose entries are in $k[X_1, \ldots, X_t]$ and $r \geq 0$ is an integer, then define $\mathcal{R}(A,r)$, the *rank-variety* (also called a

determinantal variety),

$$\mathcal{R}(A, r) = \begin{cases} \mathbb{A}^t & \text{if } r \geq \min(n, m) \\ p \in \mathbb{A}^t & \text{such that } \operatorname{rank}(A(p)) \leq r \end{cases}$$

This is an algebraic set because the statement that the rank of A is $\leq r$ is the same as saying the determinants of all $(r+1) \times (r+1)$ sub-matrices are 0.

Here are some basic properties of algebraic sets and the ideals that generate them:

PROPOSITION 12.1.2. *Let* $\mathfrak{a}, \mathfrak{b} \subset k[X_1, \ldots, X_n]$ *be ideals. Then*

(1) $\mathfrak{a} \subset \mathfrak{b} \implies \mathcal{V}(\mathfrak{a}) \supset \mathcal{V}(\mathfrak{b})$
(2) $\mathcal{V}(\mathfrak{a}\mathfrak{b}) = \mathcal{V}(\mathfrak{a} \cap \mathfrak{b}) = \mathcal{V}(\mathfrak{a}) \cup \mathcal{V}(\mathfrak{b})$
(3) $\mathcal{V}(\sum \mathfrak{a}_i) = \bigcap \mathcal{V}(\mathfrak{a}_i)$

PROOF. For statement 2 note that

$$\mathfrak{a}\mathfrak{b} \subset \mathfrak{a} \cap \mathfrak{b} \subset \mathfrak{a}, \mathfrak{b} \implies \mathcal{V}(\mathfrak{a} \cap \mathfrak{b}) \supset \mathcal{V}(\mathfrak{a}) \cup \mathcal{V}(\mathfrak{b})$$

For the reverse inclusions, let $x \notin \mathcal{V}(\mathfrak{a}) \cup \mathcal{V}(\mathfrak{b})$. Then there exist $f \in \mathfrak{a}$ and $g \in \mathfrak{b}$ such that $f(x) \neq 0$ and $g(x) \neq 0$. Then $fg(x) \neq 0$ so $x \notin \mathcal{V}(\mathfrak{a}\mathfrak{b})$. □

It follows that the algebraic sets in $\mathbb{A}^n$ satisfy the axioms of the *closed sets* in a *topology*.

DEFINITION 12.1.3. The *Zariski topology* on $\mathbb{A}^n$ has closed sets that are algebraic sets. Complements of algebraic sets will be called *distinguished open sets*.

REMARK. Oscar Zariski originally introduced this concept in [110]. This topology has some distinctive properties:
  ▷ every algebraic set is compact in this topology.
  ▷ algebraic maps (called *regular* maps) are continuous. The converse is not necessarily true, though. See exercise 4 on page 426.

The Zariski topology is also *extremely coarse* i.e, has very "large" open sets. To see this, recall that the *closure*, $\bar{S}$ of a subset $S \subset X$ of a space is the smallest closed set that contains it — i.e., the intersection of all closed sets that contain S.

Now suppose $k = \mathbb{C}$ and $S \subset \mathbb{A}^1 = \mathbb{C}$ is an arbitrarily line segment, as in figure 12.1.2 on the next page. Then we claim that $\bar{S} = \mathbb{C}$ in the Zariski topology.

Let $\mathfrak{I} \subset \mathbb{C}[X]$ be the ideal of all polynomials that vanish on S. Then the closure of S is the set of points where the polynomials in $\mathfrak{I}$ all vanish — i.e., $\mathcal{V}(\mathfrak{I})$. But nonzero polynomials vanish on finite sets of points and S is infinite. It follows that $\mathfrak{I} = (0)$ i.e., the only polynomials that vanish on S are identically zero. Since $\mathcal{V}((0)) = \mathbb{C}$, we get that the closure of S is *all* of $\mathbb{C}$, as is the closure of *any* infinite set of points.

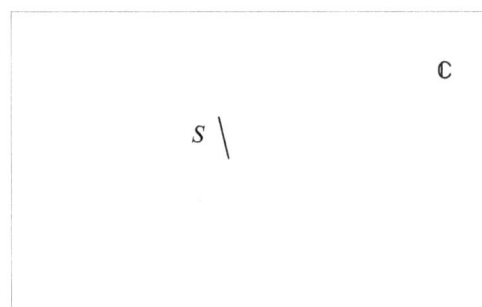

FIGURE 12.1.2.   Closure in the Zariski topology

DEFINITION 12.1.4. For a subset $W \subseteq \mathbb{A}^n$, define
$$\mathcal{I}(W) = \{f \in k[X_1, \ldots, X_n] | f(P) = 0 \text{ for all } P \in W\}$$
It is not hard to see that:

PROPOSITION 12.1.5. *The set $\mathcal{I}(W)$ is an ideal in $k[X_1, \ldots, X_n]$ with the properties:*
  (1) $V \subset W \implies \mathcal{I}(V) \supset \mathcal{I}(W)$
  (2) $\mathcal{I}(\emptyset) = k[X_1, \ldots, X_n]$; $\mathcal{I}(k^n) = 0$
  (3) $\mathcal{I}(\bigcup W_i) = \bigcap \mathcal{I}(W_i)$
  (4) *The Zariski closure of a set $X \subset \mathbb{A}^n$ is exactly $\mathcal{V}(\mathcal{I}(X))$.*

EXERCISES.

1. Show that the Zariski topology on $\mathbb{A}^2$ does not coincide with the product-topology of $\mathbb{A}^1 \times \mathbb{A}^1$ (the Cartesian product).

2. If $V \subset \mathbb{A}^n$ is an algebraic set and $p \notin V$ is a point of $\mathbb{A}^n$, show that any line, ℓ, through p intersects V in a finite number of points (if it intersects it at all).

3. If
$$0 \to M_1 \to M_2 \to M_3 \to 0$$
is a short exact sequence of modules over $k[X_1, \ldots, X_n]$, show that
$$\mathcal{V}(\text{Ann}(M_2)) = \mathcal{V}(\text{Ann}(M_1)) \cup \mathcal{V}(\text{Ann}(M_3))$$
(see definition 6.3.27 on page 241 for $\text{Ann}(*)$).

4. If $V = \mathcal{V}\left((X_1^2 + X_2^2 - 1, X_1 - 1)\right)$, what is $\mathcal{I}(V)$?

5. If $V = \mathcal{V}\left((X_1^2 + X_2^2 + X_3^2)\right)$, determine $\mathcal{I}(V)$ when the characteristic of k is 2.

6. Find the ideal $\mathfrak{a} \subset k[X, Y]$ such that $\mathcal{V}(\mathfrak{a})$ is the union of the coordinate-axes.

7. Find the ideal $\mathfrak{a} \subset k[X,Y,Z]$ such that $\mathcal{V}(\mathfrak{a})$ is the union of the three coordinate-axes.

8. If $V \subset \mathbb{A}^2$ is defined by $Y^2 = X^3$, show that every element of $k[V]$ can be uniquely written in the form $f(X) + g(X)Y$.

12.2. Hilbert's Nullstellensatz

12.2.1. The weak form. Hilbert's *Nullstellensatz* (in English, "zero-locus theorem") was a milestone in the development of algebraic geometry, making precise the connection between algebra and geometry.

> David Hilbert (1862–1943) was one of the most influential mathematicians in the 19th and early 20th centuries, having contributed to algebraic and differential geometry, physics, and many other fields.

The Nullstellensatz completely characterizes the correspondence between algebra and geometry of affine varieties. It is usually split into two theorems, called the *weak* and *strong* forms of the Nullstellensatz. Consider the question:

When do the equations
$$g(X_1, \ldots, X_n) = 0, g \in \mathfrak{a}$$
have a common zero (or are consistent)?

This is clearly impossible if there exist $f_i \in k[X_1, \ldots, X_n]$ such that $\sum f_i g_i = 1$ — or $1 \in \mathfrak{a}$, so $\mathfrak{a} = k[X_1, \ldots, X_n]$. The weak form of Hilbert's Nullstellensatz essentially says that this is the *only* way it is impossible. Our presentation uses properties of *integral extensions* of rings (see section 6.5 on page 251).

LEMMA 12.2.1. *Let F be an infinite field and suppose $f \in F[X_1, \ldots, X_n]$, $n \geq 2$ is a polynomial of degree $d > 0$. Then there exist $\lambda_1, \ldots, \lambda_{n-1} \in F$ such that the coefficient of X_n^d in*
$$f(X_1 + \lambda_1 X_n, \ldots, X_{n-1} + \lambda_{n-1} X_n, X_n)$$
is nonzero.

PROOF. If f_d is the homogeneous component of f of degree d (i.e., the sum of all monomials of degree d), then the coefficient of x_n^d in $f(X_1 + \lambda_1 X_n, \ldots, X_{n-1} + \lambda_{n-1} X_n, X_n)$ is $f_d(\lambda_1, \ldots, \lambda_{n-1}, 1)$. Since F is infinite, there is a point $(\lambda_1, \ldots, \lambda_{n-1}) \in F^{n-1}$ for which $f_d(\lambda_1, \ldots, \lambda_{n-1}, 1) \neq 0$ (a fact that is easily established by induction on the number of variables). □

The following result is called the *Noether Normalization Theorem* or Lemma. It was first stated by Emmy Noether in [83] and further developed in [84].

THEOREM 12.2.2. *Let F be an infinite field and suppose $A = F[r_1, \ldots, r_m]$ is a finitely generated F-algebra that is an integral domain with generators $r_1 \ldots, r_m$. Then for some $q \leq m$, there are algebraically independent elements $y_1, \ldots, y_q \in A$ such that the ring A is integral (see definition 6.5.3 on page 253) over the polynomial ring $F[y_1, \ldots, y_q]$.*

REMARK. Recall that an F-algebra is a vector space over F that is also a ring. The r_i generate it as a *ring* (so the vector space's dimension over F might be $> m$).

PROOF. We prove this by induction on m. If the r_i are algebraically independent, simply set $y_i = r_i$ and we are done. If not, there is a nontrivial polynomial $f \in F[x_1, \ldots, x_m]$, say of degree d such that
$$f(r_1, \ldots, r_m) = 0$$
and lemma 12.2.1 on the previous page that there a polynomial of the form
$$r_m^d + g(r_1, \ldots, r_m) = 0$$
If we regard this as a polynomial of r_m with coefficients in $F[r_1, \ldots, r_{m-1}]$ we get
$$r_m^d + \sum_{i=1}^{d-1} g_i(r_1, \ldots, r_{m-1}) r_m^i = 0$$
which implies that r_m is integral over $F[r_1, \ldots, r_{m-1}]$. By the inductive hypothesis, $F[r_1, \ldots, r_{m-1}]$ is integral over $F[y_1, \ldots, y_q]$, so statement 2 of proposition 6.5.5 on page 253 implies that r_m is integral over $F[y_1, \ldots, y_q]$ as well. □

We are now ready to prove:

THEOREM 12.2.3 (Hilbert's Nullstellensatz (weak form)). *The maximal ideals of $k[X_1, \ldots, X_n]$ are precisely the ideals*
$$\mathfrak{I}(a_1, \ldots, a_n) = (X_1 - a_1, X_2 - a_2, \ldots, X_n - a_n)$$
for all points
$$(a_1, \ldots, a_n) \in \mathbb{A}^n$$
Consequently every proper ideal $\mathfrak{a} \subset k[X_1, \ldots, X_n]$ has a 0 in $\mathbb{A}^n$.

REMARK. See proposition 5.2.4 on page 112 and lemma 5.3.2 on page 117 for a discussion of the properties of maximal ideals.

PROOF. Clearly
$$k[X_1, \ldots, X_n]/\mathfrak{I}(a_1, \ldots, a_n) = k$$
The projection
$$k[X_1, \ldots, X_n] \to k[X_1, \ldots, X_n]/\mathfrak{I}(a_1, \ldots, a_n) = k$$
is a homomorphism that evaluates polynomial functions at the point $(a_1, \ldots, a_n) \in \mathbb{A}^n$. Since the quotient is a field, the ideal $\mathfrak{I}(a_1, \ldots, a_n)$ is maximal (see lemma 5.3.2 on page 117).

We must show that *all* maximal ideals are of this form, or equivalently, if
$$\mathfrak{m} \subset k[X_1, \ldots, X_n]$$

is any maximal ideal, the quotient field is k.

Suppose $\mathfrak{m}$ is a maximal ideal and
$$K = k[X_1, \ldots X_n]/\mathfrak{m}$$
is a field. If the transcendence degree of K over k is d, the Noether Normalization Theorem 12.2.2 on the facing page implies that K is integral over
$$k[y_1, \ldots, y_d]$$
where $y_1, \ldots, y_d$ are a transcendence basis. Proposition 6.5.7 on page 254 implies that $k[y_1, \ldots, y_d]$ must also be a field. The only way for this to happen is for $d = 0$. So K must be an algebraic extension of k, which implies that it must equal k because k is algebraically closed.

The final statement follows from the fact that every proper ideal is contained in a maximal one, say $\mathcal{I}(a_1, \ldots, a_n)$ so its zero-set contains at least the point $(a_1, \ldots, a_n)$. □

12.2.2. The strong form. The strong form of the Nullstellensatz gives the precise correspondence between ideals and algebraic sets. It implies the weak form of the Nullstellensatz, but the two are usually considered separately.

DEFINITION 12.2.4. If $\mathfrak{a}$ is an ideal in a ring K, define the *radical* of $\mathfrak{a}$, $\sqrt{\mathfrak{a}}$ to be
$$\{f | f^r \in \mathfrak{a}, \text{ for some } r > 0\}$$

PROPOSITION 12.2.5. *The radical of an ideal has the following properties*

▷ $\sqrt{\mathfrak{a}}$ *is an ideal*
▷ $\sqrt{\sqrt{\mathfrak{a}}} = \sqrt{\mathfrak{a}}$

PROOF. If $a \in \sqrt{\mathfrak{a}}$, then $a^r \in \mathfrak{a}$ so $f^r a^r = (fa)^r \in \mathfrak{a}$ so $fa \in \sqrt{\mathfrak{a}}$ for all $f \in K$. If $a, b \in \sqrt{\mathfrak{a}}$ and $a^r, b^s \in \mathfrak{a}$. The binomial theorem expands $(a+b)^{r+s}$ to a polynomial in which every term has a factor of a^r or b^s.

If $a^r \in \sqrt{\mathfrak{a}}$ then $a^{rs} \in \mathfrak{a}$. □

DEFINITION 12.2.6. An ideal is called *radical* if it equals its own radical.

Equivalently, $\mathfrak{a}$ is radical if and only if $K/\mathfrak{a}$ is a *reduced ring* — a ring without nonzero nilpotent elements. Since integral domains are reduced, prime ideals (and maximal ideals) are radical.

It is not hard to see that intersections of radical ideals are radical. Since $f^r(P) = (f(P))^r$, f^r vanishes wherever f vanishes. It follows that $\mathcal{IV}(\mathfrak{a}) \supset \sqrt{\mathfrak{a}}$. We conclude this section with study of the *nilpotent elements* of a ring.

DEFINITION 12.2.7. An element $x \in R$ of a commutative ring is called *nilpotent* if $x^k = 0$ for some integer k. The set of all nilpotent elements of a ring forms an ideal, $\mathfrak{N}(R) = \sqrt{(0)}$, called the *nilradical*.

REMARK. We leave the proof that the set of all nilpotent element forms an ideal as an exercise.

THEOREM 12.2.8. *If $\mathcal{I} \subset R$ is an ideal in a commutative ring, then*
$$\sqrt{\mathcal{I}} = \bigcap \mathfrak{p}_i$$

where the intersection is taken over all prime ideals that contain $\mathfrak{J}$. Consequently, $\mathfrak{N}(R)$ is equal to the intersection of all prime ideals.

REMARK. Every ideal is contained in a maximal ideal (see proposition 5.2.11 on page 115) which is prime by proposition 5.2.4 on page 112, so there is always at least one prime in this intersection.

PROOF. Suppose $x \in \sqrt{\mathfrak{J}}$ and $\mathfrak{J} \subset \mathfrak{p}$ where $\mathfrak{p}$ is prime. Then $x^n = x \cdot x^{n-1} \in \mathfrak{J} \subset \mathfrak{p}$. If $x^{n-1} \notin \mathfrak{p}$ then $x \in \mathfrak{p}$. Otherwise, a simple downward induction on n proves that $x \in \mathfrak{p}$. It follows that

$$\sqrt{\mathfrak{J}} \subseteq \bigcap \mathfrak{p}_i$$

where we take the intersection over all prime ideals of R.

If $x \in R \setminus \sqrt{\mathfrak{J}}$, we will construct a prime ideal that *does not* contain x. Note that $S = \{x^n, n = 1, \dots\}$ is a multiplicative set. Proposition 6.4.2 on page 248 show that the maximal ideal that does *not* intersect S is prime. □

Hilbert's strong Nullstellensatz describes which ideals in $k[X_1, \dots, X_n]$ occur as $\mathcal{I}(P)$ when P is an algebraic set.

PROPOSITION 12.2.9. *For any subset $W \subset \mathbb{A}^n$, $\mathcal{V}(\mathcal{I}W)$ is the smallest algebraic subset of $\mathbb{A}^n$ containing W. In particular, $\mathcal{V}(\mathcal{I}W) = W$ if W is algebraic.*

REMARK. In fact, $\mathcal{V}(\mathcal{I}W)$ is the *Zariski closure* of W.

PROOF. Let $V = \mathcal{V}(\mathfrak{a})$ be an algebraic set containing W. Then $\mathfrak{a} \subset \mathcal{I}(W)$ and $\mathcal{V}(\mathfrak{a}) \supset \mathcal{V}(\mathcal{I}W)$. □

THEOREM 12.2.10 (Hilbert's Nullstellensatz). *For any ideal $\mathfrak{a} \in k[X_1, \dots, X_n]$, $\mathcal{I}\mathcal{V}(\mathfrak{a}) = \sqrt{\mathfrak{a}}$ (see definition 12.2.4 on the previous page). In particular, $\mathcal{I}\mathcal{V}(\mathfrak{a}) = \mathfrak{a}$ if $\mathfrak{a}$ is radical.*

PROOF. If f^n vanishes on $\mathcal{V}(\mathfrak{a})$, then f vanishes on it too so that $\mathcal{I}\mathcal{V}(\mathfrak{a}) \supset \sqrt{\mathfrak{a}}$. For the reverse inclusion, we have to show that if h vanishes on $\mathcal{V}(\mathfrak{a})$, then $h^r \in \mathfrak{a}$ for some exponent r.

Suppose $\mathfrak{a} = (g_1, \dots, g_m)$ and consider the system of $m+1$ equations in $n+1$ variables, $X_1, \dots, X_m, Y$:

$$g_i(X_1, \dots, X_n) = 0$$
$$1 - Yh(X_1, \dots, X_n) = 0$$

If $(a_1, \dots, a_n, b)$ satisfies the first m equations, then $(a_1, \dots, a_m) \in V(\mathfrak{a})$. Consequently $h(a_1, \dots, a_n) = 0$ and the equations are *inconsistent*.

According to the weak Nullstellensatz (see theorem 12.2.3 on page 420), the ideal generated by the left sides of these equations generate the whole ring $k[X_1, \dots, X_n, Y]$ and there exist $f_i \in k[X_1, \dots, X_n, Y]$ such that

$$1 = \sum_{i=1}^{m} f_i g_i + f_{m+1}(1 - Yh)$$

Now regard this equation as an identity in $k(X_1, \dots, X_n)[Y]$ — polynomials in Y with coefficients in the field of fractions of $k[X_1, \dots, X_n]$. After

substituting h^{-1} for Y, we get
$$1 = \sum_{i=1}^m f_i(X_1,\ldots,X_n,h^{-1})g_i(X_1,\ldots X_n)$$
Clearly
$$f(X_1,\ldots,X_n,h^{-1}) = \frac{\text{polynomial in } X_1,\ldots,X_n}{h^{N_i}}$$
for some N_i.

Let N be the largest of the N_i. On multiplying our equation by h^N, we get
$$h^N = \sum(\text{polynomial in } X_1,\ldots,X_n) \cdot g_i$$
so $h^N \in \mathfrak{a}$. □

Hilbert's Nullstellensatz precisely describes the correspondence between algebra and geometry:

COROLLARY 12.2.11. *The map $\mathfrak{a} \mapsto \mathcal{V}(\mathfrak{a})$ defines a 1-1 correspondence between the set of radical ideals in $k[X_1,\ldots,X_n]$ and the set of algebraic subsets of $\mathbb{A}^n$.*

PROOF. We know that $\mathcal{IV}(\mathfrak{a}) = \mathfrak{a}$ if $\mathfrak{a}$ is a radical ideal and that $\mathcal{V}(\mathcal{I}W) = W$ if W is an algebraic set. It follows that $\mathcal{V}(*)$ and $\mathcal{I}(*)$ are inverse maps. □

COROLLARY 12.2.12. *The radical of an ideal in $k[X_1,\ldots,X_n]$ is equal to the intersection of the maximal ideals containing it.*

REMARK. In general rings, the radical is the intersections of all *prime* ideals that contain it (corollary 12.2.12). The statement given here is true for algebras over algebraically closed fields.

PROOF. Let $\mathfrak{a} \subset k[X_1,\ldots X_n]$ be an ideal. Because maximal ideals are radical, every maximal ideal containing $\mathfrak{a}$ also contains $\sqrt{\mathfrak{a}}$, so
$$\sqrt{\mathfrak{a}} \subset \bigcap_{\mathfrak{m} \supset \mathfrak{a}} \mathfrak{m}$$
For each $P = (a_1,\ldots,a_n) \in k^n$, $\mathfrak{m}_P = (X_1 - a_1,\ldots,X_n - a_n)$ is a maximal ideal in $k[X_1,\ldots,X_n]$ and
$$f \in \mathfrak{m}_P \Leftrightarrow f(P) = 0$$
so
$$\mathfrak{m}_P \supset \mathfrak{a} \Leftrightarrow P \in V(\mathfrak{a})$$
If $f \in \mathfrak{m}_P$ for all $P \in \mathcal{V}(\mathfrak{a})$, then f vanishes on $V(\mathfrak{a})$ so $f \in \mathcal{IV}(\mathfrak{a}) = \sqrt{\mathfrak{a}}$.

It follows that
$$\sqrt{\mathfrak{a}} \supset \bigcap_{P \in \mathcal{V}(\mathfrak{a})} \mathfrak{m}_P$$
□

REMARK. This result allows us to directly translate between geometry and algebra:

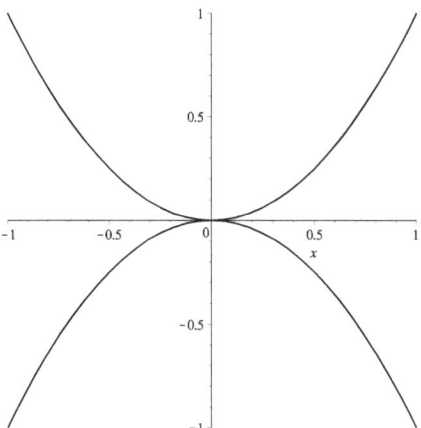

FIGURE 12.2.1.   An intersection of multiplicity 2

(1) "Since $\mathcal{V}(\mathfrak{a})$ is the *union* of the points *contained in* it, $\sqrt{\mathfrak{a}}$ is the *intersection* of the maximal ideals *containing* it."
(2) Because $\mathcal{V}((0)) = k^n$
$$\mathcal{I}(k^n) = \mathcal{I}\mathcal{V}((0)) = \sqrt{0} = 0$$
— only the zero polynomial vanishes on all of k^n.
(3) The 1-1 correspondence is order-inverting so the maximal proper radical ideals correspond to the minimal nonempty algebraic sets.
(4) But the maximal proper radical ideals are the maximal ideals and the minimal nonempty algebraic sets are one-point sets.
(5) Let W and W' be algebraic sets. Then $W \cap W'$ is the largest algebraic subset contained in W and W' — so $\mathcal{I}(W \cap W')$ must be the smallest radical ideal containing both $\mathcal{I}(W)$ and $\mathcal{I}(W')$. It follows that
$$\mathcal{I}(W \cap W') = \sqrt{\mathcal{I}(W) + \mathcal{I}(W')}$$

EXAMPLE 12.2.13. Let $W = \mathcal{V}(X^2 - Y)$ and $W' = \mathcal{V}(X^2 + Y)$.

Then $I(W \cap W') = \sqrt{(X^2, Y)} = (X, Y)$ (assuming the characteristic of k is $\neq 2$).

So $W \cap W' = \{(0,0)\}$.

When considered at the intersection of $Y = X^2$ and $Y = -X^2$ it has multiplicity 2.

LEMMA 12.2.14. *If V is an algebraic subset of $\mathbb{A}^n$, then*
(1) *The points of V are closed in the Zariski topology (thus V is a T_1-space).*
(2) *Every ascending chain of open subsets $U_1 \subset U_2 \subset \cdots$ of V eventually becomes constant — hence every descending chain of closed sets eventually becomes constant.*
(3) *Every open covering has a finite subcovering.*

REMARK. Topological spaces satisfying Condition 2 above are called *noetherian*. This is equivalent to:

> "Every nonempty set of closed subsets of V has a minimal element."

Spaces satisfying condition 3 are called compact (although the Bourbaki group requires compact spaces to be Hausdorff, so they call such spaces *quasicompact*).

PROOF. Let $\{(a_1, \ldots, a_n)\}$ be the algebraic set defined by the ideal $(X_1 - a_1, \ldots, X_n - a_n)$.

A sequence $V_1 \supset V_2 \supset \ldots$ gives rise to a sequence of radical ideals $\mathcal{I}(V_1) \subset \mathcal{I}(V_2) \subset \cdots$ which must eventually become constant by theorem 5.4.4 on page 123.

Let $V = \bigcup_{i \in I} U_i$. If $V \neq U_1$, there exists an $i_1 \in I$ such that $U_1 \subsetneq U_1 \cup U_{i_1}$. If $V \neq U_1 \cup U_{i_1}$, continue this process. By statement 2, this must stop in a finite number of steps. □

DEFINITION 12.2.15. A function $f: \mathbb{A}^n \to \mathbb{A}^m$ is a *regular mapping* if it is of the form
$$f(X_1, \ldots X_n) = \begin{bmatrix} F_1(X_1, \ldots, X_n) \\ \vdots \\ F_m(X_1, \ldots, X_n) \end{bmatrix}$$
for $F_1, \ldots, F_m \in k[X_1, \ldots, X_n]$.

If $V \subset \mathbb{A}^n$ and $W \subset \mathbb{A}^m$ are algebraic sets and $f: \mathbb{A}^n \to \mathbb{A}^m$ is a regular mapping such that
$$f(V) \subset W$$
then we call $\bar{f} = f|V: V \to W$ a *regular mapping from V to W*.

Although the Zariski topology is very coarse — implying that it is difficult for a map to be continuous in this topology — there is an important class of continuous maps:

PROPOSITION 12.2.16. *If $f: V \subset \mathbb{A}^n \to W \subset \mathbb{A}^m$ is a regular map of algebraic sets, then f is continuous in the Zariski topology.*

PROOF. The map, f, is continuous if $f^{-1}(K) \subset \mathbb{A}^n$ is a closed set whenever $K \subset \mathbb{A}^m$ is closed. Let
$$f(X_1, \ldots X_n) = \begin{bmatrix} F_1(X_1, \ldots, X_n) \\ \vdots \\ F_m(X_1, \ldots, X_n) \end{bmatrix}$$
A closed set $K \subset \mathbb{A}^m$, in the Zariski topology, is defined by a finite set of equations
$$g_1(X_1, \ldots, X_m) = 0$$
$$\vdots$$
$$g_t(X_1, \ldots, X_m) = 0$$

where the g_i are polynomials. $f^{-1}(K)$ is defined by
$$g_1(F_1,\ldots,F_m)(X_1,\ldots,X_n) = 0$$
$$\vdots$$
$$g_t(F_1,\ldots,F_m)(X_1,\ldots,X_n) = 0$$
which is a closed set in $\mathbb{A}^n$. □

EXERCISES.

1. if R is a principal ideal domain and $x = p_1^{n_1} \cdots p_k^{n_k}$ is a factorization into primes, show that
$$\sqrt{(x)} = (p_1 \cdots p_k)$$

2. Show that prime ideals are radical.

3. Show that the strong form of the Nullstellensatz implies the weak form.

4. Give an example of a map $f\colon \mathbb{A}^n \to \mathbb{A}^m$ that is continuous in the Zariski topology but not regular.

5. Suppose $f = \begin{bmatrix} f_1(X_1,\ldots,X_n) \\ \vdots \\ f_n(X_1,\ldots,X_n) \end{bmatrix} \colon \mathbb{A}^n \to \mathbb{A}^n$ is a regular map and
$$A_{i,j} = \frac{\partial f_i}{\partial X_j}$$
suppose that $z = \det A_{i,j}$ is never 0. Show that it must be a nonzero constant.

The Inverse function theorem in calculus implies that f has a smooth inverse in a neighborhood of every point.

Jacobi's Conjecture states that such an f has an *global inverse* that is a *regular map*.

The only cases that have been proved are when $k = \mathbb{C}$ and $n = 2$. It has been shown that proving it for $n = 3$ would prove it for all n when $k = \mathbb{C}$ — see [35] as a general reference.

6. Find the irreducible components of the algebraic set $X^2 - YZ = XZ - Z = 0$ in $\mathbb{A}^3$.

7. If $\mathfrak{a} = (Y^3, X - Y)$, is $X + Y \in \sqrt{\mathfrak{a}}$? If so, what power of it is in $\mathfrak{a}$?

12.3. The coordinate ring

The coordinate ring is one of the central concepts of algebraic geometry — particularly the theory of *affine* algebraic sets. It is the ring of algebraic functions on an algebraic set. and it determines *all* geometric properties.

DEFINITION 12.3.1. Let $V \subset \mathbb{A}^n$ be an algebraic set and let $\mathfrak{a} = \mathcal{I}(V)$. Then the *coordinate ring* of V is defined by
$$k[V] = k[X_1, \ldots, X_n]/\mathfrak{a}$$
(where the X_i are indeterminates). It is the ring of polynomial functions of $\mathbb{A}^n$ *restricted* to V (or the algebraic functions *on* V).

EXAMPLE 12.3.2. Let $V \subset \mathbb{A}^2$ be the hyperbola defined by $XY = 1$ or $XY - 1 = 0$. It is easily checked that $\sqrt{(XY-1)} = (XY-1)$ so the defining ideal is $(XY - 1)$. The coordinate ring is
$$k[X, Y]/(XY - 1) = k[X, X^{-1}]$$
the ring of so-called Laurent polynomials.

PROPOSITION 12.3.3. *The coordinate ring, $k[V]$, of an algebraic set, V, has the following properties:*
  (1) *The points of V are in a 1-1 correspondence with the maximal ideals of $k[V]$.*
  (2) *The closed sets of V are in a 1-1 correspondence with the radical ideals of $k[V]$.*
  (3) *If $f \in k[V]$ and $p \in V$ with corresponding maximal ideal $\mathfrak{m}_p$, then the result of evaluating f at p is the same as the image of f under the canonical projection*
$$\pi \colon k[V] \to k[V]/\mathfrak{m}_p = k$$
  In other words, $f(p) = \pi(f)$.

PROOF. Let $V \subset \mathbb{A}^n$ be an algebraic set. If
$$\pi \colon k[X_1, \ldots, X_n] \to k[V]$$
is the canonical projection, and $\mathfrak{b} \subset k[V]$ is an ideal, then lemma 5.2.9 on page 114 implies that
$$\mathfrak{b} \mapsto \pi^{-1}(\mathfrak{b})$$
is a bijection from the set of ideals of $k[V]$ to the set of ideals of $k[X_1, \ldots, X_n]$ containing $\mathfrak{a}$. Prime, and maximal ideals in $k[V]$ correspond to prime, and maximal ideals in $k[X_1, \ldots, X_n]$ containing $\mathfrak{a}$. The fact that radical ideals are intersections of maximal ideals (see corollary 12.2.12 on page 423) implies that this correspondence respects radical ideals too.

If $p = (a_1, \ldots, a_n) \in V \subset \mathbb{A}^n$ is a point, the maximal ideal of functions in $k[X_1, \ldots, X_n]$ that vanish at p is
$$\mathfrak{P} = (X_1 - a_1, \ldots, X_n - a_n) \subset k[X_1, \ldots, X_n]$$
and this gives rise to the maximal ideal $\pi(\mathfrak{P}) \subset k[V]$.

Clearly
$$\mathcal{V}\left(\pi^{-1}(\mathfrak{b})\right) = \mathcal{V}(\mathfrak{b}) \subset V$$

so $\mathfrak{b} \mapsto \mathcal{V}(\mathfrak{b})$ is a bijection between the set of radical ideals in $k[V]$ and the algebraic sets contained within V.

To see that $f(p) = \pi(f)$, note that the corresponding statement is true in $k[X_1, \ldots, X_n]$, i.e., the image of $f(X_1, \ldots, X_n)$ under the map

$$g: k[X_1, \ldots, X_n] \to k[X_1, \ldots, X_n]/\mathfrak{P} = k$$

is just $f(a_1 \ldots, a_n)$.

Let $h: k[X_1, \ldots, X_n] \to k[X_1, \ldots, X_n]/\mathfrak{a} = k[V]$ be the canonical projection. Then $\mathfrak{m}_p = h(\mathfrak{P})$ and the diagram

$$\begin{array}{ccc} k[X_1, \ldots, X_n] & \xrightarrow{h} & k[V] \\ {\scriptstyle g}\downarrow & & \downarrow{\scriptstyle \pi} \\ k & = & k \end{array}$$

commutes by lemma 5.2.10 on page 114. $\square$

PROPOSITION 12.3.4. *Let $V \in \mathbb{A}^n$ and $W \in \mathbb{A}^m$ be algebraic sets and let $f: V \to W$ be a regular map. Then f induces a homomorphism*

$$f^*: k[W] \to k[V]$$

of coordinate rings (as k-algebras).

PROOF. The fact that f is regular implies

$$f = \begin{bmatrix} F_1 \\ \vdots \\ F_m \end{bmatrix}$$

for $F_1, \ldots, F_m \in k[Y_1, \ldots, Y_n]$ and these polynomials induce a map

$$\begin{array}{rcl} F^*: k[X_1, \ldots, X_m] & \to & k[Y_1, \ldots, Y_n] \\ g(X_1, \ldots, X_m) & \mapsto & g(F_1, \ldots, F_m) \end{array}$$

Since $f(V) \subset W$ we must have $F^*(\mathcal{I}(W)) \subset \mathcal{I}(V)$. But this means that F^* induces a homomorphism of k-algebras

$$f^*: k[X_1, \ldots, X_m]/\mathcal{I}(W) = k[W] \to k[Y_1, \ldots, Y_n]/\mathcal{I}(V) = k[V]$$

$\square$

EXAMPLE 12.3.5. Suppose $V \subset \mathbb{A}^2$ is the parabola $y = x^2$. Then projection to the x-axis

$$\begin{array}{rcl} f: \mathbb{A}^2 & \to & \mathbb{A}^1 \\ (x, y) & \mapsto & x \end{array}$$

is a regular map. There is also a regular map $g: \mathbb{A}^1 \to V$

$$\begin{array}{rcl} g: \mathbb{A}^1 & \to & \mathbb{A}^2 \\ x & \mapsto & (x, x^2) \end{array}$$

It is interesting that we have a converse to proposition 12.3.4:

PROPOSITION 12.3.6. *Let $V \subset \mathbb{A}^n$ and $W \subset \mathbb{A}^m$ be algebraic sets. Any homomorphism of k-algebras*

$$f: k[W] \to k[V]$$

induces a unique regular map

$$\bar{f}: V \to W$$

REMARK. This and proposition 12.3.4 on the preceding page imply that the coordinate ring is a contravariant functor (see definition 10.2.1 on page 345) from the category of algebraic sets to that of k-algebras.

PROOF. We have a diagram

$$\begin{array}{ccc}
k[X_1,\ldots,X_m] & & k[Y_1,\ldots,Y_n] \\
\downarrow & & \downarrow \\
k[X_1,\ldots,X_m]/I(W) & & k[y_1,\ldots,y_n]/I(V) \\
\| & & \| \\
k[W] & \xrightarrow{f} & k[V]
\end{array}$$

and we can map each $X_i \in k[X_1,\ldots,X_m]$ to $k[Y_1,\ldots,Y_n]$ to make

(12.3.1)
$$\begin{array}{ccc}
k[X_1,\ldots,X_m] & \xrightarrow{r} & k[Y_1,\ldots,Y_n] \\
\downarrow & & \downarrow \\
k[X_1,\ldots,X_m]/I(W) & & k[Y_1,\ldots,Y_n]/I(V) \\
\| & & \| \\
k[W] & \xrightarrow{f} & k[V]
\end{array}$$

commute as a diagram of k-algebras. Suppose $r(X_i) = g_i(Y_1,\ldots,Y_n)$. We claim that

$$\bar{f} = \begin{bmatrix} g_1 \\ \vdots \\ g_m \end{bmatrix} : \mathbb{A}^n \to \mathbb{A}^m$$

is the required regular map. If $p = (k_1,\ldots,k_n) \in V \subset \mathbb{A}^n$ so

$$v(p) = 0$$

for any $v \in \mathcal{I}(V)$, then $w(f(p)) = f(w)(p) = 0$ for any $w \in \mathcal{I}(W)$ implying that $\bar{f}(V) \subset W$.

If we replace r in diagram 12.3.1 by a map r' that still makes it commute, the induced $\bar{f}'$ will differ from $\bar{f}$ by elements of $\mathcal{I}(V)$ so

$$\bar{f}|V = \bar{f}'|V$$

implying that the map $\bar{f}: V \to W$ is unique. □

DEFINITION 12.3.7. Let $V \subset \mathbb{A}^n$ and $W \subset \mathbb{A}^m$ be algebraic sets. Then V and W are said to be *isomorphic* if there exist regular maps
$$f: V \to W$$
$$g: W \to V$$
such that $f \circ g = 1: W \to W$ and $g \circ F = 1: V \to V$.

REMARK. We may regard isomorphic algebraic sets as equivalent in every way. Then example 12.3.5 shows that parabola $y = x^2 \subset \mathbb{A}^2$ is isomorphic to $\mathbb{A}^1$.

We have proved:

COROLLARY 12.3.8. *Algebraic sets $V \subset \mathbb{A}^n$ and $W \subset \mathbb{A}^m$ are isomorphic if and only if $k[V]$ and $k[W]$ are isomorphic as k-algebras.*

REMARK. This proves the claim made earlier: the coordinate ring defines *all* of the significant geometric properties of an algebraic set, including its isomorphism class.

We can characterize the kinds of rings that can be coordinate rings of algebraic sets:

DEFINITION 12.3.9. Given an algebraically closed field k, an *affine k-algebra* is defined to be a finitely generated k-algebra that is reduced, i.e. $\sqrt{(0)} = (0)$. If A and B are affine k-algebras, the set of homomorphisms $f: A \to B$ is denoted
$$\hom_{k-\mathrm{alg}}(A, B)$$

REMARK. The requirement that the ring be *reduced* is equivalent to saying that it has no nilpotent elements. This is equivalent to saying that the intersection of its maximal ideals is 0 — see theorem 12.2.8 on page 421.

If k is an algebraically closed field, Hilbert's Nullstellensatz (theorem 12.2.10 on page 422) implies that affine k-algebras (see definition 12.3.9) are Jacobson rings (see definition 5.7.2 on page 154).

EXERCISES.

1. Suppose the characteristic of the field k is $\neq 2$ and V is an algebraic set in $\mathbb{A}^3$ defined by the equations
$$X^2 + Y^2 + Z^2 = 0$$
$$X^2 - Y^2 - Z^2 + 1 = 0$$
Decompose V into its irreducible components.

2. Prove that the statements:
   a. X is connected if and only if the only subsets of X that are open and closed are $\emptyset$ and X,
   b. X is connected if, whenever $X = X_1 \cup X_2$ with X_1, X_2 closed nonempty subsets of X, then $X_1 \cap X_2 \neq \emptyset$.

are equivalent.

CHAPTER 13

Cohomology

"About a thousand years ago, the proof of Pythagoras's theorem was thought to be difficult and abstract and suited only for the 'scholastic' geometers. Today the same theorem is considered an intrinsic component of school education. Perhaps one day algebra, sheaf theory and cohomology will be so well understood that we will be able to explain the proof of Wiles' theorem to school children as well."
— Kapil Hari Paranjape, essay, *On Learning from Arnold's talk on the Teaching of Mathematics.*

13.1. Chain complexes and cohomology

Homology theory is one of the pillars of algebraic topology and a variant called *sheaf cohomology* is widely used in algebraic geometry. The first step to developing this theory involves defining cochain complexes — a purely algebraic construct that will be coupled to geometry later.

We will assume all objects here are in a fixed abelian category (see section 10.5 on page 361), $\mathscr{A}$. For instance, they could be abelian groups or modules over any ring, or even certain types of sheaves.

We begin with the most basic construct:

DEFINITION 13.1.1. A *chain complex* (C_i, ∂_i) is a sequence of objects of $\mathscr{A}$ and homomorphisms

$$\cdots \to C_{i+1} \xrightarrow{\partial_{i+1}} C_i \xrightarrow{\partial_i} C_{i-1} \to \cdots$$

where, for all i, $\partial_i \circ \partial_{i+1} = 0$. A *morphism* of cochain complexes $\{f_i\}: (C_i, \partial_i) \to (D_i, \partial'_i)$ (or *chain-map*) is a sequence of homomorphisms

$$f_i: C_i \to D_i$$

such that the diagrams

(13.1.1)
$$\begin{array}{ccc} C_i & \xrightarrow{f_i} & D_i \\ \partial_i \downarrow & & \downarrow \partial'_i \\ C_{i-1} & \xrightarrow{f_{i-1}} & D_{i-1} \end{array}$$

commute for all i. The maps, ∂_i, are called the *boundary maps* or *differentials* of the chain-complex. The category of chain-complexes with chain-maps as morphisms is denoted $\mathscr{C}h$.

REMARK. The condition $\partial_{i-1} \circ \partial_i = 0$ implies that $\operatorname{im} \partial_i \subseteq \ker \partial_{i-1}$. In algebraic topology, chain-complexes are geometrically defined objects that contain a great deal of topological information.

Now we define a dual concept that is very similar:

DEFINITION 13.1.2. A *cochain complex* (C^i, δ_i) is a sequence of objects of $\mathscr{A}$ and homomorphisms

$$\cdots \to C^{i-1} \xrightarrow{\delta_{i-1}} C^i \xrightarrow{\delta_i} C^{i+1} \to \cdots$$

where, for all i, $\delta_{i+1} \circ \delta_i = 0$. A *morphism* of cochain complexes $\{f_i\}\colon (C^i, \delta_i) \to (D^i, \delta_i')$ (or *chain-map*) is a sequence of homomorphisms

$$f_i \colon C^i \to D^i$$

such that the diagrams

(13.1.2)
$$\begin{array}{ccc} C^{i+1} & \xrightarrow{f_{i+1}} & D^{i+1} \\ \delta_i \uparrow & & \uparrow \delta_i' \\ C^i & \xrightarrow{f_i} & D^i \end{array}$$

commute for all i. The maps, δ_i, are called the *coboundary maps* or *codifferentials* of the cochain-complex. The category of cochain-complexes with chain-maps as morphisms is denoted $\mathscr{C}o$.

REMARK. The superscripts are not exponents! At this point, the reader may wonder what essential difference exists between chain complexes and cochain complexes. The answer is "none!" We can define cochain-complexes as chain-complexes with *negative subscripts*:

$$C^i = C_{-i}$$

(or equivalently, defining chain-complexes as cochain-complexes with negative superscripts). Anything we can prove for one is valid for the other under this equivalence.

Historically, chain-complexes appeared first and were geometrically defined. The generators of the C_i were i-dimensional building blocks for a topological space and the ∂_i mapped one of these to its boundary. Cochain complexes appeared later as sets of functions one could define on these building blocks.

In actual applications (in the next section), this symmetry will break down to some extent and they both will express complementary information. We will give greater emphasis to *cochain* complexes because they are the ones that are most significant in algebraic geometry.

The condition $\delta_{i+1} \circ \delta_i = 0$ implies that $\operatorname{im} \delta_i \subseteq \ker \delta_{i+1}$ for all i. With this in mind, we can define

DEFINITION 13.1.3. Given:
▷ a chain-complex, (C_i, ∂_i), we can define its *homology groups*, $H_i(C)$ via

$$H_i(C) = \frac{\ker \partial_i}{\operatorname{im} \partial_{i+1}}$$

▷ a cochain complex (C^i, δ_i), we can define its associated *cohomology groups*, $H^i(C)$, via

$$H^i(C) = \frac{\ker \delta_i}{\operatorname{im} \delta_{i-1}}$$

REMARK. These will also be objects in the category $\mathscr{A}$. If $H_i(C) = 0$, then the original chain complex was an exact sequence. Such chain-complexes are said to be *exact* or *acyclic*. A similar definition exists for cochain complexes.

Historically, $H_i(C)$ measured the number of i-dimensional "holes" a topological space had (so an n-sphere has $H_n = \mathbb{Z}$ and $H_i = 0$ for $0 < i < n$).

Note that the diagrams 13.1.2 on the preceding page imply that chain maps preserve images and kernels of the boundary or coboundary homomorphisms. This implies that

PROPOSITION 13.1.4. *A chain map or morphism:*

▷ *of chain complexes* $\{f_i\}: (C_i, \partial_i) \to (D_i, \partial'_i)$ *induces homomorphisms of homology*

$$f^i_*: H_i(C) \to H_i(D)$$

or

▷ *of cochain-complexes* $\{f_i\}: (C^i, \delta_i) \to (D^i, \delta'_i)$ *induces homomorphisms of cohomology groups*

$$f^*_i H^i(C) \to H^i(D)$$

Next, we consider a property of chain maps:

DEFINITION 13.1.5. Two

▷ chain maps $f, g: (C, \partial_C) \to (D, \partial_D)$ of chain-complexes are said to be *chain-homotopic* if there exists a set of homomorphisms

$$\Phi_i: C^i \to D^{i+1}$$

for all $i > 0$ called a *homotopy*, such that

$$f_i - g_i = \Phi_{i-1} \circ \partial_C + \partial_D \circ \Phi_i$$

▷ chain maps $f, g: (C, \delta_C) \to (D, \delta_D)$ of cochain-complexes are said to be chain-homotopic if there exists a set of homomorphisms

$$\Phi_i: C^i \to D^{i-1}$$

called a *cohomotopy* for all $i > 0$, such that

$$f_i - g_i = \Phi_{i+1} \circ \delta_C + \delta_D \circ \Phi_i$$

REMARK. Chain-homotopy clearly defines an equivalence relation on chain-maps. Although the definition seems odd, the maps Φ arise naturally in certain topological settings.

The main significance of chain-homotopy is that:

PROPOSITION 13.1.6. *If* $f, g: (C, \delta_C) \to (D, \delta_D)$ *are chain-homotopic chain-maps of cochain complexes then*

$$f^* = g^*: H^i(C) \to H^i(D)$$

REMARK. A corresponding result exists for chain-complexes, by "reversing all the arrows." The symmetry between chain and cochain complexes persists.

PROOF. If $x \in H^i(C)$, then there exists an element $y \in \ker(\delta_C)_i \subset C^i$ such that $x \equiv y \pmod{\operatorname{im}(\delta_C)_{i-1}}$. If we evaluate $(f-g)(y)$, we get

$$(f-g)(y) = (\Phi \circ \delta_C + \delta_D \circ \Phi)(y)$$
$$= \delta_D \circ \Phi(y) \qquad \text{because } y \in \ker(\delta_C)_i$$

It follows that $f(y) \equiv g(y) \pmod{\operatorname{im}(\delta_D)_i}$ and $f^*(x) = g^*(x) \in H^i(D)$. □

We can also define an equivalence relation on cochain-complexes:

DEFINITION 13.1.7. Two cochain-complexes (C, δ_C) and (D, δ_D) are *chain-homotopy equivalent* if there exist chain maps

$$f:(C,\delta_C) \to (D,\delta_D)$$
$$g:(D,\delta_D) \to (C,\delta_C)$$

such that $f \circ g: (D, \delta_D) \to (D, \delta_D)$ and $g \circ f: (C, \delta_C) \to (C, \delta_C)$ are both chain-homotopic to their respective identity maps.

REMARK. Clearly, homotopy equivalent cochain-complexes have isomorphic cohomology groups

$$H^i(C) \cong H^i(D)$$

for all i. Chain-homotopy equivalence is a much sharper relationship than simply having isomorphic cohomology groups. In a certain sense, (C, δ_C) and (D, δ_D) may be regarded as equivalent in every important respect.

Our final topic in the basic algebra of chain-complexes is:

DEFINITION 13.1.8. If (C, δ_C), (D, δ_D), and (E, δ_E) are cochain complexes, an *exact sequence*

$$0 \to (C, \delta_C) \xrightarrow{f} (D, \delta_D) \xrightarrow{g} (E, \delta_E) \to 0$$

are chain-maps f, g such that

$$0 \to C^i \xrightarrow{f_i} D^i \xrightarrow{g_i} E^i \to 0$$

are exact for all i.

REMARK. Exact sequences of cochain complexes arise in many natural settings and can be used to compute cohomology because of the next result:

PROPOSITION 13.1.9. *An exact sequence*

$$0 \to (C, \delta_C) \xrightarrow{f} (D, \delta_D) \xrightarrow{g} (E, \delta_E) \to 0$$

of cochain-complexes induces a homomorphism

$$c: H^i(E) \to H^{i+1}(C)$$

for all i, called the connecting map, that fits into a long exact sequence in cohomology:
$$\cdots \to H^i(C) \xrightarrow{f^*} H^i(D) \xrightarrow{g^*} H^i(E) \xrightarrow{c} H^{i+1}(C) \to \cdots$$

Here f^* and g^* are the induced maps and $c\colon H^i(E) \to H^{i+1}(C)$, called the connecting map is defined as
$$c = f^{-1} \circ \delta_D \circ g^{-1}$$
or in more detail by

(1) If $x \in H^i(E)$, then there exists $y \in \ker(\delta_C)_i$ such that $x \equiv y$ (mod im $(\delta_C)_{i-1}$).
(2) Since $g\colon D^i \to E^i$ is surjective, there exists $z \in D^i$ with $g(z) = y$.
(3) Now take $(\delta_D)_i(z) = w \in D^{i+1}$. Since $y \in \ker(\delta_C)_i$ and chain-maps commute with coboundaries, $w \in \ker g$.
(4) Since the sequence is exact, this w is in the image of (the injective map) f so we may regard $w \in C^{i+1}$.
(5) This $w \in \ker(\delta_C)_{i+1}$ because it is in the image of δ_C and its image in D is in $\ker(\delta_D)_{i+1}$ since $(\delta_D)_{i+1} \circ (\delta_D)_i = 0$.

REMARK. This will turn out to be very useful for *computing* cohomology groups.

PROOF. The proof follows by analyzing the commutative diagram

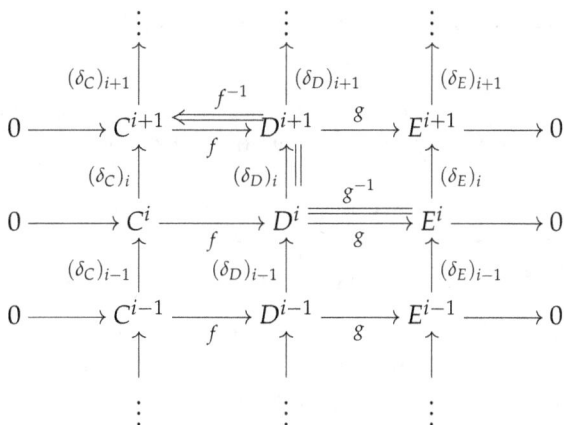

in a visual process affectionately (or angrily!) called a "diagram chase".

We show that c is well-defined: any two distinct lifts of y to D^i will differ by an element of C^i. Since the right square commutes, the final result will differ by an element of $(\delta_C)^i$, hence define the same element of $H^{i+1}(C)$. If y, y' both represent the same x, they will differ by an element of $(\delta_C)_{i-1}$ and their lifts to D^i will differ by an element of $(\delta_D)_{i-1}$, which will be annihilated when we plug it into $(\delta_D)_i$.

The proof of the remaining assertions (about the sequence being exact) follows by similar arguments and is left to the reader. □

If we define $C_i = C^{-i}$ and $H_i(C) = H^{-i}(C)$, we *immediately* get the corresponding result for exact sequences of *chain*-complexes:

PROPOSITION 13.1.10. *Given a short exact sequence of chain-complexes*
$$0 \to (C, \partial_C) \xrightarrow{f} (D, \partial_D) \xrightarrow{g} (E, \partial_E) \to 0$$
of chain-complexes, there exists a homomorphism
$$c : H_i(E) \to H_{i-1}(C)$$
for all i, called the connecting map, that fits into a long exact sequence in cohomology:
$$\cdots \to H_i(C) \xrightarrow{f^*} H_i(D) \xrightarrow{g^*} H_i(E) \xrightarrow{c} H_{i-1}(C) \to \cdots$$

We also need two more basic concepts:

DEFINITION 13.1.11. Given a chain-map of cochain complexes
$$f : (C, \delta_C) \to (D, \delta_D)$$
the *algebraic mapping cone* of f is a cochain complex defined by
$$\mathbf{A}(f)^n = C^{n+1} \oplus D^n$$

with a differential
$$\delta_A^n = \begin{bmatrix} -\delta_C^{n+1} & 0 \\ f^{n+1} & \delta_D^n \end{bmatrix} : \begin{bmatrix} C^{n+1} \\ D^n \end{bmatrix} = \mathbf{A}(f)^n \to \begin{bmatrix} C^{n+2} \\ D^{n+1} \end{bmatrix} = \mathbf{A}(f)^{n+1}$$

and giving a short exact sequence of cochain complexes

(13.1.3) $$0 \to D \to \mathbf{A}(f) \to C[+1] \to 0$$

where $C[+1]$ is C shifted upwards by one degree so $C[+1]^n = C^{n+1}$ with $\delta_{C[+1]}^n = -\delta_C^{n+1}$.

REMARK. It is left as an exercise to the reader to verify that $\delta_A^2 = 0$. As the name hints, this was originally an algebraic version of a geometric construction.

The short exact sequence in 13.1.3 induces a long exact sequence in cohomology (proposition 13.1.9 on page 436):
$$\cdots \to H^i(D) \to H^i(\mathbf{A}(f)) \to H^i(C[+1]) \to H^{i+1}(D) \to \cdots$$

with $H^i(C[+1]) = H^{i+1}(C)$. Analysis of the connecting map $H^i(C[+1]) \to H^{i+1}(D)$ shows that is identical to the map in cohomology induced by f so we can rewrite the long exact sequence as

(13.1.4) $$\cdots \to H^i(C) \xrightarrow{f^*} H^i(D) \to H^i(\mathbf{A}(f))$$
$$\to H^{i+1}(C) \xrightarrow{f^*} H^{i+1}(D) \to \cdots$$

13.1.1. "Topological" homology and cohomology. In this section, we will give a crude and (very!) non-rigorous overview of how homology and cohomology were originally developed and what they "mean" — see [51] for rigor and more details.

One way of studying topological spaces involved breaking them up into a union of discrete pieces of every dimension, called *simplices*[1], and a

---

[1] Simplices are essentially polyhedral pieces of Euclidean space. Singular homology and cohomology involves *mappings* of simplices into a space.

chain-complex was constructed from these simplices. The boundary operator actually represented taking the boundary of n-dimensional simplices and expressing them as $n-1$-dimensional simplices. Although the chain-complex one gets from this construction is not unique (far from it!), it can be proved that its homology is.

In dimension 0, $H_0(X;\mathbb{C}) = \mathbb{C}^k$ where k is the number of components of X. Higher dimensional homology encodes the number of d-dimensional "holes" a space has. For instance, if S^n is an n-sphere, then $H_i(S^n;\mathbb{C}) = 0$ for $0 < i < n$ and $H_n(S^n;\mathbb{C}) = \mathbb{C}$.

Cohomology originally studied the behavior of functions on a topological space — C^i was the set of functions on i-dimensional simplices and the coboundary operator $\delta^i\colon C^i \to C^{i+1}$ determined a function on $i+1$-dimensional simplices by taking its value on the boundary. For instance $H^0(X;\mathbb{C})$ is the set of *locally-constant* functions on X. If X has k components, this is $\mathbb{C}^k$.

In higher dimensions, $H^i(X;\mathbb{C})$ measures the extent to which certain functions on simplices are determined by their behavior on the on the boundaries of those simplices. Roughly speaking, $H^1(\mathbb{R}^2;\mathbb{C}) = 0$ is equivalent to Green's Theorem in multivariate calculus, and $H^2(\mathbb{R}^3;\mathbb{C}) = 0$ is equivalent to the Divergence Theorem.

EXERCISES.

1. If
$$0 \to (C,\delta_C) \xrightarrow{f} (D,\delta_D) \xrightarrow{g} (E,\delta_E) \to 0$$
is an exact sequence of cochain complexes, and two out of the three complexes are acyclic, show that the third must be acyclic also.

2. If
$$0 \to (C,\delta_C) \xrightarrow{f} (D,\delta_D) \xrightarrow{g} (E,\delta_E) \to 0$$
is an exact sequence of cochain-complexes and D is acyclic, show that
$$H^i(E) \cong H^{i+1}(C)$$
for all i.

3. Show that, if two chain-maps $f,g\colon (C,\delta_C) \to (D,\delta_D)$ are chain-homotopic and $F\colon \mathscr{A} \to \mathscr{A}'$ is any additive functor (for instance, $\hom_{\mathscr{A}}(M,*)$ for any $M \in \mathscr{A}$), then the induced chain-maps
$$F(f), F(G)\colon (F(C^i), F(\delta_C)) \to (F(D^i), F(\delta_D))$$
are also chain-homotopic. It follows that, if (C,δ_C) and (D,δ_D) are chain-homotopy *equivalent*, then $(F(C^i), F(\delta_C))$ and $(F(D^i), F(\delta_D))$ also are.

4. Given a commutative diagram

$$\begin{array}{ccccccccc} 0 & \longrightarrow & A & \xrightarrow{r} & B & \xrightarrow{s} & C & \longrightarrow & 0 \\ & & \downarrow u & & \downarrow v & & \downarrow w & & \\ 0 & \longrightarrow & A' & \xrightarrow{r'} & B' & \xrightarrow{s'} & C' & \longrightarrow & 0 \end{array}$$

with exact rows, do a diagram-chase to show that, if u and w are isomorphisms, then so is v. This is a watered-down version of what is called the 5-Lemma.

13.1.2. Resolutions and Derived functors. Now we will consider special types of chain- and cochain-complexes called *resolutions*. We will assume that our abelian category, $\mathscr{A}$, has enough projectives and injectives (see definition 10.5.3 on page 362). This is true for abelian groups and modules over any ring, for instance.

Resolutions are used to compute constructs called *derived functors*. *Roughly speaking*, given a functor $\mathscr{F}: \mathscr{A} \to \mathscr{B}$ between abelian categories, the *first* derived functor measures the extent to which $\mathscr{F}$ fails to be *exact* (i.e. map exact sequences to exact sequence) — if it *vanishes*, then the functor is exact. The *second* derived functor (again, roughly speaking) measures the extent to which the *first* derived functor fails to be exact, and so on. See corollary 13.1.21 on page 444 for a more precise statement.

DEFINITION 13.1.12. If $M \in \mathscr{A}$ is an object, a *right resolution*, I^*, of M is a cochain-complex

$$I_0 \xrightarrow{\delta_0} I_1 \xrightarrow{\delta_1} \cdots$$

where there exists a monomorphism $M \to I_0$ that makes the complex

$$0 \to M \to I_0 \xrightarrow{\delta_0} I_1 \xrightarrow{\delta_1} \cdots$$

exact or acyclic. If all of the I_j are injective (see definition 10.5.3 on page 362), this is called an *injective resolution*.

The *injective dimension* of an object, $M \in \mathscr{A}$, denoted inj-dim M, is the largest subscript of the shortest possible injective resolution of M — if M has a finite injective resolution — or ∞.

REMARK. The definition immediately implies that

$$H^i(I^*) = \begin{cases} M & \text{if } i = 0 \\ 0 & \text{otherwise} \end{cases}$$

Since $\mathscr{A}$ has enough injectives, every object has *some* injective resolution:

(1) set I_0 to some injective containing M and I_1 to an injective object containing I_0/M.
(2) set I_{j+1} to an injective object containing $I_j/\delta_{j-1}(I_{j-1})$.

We also have projective resolutions:

DEFINITION 13.1.13. If $M \in \mathscr{A}$ is an object, a *(left) resolution*, P_*, of M is a chain-complex
$$\cdots \to P_1 \to P_0 \to 0$$
where there exists a epimorphism $P_0 \to M$ that makes the chain complex
$$\cdots \to P_1 \to P_0 \to M \to 0$$
exact or acyclic. A resolution is called a projective resolution if all of the P_i are projective objects (see definition 10.5.2 on page 362). The *projective dimension* of an object, $M \in \mathscr{A}$, denoted proj-dim M is the largest subscript that occurs in a minimal projective resolution — if M has a finite projective resolution — or ∞.

Injective resolutions are by no means unique although they have an interesting property:

PROPOSITION 13.1.14. *Suppose $M, N \in \mathscr{A}$ are two objects with right resolutions I^* and J^*, respectively. If J^* is an injective resolution, then any morphism*
$$f: M \to N$$
induces a chain-map
$$\hat{f}: I^* \to J^*$$
Although $\hat{f}$ is not unique, any two such induced chain-maps are chain-homotopic.

REMARK. This implies that injective resolutions are unique up to chain-homotopy type.

A similar statement can be proved for projective resolutions (reverse all the arrows!).

PROOF. We make extensive use of the property of injective modules described in exercise 3 on page 365. In the diagram

$$\begin{array}{ccc} I_0 & \dashrightarrow & J_0 \\ \uparrow & & \uparrow \\ M & \xrightarrow{f} & N \end{array}$$

it is clear that a portion of I_0 maps to J_0 (namely the portion in the image of M). The injective property of J_0 implies that this extends to all of I_0. In a similar fashion, we inductively construct the chain-map $\hat{f}$ in all higher dimensions.

Suppose g_1 and g_2 are two chain-maps $g_1, g_2: (I^*, \delta) \to (J^*, \sigma)$ that cover the same map $f: M \to N$. It follows that $g = g_1 - g_2$ is chain-map that covers the *zero* map. We will show that it is homotopic to zero, i.e. there exists a map
$$\Phi_i: I_i \to J_{i-1}$$
such that

(13.1.5) $$g_i = \Phi_{i+1} \circ \delta_i + \sigma_{i-1} \circ \Phi_i$$

Since M maps to 0, we have that $g_0\colon I_0 \to J_0$ maps the kernel of δ_0 to 0, which means that it maps $\operatorname{im}\delta_0 \subset I_1$ to J_0. The injective property of J_1 implies that this extends to *all* of I_1, giving

$$\Phi_1\colon I_1 \to J_0$$

with $g_0 = \Phi_1 \circ \delta_0$. Suppose equation 13.1.5 on the previous page is true for degrees $< t$. In degree t, consider

$$g_t - \sigma_{t-1} \circ \Phi_t\colon I_t \to J_t$$

$$
\begin{aligned}
(g_t - \sigma_{t-1} \circ \Phi_t) \circ \delta_{t-1} &= g_t \circ \delta_{t-1} - \sigma_{t-1} \circ \Phi_t \circ \delta_{t-1} \\
&= \sigma_{t-1} \circ g_{t-1} - \sigma_{t-1} \circ \Phi_t \circ \delta_{t-1} \\
&= \sigma_{t-1} \circ \Phi_t \circ \delta_{t-1} + \sigma_{t-1} \circ \sigma_{i-2} \circ \Phi_{i-1} \\
&\quad - \sigma_{t-1} \circ \Phi_t \circ \delta_{t-1} \\
&= \sigma_{t-1} \circ \Phi_t \circ \delta_{t-1} - \sigma_{t-1} \circ \Phi_t \circ \delta_{t-1} \\
&= 0
\end{aligned}
$$

So $(g_t - \sigma_{t-1} \circ \Phi_t)|\operatorname{im}\delta_{t-1} = 0$ which means that $(g_t - \sigma_{t-1} \circ \Phi_t)|\ker \delta_t = 0$. The argument used above implies that $g_t - \sigma_{t-1} \circ \Phi_t$ defines a map $\Phi_{t+1}\colon I_{t+1} \to J_t$ such that

$$g_t - \sigma_{t-1} \circ \Phi_t = \Phi_{t+1} \circ \delta_t$$

The conclusion follows. □

Reversing the arrows proves the chain-complex version

PROPOSITION 13.1.15. *Suppose $M, N \in \mathscr{A}$ are two objects with left resolutions P_* and Q_*, respectively. If P_* is a projective resolution, then any morphism*

$$f\colon M \to N$$

induces a chain-map

$$\hat{f}\colon P_* \to Q_*$$

Although $\hat{f}$ is not unique, any two such induced chain-maps are chain-homotopic.

REMARK. This implies that projective resolutions are unique up to chain-homotopy type.

We will be interested in functors $F\colon \mathscr{A} \to \mathscr{A}'$ to other abelian categories:

DEFINITION 13.1.16. A functor $F\colon \mathscr{A} \to \mathscr{A}'$ is *left-exact* if an exact sequence

(13.1.6) $$0 \to A \xrightarrow{r} B \xrightarrow{s} C \to 0$$

in $\mathscr{A}$ implies that the sequence

$$0 \to F(A) \xrightarrow{F(r)} F(B) \xrightarrow{F(s)} F(C)$$

is exact. The functor, $F\colon \mathscr{A} \to \mathscr{A}'$, is *right-exact* if the sequence 13.1.6 implies the exactness of

$$F(A) \xrightarrow{F(r)} F(B) \xrightarrow{F(s)} F(C) \to 0$$

REMARK. Exercise 4 on page 365 shows that $\hom_{\mathscr{A}}(A, *)$ is left-exact.

PROPOSITION 13.1.17. *For a module M over a ring, R, the functor $* \otimes_R M$ is right exact. In other words, the exact sequence*

$$0 \to A \xrightarrow{r} B \xrightarrow{s} C \to 0$$

induces an exact sequence

$$A \otimes_R M \to B \otimes_R M \to C \otimes_R M \to 0$$

PROOF. Part of this has already been proved in Proposition 10.6.14 on page 371.

The rest follows from the left-exactness of Hom and proposition 10.6.10 on page 369. The exact sequence in 13.1.6 on the facing page induces the exact sequence

$$0 \to \mathrm{Hom}_R(C, \mathrm{hom}_R(M, D)) \to \mathrm{Hom}_R(B, \mathrm{hom}_R(M, D))$$
$$\to \mathrm{Hom}_R(A, \mathrm{hom}_R(M, D))$$

which, by proposition 10.6.10 on page 369, is isomorphic to

$$0 \to \mathrm{Hom}_R(C \otimes_R M, D) \to \mathrm{Hom}_R(B \otimes_R M, D)$$
$$\to \mathrm{Hom}_R(A \otimes_R M, D)$$

for an *arbitrary* module D. This implies

$$A \otimes_R M \to B \otimes_R M \to C \otimes_R M \to 0$$

is exact (the details are left to the reader as an exercise). □

Since injective resolutions are *unique* up to chain-homotopy type, the solution to exercise 3 on page 439 that the following constructs will be well-defined:

DEFINITION 13.1.18. If $F: \mathscr{A} \to \mathscr{A}'$ is a left-exact functor and $C \in \mathscr{A}$ has an injective resolution $I_0 \to \cdots$, then the *right derived functors* of F are

$$R^i F(C) = H^i(F(I^*)) \in \mathscr{A}'$$

for $i \geq 0$.

REMARK. It is not hard to see that $R^0 FC = C$. The $\{R^i F(C)\}$ for $i > 0$ essentially measure how much F fails to be right-exact.

DEFINITION 13.1.19. If $M, N \in \mathscr{A}$ and I^* is an injective resolution of N, the cohomology *groups* (in $\mathscr{A}b$)

$$\mathrm{Ext}^i_R(M, N) = H^i(\mathrm{hom}_{\mathscr{A}}(M, I))$$

depend only on M and N and are functorial.

REMARK. Note that we use hom here rather than Hom. The latter is used for morphisms of modules over a ring, whereas hom is used for morphisms in a category (so it is more general than Hom), in this case $\mathscr{A}$.

Exercise 14 on page 446 shows that we could've used a *projective resolution*, P_*, of M to compute $\mathrm{Ext}^i_R(M, N)$:

$$\mathrm{Ext}^i_R(M, N) = H^i(\mathrm{hom}_{\mathscr{A}}(P_*, N))$$

To analyze the behavior of derived functors, we need the following result

LEMMA 13.1.20 (Injective Horseshoe Lemma). *Suppose*
$$0 \to A \xrightarrow{r} B \xrightarrow{s} C \to 0$$
is a short exact sequence in $\mathscr{A}$ and I^* and J^* are injective resolutions of A and C, respectively. Then there exists an injective resolution W^* of B fitting into a commutative diagram

$$\begin{array}{ccccccccc} 0 & \to & A & \xrightarrow{r} & B & \xrightarrow{s} & C & \to & 0 \\ & & \epsilon_A \downarrow & & \epsilon_B \downarrow & & \epsilon_C \downarrow & & \\ 0 & \to & I^* & \xrightarrow{u} & W^* & \xrightarrow{v} & J^* & \to & 0 \end{array}$$

where the bottom row is a short exact sequence of chain-complexes.

PROOF. Clearly, $W^n = I^n \oplus J^n$ for all n. The map $\epsilon_A: A \to I^0$ shows that a *sub*-object of B maps to I^0. Injectivity implies that this map extends to *all* of B, so we get $\iota: B \to I^0$ and can define
$$\epsilon_B = \iota \oplus \epsilon_C \circ s: B \to I^0 \oplus J^0 = W^0$$
We claim that this is injective: if $b \in B$ maps to zero, it must map to zero in J^0 so that $s(b) = 0$. This means that $b \in \text{im } A$ which maps to I^0 via the injective map, ϵ_A.

Suppose this exact sequence of resolutions has been constructed up to degree n so we have

$$\begin{array}{ccccccccc} 0 & \to & I^n/\text{im } \delta_A & \xrightarrow{r} & W^n/\text{im }\delta_B & \xrightarrow{s} & J^n/\text{im }\delta_C & \to & 0 \\ & & \delta_A^n \downarrow & & f \downarrow & & \delta_C^n \downarrow & & \\ 0 & \to & I^{n+1} & \xrightarrow{u} & W^{n+1} & \xrightarrow{v} & J^{n+1} & \to & 0 \end{array}$$

where the vertical maps are inclusions. Now construct f *exactly* the way ϵ_B was constructed above. □

This immediately implies that

COROLLARY 13.1.21. *If*
$$0 \to A \xrightarrow{r} B \xrightarrow{s} C \to 0$$
is a short exact sequence in $\mathscr{A}$, and $F: \mathscr{A} \to \mathscr{A}'$ is a left-exact additive functor, there exists a natural long exact sequence

$$0 \to F(A) \to F(B) \to F(C) \to R^1 F(A) \to$$
$$\cdots \to R^i F(A) \to R^i F(B) \to R^i F(C) \to R^{i+1} F(A) \to \cdots$$

REMARK. This long exact sequence is often useful for computing the $R^i F(A)$. For instance, if $R^1 F(A) = 0$ then the sequence
$$0 \to F(A) \to F(B) \to F(C) \to 0$$
is exact. If $R^1 F(*)$ is *always* 0, then F is an exact functor.

Here is an application:

DEFINITION 13.1.22. If $F\colon \mathscr{A} \to \mathscr{A}'$ is a left-exact additive functor, an object $M \in \mathscr{A}$ is called *F-acyclic* if
$$R^i F(M) = 0$$
for $i > 0$.

The long exact sequence in corollary 13.1.21 on the facing page implies that

COROLLARY 13.1.23. *Let $F\colon \mathscr{A} \to \mathscr{A}'$ be a left-exact additive functor, and let M be injective or F-acyclic. Then*

(1) *If*
$$0 \to A \to M \to B \to 0$$
is a short exact sequence in $\mathscr{A}$, then
$$R^i F(A) \cong R^{i-1} F(B)$$
for $i > 1$ and $R^1 F(A)$ is $\operatorname{coker} F(M) \to R(B)$.

(2) *if*
$$0 \to A \to M_0 \to \cdots \to M_{n-1} \to B \to 0$$
with the M_i injective or F-acyclic then
$$R^i F(A) \cong R^{i-n} R(B)$$
and $R^n F(A)$ is $\operatorname{coker} F(M_{n-1}) \to F(B)$.

(3) *If*
$$0 \to A \to M_0 \to \cdots$$
is a resolution by F-acyclic objects, then $R^i F(A) = H^i(F(M))$.

PROOF. To prove the first statement, note that the long exact sequence in corollary 13.1.21 on the preceding page reduces to
$$0 \to F(A) \to F(M) \to F(B) \to R^1 F(A) \to 0$$
and
$$0 \to R^n F(B) \xrightarrow{\delta} R^{n+1} F(A) \to 0$$
for all $n > 0$. The second statement follows from the first applied to short exact sequences
$$0 \to A \to M_0 \to K_1 \to 0$$
$$0 \to K_i \to M_i \to K_{i+1} \to 0$$
$$0 \to K_{n-1} \to M_{n-1} \to B \to 0$$
and induction on n.

To prove the third statement, note that we can truncate the resolution by F-acyclic objects at any point to get
$$0 \to A \to M_0 \to \cdots \to M_{n-1} \to \ker \delta^n \to 0$$
and
$$R^n F(A) = \operatorname{coker} F(\delta^{n-1}) \colon F(M_{n-1}) \to F(\ker \delta^n) = \ker F(\delta^n) = H^n(F(M))$$
where $F(\ker \delta^n) = \ker F(\delta^n)$ is due to the left-exactness of F. $\square$

EXERCISES.

5. Fill in the details of the proof of proposition 10.6.14 on page 371: If
$$0 \to \hom_R(C \otimes_R M, D) \xrightarrow{\hom(\beta \otimes 1,1)} \hom_R(B \otimes_R M, D)$$
$$\xrightarrow{\hom(\alpha \otimes 1,1)} \hom_R(A \otimes_R M, D)$$
is exact for an *arbitrary* module, D, then
$$A \otimes_R M \xrightarrow{\alpha \otimes 1} B \otimes_R M \xrightarrow{\beta \otimes 1} C \otimes_R M \to 0$$
is exact.

6. If $A \in \mathscr{A}b$ is a finite abelian group, show that there is an (unnatural!) isomorphism
$$A \cong \hom_{\mathscr{A}b}(A, \mathbb{Q}/\mathbb{Z})$$
and that $\operatorname{Ext}^1_\mathbb{Z}(A, \mathbb{Z}) = A$.

7. Show that $\mathbb{Q}$ is an injective object in the category of abelian groups.

8. Show that $\operatorname{Ext}^i_R(A \oplus B, C) = \operatorname{Ext}^i_R(A, C) \oplus \operatorname{Ext}^i_R(B, C)$

9. Show that $\operatorname{Ext}^i_R\left(\bigoplus_j A_j, C\right) = \prod_j \operatorname{Ext}^i_R(A_j, C) \oplus \operatorname{Ext}^i_R(B, C)$

10. Show that $\operatorname{Ext}^i_R(\varinjlim A_j, C) = \varprojlim \operatorname{Ext}^i_R(A_j, C)$

11. If N is an injective object of $\mathscr{A}$, show that
$$\operatorname{Ext}^i_R(M, N) = 0$$
for $i > 0$ and *any* object $M \in \mathscr{A}$. Conclude that $\hom(*, N)$ is an exact functor.

12. If $\mathscr{A} = R\text{-mod}$, the category of modules over a commutative ring, R, show that
$$\operatorname{Ext}^i_R(R, M) = 0$$
for $i > 0$ and any R-module, M.

13. If $\mathscr{A} = R\text{-mod}$, the category of modules over a commutative ring, R, and P is any projective module over M, show that
$$\operatorname{Ext}^i_R(P, M) = 0$$
for $i > 0$ and any R-module, M, so that $\hom_R(P, *)$ is an exact functor.

14. Suppose $\mathscr{A} = R\text{-mod}$, M and N are R-modules, and
$$\cdots \to P_1 \to P_0 \to M \to 0$$
is a projective resolution (see definition 13.1.13 on page 441). Show that
$$\operatorname{Ext}^i_R(M, N) = H^i(\hom_R(P_*, N))$$
so *projective* resolutions could be used to compute the Ext^i_R-groups.

15. Compute $\operatorname{Ext}^1_\mathbb{Z}(\mathbb{Z}_n, \mathbb{Z})$ using the exact sequence
$$0 \to \mathbb{Z} \xrightarrow{\times n} \mathbb{Z} \to \mathbb{Z}_n \to 0$$
of $\mathbb{Z}_n$.

16. Find an injective resolution for $\mathbb{Z}$ in $\mathscr{A}b$.

17. If $\mathscr{A} = \mathscr{A}b$ show that every abelian group, A, has an injective resolution that ends in degree 1 — i.e. is of the form
$$I_0 \to I_1$$
so that $\operatorname{Ext}^i_{\mathbb{Z}}(A, B) = 0$ for $A, B \in \mathscr{A}b$ and $i > 1$.

18. Prove a *projective* version of corollary 13.1.22 on page 445:
If
$$0 \to U \to P_n \to \cdots \to P_1 \to A \to 0$$
is an exact sequence of modules over a ring, R, with all of the P_i *projective*, then
$$\operatorname{Ext}^i_R(U, M) = \operatorname{Ext}^{i+n}_R(A, M)$$
if $i > 0$.

13.2. Rings and modules

Since the category of modules over a ring has enough projectives (just map a suitable free module to a module) and enough injectives (proposition 10.5.10 on page 364), we can define homology and cohomology-based functors.

Suppose $\mathscr{A} = \mathscr{M}_R$, the category of modules over a commutative ring, R.

When R is not commutative (for instance, when $R = \mathbb{Z}G$, a group-ring of a nonabelian group), most of these results still go through but become slightly more complicated: for instance, we must make a distinction between right-modules and left-modules.

13.2.1. Tor^i_R-functors.
In this section, we study the derived functors of $\otimes$.

DEFINITION 13.2.1. Given a ring, R, and R-modules, M and N, let
$$\cdots \to P_2 \to P_1 \to P_0 \to M \to 0$$
be an R-projective resolution of M and define
$$\operatorname{Tor}^i_R(M, N) = H_i(P_* \otimes N)$$

REMARK. Proposition 13.1.15 on page 442 implies that $\operatorname{Tor}^i_R(M, N)$ does not depend on the resolution used. The right-exactness of $* \otimes N$ (see proposition 13.1.17 on page 443) implies that
$$\operatorname{Tor}^0_R(M, N) = M \otimes_R N$$

PROPOSITION 13.2.2. *If M is and R-module and*
$$0 \to A \to B \to C \to 0$$
is an exact sequence of R-modules, there exist homomorphisms
$$\partial_{i+1} \colon \operatorname{Tor}^{i+1}_R(C, M) \to \operatorname{Tor}^i_R(A, M)$$

that fit into a long exact sequence

$$\cdots \xrightarrow{\partial_{n+1}} \operatorname{Tor}_R^n(A,M) \to \operatorname{Tor}_R^n(B,M) \to \operatorname{Tor}_R^n(C,M) \to \cdots$$
$$\xrightarrow{\partial_2} \operatorname{Tor}_R^1(A,M) \to \operatorname{Tor}_R^1(B,M) \to \operatorname{Tor}_R^1(C,M)$$
$$\xrightarrow{\partial_1} A \otimes_R M \to B \otimes_R M \to C \otimes_R M \to 0$$

REMARK. This implies that $\operatorname{Tor}_R^1(C,M)$ measures how much $* \otimes_R M$ fails to be right-exact.

PROOF. Let P_* be a projective resolution of M. Since projective modules are flat (see proposition 10.6.20 on page 374) and tensor products with flat modules preserve exact sequences (see proposition 10.6.18 on page 373), we get an exact sequence of chain-complexes

$$0 \to A \otimes_R P_* \to B \otimes_R P_* \to C \otimes_R P_* \to 0$$

which induces a long exact sequence in homology — see proposition 13.1.10 on page 438. This is the conclusion. □

EXAMPLE. If $r \in R$ is not a zero-divisor the exact sequence

$$0 \to R \xrightarrow{\times r} R \to R/r \cdot R \to 0$$

induces

$$0 \to \operatorname{Tor}_R^1(R/r \cdot R, M) \to M \xrightarrow{\times r} M \to M/r \cdot M \to 0$$

since $R \otimes_R M = M$ and $(R/r \cdot R) \otimes_R M = M/r \cdot M$. It follows that

$$\operatorname{Tor}_R^1(R/r \cdot R, M) = \{m \in M | r \cdot m = 0\}$$

i.e., the *r-torsion* elements of M. That's the basis of the name Tor_R^i.

13.2.2. Ext_R^1 and extensions. In this section and the next, we will consider properties of the Ext_R^i-functors in definition 13.1.19 on page 443.

DEFINITION 13.2.3. If A and B are R-modules, an *extension* of A by B is a short exact sequence

$$0 \to B \to E \to A \to 0$$

where E is some module. Two such extensions are considered *equivalent* if they fit into a commutative diagram

$$\begin{array}{ccccccccc}
0 & \longrightarrow & B & \xrightarrow{r} & E_1 & \xrightarrow{s} & A & \longrightarrow & 0 \\
& & \| & & \downarrow v & & \| & & \\
0 & \longrightarrow & B & \xrightarrow{r'} & E_2 & \xrightarrow{s'} & A & \longrightarrow & 0
\end{array}$$

Exercise 4 on page 439 implies that v is an isomorphism.

Regard an extension as *equivalent to* 0 if it is *split*, i.e. of the form

$$0 \to B \to B \oplus A \to A \to 0$$

Given an extension, Corollary 13.1.21 on page 444 shows it induces a long exact sequence

$$0 \to \hom_{\mathscr{A}}(A,B) \to \hom_{\mathscr{A}}(A,E) \to \hom_{\mathscr{A}}(A,A) \xrightarrow{\delta} \mathrm{Ext}^1_R(A,B)$$

We will associate the extension to

(13.2.1) $$\delta(1) \in \mathrm{Ext}^1_R(A,B)$$

where $1 \in \hom_{\mathscr{A}}(A,A)$ is the identity map. The fact that the long exact sequence is *natural* means that an equivalence of extensions gives rise to a commutative diagram

$$\begin{array}{ccccc}
\hom_{\mathscr{A}}(A,E_1) & \longrightarrow & \hom_{\mathscr{A}}(A,A) & \xrightarrow{\delta} & \mathrm{Ext}^1_R(A,B) \\
\downarrow & & \parallel & & \parallel \\
\hom_{\mathscr{A}}(A,E_2) & \longrightarrow & \hom_{\mathscr{A}}(A,A) & \xrightarrow[\delta]{} & \mathrm{Ext}^1_R(A,B)
\end{array}$$

so equivalent extensions give rise to the *same* element of $\mathrm{Ext}^1_R(A,B)$.

Given $x \in \mathrm{Ext}^1_R(A,B)$ and an injective resolution for B, I^*, represent x by a homomorphism $x \colon A \to I^1$ whose image lies in the kernel of $\delta^1 \colon I^1 \to I^2$. This means it is in the image of I^0 and this *image* is isomorphic to I^0/B. We get a commutative diagram

$$\begin{array}{ccccccccc}
0 & \longrightarrow & B & \xrightarrow{r} & E_1 & \xrightarrow{s} & A & \longrightarrow & 0 \\
& & \parallel & & \downarrow v & & \downarrow x & & \\
0 & \longrightarrow & B & \xrightarrow[r']{} & I^0 & \xrightarrow[s']{} & I^0/B & \longrightarrow & 0
\end{array}$$

where E_1 is a pull-back (see exercise 6 on page 345) of

(13.2.2) $$\begin{array}{ccc}
I^0/B & \xleftarrow{x} & A \\
{\scriptstyle s'}\uparrow & & \\
I^0 & &
\end{array}$$

so

$$E_1 = \{(\alpha,\beta) \in I^0 \oplus A \mid s'(\alpha) = x(\beta)\}$$

and

$$s \colon E_1 \to A$$

is the composite $E_1 \hookrightarrow I^0 \oplus A \xrightarrow{p} A$. The kernel of s is the pull-back

$$\begin{array}{ccc}
I^0/B & \longleftarrow & 0 \\
{\scriptstyle s'}\uparrow & & \\
I^0 & &
\end{array}$$

which is isomorphic to B.

This induces a diagram

$$\begin{array}{ccccccc} \hom_{\mathscr{A}}(A,E_1) & \longrightarrow & \hom_{\mathscr{A}}(A,A) & \xrightarrow{\delta} & \text{Ext}^1_R(A,B) & \longrightarrow & \text{Ext}^1_R(A,E_1) \\ \downarrow & & \downarrow \hom_{\mathscr{A}}(1,x) & & \| & & \downarrow \\ \hom_{\mathscr{A}}(A,I^0) & \longrightarrow & \hom_{\mathscr{A}}(A,I^0/B) & \xrightarrow{\delta} & \text{Ext}^1_R(A,B) & \longrightarrow & \text{Ext}^1_R(A,I^0) \end{array}$$

Since $\text{Ext}^1_R(A,I^0) = 0$, it is clear that the identity map of A maps to x.

The proof that split exact sequences give 0 is left to the reader.

Suppose we vary x by a coboundary — i.e. $x' = x + \partial f$ where $f: A \to I^0$ is a homomorphism. In this case, we get an isomorphism of pull-backs

$$F = \begin{bmatrix} 1 & f \\ 0 & 1 \end{bmatrix} : E_1 \subset I^0 \oplus A \to E_2 \subset I^0 \oplus A$$

that fits into a commutative diagram

$$\begin{array}{ccccccccc} 0 & \longrightarrow & B & \xrightarrow{r} & E_1 & \xrightarrow{s} & A & \longrightarrow & 0 \\ & & \| & & F \downarrow & & \| & & \\ 0 & \longrightarrow & B & \xrightarrow{r'} & E_2 & \xrightarrow{s'} & A & \longrightarrow & 0 \end{array}$$

It follows that the *class* of $x \in \text{Ext}^1_R(A,B)$ is what is significant.

We summarize this discussion

THEOREM 13.2.4. *Given a ring R and modules A and B, let $E(A,B)$ denote the equivalence classes of extensions of A by B. The process in equation 13.2.1 on the previous page gives a 1-1 correspondence*

$$E(A,B) \leftrightarrow \text{Ext}^1_R(A,B)$$

REMARK. This gives us an *interpretation* of $\text{Ext}^1_R(A,B)$: it the set of equivalence classes of extensions of A by B, and is the reason for the name of the functor.

Suppose $A = \mathbb{Z}_n$ and $B = \mathbb{Z}$ over the ring $\mathbb{Z}$. In this case

$$0 \to \mathbb{Z} \to \mathbb{Q} \to \mathbb{Q}/\mathbb{Z} \to 0$$

is an injective resolution of $\mathbb{Z}$. A homomorphism

$$x_k : \mathbb{Z}_n \to \mathbb{Q}/\mathbb{Z}$$

is defined by

$$1 \mapsto k/n \in \mathbb{Q}/\mathbb{Z}$$

where

$$0 \le k < n$$

The extension defined by this is the pullback

$$\begin{array}{ccc} \mathbb{Q}/\mathbb{Z} & \xleftarrow{x_k} & \mathbb{Z}_n \\ s' \uparrow & & \\ \mathbb{Q} & & \end{array}$$

If $x = 0$, this pullback is $\mathbb{Z} \oplus \mathbb{Z}_n$; otherwise it is $\mathbb{Z}$. Note that extensions

$$0 \to \mathbb{Z} \to \mathbb{Z} \to \mathbb{Z}_n \to 0$$

for *distinct* nonzero values of k are *inequivalent* although their extended modules are all $\mathbb{Z}$.

DEFINITION 13.2.5 (Baer Sum). Given modules A and B over a ring, R, and two extensions

$$0 \to B \to E_1 \to A \to 0$$
$$0 \to B \to E_2 \to A \to 0$$

of A by B, we define the *Baer Sum* of the extensions to be

$$0 \to B \to E \to A \to 0$$

where E is defined via:

Let $\bar{E}$ be the pull-back

$$\begin{array}{ccc} A & \leftarrow & E_1 \\ \uparrow & & \uparrow \\ E_2 & \leftarrow & \bar{E} \end{array}$$

This contains three embedded copies of B, namely $(B, 0)$, $(0, B)$, and the diagonally embedded copy:

$$\iota : B \hookrightarrow \bar{E} \quad \subset \quad E_1 \oplus E_2$$
$$b \mapsto (b, -b)$$

Define

$$E = \frac{\bar{E}}{\iota(B)}$$

This fits into an extension

$$0 \to B \to E \to A \to 0$$

EXERCISES.

1. Show that a module M is projective if and only if

$$\operatorname{Ext}^1_R(M, N) = 0$$

for all R-modules, N.

2. Formulate the correspondence in theorem 13.2.4 on the facing page using a *projective* resolution of A rather than an *injective* resolution of B.

3. Shows that the Baer sum in definition 13.2.5 makes the correspondence in theorem 13.2.4 on the facing page a homomorphism of abelian groups.

4. If p is a prime, we know that $\text{Ext}^1_{\mathbb{Z}}(\mathbb{Z}_p, \mathbb{Z}) = \mathbb{Z}_p$ implying that there are p inequivalent extensions

$$0 \to \mathbb{Z} \xrightarrow{\iota} M \xrightarrow{\pi} \mathbb{Z}_p \to 0$$

and the *split* extension ($M = \mathbb{Z} \oplus \mathbb{Z}_p$) corresponds to $0 \in \text{Ext}^1_{\mathbb{Z}}(\mathbb{Z}_p, \mathbb{Z})$. Describe the other $p - 1$ extensions.

5. Extend exercise 4 to the case

$$0 \to \mathbb{Z} \xrightarrow{\iota} M \xrightarrow{\pi} \mathbb{Z}_{pq} \to 0$$

where p and q are two different primes. Hint: Note that $\mathbb{Z}_{pq} = \mathbb{Z}_p \oplus \mathbb{Z}_q$ and we have

▷ the split extension

$$0 \to \mathbb{Z} \xrightarrow{\iota} \mathbb{Z} \oplus \mathbb{Z}_{pq} \xrightarrow{\pi} \mathbb{Z}_{pq} \to 0$$

▷ *partially* split extensions

$$0 \to \mathbb{Z} \xrightarrow{\iota} \mathbb{Z} \oplus \mathbb{Z}_p \xrightarrow{\pi} \mathbb{Z}_{pq} \to 0$$

$$0 \to \mathbb{Z} \xrightarrow{\iota} \mathbb{Z} \oplus \mathbb{Z}_q \xrightarrow{\pi} \mathbb{Z}_{pq} \to 0$$

▷ and the totally *nonsplit* extensions

$$0 \to \mathbb{Z} \xrightarrow{\iota} \mathbb{Z} \xrightarrow{\pi} \mathbb{Z}_{pq} \to 0$$

13.2.3. n-fold extensions. We can continue the reasoning in the last section to so-called Yoneda extensions.

DEFINITION 13.2.6. If R is a ring and A and B are R-modules, an *n-fold extension* of A by B is an exact sequence

$$0 \to B \to E_n \to \cdots \to E_1 \to A \to 0$$

Two such extensions

$$0 \to B \to E_n \to \cdots \to E_1 \to A \to 0$$

and

$$0 \to B \to E'_n \to \cdots \to E'_1 \to A \to 0$$

are defined to be *equivalent* if there exists a commutative diagram

$$\begin{array}{ccccccccc}
0 & \to & B & \to & E_n & \to \cdots \to & E_1 & \to & A & \to & 0 \\
& & \parallel & & \downarrow & & \downarrow & & \parallel & & \\
0 & \to & B & \to & E'_n & \to \cdots \to & E'_1 & \to & A & \to & 0
\end{array}$$

It is possible to "standardize" n-fold extensions:

13.2. RINGS AND MODULES

LEMMA 13.2.7. *Let*
$$\cdots \to P_{n+1} \to P_n \to \cdots \to P_1 \to A \to 0$$
(note that we are indexing the projective modules in this resolution from 1 rather than from 0) be a projective resolution of A. An n-fold extension of A by B is equivalent to one in which $n - 1$ modules in the extension are projective and of the form:

$$
\begin{array}{ccccccccccc}
0 & \to & B & \to & M & \to & P_{n-1} & \to & \cdots \to P_1 & \to & A & \to & 0 \\
& & \| & & \downarrow & & \downarrow & & \downarrow & & \| & & \\
0 & \to & B & \to & E_n & \to & E_{n-1} & \to & \cdots \to E_1 & \to & A & \to & 0
\end{array}
$$

REMARK. The significant thing about this is that most of our "standardized" extension is fixed.

PROOF. Proposition 13.1.15 on page 442 implies that we get a chain-map:

(13.2.3)
$$
\begin{array}{ccccccccccc}
\cdots \to & P_{n+1} & \xrightarrow{v} & P_n & \xrightarrow{q} & P_{n-1} & \to \cdots \to & P_0 & \to & A & \to 0 \\
& u \downarrow & & \downarrow & & \downarrow & & \downarrow & & \| & \\
0 \to & B & \to & E_n & \to & E_{n-1} & \to \cdots \to & E_1 & \to & A & \to 0
\end{array}
$$

Given a homomorphism $x \in \hom(P_n, B)$, we can form the push-out

$$
\begin{array}{ccc}
P_{n+1} & \longrightarrow & P_n \\
\downarrow & & \vdots \\
B & \cdots\cdots\cdots & M
\end{array}
$$

or
$$M = \frac{B \oplus P_n}{f(P_{n+1})}$$
where
$$f = (u, -v) \colon P_{n+1} \to B \oplus P_n$$

We have a homomorphism
$$\iota \colon B \to M$$
defined as the composite
$$B \hookrightarrow B \oplus P_n \to M$$
We claim that this is injective. The kernel is given by
$$\ker \iota = B \oplus 0 \cap \operatorname{im} f$$
Commutativity of
$$
\begin{array}{ccc}
P_{n+1} & \xrightarrow{v} & P_n \\
u \downarrow & & \downarrow p_n \\
B & \xrightarrow{a} & E_n
\end{array}
$$

— the leftmost-square of diagram 13.2.3 on the previous page — means that $v(x) = 0$ implies $p_n \circ v(x) = 0$ which implies that $a \circ u(x) = 0$. Since a is *injective*, this means that $u(x) = 0$ so that $\ker \iota = 0$.

We define
$$M \to P_{n-1}$$
to be induced by
$$(0, q) \colon B \oplus P_n \to P_{n-1}$$
which is well-defined if $\ker f \subset B \oplus \ker q$. This follows from $q \circ v = 0$.

The map $M \to E_n$ is induced by the defining property of a push-out. $\square$

This leads to the result

THEOREM 13.2.8. *Let R be a ring and let A and B be R-modules. Then the set of equivalence classes of n-fold extensions of A by B is in a one to one correspondence with*
$$\operatorname{Ext}_R^n(A, B)$$

PROOF. Two n-fold extensions are equivalent if and only if their standardized forms
$$0 \to B \to M_1 \to \operatorname{im} P_n \to 0$$
and
$$0 \to B \to M_2 \to \operatorname{im} P_n \to 0$$
are equivalent. These equivalence classes are in a one-to-one correspondence with $\operatorname{Ext}_R^1(\operatorname{im} P_n, B)$. Exercise 18 on page 447 implies that
$$\operatorname{Ext}_R^1(\operatorname{im} P_n, B) = \operatorname{Ext}_R^n(A, B)$$
and the conclusion follows. $\square$

EXERCISES.

6. Define a generalization of the Baer sum for n-fold extensions.

13.3. Cohomology of groups

13.3.1. Introduction. The cohomology of groups is a vast subject suitable for its own book (for instance, the excellent [17]). We will only touch on it here.

If G is a group, we can construct the *group-ring*, $\mathbb{Z}G$ — a ring of formal linear combinations of elements of G. Let G act trivially on $\mathbb{Z}$ so that $\mathbb{Z}$ is a $\mathbb{Z}G$-module and let
$$\cdots \to P_2 \to P_1 \to P_0 \to \mathbb{Z} \to 0$$
be a projective resolution of $\mathbb{Z}$.

DEFINITION 13.3.1. If A is a left $\mathbb{Z}G$-module define
$$H^i(G, A) = H^i(\hom_{\mathbb{Z}G}(P_*, A)) = \text{Ext}^i_{\mathbb{Z}G}(\mathbb{Z}, A)$$

REMARK. In a few cases, it's easy to say what these groups are:
$$H^0(G, A) = \hom_{\mathbb{Z}G}(\mathbb{Z}, A) = A^G$$
the submodule of A *fixed* by G — definition 11.1.7 on page 390.

In order to proceed further, we must construct explicit resolutions of $\mathbb{Z}$.

PROPOSITION 13.3.2. *If* $G = \mathbb{Z}_n$, *generated by* T *then*
$$\mathbb{Z}G = \mathbb{Z}[T]/(T^n - 1)$$
a quotient of a polynomial ring and
$$\mathbb{Z} = \frac{\mathbb{Z}G}{(T-1)}$$
If $x \in \mathbb{Z}G$ *has the property that* $(T-1) \cdot x = 0$, *then there exists* $y \in \mathbb{Z}G$ *such that* $x = N \cdot y$ *where*
$$N = 1 + T + T^2 + \cdots + T^{n-1}$$
It follows that
$$\cdots \to \mathbb{Z}G \xrightarrow{\cdot(T-1)} \mathbb{Z}G \xrightarrow{\cdot N} \mathbb{Z}G \xrightarrow{\cdot(T-1)} \mathbb{Z}G \to \mathbb{Z} \to 0$$
is a free $\mathbb{Z}G$-*resolution of* $\mathbb{Z}$.

REMARK. Note that this resolution is *periodic* so that, if A is a $\mathbb{Z}G$-module with trivial G-action:
$$H^0(G, A) \cong A$$
and
$$H^{2i}(G, A) \cong A/n \cdot A$$
where $i \geq 1$, and
$$H^{2i-1}(G, A) \cong H^1(G, A) = A[n]$$
the n-torsion subgroup (see definition 4.6.23 on page 71).

PROOF. The statement that
$$\mathbb{Z} = \frac{\mathbb{Z}G}{(T-1)}$$
is clear. If $(T-1) \cdot x = 0$, and $x = \sum_{i=0}^{n-1} a_i T^i$, then
$$(T-1) \cdot x = \sum_{i=0}^{n-1} (a_{(i+1) \bmod n} - a_i) T^i$$
so $(T-1) \cdot x = 0$ implies that all of the coefficients are the *same* — i.e., $x = N \cdot a_0$. □

The definition of cohomology groups immediately implies that:
If $G = \mathbb{Z}_n$, then
$$H^1(G, A) = \frac{\ker(N \colon A \to A)}{(T-1) \cdot A}$$
From this, we get a *cohomological* statement of Hilbert's Theorem 90 (see theorem 8.8.2 on page 318):

THEOREM 13.3.3 (Hilbert's Theorem 90). *If $G = \mathbb{Z}_n$ for some integer n and G is the Galois group of a field extension*

$$\begin{array}{c} H \\ | \\ F \end{array}$$

then
$$H^1(G, H^\times) = \{1\}$$
where $H^\times = H \setminus \{0\}$ is the multiplicative (abelian) group of H, regarded as a $\mathbb{Z}G$-module.

REMARK. Note: we haven't used group cohomology to *prove* Hilbert's Theorem 90; we are merely *stating* it in terms of group-cohomology.

PROOF. If $x \in A$, the statement that $N \cdot x = 0$ is equivalent (when writing the operation *multiplicatively*) to the statement
$$x \cdot \theta(x) \cdots \theta^{n-1}(x) = 1$$
where θ generates the Galois group, G. This is equivalent to saying the norm of x is 1, by theorem 8.8.1 on page 318. The statement that $H^1(G, H^\times) = \{1\}$ means that $x = \theta(y)y^{-1}$ for some $y \in H$ — or $x = z\theta(z)^{-1}$, where $z = y^{-1}$ — which is the statement of theorem 8.8.2 on page 318. □

EXAMPLE 13.3.4. If $G = \mathbb{Z}$, $\mathbb{Z}G = \mathbb{Z}[T, T^{-1}]$ and
$$0 \to \mathbb{Z}G \xrightarrow{\cdot(T-1)} \mathbb{Z}G \to \mathbb{Z} \to 0$$
is a projective resolution of $\mathbb{Z}$, so
 (1) $H^0(\mathbb{Z}, A) = A^{\mathbb{Z}}$, the largest *submodule* fixed by $\mathbb{Z}$.
 (2) $H^1(\mathbb{Z}, A) = A/((T-1) \cdot A)$, the largest *quotient* fixed by $\mathbb{Z}$.
 (3) $H^i(\mathbb{Z}, A) = 0$ for $i \geq 2$.

In order to proceed further, we need a description of projective resolutions for *all* groups.

DEFINITION 13.3.5. If G is a group, its *bar resolution*, $\mathcal{B}G$, is a $\mathbb{Z}G$-chain complex
$$\cdots \to \mathcal{B}G_3 \xrightarrow{d_3} \mathcal{B}G_2 \xrightarrow{d_2} \mathcal{B}G_1 \xrightarrow{d_1} \mathcal{B}G_0 \xrightarrow{\epsilon} \mathbb{Z} \to 0$$
of free $\mathbb{Z}G$-modules where
 (1) The basis for $\mathcal{B}G_n$ is symbols $[g_1|\cdots|g_n]$ for all elements $g_i \neq 1 \in G$,
 (2) $\mathcal{B}G_0 = \mathbb{Z}G \cdot [\,]$, where $[\,]$ is a symbol that is its only basis element.

(3) $\epsilon\colon \mathbb{Z}G \cdot [\,] \to \mathbb{Z}$ sends all elements of G to $1 \in \mathbb{Z}$
(4) for $i > 0$, P_i is the free $\mathbb{Z}G$-module on a basis that consists of all possible symbols $[g_1|\cdots|g_n]$ where $1 \neq g_i \in G$ for all i.
(5) $d_n\colon \mathcal{B}G_n \to \mathcal{B}G_{n-1}$ is defined by

$$d_n([g_1|\cdots|g_n]) = g_1 \cdot [g_2|\cdots|g_n]$$
$$+ \sum_{i=1}^{n-1}(-1)^i[g_1|\cdots|g_ig_{i+1}|\cdots|g_n]$$
$$\cdots + (-1)^n[g_1|\cdots|g_{n-1}]$$

Here, if a term has a 1 in it, we identify it with 0, so

$$d_1([g|g^{-1}]) = g[g^{-1}] - [1] + [g] = g[g^{-1}] + [g]$$

This is a *resolution* because it has a contracting homotopy given by

$$\Phi_i\colon \mathcal{B}G_i \to \mathcal{B}G_{i+1}$$
$$g \cdot [g_1|\cdots|g_i] \mapsto [g|g_1|\cdots|g_i] \text{ if } g \neq 1$$
$$1 \cdot [g_1|\cdots|g_i] \mapsto 0$$

for $i \geq 0$ and

$$\Phi_{-1}(1) = [\,]$$

such that

(13.3.1) $$1 - \epsilon_i = d_{i+1} \circ \Phi_i + \Phi_{i-1} \circ d_i$$

where $\epsilon_i = 0$, if $i > 0$, and $\epsilon_0 = \epsilon$, as defined above. It follows that the 0-map is chain-homotopic to the identity (see definition 13.1.5 on page 435 and proposition 13.1.6 on page 435) above dimension 0, and the homology of this complex must vanish.

REMARK. This was originally defined by Eilenberg and Maclane in [32, 33] and MacLane's book [70]. The term "bar resolution" comes from their use of vertical bars rather than $\otimes$'s because they take up less space[2].

This is what is called the *normalized* bar resolution; it is slightly smaller than the unnormalized bar resolution, which allow 1's in the bar-symbols.

EXERCISES.

1. Show that the resolution in definition 13.3.5 on the facing page is a *chain-complex*, i.e., that $d_i \circ d_{i+1} = 0$ for all i.

2. Verify equation 13.3.1.

---

[2]Typographically!

13.3.2. Extensions of groups. In this section we will apply the cohomology of groups to study the *extension problem for groups:*

> Given two groups H and N, which groups, G, have the property that
> (1) $N \triangleleft G$
> (2) there exists an exact sequence

(13.3.2) $$1 \to N \xrightarrow{\iota} G \xrightarrow{\pi} H \to 1$$

> ?
> Such groups, G, are called *extensions of H by N* (compare definition 13.2.3 on page 448).

This is an extremely difficult problem in general but we know solutions in special cases:

If N is abelian and the extension is *split* — i.e., if there's a homomorphism $g: H \to G$ with the property that $\pi \circ g = 1: H \to H$ in 13.3.2, then exercise 5 on page 80 shows that

$$G = N \rtimes_\varphi H$$

the semidirect product.

PROPOSITION 13.3.6. *An extension of a group H by an abelian group, N*

$$1 \to N \xrightarrow{\iota} G \xrightarrow{\pi} H \to 1$$

defines a $\mathbb{Z}H$-module structure on N by conjugation

$$\varphi: H \to \mathrm{Aut}(N)$$

PROOF. We define a *function* (not a homomorphism!)

$$g: H \to G$$

such that $\pi \circ g = 1: H \to H$. The H-action on N is defined by conjugation

$$\varphi(h)(n) = g(h) \cdot n \cdot g(n)^{-1}$$

Note: since N is *abelian*, varying $g(h)$ by an element of N doesn't change the result of conjugation so we actually get a homomorphism

$$H \to \mathrm{Aut}(N)$$

□

Compare this with 13.2.3 on page 448

DEFINITION 13.3.7. Two extensions, G_1 and G_2 are defined to be *equivalent* if there exists a commutative diagram

$$\begin{array}{ccccccccc}
1 & \longrightarrow & N & \xrightarrow{\iota_1} & G_1 & \xrightarrow{\pi_1} & H & \longrightarrow & 1 \\
& & \| & & \downarrow & & \| & & \\
1 & \longrightarrow & N & \xrightarrow{\iota_2} & G_2 & \xrightarrow{\pi_2} & H & \longrightarrow & 1
\end{array}$$

The results of section 13.2.2 on page 448 suggest that the cohomology of groups might be useful — at least in the case where N is abelian.

Throughout the remainder of this section N denotes an abelian group that is a module over $\mathbb{Z}H$ and we will study extensions like

(13.3.3) $$1 \to N \xrightarrow{\iota} G \xrightarrow{\pi} H \to 1$$

We will consider functions

(13.3.4) $$g: H \to G$$

such that $\pi \circ g = 1: H \to H$. Given $h_1, h_2 \in H$, consider $x(h_1, h_2) = g(h_1)g(h_2)g(h_1h_2)^{-1}$ and we require $g(1) = 1$.

DEFINITION 13.3.8. Given a group-extension as in 13.3.3 and a map $g: H \to G$ with the property $g(1) = 1$, the function

$$x: H \times H \to N$$

defined as $x(h_1, h_2) = g(h_1)g(h_2)g(h_1h_2)^{-1}$ is called a *factor-set* for the extension.

Since the extension in 13.3.3 is not split, g in 13.3.4 is not a homomorphism so $x(h_1, h_2) \neq 1$. Since it maps to 1 in H, we conclude that $x(h_1, h_2) \in N$.

If we identify G with $N \times g(H)$ (as sets), a factor-set for an extension *determines* it because

$$\begin{aligned}(n_1 g(h_1))(n_2 g(h_2)) &= (n_1 g(h_1) n_2 g(h_1)^{-1} g(h_1) g(h_2)) \\ &= (n_1 \varphi(h_1)(n_2) g(h_1) g(h_2)) \\ &= (n_1 \varphi(h_1)(n_2) g(h_1) g(h_2) g(h_1 h_2)^{-1} g(h_1 h_2)) \\ &= (n_1 \varphi(h_1)(n_2) \cdot x(h_1, h_2) g(h_1 h_2))\end{aligned}$$

or

(13.3.5) $$(n_1, h_1)(n_2, h_2) = (n_1 + \varphi(h_1)(n_2) + x(h_1, h_2), h_1 h_2)$$

since the N-group is abelian. Note that, if the factor-set is zero, this becomes the product-formula for the semidirect product.

PROPOSITION 13.3.9. *A function, $x: H \times H \to N$, is a factor-set for the extension in 13.3.3 if and only if it defines a 2-cocycle*

$$\mathcal{B}H_2 \to N$$

with respect to the bar-resolution (see definition 13.3.5 on page 456).

PROOF. This is the same as saying the composite

$$\mathcal{B}H_3 \xrightarrow{d_3} \mathcal{B}H_2 \xrightarrow{x} N$$

is zero, i.e.

(13.3.6) $$\varphi(h_1)\left(x(h_2, h_3)\right) - x(h_1 h_2, h_3) + x(h_1, h_2 h_3) - x(h_1, h_2) = 0$$

If x is a factor-set, this follows from the associativity of the product in equation 13.3.5 on the preceding page:

$$((0,h_1)(0,h_2))(0,h_3) = (x(h_1,h_2) + x(h_1h_2,h_3), h_1h_2h_3)$$
$$(0,h_1)((0,h_2)(0,h_3)) = (\varphi(h_1)(x(h_2,h_3)) + x(h_1h_2,h_3), h_1h_2h_3)$$

and setting the difference between these to 0 gives equation 13.3.6 on the previous page.

Conversely, if x is a 2-cocycle, equation 13.3.5 on the preceding page gives an associative product on $N \times H$ and

$$(n,h)^{-1} = \left(-\varphi(h^{-1})(n) - \varphi(h^{-1})\left(x(h,h^{-1})\right), h^{-1}\right)$$

so we can define an extension group from x. $\square$

LEMMA 13.3.10. *If $x_1, x_2 \colon H \times H \to N$ are two factor-sets giving rise to the same extension*

$$1 \to N \to E \to H \to 1$$

then

$$x_1 - x_2$$

is a coboundary in the bar-resolution of N. It follows that a factor-set corresponds to a well-defined element of

$$H^2(H,N)$$

PROOF. We will write the group-operation *multiplicatively*. If the x_i are induced by functions

$$g_i \colon H \to E$$

then $g_1(h)g_2(h)^{-1} = \theta(h) \in N$ for all $h \in H$, so $g_1(h) = \theta(h) \cdot g_2(h)$.

$$x_1(h_1,h_2) = g_1(h_1)g_1(h_2)g_1(h_1h_2)^{-1}$$
$$= \theta(h_1)g_2(h_1)\theta(h_2)g_2(h_2)[\theta(h_1h_2)g_2(h_1h_2)]^{-1}$$
$$= \theta(h_1)g_2(h_1)\theta(h_2)g_2(h_2)g_2(h_1h_2)^{-1}\theta(h_1h_2)^{-1}$$
$$= \theta(h_1)g_2(h_1)\theta(h_2)g_2(h_1)^{-1}g_2(h_1)g_2(h_2)g_2(h_1h_2)^{-1}\theta(h_1h_2)^{-1}$$
$$= \theta(h_1)\theta(h_2)^{g_2(h_1)}g_2(h_1)g_2(h_2)g_2(h_1h_2)^{-1}\theta(h_1h_2)^{-1}$$
$$= \theta(h_1)\theta(h_2)^{g_2(h_1)}x_2(h_1,h_2)\theta(h_1h_2)^{-1}$$

Since the equation on the bottom line takes place entirely in the abelian group, N, we may write it *additively*

$$x_1(h_1,h_2) = \theta(h_1) + \theta(h_2)^{g_2(h_1)} - \theta(h_1h_2) + x_2(h_1,h_2)$$

or

$$x_1(h_1,h_2) - x_2(h_1,h_2) = \theta(h_1) + \theta(h_2)^{g_2(h_1)} - \theta(h_1h_2)$$
$$= \varphi(h_1)(\theta(h_2)) - \theta(h_1h_2) + \theta(h_1)$$
$$= \theta(d_2([h_1|h_2]))$$

where

$$d_2 \colon \mathcal{B}H_2 \to \mathcal{B}H_1$$

is the *boundary map* of the bar-resolution of H (see definition 13.3.5 on page 456). □

Considering definition 13.3.7 on page 458, we conclude that

THEOREM 13.3.11. *If H is a group and N is an abelian group with the structure of a $\mathbb{Z}H$-module, the set of equivalence classes of extensions of N by H is in a one-to-one correspondence with elements of*

$$H^2(H,N)$$

The split extension corresponds to $0 \in H^2(H,N)$ (the semi-direct product — see 4.7.15 on page 79 and exercise 5 on page 80).

Extensions in which N is *not* abelian are much more complex and their study involves *nonabelian cohomology* and algebraic topology. See [18].

EXERCISES.

3. If F is a free group and A is any $\mathbb{Z}F$ module, show that
$$H^2(F,A) = 0$$

4. If F is a free group and A is any $\mathbb{Z}F$ module, show that
$$H^i(F,A) = 0$$
for *all $i > 1$*.

CHAPTER 14

Axiomatic Set Theory

"Either mathematics is too big for the human mind or the human mind is more than a machine."
— Kurt Gödel.

14.1. Introduction

In this section, we will give a rigorous set of axioms defining set theory. The reader might wonder why this is necessary: after all sets are just collections of stuff.

In the 19th century, sets were regarded as any collection of objects that could be described in *any terms* — and elements of sets could be any objects that one could describe.

In [108], the philosopher Bertrand Russell pointed out a problem with this very loose definition of a set:

> Since sets could theoretically be elements of themselves, let S be the set of all sets that do *not* contain themselves as an element. The question is: Is S an element of S? If the answer is *yes*, then S *cannot* be an element of S. If the answer is *no*, then S *must* be an element of S.

> Bertrand Arthur William Russell, 3rd Earl Russell, (1872 – 1970) was a British philosopher, logician, mathematician, historian, writer, social critic and political activist. He is well-known for his political leaning toward pacifism, initiating the Russell–Einstein Manifesto opposing the development of atomic weapons. His main mathematical work is [108].

The result of this is that set theory today is defined in terms of *axioms:*

14.2. Zermelo-Fraenkel Axioms

We resolve questions like the Russell Paradox via a system of axioms that define what we mean by sets. We present an abbreviated version of these axioms. In these axioms, symbols like $\in$ are boolean-valued operations so, if x is *not* a set, $y \in x$ is false for any y. The axioms describe how we are allowed to use these symbols in equations[1].

AXIOM 14.2.1 (Extensionality). *Two sets are equal (are the same set) if they have the same elements:*

$$\forall_{x,y} [\forall_z (z \in x \Leftrightarrow z \in y) \implies x = y]$$

This is straightforward but has to be said.

---

[1]Roughly speaking, concrete "definitions" of these symbols in terms of other objects are called *models* of set theory. In [24], P. Cohen shocked the mathematical world by showing that multiple *inequivalent* models exist for set theory (in one of which all sets are *countable*).

AXIOM 14.2.2 (Regularity). *Every nonempty set x contains an element y such that x and y are disjoint.*

$$\forall_x \left[\exists_a (a \in x) \implies \exists_y (y \in x \land \forall_z (z \in x \implies z \notin y)) \right]$$

The element $y \in x$ with this property is said to be $\in$-*minimal*. This axiom has a number of interesting consequences:

(1) No set can be a member of *itself* (preventing the Russell Paradox). Suppose A is a set and consider the set $\{A\}$. The Regularity Axiom implies that there exists a $y \in \{A\}$ that is disjoint from it. Since the *only* element of $\{A\}$ is A, we have $y = A$ and

$$A \in \{A\} \implies A \notin y = A$$

(2) No *infinite descending chain* of set-containments[2] exists. Let f be a function defined on the natural numbers such that
   (a) $f(n)$ is a set for all n and
   (b) $f(n+1) \in f(n)$ for all n
Let $S = \{f(n) | n \in \mathbb{N}\}$. If $y \in S$ is any element, $y = f(k)$ for some k and $f(k+1) \in y$, so $f(k+1) \in f(k) \cap S$, violating the Regularity Axiom. This only happens for *infinite* chains — something that might crudely be represented by

$$\left\{ \{^\infty x\}^\infty, \{^{\infty-1} x\}^{\infty-1}, \dots \right\}$$

where one is to imagine an infinite number of brackets enclosing the x.

Given a *finite chain* like

$$\{\{\{\{x\}\}\}, \{\{x\}\}, \{x\}, x\}$$

we have

$$x \in \{x\} \in \{\{x\}\} \in \{\{\{x\}\}\}$$

and there's a minimal element (x, in this case) that doesn't contain any of the others.

AXIOM 14.2.3 (Specification). *Given a set, one can define a subset by imposing a condition on its elements. In other words, one is allowed to use constructions like*

$$\{x \in S | \phi(x)\}$$

to define sets, where ϕ is some logical function and S is a set.

This is what is called an axiom *schema* since there are an infinite number of possible functions ϕ and this gives rise to an axiom for each of them.

One consequence of this is that *empty sets* exist. If x is any set, we can define a set

$$\{y \in x | y \neq y\}$$

and the Axiom of Extensionality implies that any two such sets are equal, so we have a *unique* empty set, denoted $\emptyset$.

AXIOM 14.2.4 (Pairing). *Given two sets x and y there exists a set, z, that contains x and y as elements*

$$\forall_{x,y} \exists_z (x \in z \land y \in z)$$

AXIOM 14.2.5 (Union). *The union over the elements of a set exists and is a set.*

$$\forall_z \exists_u \forall_{x,y} ((x \in y) \land (y \in z)) \implies x \in u$$

---

[2] In this context "set-containment" means one set is an *element* of another — not a subset.

AXIOM 14.2.6 (Replacement). *The image of a set under any definable function lies in a set.*

This is also an axiom schema since the set of possible functions is infinite.

AXIOM 14.2.7 (Infinity). *There exists a set with infinitely many elements.*

$$\exists_x \left[\varnothing \in x \wedge \left(\forall_y y \in x \implies (y \cup \{y\}) \in x \right) \right]$$

AXIOM 14.2.8 (Power set). *Given a set, S, there exists a set whose elements are all of the subsets of S (where a subset of S is defined as a set whose elements are contained in S)*

AXIOM 14.2.9 (Well-ordering). *Any set, S, can be well-ordered. This means that there exists a binary order-relation, $\prec$, such that*
- *if $x, y \in S$ with $x \neq y$, then $x \prec y$ or $y \prec x$ (but not both),*
- *if $T \subset S$ is a subset then T has a minimal element — i.e., there exists $x \in T$ such that $x \prec y$ for all $y \in T$ with $y \neq x$.*

REMARK 14.2.10. The well-ordering axiom is completely trivial for finite sets.

For *infinite* sets, it has many subtle ramifications. For instance, if $S = \mathbb{Q}$, the well-ordering cannot possibly be the *usual* ordering, $<$, of rational numbers because an open interval like $(0,1)$ has no minimal element. We can construct a well-ordering of the positive rationals: If $p/q \in \mathbb{Q}$ is *reduced* (i.e. $\gcd(p,q) = 1$), define $h(p/q) = |p| + |q|$ and order the rationals by this h-function. When two numbers have the same value of h, order them by the denominator. In this ordering, the *minimal* element of $(0,1)$, is $1/2$.

At the time of this writing, there is *no known* well-ordering of the *real* numbers.

The well-ordering axiom implies another

THEOREM 14.2.11 (Axiom of choice). *Given a set S and a collection of nonempty sets indexed by S: $\{G_\alpha\}$, $\alpha \in S$, there exists a function*

$$f: S \to G = \bigcup_{\alpha \in S} G_\alpha$$

that selects an element $f(\alpha) \in G_\alpha \subset G$.

REMARK. This is often stated as

The Cartesian product

$$\prod_{\alpha \in S} G_\alpha$$

is nonempty.

Although this also seems self-evident, it has bizarre implications for infinite families of sets. It is possible to prove the well-ordering axiom from the axiom of choice and the other axioms of set theory, so the two are *equivalent*. This proof is beyond the scope of this book.

PROOF. Simply well-order G and define $f(\alpha)$ to be a minimal element in $G_\alpha \subset G$. □

The well-ordering axiom also implies Zorn's lemma, a classic result found in [65, 111]:

LEMMA 14.2.12. *If S is a partially-ordered set with the property that every increasing sequence of elements*

$$e_1 \preceq e_2 \preceq \cdots$$

has an least upper bound, then S contains a maximal element ("increasing" means $e_i \neq e_{i+1}$ for all i).

If every decreasing sequence of elements

$$e_1 \succeq e_2 \succeq \cdots$$

has a lower bound, then S has a minimal element.

REMARK. In this context,
(1) "least upper bound" means an element f such that $f \succeq e_i$ for all i and for any other $g \succeq e_i$, $f \preceq g$.
(2) "maximal" means "there is an element $e \in S$ such that there does *not* exist an element $e' \in S$ with $e \preceq e'$ and $e' \neq e$."

Zorn's lemma is *equivalent* to the axiom of choice and the well-ordering axiom (which are equivalent to each other) in set theory. Proving *all* of these equivalences is beyond the scope of this book[3] — see [55] as a general reference.

PROOF. We give a non-rigorous proof that can be made rigorous using transfinite induction and ordinals. Let $P = \mathcal{P}(S) \setminus \{\emptyset\}$ and let

$$\bar{S} = \bigcup_{A \in P} A$$

Now use the Axiom of Choice (14.2.11 on the preceding page) to define a function

$$f: P \to \bar{S}$$

with $f(A) = a \in A$ — this just maps every nonempty subset of S to one of its elements. Define the function

$$F: I \to A \cup \{A\}$$

where I is some ordered collection of indices to be defined later, via

$$F(\alpha) = \begin{cases} f(\{s \in S | F(\beta) \preceq s, \text{ for all } \beta < \alpha\}) & \text{if this set is nonempty} \\ A & \text{otherwise} \end{cases}$$

If S is *finite* and $I = \mathbb{Z}$, then $F(n)$ will be A for some value of n — we will have constructed an *ascending chain* that exhausts all of the elements of S with elements $\preceq$ all other elements of S. Since this chain has an upper bound, $F(n-1)$ will be maximal.

If S is *infinite*, the idea is basically the same except that I consists of *infinite ordinals*[4] — the totality of which is known *not* to constitute a set. We also need transfinite induction to properly define F. It turns out that there exists an ordinal α such that $F(\alpha) = A$ and $\alpha = \beta + 1$ and $F(\beta)$ is the maximal element. □

---

[3] We proved the "easy" one in theorem 14.2.11 on the previous page.
[4] See [55] as a general reference, again.

CHAPTER 15

Further topics in ring theory

"Geometry is one and eternal shining in the mind of God. That share in it accorded to men is one of the reasons that Man is the image of God."
— Johannes Kepler, *Conversation with the Sidereal Messenger (an open letter to Galileo Galilei),* [94].

This appendix deals with topics that are more specialized than those considered in chapter 5 on page 107. They have applications to Algebraic Geometry and other areas.

15.1. Discrete valuation rings

In this section we define a class of rings that is important in algebraic geometry. Their main property is that they have an especially simple ideal-structure.

Krull introduced them with the concept of *valuation* in his work on algebraic number theory in [62]. A *valuation* on a field is a function that can be used to define a metric on this field. We have already seen an example of this in claim 5.1.8 on page 109 — the function $v(x)$, there, is an example of a valuation.

DEFINITION 15.1.1. Let F be a field and let $F^\times \subset F$ denote the subset of nonzero elements. A *discrete valuation* on F is a surjective function

$$v: F^\times \to \mathbb{Z}$$

with the properties:
(1) $v(x \cdot y) = v(x) + v(y)$ for all $x, y \in F^\times$
(2) $v(x + y) \geq \min(v(x), v(y))$

REMARK. Statement 2 implies that $v(1) = v(1 \cdot 1) = v(1) + v(1) = 0$. If $0 < \alpha < 1$ is some real number, it is not hard to see that

$$\alpha^{v(*)}: F^\times \to [0, 1]$$

defines a *metric* on F, where we define the metric of 0 to be 0.

The definition of valuation easily implies the following properties:

PROPOSITION 15.1.2. *Let F be a field with valuation $v: F^\times \to \mathbb{Z}$. Then*
(1) $v(x^{-1}) = -v(x)$ *for all $x \in F^*$, since $v(x \cdot x^{-1}) = v(1) = 0 = v(x) + v(x^{-1})$.*
(2) $v(-1) = 0$, *because $v(1) = v((-1) \cdot (-1)) = v(-1) + v(-1)$ if the characteristic of F is $\neq 2$. If it is 2, then $-1 = 1$, so the statement still holds.*
(3) $v(-x) = v(x)$, *because $-x = x \cdot (-1)$.*
(4) *if $v(x) > v(y)$ then $v(x + y) = v(y)$. Certainly, it must be $\geq v(y)$ but, if we write $y = x + y - x$, we get $v(y) \geq \min(v(x + y), v(x))$.*

It is easy to find examples of valuations:

EXAMPLE 15.1.3. If $F = \mathbb{Q}$ and $p \in \mathbb{Z}$ is any prime then we can define the p-adic valuation, v_p, as follows:

For any $q \in \mathbb{Q}$ we have a unique representation
$$q = \prod p_i^{n_i}$$
where the p_i are primes and $n_i \in \mathbb{Z}$ are integers (which are negative if a prime only occurs in the denominator of q). If $p = p_j$, define
$$v_p(q) = n_j$$

It is well-known that the p-adic valuations constitute *all* of the discrete valuations on $\mathbb{Q}$ — see [61].

If a field, F, has a valuation
$$v \colon F^\times \to \mathbb{Z}$$
proposition 15.1.2 on the previous page implies that the set of elements $x \in F$ with $v(x) \geq 0$ form a ring, i.e., are closed under addition and multiplication.

Other interesting examples are provided by power-series rings and variants

EXAMPLE 15.1.4. If k is a field, $R = k[[X]]$ is the ring power-series in X and $F = k((X))$ is the field of fractions of R, exercise 6 on page 251 implies that every element in F can be written uniquely in the form
$$f = X^\alpha \cdot r$$
with $r \in R$. It follows that F is a field with valuation given by $v(f) = \alpha$. The subring of elements with a valuation ≥ 0 is precisely $R \subset F$.

There are valuations that are *not* discrete in the sense above:

EXAMPLE 15.1.5. We can also define the field of *Puiseux series*, discovered by Isaac Newton in 1676 ([81]) and rediscovered by Victor Puiseux ([91]) in 1850:
$$k\{\{X\}\} = \bigcup_{n=1}^\infty k((X^{1/n}))$$

An argument analogous to that used in the power-series case implies that every element of $k\{\{X\}\}$ can be uniquely written in the form
$$f = X^q \cdot \left(a_0 + a_1 X^{1/n} + a_2 X^{2/n} + \cdots\right)$$
for some $n \in \mathbb{Z}$, $n \geq 1$, some $q \in \mathbb{Q}$, and $a_0 \neq 0 \in k$. We can define the *valuation* of f to be q.

REMARK. If k is algebraically closed and of characteristic 0, it turns out that $k\{\{X\}\}$ is the *algebraic closure* of $k((X))$. Newton sketched a proof in a letter he wrote in 1676. See [86] for a short modern proof.

DEFINITION 15.1.6. Let R be an integral domain with field of fractions F. Then R is a *discrete valuation ring* if there exists a valuation
$$v \colon F^\times \to \mathbb{Z}$$
such that
$$R = \{x \in F \mid v(x) \geq 0\}$$

This ring has an ideal
$$\mathfrak{m} = \{x \in F \mid v(x) > 0\}$$

The notation for a discrete valuation ring is $(R, \mathfrak{m})$.

REMARK. The properties of a valuation (in proposition 15.1.2 on page 467) imply that m is an ideal and that all $x \in R \setminus \mathfrak{m}$ are units, so m is the unique maximal ideal, and R is a local ring.

For the p-adic valuation on $\mathbb{Q}$, the corresponding discrete valuation ring is $R_p \subset \mathbb{Q}$ of fractions whose denominator is relatively prime to p (when it is reduced to the lowest form). The maximal ideal is $p \cdot R_p$.

As mentioned above, discrete valuation rings have an extremely simple ideal-structure:

LEMMA 15.1.7. *Let $(R, \mathfrak{m})$ be a discrete valuation ring defined by a valuation*
$$v: F^\times \to \mathbb{Z}$$
on the field of fractions, F, of R. Then there exists an element $r \in R$ such that $\mathfrak{m} = (r)$ and all ideals of R are of the form (r^n) for $n \in \mathbb{Z}^+$.

PROOF. Suppose $u \in R$ has $v(u) = 0$. Then $v(u^{-1}) = 0$ also, so u is a unit. If $\mathfrak{I} \subset R$ is an ideal, let $x \in \mathfrak{I}$ be the element with the smallest valuation. If $y \in \mathfrak{I}$, then $x^{-1}y \in F$ has a valuation $v(y) - v(x) \geq 0$ so $x^{-1}y \in R$ and $y = x \cdot x^{-1}y$ and $\mathfrak{I} = (x)$, and all ideals are principal. It follows that $\mathfrak{m} = (r)$ and $v(r) = 1$ (since the valuation-map is surjective).

Suppose $y \in R$ has the property that $v(y) = n$. Then $r^{-n}y$ has valuation 0 so it is a unit and $(y) = (r^n)$. □

It is interesting to determine the properties a general ring must have to be a discrete valuation ring:

LEMMA 15.1.8. *Let R be a noetherian local domain with maximal ideal $\mathfrak{m} \in R$ and suppose that this is the only prime ideal (other than the trivial prime ideal, (0)). Then R is a discrete valuation ring if and only if it is integrally closed in its field of fractions.*

PROOF. First, we show that a discrete valuation ring is integrally closed. If F is the field of fractions of R and $x/y \in F$ is integral over R, then
$$(x/y)^n + a_{n-1}(x/y)^{n-1} + \cdots + a_0 = 0$$
with the $a_i \in R$. If $v(x) < v(y)$ then
$$v((x/y)^n + a_{n-1}(x/y)^{n-1} + \cdots + (x/y)a_1) = v(-a_0) \geq 0$$
Proposition 15.1.2 on page 467 implies that
$$v((x/y)^n + a_{n-1}(x/y)^{n-1} + \cdots + (x/y)a_1) = v((x/y)^n) < 0$$
which is a contradiction. It follows that $v(x) \geq v(y)$ and $x/y \in R$.

Now we work in the *other direction:* assume R satisfies the hypotheses (i.e., it is a noetherian local domain with a unique prime ideal) and show that it is a discrete valuation ring if it is integrally closed.

For every $u \in \mathfrak{m}$ and $v \in R \setminus (u)$ define

(15.1.1) $\qquad (u{:}v) = \{r \in R | rv \in (u)\}$

This is easily verified to be an ideal and nonempty (since it contains u at least). Let $(a{:}b)$ be the maximal such ideal (with respect to inclusion). We claim that it is a prime ideal. If $xy \in (a{:}b)$, then $xyb \in (a)$. Note that $(a{:}yb) \supseteq (a{:}b)$. If $x, y \notin (a{:}b)$, then $yb \notin (a)$ and $x \in (a{:}yb)$ so $(a{:}yb) \supsetneq (a{:}b)$, which contradicts the maximality of $(a{:}b)$.

Since m is the only prime ideal of R, we have $\mathfrak{m} = (a{:}b)$. We claim that $\mathfrak{m} = (a/b)$ (so $b|a$). Equation 15.1.1 for $(a{:}b)$ implies that $(b/a) \cdot (a{:}b) = (b/a) \cdot \mathfrak{m} \subset R$.

If $(b/a) \cdot \mathfrak{m} \neq R$ then $(b/a) \cdot \mathfrak{m}$ must be an *ideal* of R, hence $(b/a) \cdot \mathfrak{m} \subset \mathfrak{m}$. Since R is noetherian, $\mathfrak{m}$ must be a finitely generated R-module. Since (b/a) maps a finitely-generated R-module to itself, proposition 6.5.2 on page 252 implies that b/a is integral over R, hence *in* R (because R is integrally closed). This is a contradiction (by the condition above equation 15.1.1 on the previous page), so we conclude that $(b/a)\mathfrak{m} = R$ and $\mathfrak{m} = R \cdot (a/b) = (a/b)$.

We claim that *all* ideals in R are principal. If not, there is a maximal non-principal ideal $\mathfrak{J}$ (because R is noetherian). We must have

$$\mathfrak{J} \subset \mathfrak{m} = (a/b)$$

Now consider

$$\mathfrak{J} \subset (b/a) \cdot \mathfrak{J} \subset (b/a) \cdot \mathfrak{m} = R$$

If $\mathfrak{J} = (b/a) \cdot \mathfrak{J}$, then by the reasoning above and proposition 6.5.2 on page 252, we conclude that (b/a) is integral over R, hence in R. This is the same contradiction as before (with $\mathfrak{m}$) and we conclude that

$$\mathfrak{J} \subsetneq (b/a) \cdot \mathfrak{J}$$

which implies that the ideal $(b/a) \cdot \mathfrak{J}$ is principal, say $(b/a) \cdot \mathfrak{J} = (x)$. Then we get $\mathfrak{J} = (x \cdot a/b)$ which is a contradiction.

We conclude that all ideals are principal, and that R is a unique factorization domain by remark 5.3.12 on page 119. The element $\pi = a/b$ that defines $\mathfrak{m}$ must be irreducible and a prime, so we can define a function

$$v: R \setminus \{0\} \to \mathbb{Z}$$

by setting $v(x)$ to the highest power of π that divides x. This extends to a valuation

$$\begin{aligned} v: F^\times &\to \mathbb{Z} \\ v(x/y) &= v(x) - v(y) \end{aligned}$$

and R is a discrete valuation ring. □

15.2. Metric rings and completions

Rings with metrics arise naturally in areas of analysis and number theory. We begin by defining a metric space.

DEFINITION 15.2.1. A metric space, M, is a set of points equipped with a function, $d(x,y) \in \mathbb{R}$, for every $x, y \in M$ with the properties
  (1) $d(x,y) \geq 0$ for all $x, y \in M$ and $d(x,y) = 0$ if and only if $x = y$
  (2) $d(x,y) = d(y,x)$ for all $x, y \in M$.
  (3) Given $x, y, z \in M$, $d(x,z) \leq d(x,y) + d(y,z)$ (Triangle inequality).

REMARK. It is well-known that $\mathbb{Q}$ and $\mathbb{R}$ are metric spaces with $d(x,y) = |x - y|$. We have many more examples of metric rings because:

PROPOSITION 15.2.2. *If R is a discrete valuation ring with valuation $v: F^\times \to \mathbb{Z}$ and $\alpha \in (0,1)$, then $v(x,y)$ defined by $d(x,y) = |x - y|$, where $|*|$ is defined by*
  (1) $|0| = 0$ *for all* $x \in R$
  (2) *if* $x \in R$, *and* $x \neq 0$, *then* $d(x,y) = \left(\frac{1}{2}\right)^{v(x)}$

is a metric on R.

REMARK. Note that this "absolute value" has the property that $|x \cdot y| = |x| \cdot |y|$.

PROOF. This follows by the properties of a valuation in proposition 15.1.2 on page 467: □

DEFINITION 15.2.3. An infinite sequence of points, $\{a_n\}$, in a metric space approaches a *limit a* if, for every $\epsilon > 0$, there exists an integer $N(\epsilon)$ such that
$$|a_i - a| < \epsilon$$
for all $i > N(\epsilon)$ (where $|*|$ denotes distance). A sequence that has a limit is said to *converge*.

We need a criterion for a limit to exist. It's certainly necessary for the terms of the sequence to get arbitrarily close to each other:

DEFINITION 15.2.4. An infinite sequence of points in a metric space $\{a_n\}$ is said to be *Cauchy* if, for every $\epsilon > 0$, there exists an integer $N(\epsilon)$ such that
$$|a_i - a_j| < \epsilon$$
for all $N(\epsilon) < i < j$. If every Cauchy sequence in a metric space, M, converges, we say that M is *complete*.

REMARK. Clearly, any convergent sequence is Cauchy.

15.3. Graded rings and modules

A graded ring is a kind of ring subdivided into distinct direct summands. These appear in the study of projective varieties and sheaf-cohomology.

DEFINITION 15.3.1. A ring, G, is called a *graded ring* over k if there exists a decomposition
$$G = G_0 \oplus G_1 \oplus \cdots$$
such that $G_i \cdot G_j \subset G_{i+j}$ for all $i, j \geq 0$. If $\mathfrak{I} \subset G$ is an ideal, we say that $\mathfrak{I}$ is a *graded ideal* if
$$\mathfrak{I} = \mathfrak{I}_0 \oplus \mathfrak{I}_1 \oplus \cdots$$
where $\mathfrak{I}_j = \mathfrak{I} \cap G_j$ for all $j \geq 0$. An ideal $\mathfrak{H} \subset G$ is *homogeneous* if all of its generators come from the same G_n for some n.

REMARK. Any ring, R, can be regarded as a graded ring, if we define $R_0 = R$, $R_i = 0$ for $i > 0$. A polynomial ring, $k[X_1, \ldots, X_t]$ is naturally graded with gradation given by the total degree of a monomial (where we have specified the degree of *each* of the X_i):
$$k[X_1, \ldots, X_t] = k \oplus K_1 \oplus K_2 \oplus \cdots$$
where K_n is the vector space generated by all monomials of total degree n. For instance, let $G = k[X, Y]$ where X and Y are of degree 1 and let $H = k[X, Y]$ where X is of degree 1 and Y is of degree 2. Then:
$$\begin{aligned} G_0 &= k \\ G_1 &= k \cdot \{X, Y\} \\ G_2 &= k \cdot \{X^2, XY, Y^2\} \\ &\vdots \end{aligned}$$
and
$$\begin{aligned} H_0 &= k \\ H_1 &= k \cdot X \\ H_2 &= k \cdot Y \\ &\vdots \end{aligned}$$
so G and H are isomorphic as rings but not as *graded* rings.

Given graded algebras over a ring R, we can define the graded tensor product

DEFINITION 15.3.2. If A, B are graded algebras over a (non-graded) ring, R, the tensor product is the graded algebra $A \otimes_R B$ defined by

$$(A \otimes_R B)_n = \bigoplus_{i+j=n} A_i \otimes_R B_j$$

REMARK. This definition is consistent with the convention

$$(a \otimes b) \cdot (c \otimes d) = (ac \otimes bd)$$

It is not hard to see that:

PROPOSITION 15.3.3. *If $\mathfrak{H} \subset G$ is a homogeneous ideal of a graded ring G, it is not hard to see that*

(15.3.1) $$\frac{G}{\mathfrak{H}} = \bigoplus_{i=0}^{\infty} \frac{G_i + \mathfrak{H}}{\mathfrak{H}}$$

is also a graded ring.

Here is a standard construction of a graded ring:

DEFINITION 15.3.4. If $\mathfrak{a} \subset R$ is an ideal in a ring, we can define a graded ring

$$\Gamma(\mathfrak{a}) = R \oplus \mathfrak{a} \oplus \mathfrak{a}^2 \oplus \cdots = R[T \cdot \mathfrak{a}] \subset R[T]$$

by giving R a degree of 0, and $\mathfrak{a}$ a degree of 1 — or, equivalently, giving T a degree of 1. This is called the *Rees algebra* of $\mathfrak{a}$.

We begin with

DEFINITION 15.3.5. If G is a graded ring (see definition 15.3.1 on the previous page), a module, M, over G is a *graded module* if

$$M = M_0 \oplus M_1 \oplus \cdots$$

with the property $G_i \cdot M_j \subset M_{i+j}$ for all $i, j \geq 0$.

If ℓ is an integer and M is a graded module over a graded ring, we define the ℓ-twist of M, denoted $M(\ell)$ is defined by

$$M(\ell)_i = M_{i+\ell}$$

REMARK. Although any k-algebra could be regarded as a graded algebra (put it all in G_0), some have a natural grading. For instance,

$$G = k[X_0, \ldots, X_n]$$

is naturally a graded ring by degrees of monomials, i.e., G_k consists of homogeneous polynomials of degree k. This grading is geometrically significant.

It is not hard to see

PROPOSITION 15.3.6. *If $\mathfrak{J} \subset G$ is a graded ideal in a graded algebra, the quotient*

$$R = \frac{G}{\mathfrak{J}}$$

is naturally a graded algebra with

(15.3.2) $$R_j = \frac{G_j}{\mathfrak{J}_j}$$

for all j.

REMARK. Graded ideals are just graded submodules of G, regarding it as a graded module over itself. In general, all of the results of section 6 on page 163 have versions for graded modules over graded algebras, given the following

LEMMA 15.3.7. *Let $R = R_0 \oplus R_1 \oplus \cdots$ be a graded ring and let M be a graded module over R. If $m \in M$ and $\mathfrak{p} = \mathrm{ann}(m) \subset R$ is prime, then $\mathfrak{p}$ is homogeneous and $\mathfrak{p}$ is the annihilator of a homogeneous element of M.*

PROOF. If $r \in \mathfrak{p}$, we have a unique expression $r = \sum_{i=1}^s r_i$ where r_i is homogeneous of degree d_i, with $d_1 < d_2 < \cdots < d_s$. We will prove that $\mathfrak{p}$ is homogeneous by showing that $r \in \mathfrak{p}$ implies that all of the $r_i \in \mathfrak{p}$. By induction on s, it suffices to show that $r_1 \in \mathfrak{p}$.

Similarly, we have a unique expression $m = \sum_{j=1}^t m_j$ with m_j homogeneous of degree e_i with $e_1 < \cdots < e_t$. We claim that $r_1 \cdot m_1 = 0$, since this is the term in $r \cdot m = 0$ of lowest degree. This proves the result in the case where $t = 1$. Now suppose it has been proved for all smaller values of t. The element

$$r_1 \cdot m = \sum_{j=2}^t r_1 \cdot m_j$$

is a sum of $< t$ homogeneous components. Let $\mathfrak{q} = \mathrm{ann}(r_1 \cdot m)$. By induction, we conclude that $\mathfrak{q}$ is homogeneous if it is prime, and $\mathfrak{p} \subseteq \mathfrak{q}$. If $\mathfrak{p} = \mathfrak{q}$, we are done. Otherwise, let $g \in \mathfrak{q} \setminus \mathfrak{p}$. Then $g \cdot r_1 \cdot m = 0$ so $gr_1 \in \mathrm{ann}(m) = \mathfrak{p}$. Since $\mathfrak{p}$ is prime and $g \notin \mathfrak{p}$, we conclude that $r_1 \in \mathfrak{p}$, and $\mathfrak{p}$ is homogeneous.

Now, since $\mathfrak{p}$ is homogeneous, $\mathfrak{p} \cdot m_j = 0$ for all j, so

$$\mathfrak{p} = \mathrm{ann}(m) \supset \bigcap_{j=1}^t \mathrm{ann}(m_j) \supset \mathfrak{p}$$

which implies that $\mathfrak{p} = \bigcap_{j=1}^t \mathrm{ann}(m_i) \supset \prod_{j=1}^t \mathrm{ann}(m_j)$. The fact that $\mathfrak{p}$ is prime implies that $\mathfrak{p} \supset \mathrm{ann}(m_j)$ (see exercise 5 on page 115) for some j, which means that $\mathfrak{p} = \mathrm{ann}(m_j)$. □

With this in hand, we can easily generalize prime filtrations of modules to *graded* modules:

LEMMA 15.3.8. *Let M be a graded module over a graded ring, R. Then there exists a finite ascending chain of graded-submodules*

$$0 = M_0 \subset M_1 \subset \cdots \subset M_t = M$$

with the property that for each i

$$\frac{M_{i+1}}{M_i} \cong \frac{R}{\mathfrak{p}_i}(\ell_i)$$

where $\mathfrak{p}_i \subset R$ is a homogeneous prime ideal and ℓ_i is an integer.

PROOF. The proof is exactly the same as that of theorem 6.3.30 on page 242, except that we use lemma 15.3.7 to guarantee that the ideals $\{\mathfrak{p}_i\}$ are all homogeneous so that the quotients $R/\mathfrak{p}_i$ are now graded rings. The ℓ_i occur because the natural grading of $R/\mathfrak{p}_i$ may be shifted in forming iterated quotients. □

We can also conclude something about other filtrations of modules:

DEFINITION 15.3.9. *If M is a module over a ring R and $\mathfrak{a} \subset R$ is an ideal, a filtration*

$$\cdots \subset M_t \subset \cdots \subset M_0 = M$$

if called an $\mathfrak{a}$-filtration if $\mathfrak{a} \cdot M_n \subseteq M_{n+1}$ for all $n \geq 0$. It is called a stable $\mathfrak{a}$-filtration if $\mathfrak{a} \cdot M_n = M_{n+1}$ for all $n > n_0$.

REMARK. Note that these conditions only apply from some finite point on.

We can define a kind of module-analogue of the Rees algebra:

DEFINITION 15.3.10. If M is a module over a ring, R, with a filtration
$$\cdots \subset M_t \subset \cdots \subset M_0 = M$$
define the Rees module of this to be
$$\Gamma(M) = M_0 \oplus M_1 \oplus \cdots$$

REMARK. If the filtration is an $\mathfrak{a}$-filtration for some ideal $\mathfrak{a} \subset R$ then $\Gamma(M)$ is naturally a graded-module over $\Gamma(\mathfrak{a})$, since $\mathfrak{a}^t \cdot M_n \subseteq M_{n+t}$.

One of our main results is:

LEMMA 15.3.11. *If M is a finitely generated module over a ring R with an $\mathfrak{a}$-filtration*
$$\cdots \subset M_t \subset \cdots \subset M_0 = M$$
by finitely-generated submodules, for some ideal $\mathfrak{a} \subset R$, then $\Gamma(M)$ is finitely generated over $\Gamma(\mathfrak{a})$ if and only if this filtration is stable.

PROOF. If $\Gamma(M)$ is finitely generated over $\Gamma(\mathfrak{a})$, then for some n
$$M_0 \oplus \cdots \oplus M_n$$
generates all of $\Gamma(M)$. Since the filtration is an $\mathfrak{a}$-filtration, we have $\mathfrak{a}^i \cdot M_{n-i} \subseteq M_n$, so we can really say that M_n generates
$$M_n \oplus M_{n+1} \oplus \cdots$$
and considerations of grading imply that $M_{n+i} = \mathfrak{a}^i \cdot M_n$ so the filtration is stable from degree n on.

Conversely, if the filtration is stable from degree n on, then $M_{n+i} = \mathfrak{a}^i \cdot M_n$ so that $\Gamma(M)$ generated by $M_0 \oplus \cdots \oplus M_n$ over $\Gamma(\mathfrak{a})$. □

The main result of this section is

THEOREM 15.3.12 (Artin-Rees Theorem). *Suppose R is a noetherian ring with ideal $\mathfrak{a} \subset R$. If $N \subset M$ are finitely generated R-modules and M has a stable $\mathfrak{a}$-filtration*
$$\cdots \subset M_t \subset \cdots \subset M_0 = M$$
then the filtration
$$\cdots \subset M_t \cap N \subset \cdots \subset M_0 \cap N = N$$
is also a stable $\mathfrak{a}$-filtration. In other words there exists an integer n such that
$$\left(\mathfrak{a}^i \cdot M_n\right) \cap N = \mathfrak{a}^i \cdot (M_n \cap N)$$

PROOF. Since R is noetherian, $\mathfrak{a}$ is finitely generated and $\Gamma(\mathfrak{a})$ is a finitely-generated R-algebra, hence noetherian. Since the filtration on M is stable, $\Gamma(M)$ is finitely-generated over $\Gamma(\mathfrak{a})$. It is not hard to see that, computed with respect to the induced filtration, $\Gamma(N) \subset \Gamma(M)$, which means that it is *also* finitely generated (see lemma 6.3.14 on page 227). The conclusion follows from lemma 15.3.11). □

Solutions to Selected Exercises

Chapter 3, 3.1 Exercise 1 (p. 18) They are numbers k such that $k \cdot \ell \equiv 1 \pmod{m}$ for some ℓ such that $0 < \ell < m$, or

$$k \cdot \ell = 1 + n \cdot m$$

or

$$k \cdot \ell - n \cdot m = 1$$

The proof of lemma 3.1.5 on page 14 implies that the smallest positive value attained by linear combinations of ℓ and m is their greatest common divisor — in this case, 1. It follows that an integer $0 < k < m$ is a unit in $\mathbb{Z}_m$ if and only if it is relatively prime to m.

Chapter 3, 3.1 Exercise 2 (p. 18) We compute $123 = 27 q_1 + r_1$ with

$$\begin{aligned} q_1 &= 4 \\ r_1 &= 15 \end{aligned}$$

Now $27 = 15 q_2 + r_2$ with

$$\begin{aligned} q_2 &= 1 \\ r_2 &= 12 \end{aligned}$$

In stage 3, $15 = 12 q_3 + r_3$ with

$$\begin{aligned} q_3 &= 1 \\ r_3 &= 3 \end{aligned}$$

The process terminates at this point so $\gcd(27, 123) = 3$. Now we apply 3.1.12 on page 17 to calculate a and b: $x_0 = 0$, $y_0 = 1$, $x_1 = 1$, $y_1 = -q_1 = -4$ and

$$\begin{aligned} x_2 &= x_0 - q_2 x_1 = -1 \\ y_2 &= y_0 - q_2 y_1 = 5 \\ x_3 &= x_1 - q_3 x_2 = 2 \\ y_3 &= y_1 - q_3 y_2 = -9 \end{aligned}$$

So

$$3 = 2 \cdot 123 - 9 \cdot 27$$

Chapter 3, 3.1 Exercise 3 (p. 18) This follows immediately from the fact that integers uniquely factor into primes. Suppose

$$x = \prod p_i^{\alpha_i}$$

is an expression where the p_i are primes. Clearly x^y can be an integer: $64^{1/2} = 8$. Suppose x^y is not an integer but rational, i.e.

$$\left(\prod p_i^{\alpha_i} \right)^{\frac{R}{S}} = \frac{\prod a_k^{m_k}}{\prod b_\ell^{n_\ell}}$$

where $\prod b_\ell^{n_\ell} \neq 1$ and where the a_j and the b_ℓ are disjoint sets of primes. Then

$$(\prod p_i^{\alpha_i})^R = \frac{\prod a_k^{m_k \cdot S}}{\prod b_\ell^{n_\ell \cdot S}}$$

or

$$\prod p_i^{\alpha_i \cdot R} = \frac{\prod a_k^{m_k \cdot S}}{\prod b_\ell^{n_\ell \cdot S}}$$

$$\prod p_i^{\alpha_i \cdot R} \cdot \prod b_\ell^{n_\ell \cdot S} = \prod a_k^{m_k \cdot S}$$

This is a contradiction because the set of primes $\{b_\ell\}$ is disjoint from the $\{a_k\}$.

Chapter 3, 3.2 Exercise 1 (p. 21) It is not hard to see that $7 \equiv 2 \pmod{5}$. The fact that modulo arithmetic preserves products implies that

$$7^k \equiv 2^k \pmod{5}$$

for all k.

Chapter 3, 3.2 Exercise 3 (p. 21) Suppose k has the property that $d \cdot k \equiv 0 \pmod{n}$ but k is *not* a multiple of n/d. Proposition 3.1.1 on page 13 implies that we can write

$$k = q \cdot \frac{n}{d} + r$$

where $0 < r < n/d$. If we multiply this by d we get

$$dk = qn + dr$$
$$\equiv dr \pmod{n}$$

But $dr < n$ since $r < n/d$ so that $dr \not\equiv 0 \pmod{n}$, a contradiction.

Chapter 3, 3.2 Exercise 4 (p. 21) Note that

$$p - \left(\frac{p-1}{2} - k\right) = \frac{p+1}{2} + k$$

so

$$\frac{p-1}{2} - k \equiv -\left(\frac{p+1}{2} + k\right) \pmod{p}$$

so

$$(z!)^2 = 1 \cdot 2 \cdots \frac{p-1}{2} \cdot 1 \cdot 2 \cdots \frac{p-1}{2}$$
$$= 1 \cdot 2 \cdots \frac{p-1}{2} \cdot \frac{p-1}{2} \cdots 2 \cdot 1$$
$$\equiv 1 \cdot 2 \cdots \frac{p-1}{2} \cdot \left(-\frac{p+1}{2}\right)$$
$$\cdots \left(-\left(\frac{p+1}{2} + k\right)\right) \cdots - (p-1) \pmod{p}$$

Chapter 3, 3.3 Exercise 1 (p. 25) Since $52 = 4 \times 13$

$$\phi(52) = \phi(4)\phi(13)$$
$$= \phi(2^2) \cdot 12$$
$$= (4-2) \cdot 12$$
$$= 24$$

Chapter 3, 3.3 Exercise 2 (p. 25) This is just 7^{1000} (mod 100) — so we must compute $\phi(100)$. Since $100 = 2^2 \cdot 5^2$, we get
$$\phi(100) = (2^2 - 2)(5^2 - 5) = 40$$
Since $40 \mid 1000$, we conclude that
$$7^{1000} \equiv 7^0 = 1 \pmod{100}$$
so the lowest *two* digits are '01'.

Chapter 3, 3.4 Exercise 1 (p. 26) In this case $(p-1)(q-1) = 120$ so we may pick $n = 7$. To find the corresponding m, we must find integers a and b such that
$$a \cdot 7 + b \cdot 120 = 1$$
Since $120 = 17 \cdot 7 + 1$, we conclude that $a = -17$ and $b = 1$. It follows that
$$(-17)7 \equiv (120 - 17)7 \equiv 1 \pmod{120}$$
so $m = 120 - 17 = 103$.

Chapter 4, 4.1 Exercise 1 (p. 38) This follows from the basic property of an identity element (statement 1 in definition 4.1.2 on page 34:
$$\begin{aligned} 1_1 \cdot 1_2 &= 1_1 \text{ using } g \cdot 1 = g \\ &= 1_2 \text{ using } 1 \cdot g = g \end{aligned}$$

Chapter 4, 4.1 Exercise 2 (p. 38) Left-multiply the equation by a^{-1}:
$$\begin{aligned} a^{-1}ab &= a^{-1}ac \\ 1 \cdot b &= 1 \cdot c \\ b &= c \end{aligned}$$

Chapter 4, 4.1 Exercise 3 (p. 38) If $a = R$ and $b = c_1$, then table 4.1.1 on page 37 shows that
$$\begin{aligned} Rc_1 &= d_2 \\ &= d_2^{-1} \\ R^{-1}c_1^{-1} &= d_1 \end{aligned}$$

Chapter 4, 4.1 Exercise 4 (p. 38) In the equation $ab = 1$, multiply on the left by a^{-1}. We get
$$\begin{aligned} a^{-1}ab &= a^{-1} \\ 1b &= a^{-1} \\ b &= a^{-1} \end{aligned}$$
so the conclusion follows from statement 2 in definition 4.1.2 on page 34.

Chapter 4, 4.1 Exercise 5 (p. 38) Just multiply them together
$$\begin{aligned} ab \cdot b^{-1}a^{-1} &= a \cdot 1 \cdot a^{-1} \\ &= 1 \end{aligned}$$

Chapter 4, 4.1 Exercise 6 (p. 38) These are elements $x \in \mathbb{Z}_{10}$ with the property that $n \cdot x$ ranges over all of the elements of $\mathbb{Z}_{10}$ as $n \in \mathbb{Z}$ varies. In particular, there is a value of n such that $n \cdot x = 1 \in \mathbb{Z}_{10}$. Example 3.2.6 on page 20 shows that these are the elements x such that $\gcd(x, 10) = 1$ or $\{1, 3, 7, 9\}$.

Chapter 4, 4.1 Exercise 7 (p. 38) It is easy to see that it is closed under multiplication.

Chapter 4, 4.1 Exercise 11 (p. 39) If we form all possible powers of g, we get a set $\{g, g^2, g^3, \dots\}$. Since G is finite, we must have $g^j = g^k$ for some values of j and k. If $k > j$, multiply this equation by g^{-j} to get
$$g^{k-j} = 1$$

Chapter 4, 4.1 Exercise 13 (p. 39) Besides $(0, 0)$ and the entire Klein 4-group, there are the subgroups
$$\mathbb{Z}_2 \times 0, \text{ and } 0 \times \mathbb{Z}_2$$
and the subgroup generated by $(1, 1)$.

Chapter 4, 4.2 Exercise 1 (p. 40) This follows from the fact that $f(1)$ has the property that its product with $f(g) \neq 1$ is $f(g)$:
$$f(1)f(g) = f(1g) = f(g)$$
$$f(1)f(g)f(g)^{-1} = f(g)f(g)^{-1}$$
$$f(1) = 1$$

Chapter 4, 4.2 Exercise 2 (p. 40) This follows from the fact that
$$f(g^{-1})f(g) = f(g^{-1}g) = f(1) = 1$$

Chapter 4, 4.2 Exercise 3 (p. 40) If $k_1, k_2 \in \ker f$, then $f(k_1 k_2) = f(k_1)f(k_2) = 1 \cdot 1 = 1$ so $k_1 k_2 \in \ker f$.

Chapter 4, 4.2 Exercise 7 (p. 40) These isomorphisms are both immediate consequences of the Chinese Remainder Theorem 3.3.5 on page 24.

Chapter 4, 4.4 Exercise 1 (p. 47) If $h_1 k_1$ and $h_2 k_2$ are elements of HK, $h_1 k_1 h_2 k_2 = h_1 h_2 h_2^{-1} k_2 h_2$. Since K is normal, $h_2^{-1} k_2 h_2 = k' \in K$ and $h_1 k_1 h_2 k_2 = h_1 h_2 k' \in HK$.

Chapter 4, 4.4 Exercise 2 (p. 47) Certainly, $K \subset HK$. Since it is normal in the *whole group* G, it is also normal in the smaller group HK.

Chapter 4, 4.4 Exercise 3 (p. 47) If $x \in H \cap K$ and $h \in H$, then $hxh^{-1} \in H$ because H is a subgroup, and $hxh^{-1} \in K$ because $x \in K$ and K is normal. It follows that $hxh^{-1} \in H \cap K$.

Chapter 4, 4.4 Exercise 4 (p. 48) Consider the cosets used in forming the quotient HK/K — they are of the form $hk \cdot K = h \cdot K$. It follows that the map
$$h \mapsto h \cdot K$$
defines a surjective homomorphism
$$f: H \to \frac{HK}{K}$$
An element $h \in H$ is in the kernel if and only if $h \in K$, i.e. if and only if $h \in H \cap K$. The conclusion follows from the First Isomorphism Theorem, 4.4.8 on page 45.

Chapter 4, 4.4 Exercise 5 (p. 48) If $K \subset H$, then $p(H) \subset G/K$ is a subgroup. If $W \subset G/K$ is a subgroup, then
$$p^{-1}(W) = \bigcup_{x \in G, p(x) \in W} xK$$
contains K and is closed under multiplication and the act of taking inverses, so it is a subgroup of G. If $W \triangleleft G/K$, it is not hard to see that $p^{-1}(W)$ is normal in G.

Chapter 4, 4.4 Exercise 6 (p. 48) The statement that $H/K \triangleleft G/K$ follows from the Correspondence Theorem (exercise 5 on page 48). We define a homomorphism
$$p: G \to \frac{G/K}{H/K}$$

by $p(g) = g \cdot K \cdot H$, which equals $g \cdot H$ since $K \subset H$. Since this homomorphism is clearly surjective, the conclusion follows from theorem 4.4.8 on page 45.

Chapter 4, 4.4 Exercise 8 (p. 48) We have to verify that $Z(G)$ is closed under multiplication. If $z_1, z_2 \in Z(G)$, and $g \in G$ is an arbitrary element, then

$$g z_1 z_2 = z_1 g z_2 = z_1 z_2 g$$

so $z_1 z_2 \in Z(G)$.

Chapter 4, 4.4 Exercise 9 (p. 48) Since the elements of $Z(G)$ commute with everything in G, they also commute with each other.

Chapter 4, 4.4 Exercise 10 (p. 48) The second isomorphism theorem implies that

$$p_1 \colon G \to \frac{G}{H} \cong K$$
$$p_2 \colon G \to \frac{G}{K} \cong H$$

and the map

$$(p_1, p_2) \colon G \to K \times H$$

is an isomorphism.

Chapter 4, 4.4 Exercise 11 (p. 48) If $x \in G$ then theorem 4.4.2 on page 43 implies that the order of x is a divisor of 4, in other words, 2 or 4. If $\text{ord}(x) = 4$ then $G = \{1, x, x^2, x^3\}$ and G is cyclic. If G has no elements of order 4, its elements (other than 1) must be of order 2. Suppose $G = \{1, x, y, z\}$. Then $xy \neq x$ and $xy \neq y$ so $xy = z$. Similar reasoning shows that $yx = z$, $xz = zx = y$, $yz = zy = x$ and G is abelian.

Chapter 4, 4.4 Exercise 12 (p. 48) This follows from the fact that $g h_1 g^{-1} \cdot g h_2 g^{-1} = g h_1 h_2 g^{-1}$, so H^g is closed under the group-operation.

Chapter 4, 4.4 Exercise 13 (p. 48) Fix an element $g \in G$. We must show that the map

$$\begin{aligned} f \colon G &\to G \\ x &\mapsto x^g \end{aligned}$$

is an automorphism. This means

(1) it is a *homomorphism*: $x_1^g x_2^g = g x_1 g^{-1} g x_2 g^{-1} = g(x_1 x_2) g^{-1} = (x_1 x_2)^g$.
(2) it is *injective*: $x_1^g = x_2^g \implies x_1 = x_2$.
(3) it is *surjective* $x \in G \implies x = (x^{g^{-1}})^g$.

Chapter 4, 4.4 Exercise 14 (p. 48) Every automorphism is bijective, so it has an inverse. The composition of two automorphisms is an automorphism.

Chapter 4, 4.4 Exercise 15 (p. 48) Any automorphism of $\mathbb{Z}_m$ (or any cyclic group, for that matter) is uniquely determined by its effect on 1. If $f \in \text{Aut}(\mathbb{Z}_m)$ and $f(1) = k \in \mathbb{Z}_m$ 1to $k \in \mathbb{Z}_m$ then $f(n) = k \cdot n$ and multiplication by k is an automorphism if and only if $k \in \mathbb{Z}_m^\times$.

Chapter 4, 4.4 Exercise 16 (p. 48) The composite of two inner automorphisms is an inner automorphism:

$$(x^{g_1})^{g_2} = g_2 (g_1 x g_1^{-1}) g_2^{-1} = x^{g_2 g_1}$$

Chapter 4, 4.4 Exercise 17 (p. 48) Suppose $f \colon G \to G$ is an automorphism of G. If $x, g \in G$ then conjugating by an automorphism, f, means computing the composite

$f((f^{-1}x)^g)$, given by

$$f(f^{-1}(x)^g) = f(gf^{-1}(x)g^{-1})$$
$$= f(g)xf(g^{-1})$$
$$= x^{f(g)}$$

which is still an inner automorphism.

Chapter 4, 4.4 Exercise 18 (p. 48) It is not hard to see that $[g_1, g_2]^h = [g_1^h, g_2^h]$.

Chapter 4, 4.4 Exercise 19 (p. 49) Let

$$p: G \to \frac{G}{[G,G]}$$

If $g_1, g_2 \in G$, then $p([g_1, g_2]) = 1 = p(g_1)p(g_2)p(g_1)^{-1}p(g_2)^{-1}$. Multiplying the equation

$$p(g_1)p(g_2)p(g_1)^{-1}p(g_2)^{-1} = 1$$

on the right by $p(g_2)$ and then $p(g_1)$ gives the equation

$$p(g_1)p(g_2) = p(g_2)p(g_1)$$

in the quotient.

Chapter 4, 4.4 Exercise 20 (p. 49) Let $g, h \in G$. Since A is abelian $p(g)p(h)p(g)^{-1}p(h)^{-1} = p(ghg^{-1}h^{-1}) = 1$, which implies that $ghg^{-1}h^{-1} \in K$.

Chapter 4, 4.5 Exercise 3 (p. 58) This follows from a straightforward induction on the number of cycles in g and proposition 4.5.11 on page 54 and corollary 4.5.12 on page 54.

Chapter 4, 4.5 Exercise 4 (p. 58) If $x \in Z(S_3)$, then $xy = yx$ for any $y \in S_3$, or $xyx^{-1} = y$ for all $y \in S_3$. But xyx^{-1} has the same cycle-structure as y with elements mapped via x — see proposition 4.5.11 on page 54. If $x \neq 1$ it follows that $y \neq xyx^{-1}$, so $Z(S_3) = \{1\}$, the trivial group.

Chapter 4, 4.5 Exercise 5 (p. 58) Just consider the effect each element of D_8 has on the vertices: the rotation is the permutation $(1,2,3,4)$, the reflections are $(1,2)(3,4)$, $(1,3)$, $(1,4)(2,3)$, and $(2,4)$.

Chapter 4, 4.5 Exercise 6 (p. 58) By re-indexing, if necessary, we can assume the 5-cycle is $(1,2,3,4,5)$ and that the transposition is $(1,i)$. If $i = 5$, we can reduce to the case where $i = 2$ because

(15.3.3) $$(1,5)^{(1,2,3,4,5)} = (1,2)$$

If $i = 4$ we get

$$(1,4)^{(1,2,3,4,5)^2} = (1,3)$$

Claim: we can *always* assume $(1,2) \in G$. If $(1,3) \in G$, then note that

$$(1,3)^{(1,2,3,4,5)^2} = (3,5)$$

and

$$(1,3)(3,5)(1,3) = (1,5)$$

and the conclusion follows from equation 15.3.3.

It follows that $(1,2) \in G$ and

$$(1,2)^{(1,2,3,4,5)^{i-1}} = (i, i+1)$$

implies that all the adjacent transpositions lie in G as well. The conclusion follows from corollary 4.5.6 on page 52.

Chapter 4, 4.6 Exercise 1 (p. 72) Note that $60 = 2^2 \cdot 3 \cdot 5$ so that theorem 4.6.10 on page 65 implies that there are *two* abelian groups of order 60:
$$\mathbb{Z}_2 \oplus \mathbb{Z}_2 \oplus \mathbb{Z}_3 \oplus \mathbb{Z}_5$$
$$\mathbb{Z}_4 \oplus \mathbb{Z}_3 \oplus \mathbb{Z}_5$$

Chapter 4, 4.6 Exercise 2 (p. 72) Note that $x \in \mathbb{Q}^\times$ can be *uniquely* written in the form
$$x = p_1^{k_1} \cdots p_n^{k_n}$$
where the p_i are prime numbers. It follows that the prime numbers form a basis of Q^+.

Chapter 4, 4.6 Exercise 3 (p. 73) If A has a *countable* basis, it is essentially
$$\bigoplus_{i=1}^{\infty} \mathbb{Z} = \bigcup_{n=1}^{\infty} \left(\bigoplus_{i=1}^{n} \mathbb{Z} \right)$$
which is a countable union of countable sets — therefore *countable*. It follows that an *uncountable* free abelian group has a *uncountable* basis so $A/2A$ is an uncountable direct sum of copies of $\mathbb{Z}_2$ and is uncountable itself.

Chapter 4, 4.6 Exercise 4 (p. 73) Just map the sequence
$$(n_1, \dots)$$
to
$$(m_1, \dots)$$
where $m_i = 2^i n_i$. This defines a homomorphism $B \to S$, which shows that S is uncountable.

Chapter 4, 4.6 Exercise 5 (p. 73) Since every infinite sequence in S
$$(n_1, \dots)$$
is eventually divisible by an arbitrarily high power of 2, we can eventually *divide* it by 2 to get another sequence in S. Given a sequence in S
$$x = (n_1, \dots, n_k, \dots)$$
there exists another sequence
$$(0, \dots, 0, n_{k+1}/2, n_{k+2}/2, \dots)$$
in S. When we take the quotient, x will be in the *same coset* as
$$(n_1, \dots, n_k, 0, \dots)$$
This means that
$$\frac{S}{2S} \cong \bigoplus_{i=1}^{n} \mathbb{Z}_2$$
which is countable.

Chapter 4, 4.6 Exercise 6 (p. 73) If the Baer-Specker group, B, is free abelian, proposition 4.6.6 on page 61 implies that the subgroup, S, defined in exercise 4 on page 73 is also free abelian. Then exercise 3 on page 73 implies that $S/2S$ will be uncountable. Since exercise 5 on page 73 shows that this is *not* the case, we get a contradiction.

Chapter 4, 4.7 Exercise 1 (p. 80) Since G has order p^2, proposition 4.7.14 on page 79 implies that $Z(G)$ is of order p or p^2. If it is of order p^2, we are done, since the center is abelian. If it is of order p then
$$\frac{G}{Z(G)} = \mathbb{Z}_p$$

and every element of G can be written in the form $g = z \cdot v^i$ where $z \in Z(G) = \mathbb{Z}_p$ and v maps to a generator of the quotient. All such elements commute with each other.

Chapter 4, 4.7 Exercise 3 (p. 80) The equation
$$(n_1, h_1) \cdot (n_2, h_2) = (n_1 \varphi(h_1)(n_2), h_1 h_2) = (1,1)$$
shows that $h_2 = h_1^{-1}$ and $\left(\varphi(h_1)^{-1}(n_1)\right)^{-1} = \varphi(h_1^{-1})(n_1^{-1}) = n_2$, so $(n,h)^{-1} = (\varphi(h^{-1})(n^{-1}), h^{-1})$.

Chapter 4, 4.7 Exercise 4 (p. 80) This is direct computation
$$\begin{aligned}(1,h) \cdot (n,1) \cdot (1,h)^{-1} &= (\varphi(h)(n), h_1) \cdot (1,h)^{-1} \\ &= (\varphi(h)(n), h) \cdot (\varphi(h^{-1})(1), h^{-1}) \\ &= \left(\varphi(h)(n)\varphi(h)\left(\varphi(h^{-1})(1)\right), 1\right) \\ &= (\varphi(h)(n), 1)\end{aligned}$$

Chapter 4, 4.7 Exercise 5 (p. 80) This is straight computation
$$\begin{aligned}(n_1, h_1) \cdot (n_2, h_2) &= n_1 h_1 n_2 h_2 \\ &= n_1 h_1 n_2 h_1^{-1} h_1 h_2 \\ &= n_1 (h_1 n_2 h_1^{-1}) h_1 h_2 \\ &= n_1 \varphi(h_1)(n_2) h_1 h_2\end{aligned}$$

Chapter 4, 4.8 Exercise 1 (p. 84) The first Sylow Theorem implies that G has a subgroups of order 5 and of order 7 — both of which are cyclic (because 5 and 7 are primes — see proposition 4.4.4 on page 43). The second Sylow theorem states that all 5 element subgroups are conjugate, and all 7-element subgroups are conjugate. The third Sylow theorem states that the number of distinct 7-element subgroups must divide 5 and be congruent to 1 modulo 7 — i.e. there must be only one of them, so the 7-element subgroup is normal. It also follows that the 5-element subgroup is normal. If x generates the 5-element subgroup and y generates the 7-element one, then conjugation by y induces an *automorphism* of the 5-element subgroup. The group of automorphisms of $\mathbb{Z}_5$ is $\mathbb{Z}_4$ (see exercise 15 on page 48), so if this automorphism is not the identity, its order must divide 4 — which is impossible.

Chapter 4, 4.8 Exercise 2 (p. 84) Since $70 = 2 \cdot 5 \cdot 7$ we have Sylow 2-groups, 5-groups, and 7-groups. If n_7 is the number of distinct Sylow 7-groups, the third Sylow theorem states that $n_7 \mid 10$ and $n_7 \equiv 1 \pmod{7}$. The only possible value that satisfies these conditions is $n_7 = 1$. Since all conjugates of this Sylow 7-group are 7-groups, it follows that this Sylow subgroup is *normal*.

Chapter 4, 4.9 Exercise 1 (p. 87) Let
$$\{1\} \subset G_0 \triangleleft G_1 \triangleleft \cdots \triangleleft G_k \triangleleft G$$
be a composition series with each $G_{i+1}/G_i \cong \mathbb{Z}_{p_{i+1}}$ with p_i a prime. Then G_k is a subgroup with $[G : G_k] = p_k$, a prime.

Chapter 4, 4.9 Exercise 2 (p. 87) Clearly definition 4.9.7 on page 86 implies the new definition. Conversely, if a group, G, has a subnormal series
$$\{1\} \subset G_0 \subset G_1 \subset \cdots \subset G \tag{15.3.4}$$
with $G_{i+1}/G_i = A_{i+1}$, an abelian group for all i, we can *refine* it (i.e. add more terms) to get a composition series for G that satisfies definition 4.9.7 on page 86. Fix

a value of i and let
$$p: G_i \to A_i = A$$
be the projection to the quotient. If
$$\{1\} \subset L_0 \triangleleft L_1 \triangleleft \cdots \triangleleft L_k \triangleleft A$$
is a composition series for A, let $P_j = p^{-1}(L_j)$. It is not hard to see that $P_0 = G_{i-1}$ and we replace the portion $G_{i-1} \triangleleft G_i$ of equation 15.3.4 on the preceding page by
$$G_{i-1} \triangleleft P_1 \triangleleft \cdots \triangleleft P_k \triangleleft G_i$$
The Third Isomorphism Theorem (see exercise 6 on page 48) implies that
$$\frac{P_{j+1}}{P_j} \cong \frac{L_{j+1}}{L_j}$$
so the quotients are all cyclic groups of prime order.

The conclusion follows by performing this refinement-operation on *every* stage of 15.3.4 on the preceding page.

Chapter 4, 4.9 Exercise 4 (p. 87) Since G is solvable, there exists a composition series
$$\{1\} \subset G_0 \triangleleft G_1 \triangleleft \cdots \triangleleft G_k \triangleleft G$$
with each factor G_k/G_{k+1} cyclic — in particular *abelian*. We claim that this implies that $[G_k, G_k] \subset G_{k+1}$ since $[G_k, G_k]$ is the *smallest* normal subgroup of G_k giving an abelian quotient (see exercise 20 on page 49). Since $H \subset K$ implies that $[H, H] \subset [K, K]$, a simple induction implies that *all* of the terms of the *derived* series of G are *contained* in corresponding terms of the *composition* series (allowing for the fact that they are indexed in opposite ways!).

Since the *composition* series terminates with $\{1\}$ in a finite number of steps, so must the *derived* series.

Conversely, if the derived series terminates in a finite number of steps, it is a subnormal series with abelian quotient groups, so the conclusion follows from exercise 2 on page 87.

Chapter 4, 4.9 Exercise 5 (p. 87) Both conclusions follow from exercise 4 on page 87. If G is solvable and $H \subset G$ with derived series
$$\cdots \subset G_n \subset \cdots \subset G_1 \subset G_0 = G$$
and
$$\cdots \subset H_n \subset \cdots \subset H_1 \subset H_0 = H$$
Since $H \subset K$ implies that $[H, H] \subset [K, K]$, a simple induction shows that $H_i \subset G_i$ for all i. Since the derived series of G terminates in a finite number of steps, the same is true of the derived series of H.

If
$$p: G \to H$$
is a surjective homomorphism, then $p([G, G]) = [H, H]$ and a simple induction shows that $p(G_i) = H_i$. Again, the finite termination of G's derived series implies that of H.

Chapter 4, 4.9 Exercise 6 (p. 87) Since G/H is solvable, let
$$\{1\} \subset Q_0 \triangleleft Q_1 \triangleleft \cdots \triangleleft Q_k \triangleleft Q = G/H$$
be its composition series. This "lifts" to a series in G,
$$H \triangleleft G_0 \triangleleft G_1 \triangleleft \cdots \triangleleft G_k \triangleleft G$$
and we splice this with a composition series for H to get a complete one for G.

Chapter 4, 4.10 Exercise 1 (p. 93) This follows from the basic property of a free group, theorem 4.10.5 on page 88, and exercise 5 on page 80.

Chapter 4, 4.10 Exercise 2 (p. 93) The group $(\mathbb{Q},+)$ is generated by elements $\{1/n!, n \in \mathbb{Z}^+\}$. Define a homomorphism

$$f\colon G \to (\mathbb{Q},+)$$

via

$$1 \mapsto 0$$
$$x_n \mapsto 1/n!$$
$$x_n^{-1} \mapsto -1/n!$$

where multiplication in G corresponds to addition in $(\mathbb{Q},+)$. This is compatible with the relations because $n \cdot 1/n! = 1/(n-1)!$. It is clearly surjective, so we must analyze the kernel. Note:

G is *abelian*. This follows from the fact that in a product, $x_i x_j$, both x_i and x_j are powers of $x_{\max(i,j)}$ so they commute.

Suppose

(15.3.5)
$$g = x_1^{\alpha_1} \cdots x_n^{\alpha_n}$$

is in the kernel of f (where, without loss of generality, we assume the letters are arranged in order of increasing subscripts). This means

$$\alpha_1 + \alpha_2/2! + \cdots + \alpha_n/n! = 0$$
$$(n!\alpha_1 + (n!/2!)\alpha_2 + \cdots + \alpha_n)/n! = 0$$

We can also do the same for the expression in 15.3.5 to get

$$g = x_n^{n!\alpha_1 + (n!/2!)\alpha_2 + \cdots + \alpha_n} = x_n^0 = 1$$

Chapter 4, 4.10 Exercise 5 (p. 104) We know H is free, because of corollary 4.10.26 on page 104. It cannot be free on a finite number of generators because $x^i y^i$ cannot be written in terms of $x^j y^j$ for all values of $j \neq i$.

Chapter 5, 5.2 Exercise 2 (p. 115) Since $(x,y) = R$ it follows that there exist $a, b \in R$ such that

$$ax + by = 1$$

or $ax = 1$ in $R/(b)$. This means that $a^n x^n = 1$ in $R/(b)$ so that $(x^n, y) = R$. A similar argument with the image of b in $R/(a)$ implies the conclusion.

Chapter 5, 5.2 Exercise 3 (p. 115) Suppose α has a multiplicative inverse

$$\beta = \sum_{j=0}^{\infty} b_j X^j$$

Then the product is

$$\alpha \cdot \beta = \sum_{n=0}^{\infty} c_n X^n$$

where $c_0 = a_0 \cdot b_0 = 1$.

Chapter 5, 5.2 Exercise 5 (p. 115) If $\mathfrak{a} \not\subset \mathfrak{p}$ and $\mathfrak{b} \not\subset \mathfrak{p}$ then there exists $x \in \mathfrak{a}$ with $x \notin \mathfrak{p}$ and $y \in \mathfrak{b}$ with $y \notin \mathfrak{p}$. The product, xy, will be in $\mathfrak{a} \cdot \mathfrak{b}$ so $xy \in \mathfrak{p}$. This contradicts the definition of a prime ideal (see definition 5.2.3 on page 111).

Chapter 5, 5.2 Exercise 6 (p. 116) Suppose $x \cdot y \in \mathfrak{p}$ but $x \notin \mathfrak{p}$. Then there exists an integer n such that, for $i > n$, $x \notin \mathfrak{p}_i$. The fact that $\mathfrak{p}_i$ is prime implies that $y \in \mathfrak{p}_i$ for $i > n$, so $y \in \mathfrak{p}$.

Chapter 5, 5.2 Exercise 7 (p. 116) The ideal (X) consists of polynomials such that the X-degree of every monomial is ≥ 1. If $a(X) \cdot b(X) \in (X)$, each monomial of $a(X) \cdot b(X)$ must have X-degree ≥ 1. If $a(X)$ and $b(X)$ both contain a monomial of X-degree 0, the product of those monomials will also have X-degree zero and $a(X) \cdot b(X) \notin (X)$.

This ideal is not maximal because it is contained in the proper ideal (X, Y).

Chapter 5, 5.2 Exercise 8 (p. 116) This map clearly preserves addition. It remains to show that $f(x \cdot y) = f(x) \cdot f(y)$ for all $x, y \in \mathbb{Q}[\sqrt{2}]$. If

$$x = a + b\sqrt{2}$$
$$y = c + d\sqrt{2}$$

are two elements, then

$$xy = ac + 2bd + (ad + bc)\sqrt{2}$$

and

$$f(x) = a - b\sqrt{2}$$
$$f(y) = c - d\sqrt{2}$$

so

$$f(x) \cdot f(y) = ac + 2bd - (ad + bc)\sqrt{2} = f(x \cdot y)$$

Chapter 5, 5.2 Exercise 9 (p. 116) If

$$x = a + b\sqrt{2}$$
$$y = c + d\sqrt{2}$$

are two elements of $\mathbb{Q}[\sqrt{2}]$, then

$$xy = ac + 2bd + (ad + bc)\sqrt{2}$$

If we set $c = a$ and $d = -b$, so $y = a - b\sqrt{2}$, then we get

$$xy = a^2 - 2b^2 \in \mathbb{Q}$$

and the $\sqrt{2}$ term is zero. It follows that

$$(a + b\sqrt{2})^{-1} = \frac{a - b\sqrt{2}}{a^2 - 2b^2}$$

If $a + b\sqrt{2} \neq 0$, denominator is nonzero since $\sqrt{2}$ is irrational.

Chapter 5, 5.2 Exercise 10 (p. 116) If r is *not* a unit, $(r) \subset R$ is a proper ideal and there exists a *maximal* ideal $\mathfrak{m}$ such that $(r) \subset \mathfrak{m}$. But $r - 1 \in \mathfrak{J} \subset \mathfrak{m}$ (since $\mathfrak{J}$ is the intersection of all maximal ideals), so r and $r - 1$ are *both* contained in $\mathfrak{m}$. This implies that $r - (r - 1) = 1 \in \mathfrak{m}$, which contradicts the fact that $\mathfrak{m}$ is a proper ideal of R.

Chapter 5, 5.2 Exercise 11 (p. 116) Without loss of generality assume $i = 1$ and write

$$R = (\mathfrak{a}_1 + \mathfrak{a}_2)(\mathfrak{a}_1 + \mathfrak{a}_3) \cdots (\mathfrak{a}_1 + \mathfrak{a}_n)$$

When carrying out the multiplication, all but one term in the product has a factor of $\mathfrak{a}_1$, hence is *contained* in $\mathfrak{a}_1$ (by the defining property of an ideal — see definition 5.2.3 on page 111). The *one* exception is the term $\mathfrak{a}_1 + \prod_{j=2}^{n} \mathfrak{a}_j$ — and this contains $\mathfrak{a}_1$ and *all* of the other terms. It follows that

$$R = (\mathfrak{a}_1 + \mathfrak{a}_2) \cdots (\mathfrak{a}_1 + \mathfrak{a}_n) \subseteq \mathfrak{a}_1 + \prod_{j=2}^{n} \mathfrak{a}_j$$

so
$$\mathfrak{a}_1 + \prod_{j=2}^{n} \mathfrak{a}_j = R$$

Chapter 5, 5.2 Exercise 12 (p. 116) We get a natural map

(15.3.6) $$R \to \prod_{i=1}^{n} \frac{R}{\mathfrak{a}_i}$$

that sends $x \in R$ to $(p_1(x), \ldots, p_n(x))$, where p_i is the natural projection

$$p_i \colon R \to \frac{R}{\mathfrak{a}_i}$$

The kernel of the map in 15.3.6 is clearly $\mathfrak{a}$. It only remains to show that this map is *surjective*. Use the solution to exercise 11 on page 116 to conclude that

$$\mathfrak{a}_i + \prod_{j \neq i} \mathfrak{a}_j = R$$

for all i. This means that, for each i, there is an element $u_i \in \mathfrak{a}_i$ and $v_i \in \prod_{j \neq i} \mathfrak{a}_j$ such that $u_i + v_i = 1$. It is not hard to see that

$$\begin{aligned} p_i(v_i) &= 1 \\ p_j(v_i) &= 0 \end{aligned}$$

for any $j \neq i$. If

$$(x_1, \ldots, x_n) \in \prod_{i=1}^{n} \frac{R}{\mathfrak{a}_i}$$

is an arbitrary element, set

$$x = \sum_{i=1}^{n} x_i v_i$$

Then $p_i(x) = p_i(v_i x_i) = x_i$ so the map is surjective.

Chapter 5, 5.3 Exercise 1 (p. 121) We work in the ring $\mathbb{F}[x]$. Definition 5.3.3 on page 118 and example 5.3.4 on page 118 implies that $\alpha^n = 1$ in $\mathbb{F}$ if and only if $x - \alpha | (x^n - 1)$. Each such $x - \alpha$ is an irreducible factor of $x^n - 1$ and we get

$$x^n - 1 = (x - \alpha_1)(x - \alpha_2) \cdots (x - \alpha_k) p(x)$$

where $p(x)$ is a product of the other irreducible factors. Corollary 5.3.7 on page 118 implies that this factorization is *unique*, so $k \leq n$.

Chapter 5, 5.3 Exercise 2 (p. 121) Two functions $g_1, g_2 \in C[0,1]$ map to the same element of the quotient $C[0,1]/\mathfrak{f}_a$ if and only if $g_1(a) = g_2(a)$. It follows that $C[0,1]/\mathfrak{f}_a \cong \mathbb{R}$. Since this is a field, lemma 5.3.2 on page 117 implies that $\mathfrak{f}_a$ must be maximal.

Chapter 5, 5.3 Exercise 3 (p. 121) We start by dividing the larger polynomial by the smaller one to get

$$\begin{aligned} b(X) &= q_1(X) \cdot a(X) + r_1(X) \\ q_1(X) &= X - 3 \\ r_1(X) &= 10X^3 - 7X^2 + 3X + 8 \end{aligned}$$

Now we compute

$$\begin{aligned} a(X) &= q_2(X) \cdot r_1(X) + r_2(X) \\ q_2(X) &= \frac{1}{10} X^2 - \frac{7}{100} X - \frac{81}{1000} \\ r_2(X) &= -\frac{1577}{1000} X^2 + \frac{683}{1000} X - \frac{706}{125} \end{aligned}$$

Now we divide $r_1(X)$ by $r_2(X)$ to get
$$r_1(X) = q_3(X) \cdot r_2(X) + r_3(X)$$
$$q_3(X) = \frac{10000}{1577}X + \frac{4209000}{2486929}$$
$$r_3(X) = \frac{93655000}{2486929}X - \frac{3877000}{2486929}$$

We finally divide $r_2(X)$ by $r_3(X)$ to get
$$r_2(X) = q_4(X)r_3(X) + r_4(X)$$
$$q_4(X) = -\frac{3921887033}{93655000000}X + \frac{8992165757259}{5482036890625000}$$
$$r_4(X) = \frac{6220545559984}{1096407378125}$$

Since this is a *unit* of $\mathbb{Q}[X]$, it shows that $a(X)$ and $b(X)$ are relatively prime.

Chapter 5, 5.3 Exercise 4 (p. 121) In R, $X^5 \nmid X^6$ and their common divisors are 1, X^2, and X^3 — none of which is divisible *in R* by the other two.

Chapter 5, 5.3 Exercise 5 (p. 121) We get an injective homomorphism
$$\frac{R}{f^{-1}(\mathfrak{p})} \hookrightarrow \frac{S}{\mathfrak{p}}$$
Since $\mathfrak{p}$ is prime $S/\mathfrak{p}$ is an integral domain — see lemma 5.3.2 on page 117. This means $R/f^{-1}(\mathfrak{p})$ is also an integral domain, which implies that $f^{-1}(\mathfrak{p})$ is prime.

Chapter 5, 5.4 Exercise 1 (p. 125) This is just the pigeonhole principal: Suppose R is an integral domain with n elements and $x \in R$ is nonzero. Multiply all of the nonzero elements of R by x:
$$\{x \cdot y_1, \ldots, x \cdot y_{n-1}\}$$
We claim that these products must all be distinct. If $x \cdot y_i = x \cdot y_j$ then $x \cdot (y_i - y_j) = 0$ and the only way this can happen in an integral domain is for $y_i = y_j$. It follows that 1 must be in this set of products, so $1 = x \cdot y_k$ for some k and $y_k = x^{-1}$.

Chapter 5, 5.4 Exercise 2 (p. 125) This follows immediately from the Ascending Chain Condition and lemma 5.2.9 on page 114.

Chapter 5, 5.4 Exercise 3 (p. 125) The ring $\mathbb{Q}[X,Y]$ is certainly an integral domain. So see that it is not Euclidean note that the two variables X and Y have no common divisors other than 1.

If $\mathbb{Q}[X,Y]$ was a Euclidean ring, it would be possible to find polynomials $a(X,Y)$ and $b(X,Y)$ such that
$$1 = a(X,Y) \cdot X + b(X,Y) \cdot Y$$
This is impossible since we could make the right side of the equation equal to 0 by setting $X = 0$ and $Y = 0$, so we would get
$$1 = 0$$

It is interesting that $\mathbb{Q}[X,Y]$ has unique factorization — see Lemma 5.6.2 on page 147.

Chapter 5, 5.4 Exercise 4 (p. 125) Suppose an ideal $\mathfrak{J}$ contains polynomials $p(X), q(X)$. If these polynomials are relatively prime in $\mathbb{Q}[X]$ then there is a linear combination
$$a(X)p(X) + b(X)q(X) = 1$$
in $\mathbb{Q}[X]$, and after clearing out the denominators, we get
$$n \cdot a(X)p(X) + n \cdot b(X)q(X) = n$$

so this ideal also contains an integer, n. If an ideal *does not* contain any integer then it is not maximal.

The requirement that $\mathbb{Z}[X]/\mathfrak{J}$ is a field (see lemma 5.3.2 on page 117) implies that n is a prime, p. We can compute the quotient of $\mathbb{Z}[X]/\mathfrak{J}$ in two stages:

Form the quotient with respect to (p), forming

$$\mathbb{Z}_p[X]$$

and then taking the quotient by the image of the polynomials in $\mathfrak{J}$. Since $\mathbb{Z}_p[X]$ is a principal ideal domain, we can assume that the image of $\mathfrak{J}$ in $\mathbb{Z}_p[X]$ is a principal ideal $(q(X))$. The quotient

$$\mathbb{Z}_p[X]/(q(X))$$

is a field if and only if $q(X)$ is irreducible. It follows that our maximal ideals of $\mathbb{Z}[X]$ are all of the form

$$(p, q_p(X))$$

where $p \in \mathbb{Z}$ is a prime and $q_p(X)$ has an irreducible image in $\mathbb{Z}_p[X]$. Two such ideals

$$(p, a_p(X)), (p, b_p(X))$$

will be equal if and only if $(a_p(X)) = (b_p(X)) \subset \mathbb{Z}_p[X]$.

Chapter 5, 5.4 Exercise 5 (p. 125) This follows by straightforward induction on n and proposition 5.1.9 on page 109.

Chapter 5, 5.4 Exercise 6 (p. 125) Since R is noetherian, $\mathfrak{N}(R) = (x_1, \ldots, x_n)$ for some finite set of elements of $\mathfrak{N}(R)$. Each of these elements must be nilpotent, i.e.

$$x_i^{\alpha_i} = 0$$

for suitable values of α_i. If $\alpha = \max(\alpha_1, \ldots, \alpha_n)$ then the Pigeonhole Principal implies that

$$\mathfrak{N}(R)^{n \cdot \alpha} = 0$$

Chapter 5, 5.4 Exercise 7 (p. 125) The localization, $R_\mathfrak{p}$, only has one prime ideal, $\mathfrak{p} \cdot R_\mathfrak{p}$, and theorem 12.2.8 on page 421 implies that all of the elements of $\mathfrak{p} \cdot R_\mathfrak{p}$ are nilpotent. If $x \in \mathfrak{p}$, then $x/1 \in \mathfrak{p} \cdot R_\mathfrak{p}$ is nilpotent so that there exists an element, $y \in R \setminus \mathfrak{p}$ such that $y \cdot x^n = 0$ for some n.

Chapter 5, 5.5 Exercise 1 (p. 141) We plug $x = 1/2, y = 1/2$, and $z = 1 + \sqrt{2}/2$ into the ideal $\mathfrak{P}$ in example 5.5.17 on page 139 to give

$$\mathfrak{P}' = (-1 + 2\,b_5{}^2, -b_5 + a_5, -2 - \sqrt{2} + 2\,b_4 + b_4\sqrt{2},$$
$$- 1 + b_4, -2 + 2\,b_4{}^2, a_4, b_3 + b_5 b_4, -\sqrt{2} + 2\,a_3,$$
$$\sqrt{3}b_5 + \sqrt{3}b_5\sqrt{2} - 2\,b_4\sqrt{3}b_5 + 3\,b_2,$$
$$3\,a_2 - 2\sqrt{3} - 1/2\sqrt{2}\sqrt{3} + b_4\sqrt{3})$$

If we take a Gröbner basis of *this*, we get an even simpler representation

$$\mathfrak{P}' = (-1 + 2\,b_5{}^2, -b_5 + a_5, -1 + b_4, a_4, b_3 + b_5, -\sqrt{2} + 2\,a_3,$$
$$3\,b_2 + \left(\sqrt{2}\sqrt{3} - \sqrt{3}\right)b_5, 6\,a_2 - \sqrt{2}\sqrt{3} - 2\sqrt{3})$$

from which we conclude

$$\begin{aligned} a_5 = b_5 &= \pm 1/\sqrt{2} \\ b_4 &= 1 \\ a_4 &= 0 \\ b_3 &= -b_5 \\ a_3 &= 1/\sqrt{2} \\ b_2 &= -b_5(\sqrt{6}-\sqrt{3})/3 \\ a_2 &= (\sqrt{6}+2\sqrt{3})/6 \end{aligned}$$

which gives two solutions:
(1) $\phi_1 = 45°, \theta_1 = 90°, \theta_2 = 315°, \theta_3 = 99.735°$
(2) $\phi_1 = 225°, \theta_1 = 90°, \theta_2 = 45°, \theta_3 = 80.264°$

Note that the two possible values of θ_3 sum up to 180°.

Chapter 5, 5.5 Exercise 2 (p. 141) We start with the same equations as before:

(15.3.7)
$$\begin{aligned} a_5 a_4 a_3 - a_5 b_4 b_3 + a_5 a_4 - x &= 0 \\ b_5 a_4 a_3 - b_5 b_4 b_3 + b_5 a_4 - y &= 0 \\ b_4 a_3 + a_4 b_3 + b_4 - z &= 0 \\ a_3^2 + b_3^2 - 1 &= 0 \\ a_4^2 + b_4^2 - 1 &= 0 \\ a_5^2 + b_5^2 - 1 &= 0 \end{aligned}$$

And we plug in the new directions to get

(15.3.8)
$$\begin{aligned} (a_5 a_4 a_3 - a_5 b_4 b_3) a_2 + (-a_5 a_4 b_3 - a_5 b_4 a_3) b_2 - 1 &= 0 \\ (b_5 a_4 a_3 - b_5 b_4 b_3) a_2 + (-b_5 a_4 b_3 - b_5 b_4 a_3) b_2 &= 0 \\ (b_4 a_3 + a_4 b_3) a_2 + (a_4 a_3 - b_4 b_3) b_2 &= 0 \\ a_2^2 + b_2^2 - 1 &= 0 \end{aligned}$$

The Gröbner basis with lexicographic ordering is

$(y, b_5, a_5^2 - 1,$
$4x^2 b_4^2 - 4x^2 + 2x^2 z^2 - 4x^2 z b_4 + x^4 + z^4 - 4z^3 b_4 + 4z^2 b_4^2,$
$z^2 a_5 - 2za_5 b_4 + x^2 a_5 + 2xa_4,$
$-4xa_5 + 4zb_4 a_4 - 2z^2 a_4 + 4b_4^2 xa_5 + z^2 a_5 x - 2za_5 xb_4 + x^3 a_5,$
$a_4^2 + b_4^2 - 1,$
$2b_4 a_4 - za_4 + b_3 + xa_5 b_4$
$-2 + 4b_4^2 - 4zb_4 + z^2 + 2a_3 + x^2,$
$za_5 - a_5 b_4 + b_2, -a_5 a_4 - x + a_2)$

from which we conclude that $y = 0$ and $a_5 = \pm 1$. The term next to the last implies that

$$x^2 - 4zb_4 + z^2 + 4b_4^2 = x^2 + (z - 2b_4)^2 = 2 - 2a_3$$

which means x and z lie on a circle of radius $\sqrt{2(1-a_3)}$ and center $(0, 2b_4)$. If we specify that $a_3 = c$, some constant and take a further Gröbner basis (not including c

in the list of variables), we get an additional relation between x and z (among other things):
$$(c-1)z^2 + (1+c)x^2 = 0$$
or
$$z = \pm x \sqrt{\frac{1+c}{1-c}}$$
so the reachability set is contained in this pair of lines in the xz-plane (and very small!). The possible values of z are
$$b_4 c + b_4 \pm \sqrt{1 - b_4{}^2 - c^2 + b_4{}^2 c^2}$$
It is interesting that, although the set of points that can be reached is limited, there are many ways to reach each of these points.

Chapter 5, 5.5 Exercise 3 (p. 141) We compute the intersection of the principal ideals generated by these polynomials and take their intersection, using the method of proposition 5.5.15 on page 135: we find a Gröbner basis of the ideal
$$(T(-X^3 - 2YX^2 - XY^2 + 2X),$$
$$(1-T)(4 - 4X^2 - 4Y^2 + X^4 - 2Y^2X^2 + Y^4))$$
using a lexicographic ordering $T \succ X \succ Y$ to get
$$Y^4 X - 2X^3 Y^2 + X^5 - 4XY^2 - 4X^3 + 4X,$$
$$X^4 - 3Y^2X^2 - 2X^2 + TY^2X^2 - 2TX^2 - 2XY^3$$
$$+ 4XY + 2XTY^3 - 4TXY + Y^4 T - 4TY^2 + 4T,$$
$$-X^3 - 2YX^2 - XY^2 + 2X + X^3 T + 2YTX^2 + XTY^2 - 2XT$$
Since the only term that does *not* contain T is the top one, it is the answer.

Chapter 5, 5.5 Exercise 4 (p. 141) No. The basis given for $\mathfrak{a}$ is a Gröbner basis with lexicographic ordering and
$$X + Y \to_\mathfrak{a} 2Y$$
so $X + Y \notin \mathfrak{a}$.

Chapter 5, 5.5 Exercise 5 (p. 141)
$$\begin{aligned} X + Y &\to_\mathfrak{a} X + Y \\ (X+Y)^2 &\to_\mathfrak{a} 4XY \\ (X+Y)^3 &\to_\mathfrak{a} 12XY^2 - 4Y^3 \\ (X+Y)^4 &\to_\mathfrak{a} 32XY^3 - 16Y^4 \\ (X+Y)^5 &\to_\mathfrak{a} 0 \end{aligned}$$
so $(X+Y)^5 \in \mathfrak{a}$.

Chapter 5, 5.5 Exercise 6 (p. 141) We compute the intersection of the principal ideals generated by these polynomials and take their intersection, using the method of proposition 5.5.15 on page 135: we find a Gröbner basis of the ideal
$$(T(-X^3 - 2YX^2 - XY^2 + 2X),$$
$$(1-T)(4 - 4X^2 - 4Y^2 + X^4 - 2Y^2X^2 + Y^4))$$

using a lexicographic ordering $T \succ X \succ Y$ to get

$$Y^4 X - 2X^3 Y^2 + X^5 - 4XY^2 - 4X^3 + 4X,$$
$$X^4 - 3Y^2 X^2 - 2X^2 + TY^2 X^2 - 2TX^2 - 2XY^3$$
$$+ 4XY + 2XTY^3 - 4TXY + Y^4 T - 4TY^2 + 4T,$$
$$- X^3 - 2YX^2 - XY^2 + 2X + X^3 T + 2YTX^2 + XTY^2 - 2XT$$

Since the only term that does *not* contain T is the top one, it is the answer.

Chapter 5, 5.5 Exercise 7 (p. 141) No. The basis given for $\mathfrak{a}$ is a Gröbner basis with lexicographic ordering and

$$X + Y \to_{\mathfrak{a}} 2Y$$

so $X + Y \notin \mathfrak{a}$.

Chapter 5, 5.5 Exercise 8 (p. 141)

$$\begin{aligned} X + Y &\to_{\mathfrak{a}} X + Y \\ (X+Y)^2 &\to_{\mathfrak{a}} 4XY \\ (X+Y)^3 &\to_{\mathfrak{a}} 12XY^2 - 4Y^3 \\ (X+Y)^4 &\to_{\mathfrak{a}} 32XY^3 - 16Y^4 \\ (X+Y)^5 &\to_{\mathfrak{a}} 0 \end{aligned}$$

so $(X+Y)^5 \in \mathfrak{a}$.

Chapter 5, 5.5 Exercise 9 (p. 144) The presence of X^3 implies that we should start with

$$\sigma_1^3 = X^3 + 3X^2 Y + 3X^2 Z + 3Y^2 Z + Y^3 + Z^3 + 6XYZ$$

so

$$X^3 + Y^3 + Z^3 - \sigma_1^3 = -3X^2 Y - 3X^2 Z - 3Y^2 Z - 6XYZ$$

The highest ordered monomial is $-3X^2 Y$, which is the highest ordered monomial of $-3\sigma_1 \sigma_2$ (see equation 5.5.15 on page 143). We get

$$\begin{aligned} X^3 + Y^3 + Z^3 - \sigma_1^3 + 3\sigma_1 \sigma_2 &= 3XYZ \\ &= 3\sigma_3 \end{aligned}$$

so

$$X^3 + Y^3 + Z^3 = \sigma_1^3 - 3\sigma_1 \sigma_2 + 3\sigma_3$$

Chapter 5, 5.5 Exercise 10 (p. 144) The roots of $X^n - 1$ are ω^i, for $0 \le i < n$, where $\omega = e^{2\pi i/n}$.

We start with the computation

$$\begin{aligned} \Delta &= \prod_{i<j}(\omega^i - \omega^j)^2 \\ &= \pm \prod_{i \ne j}(\omega^i - \omega^j) \\ &= \pm \prod_{i \ne j}\omega^i(1 - \omega^{j-i}) \\ &= \pm \prod_i \omega^i \left(\prod_{k \ne 0}(1 - \omega^k) \right) \\ &= \pm \prod_i \omega^i (n) \\ &= \pm n^n \end{aligned}$$

where $\prod_{k \neq 0}(1 - \omega^k) = n$ comes from the fact that it is $h(1)$ for

$$h(X) = \frac{X^n - 1}{X - 1} = X^{n-1} + \cdots + 1$$

which we get from

$$h(X) = \prod_{k \neq 0}(X - \omega^k)$$

Compare equation 5.5.12 on page 142.

Chapter 5, 5.5 Exercise 11 (p. 145) Since Δ is unchanged by any permutation of the α_i, it is a symmetric function of the α_i, therefore a polynomial function of the *elementary* symmetric functions of the α_i by theorem 5.5.18 on page 142. But elementary symmetric polynomials are just the *coefficients* of p_1, by equation 5.5.12 on page 142. A similar argument shows that Δ is a polynomial function of the coefficients of p_2.

Now , regard the α_i and β_j as indeterminates and express $\text{Res}(p_1, p_2, t)$ as a function of the α_i, β_j by plugging elementary symmetric functions in for the coefficients of p_1 and p_2. So

$$\text{Res}(p_1, p_2, t) \in k[\alpha_1, \ldots \alpha_n, \beta_1, \ldots \beta_m]$$

Since $\text{Res}(p_1, p_2, t)$ vanishes whenever an $\alpha_i = \beta_j$, it follows that $(\alpha_i - \beta_j) | \text{Res}(p_1, p_2, t)$ for all i, j. This means that $\Delta | \text{Res}(p_1, p_2, t)$ in $k[\alpha_1, \ldots \alpha_n, \beta_1, \ldots \beta_m]$, and the quotient, q, will *also* be symmetric in the α_i, β_j — hence a polynomial function of the coefficients of p_1 and p_2. Now, note that the degree of each α_i in Δ and in $\text{Res}(p_1, p_2, t)$ is m and the degree of each β_j in both is n. It follows that q is a constant.

Chapter 5, 5.5 Exercise 12 (p. 145) We use induction on n. If $n = 2$, the conclusion is clear. Now we assume the conclusion for n, and we will prove it for $n+1$. We have

$$V_{n+1} = \begin{bmatrix} 1 & \alpha_1 & \alpha_1^2 & \cdots & \alpha_1^n \\ 1 & \alpha_2 & \alpha_2^2 & \cdots & \alpha_2^n \\ 1 & \alpha_3 & \alpha_3^2 & \cdots & \alpha_3^n \\ \vdots & \vdots & \vdots & \ddots & \vdots \\ 1 & \alpha_{n+1} & \alpha_{n+1}^2 & \cdots & \alpha_{n+1}^n \end{bmatrix}$$

and replace α_1 by X. The determinant is a polynomial $p(X)$ that vanishes if $X = \alpha_2, \ldots, \alpha_{n+1}$. It follows that

$$f(X) = C \cdot (X - \alpha_2) \cdots (X - \alpha_{n+1})$$

where the coefficient of X^n is precisely C. Expanding the determinant of V_{n+1} by minors in the first row shows that the coefficient of X^n, or C, is equal to the $(-1)^n \times$ the determinant of

$$V = \begin{bmatrix} 1 & \alpha_1 & \alpha_1^2 & \cdots & \alpha_1^{n-1} \\ 1 & \alpha_2 & \alpha_2^2 & \cdots & \alpha_2^{n-1} \\ 1 & \alpha_3 & \alpha_3^2 & \cdots & \alpha_3^{n-1} \\ \vdots & \vdots & \vdots & \ddots & \vdots \\ 1 & \alpha_n & \alpha_n^2 & \cdots & \alpha_n^{n-1} \end{bmatrix}$$

so
$$\det V_{n+1} = (-1)^n \det V_n \cdot \prod_{j=2}^{n+1}(\alpha_1 - \alpha_j)$$
$$= \det V_n \cdot \prod_{j=2}^{n+1}(\alpha_j - \alpha_1)$$
$$= \prod_{1 \leq i < j \leq n+1}(\alpha_j - \alpha_i)$$

Chapter 5, 5.5 Exercise 13 (p. 145) The map
$$\frac{\mathbb{F}[X_1,\ldots,X_n]}{\mathfrak{J}} \to \frac{\mathbb{F}[X_1,\ldots,X_{n+1}]}{\mathfrak{J} + (X_{n+1} - f(X_1,\ldots,X_n))}$$
$$X_i \mapsto X_i$$
and the inverse is defined similarly, except that
$$X_{n+1} \mapsto f(X_1,\ldots,X_n)$$

Chapter 6, 6.2 Exercise 1 (p. 171) The only one that defines a linear transformation is number 2. The others are either nonlinear (i.e., xy) or have constants added to them so that $f(0,0,0) \neq 0$.

Chapter 6, 6.2 Exercise 2 (p. 171) Brute-force computation!

Chapter 6, 6.2 Exercise 3 (p. 171) This is an immediate consequence of the *associativity* of matrix-multiplication:
$$\underbrace{(A \cdots A)}_{n \text{ times}} \cdot \underbrace{(A \cdots A)}_{m \text{ times}} = A^{n+m} = \underbrace{(A \cdots A)}_{m \text{ times}} \cdot \underbrace{(A \cdots A)}_{n \text{ times}}$$
which implies that the order of the parentheses is irrelevant.

Chapter 6, 6.2 Exercise 4 (p. 171) Since powers of A commute with each other, distributivity implies that scalar linear combinations of them (i.e., *polynomials* in A) will also commute with each other.

Chapter 6, 6.2 Exercise 5 (p. 190) The Sylvester matrix of $t - x(1+t^2)$ and $t^2 - y(1-t)$ is
$$\begin{bmatrix} -x & 1 & -x & 0 \\ 0 & -x & 1 & -x \\ 1 & y & -y & 0 \\ 0 & 1 & y & -y \end{bmatrix}$$
and the determinant is
$$\text{Res}(t - x(1+t^2), t^2 - y(1-t), t) = x^2 + 2yx^2 + 2y^2x^2 + yx - y^2x - y$$
so the implicit equation is
$$x^2 + 2yx^2 + 2y^2x^2 + yx - y^2x - y = 0$$

Chapter 6, 6.2 Exercise 6 (p. 190) These parametric equations are equivalent to $t - x(1-t^2) = 0$ and $t - y(1+t^2) = 0$ with a Sylvester matrix of
$$\begin{bmatrix} x & 1 & -x & 0 \\ 0 & x & 1 & -x \\ -y & 1 & -y & 0 \\ 0 & -y & 1 & -y \end{bmatrix}$$

and resultant of
$$r = 4y^2x^2 - x^2 + y^2$$
so the implicit equation is $r = 0$.

Chapter 6, 6.2 Exercise 7 (p. 190) Our polynomials are $1 - t - x(1+t) = 0$ and $t^2 - y(1+t^2) = 0$ with a Sylvester matrix of

$$\begin{bmatrix} -1-x & 1-x & 0 \\ 0 & -1-x & 1-x \\ 1-y & 0 & -y \end{bmatrix}$$

giving the implicit equation
$$-2y + 1 - 2x - 2yx^2 + x^2 = 0$$

Chapter 6, 6.2 Exercise 8 (p. 190) The resultant in question is
$$x^4 + 2x^3 + x^2 - 4x = x(x-1)(x^2 + 3x + 4)$$
It follows that x can have one of the 4 values
$$\left\{ 0, 1, \frac{-3 \pm i\sqrt{7}}{2} \right\}$$

Each of these x-values turns out to correspond to a *unique* y-value. Our four solutions are
$$(x,y) = \left\{ (0,1), (1,0), \left(\frac{-3-i\sqrt{7}}{2}, \frac{3-i\sqrt{7}}{2} \right), \left(\frac{-3+i\sqrt{7}}{2}, \frac{3+i\sqrt{7}}{2} \right) \right\}$$

Chapter 6, 6.2 Exercise 9 (p. 190) We get
$$\begin{aligned} \text{Res}(s+t-x, s^2 - t^2 - y, s) &= -2xt + x^2 - y \\ \text{Res}(s^2 - t^2 - y, 2s - 3t^2 - z, s) &= 9t^4 + 6t^2 z - 4t^2 - 4y + z^2 \\ \text{Res}(s+t-x, 2s - 3t^2 - z, s) &= -3t^2 - 2t + 2x - z \end{aligned}$$
and
$$R = \text{Res}(-2xt + x^2 - y, -3t^2 - 2t + 2x - z, t) = \\ -3x^4 + 4x^3 + 6x^2 y - 4x^2 z + 4yx - 3y^2$$
so the implicit equation is
$$3x^4 - 4x^3 - 6x^2 y + 4x^2 z - 4yx + 3y^2 = 0$$

If we compute the resultant of $9t^4 + 6t^2 z - 4t^2 - 4y + z^2$ and $-2xt + x^2 - y$ we get
$$9x^8 - 36x^6 y + 24x^6 z - 16x^6 + 54x^4 y^2 \\ - 48x^4 yz - 32x^4 y + 16x^4 z^2 \\ - 36x^2 y^3 + 24x^2 y^2 z - 16x^2 y^2 + 9y^4$$
which turns out to be a multiple of R.

Chapter 6, 6.2 Exercise 10 (p. 194) We can put C in echelon form by doing this with A and B independently of each other, i.e.
$$\bar{C} = \left[\begin{array}{c|c} \bar{A} & 0 \\ \hline 0 & \bar{B} \end{array} \right]$$

where $\bar{C}$ is the upper-triangular echelon matrix for C and the corresponding statements apply to $\bar{A}$ and $\bar{B}$. The conclusion now follows from the fact that determinants of an upper triangular matrix is the product of its diagonal elements.

Chapter 6, 6.2 Exercise 13 (p. 195) Just form the matrix

$$P = \begin{bmatrix} 8 & -1 & 2 \\ 4 & 0 & 1 \\ 3 & -1 & 1 \end{bmatrix}$$

and compute

$$P^{-1}AP = \begin{bmatrix} -27 & 5 & -8 \\ 53 & -10 & 15 \\ 137 & -25 & 40 \end{bmatrix}$$

Chapter 6, 6.2 Exercise 14 (p. 201) Just plug $n = 1/2$ into equation 6.2.33 on page 198 to get

$$\sqrt{A} = \begin{bmatrix} -4\sqrt{2} - 3\sqrt{3} + 8 & \sqrt{3} - 1 & -\sqrt{3} - \sqrt{2} + 2 \\ 8\sqrt{2} - 8 & 1 & 2\sqrt{2} - 2 \\ 20\sqrt{2} + 12\sqrt{3} - 32 & 4 - 4\sqrt{3} & 4 \cdot \sqrt{3} + 5\sqrt{2} - 8 \end{bmatrix}$$

Chapter 6, 6.2 Exercise 15 (p. 201) If A is a diagonal matrix, then

$$\chi_A(A) = (A - d_1 I) \cdots (A - d_n I)$$

$$= \begin{bmatrix} 0 & 0 & \cdots & 0 \\ 0 & d_2 - d_1 & \ddots & 0 \\ \vdots & \ddots & \ddots & \vdots \\ 0 & 0 & \cdots & d_n - d_1 \end{bmatrix} \cdots \begin{bmatrix} d_1 - d_n & 0 & \cdots & 0 \\ 0 & d_2 - d_n & \ddots & 0 \\ \vdots & \ddots & \ddots & \vdots \\ 0 & 0 & \cdots & 0 \end{bmatrix}$$

$$= \begin{bmatrix} 0 & \cdots & 0 \\ \vdots & \ddots & \vdots \\ 0 & \cdots & 0 \end{bmatrix}$$

where the d_i are the diagonal entries.

The case where the eigenvalues of A are all distinct is almost as simple. In this case

$$A = PDP^{-1}$$

and $\chi_A(A) = P\chi_D(D)P^{-1} = 0$.

Chapter 6, 6.2 Exercise 16 (p. 201) We start with

$$\lambda I - C = \left[\begin{array}{c|c} \lambda I - A & 0 \\ \hline 0 & \lambda I - B \end{array} \right]$$

and refer to exercise 10 on page 194.

Chapter 6, 6.2 Exercise 17 (p. 201) The hypotheses imply that

$$\lambda I - B = C^{-1}(\lambda I - A)C$$

so

$$\det(\lambda I - B) = \det(C)^{-1} \det(\lambda I - A) \det(C)$$
$$= \det(\lambda I - A)$$

Chapter 6, 6.2 Exercise 18 (p. 201) This follows from exercise 17 on page 201 and the fact that the $(n-1)^{st}$ coefficient of $\chi_*(\lambda)$ is $-\operatorname{Tr}(*)$.

Chapter 6, 6.2 Exercise 19 (p. 201) Let

$$D_1 = U_1^{-1} A U_1$$
$$D_2 = U_2^{-1} B U_2$$

be the diagonal forms of the matrices, where the U_i are invertible.

$$\chi_A(\lambda) = \prod_{i=1}^{n} (\lambda - \lambda_i)$$
$$\chi_B(\lambda) = \prod_{i=1}^{n} (\lambda - \mu_i)$$

in the algebraic closure, $\mathbb{F}$, of the field where A and B live. Here $\{\lambda_i\}$ are the eigenvalues of A and $\{\mu_i\}$ are the eigenvalues of B. The hypotheses imply that

$$\prod_{i=1}^{n} (\lambda - \lambda_i) = \prod_{i=1}^{n} (\lambda - \mu_i)$$

Since $\mathbb{F}[\lambda]$ has unique factorization, it follows that the list

$$\{\lambda_1, \ldots, \lambda_n\}$$

is just a *permutation* of the list $\{\mu_1, \ldots, \mu_n\}$ and that there exists a permutation matrix, P, such that

$$D_2 = P^{-1} D_1 P$$

It follows that

$$B = U_2 P^t U_1^{-1} A U_1 P U_2^{-1}$$

so

$$C = U_1 P U_2^{-1}$$

Chapter 6, 6.2 Exercise 23 (p. 211) The product

$$MM^t = A$$

is a matrix with $A_{i,j} = \mathbf{u}_i \bullet \mathbf{u}_j$ — the *identity matrix*. See definition 6.2.84 on page 218.

Chapter 6, 6.2 Exercise 24 (p. 211) We already know that v_1 and v_2 are linearly independent (see proposition 6.2.54 on page 196). We have

$$A v_1 = \lambda_1 v_1$$

so

$$v_2^t A v_1 = v_2^t \lambda_1 v_1 = \lambda_1 v_2^t v_1 = \lambda_1 v_1 \bullet v_2$$

Since this is a scalar (a 1×1 matrix) it equals its own transpose, so we get

$$(\lambda_1 v_1 \bullet v_2)^t = \lambda_1 v_1 \bullet v_2$$
$$= (v_2^t A v_1)^t$$
$$= v_1^t A^t v_2$$
$$= v_1^t A v_2$$
$$= \lambda_2 v_1 \bullet v_2$$

So we conclude $\lambda_1 v_1 \bullet v_2 = \lambda_2 v_1 \bullet v_2$. Since $\lambda_1 \neq \lambda_2$, we get $v_1 \bullet v_2 = 0$.

Chapter 6, 6.2 Exercise 25 (p. 214) The x-component of $\mathbf{u} \times (\mathbf{v} \times \mathbf{w})$ is given by

$$\begin{aligned}(\mathbf{u} \times (\mathbf{v} \times \mathbf{w}))_1 &= u_2(v_1w_2 - v_2w_1) - u_3(v_3w_1 - v_1w_3) \\ &= v_1(u_2w_2 + u_3w_3) - w_1(u_2v_2 + u_3v_3) \\ &= v_1(u_2w_2 + u_3w_3) - w_1(u_2v_2 + u_3v_3) + (u_1v_1w_1 - u_1v_1w_1) \\ &= v_1(u_1w_1 + u_2w_2 + u_3w_3) - w_1(u_1v_1 + u_2v_2 + u_3v_3) \\ &= (\mathbf{u} \cdot \mathbf{w})v_1 - (\mathbf{u} \cdot \mathbf{v})w_1\end{aligned}$$

The other components are computed by permuting the subscripts in this formula.

Chapter 6, 6.2 Exercise 26 (p. 214)

$$\begin{aligned}\mathbf{v}' &= -\begin{bmatrix}1\\0\\-1\end{bmatrix} + \cos 30°\left(\begin{bmatrix}1\\2\\3\end{bmatrix} + \begin{bmatrix}1\\0\\-1\end{bmatrix}\right) + \sin 30°\left(\frac{1}{\sqrt{2}}\begin{bmatrix}1\\0\\-1\end{bmatrix} \times \begin{bmatrix}1\\2\\3\end{bmatrix}\right) \\ &= \begin{bmatrix}\sqrt{3}-1\\\sqrt{3}\\\sqrt{3}+1\end{bmatrix} + \frac{1}{2}\left(\frac{1}{\sqrt{2}}\begin{bmatrix}2\\-4\\2\end{bmatrix}\right) \\ &= \begin{bmatrix}\sqrt{3}-1+1/\sqrt{2}\\\sqrt{3}-\sqrt{2}\\\sqrt{3}+1/\sqrt{2}\end{bmatrix}\end{aligned}$$

Chapter 6, 6.2 Exercise 27 (p. 222) We start with the standard basis of $\mathbb{R}^3$, given in example 6.2.79 on page 216. Since that example shows that $\|e_1\| = \sqrt{2}$, we get

$$\mathbf{u}_1 = \frac{1}{\sqrt{2}}e_1$$

Now $\langle \mathbf{u}_1, e_2 \rangle = -1/\sqrt{2}$, so

$$\text{Proj}_{\mathbf{u}_1} e_2 = -\frac{e_1}{2}$$

and

$$f = e_2 - \text{Proj}_{\mathbf{u}_1} e_2 = \begin{bmatrix}1/2\\1\\0\end{bmatrix}$$

and $\|f\| = \sqrt{3/2}$ and

$$\mathbf{u}_2 = \sqrt{\frac{2}{3}}\begin{bmatrix}1/2\\1\\0\end{bmatrix} = \begin{bmatrix}1/\sqrt{6}\\\sqrt{2/3}\\0\end{bmatrix}$$

Finding $\mathbf{u}_3$ is similar.

Chapter 6, 6.3 Exercise 1 (p. 230) It $\mathfrak{A}$ is *not* principal, then any free basis of $\mathfrak{A}$ must contain at least two elements $x, y \in R$. The equation

$$y \cdot x - x \cdot y = 0$$

implies that they can't be part of a free basis (i.e., all elements of R may be used as coefficients in linear combinations). If x is a non-zero divisor, then $R \cdot x \cong R$.

Chapter 6, 6.3 Exercise 2 (p. 230) Define $\tilde{f}$ to send $m \pmod{A}$ to $f(m) \pmod{f}(A)$ for all $m \in M_1$. This is well-defined because $m \in A$ implies that $f(m) \in f(A)$.

Chapter 6, 6.3 Exercise 3 (p. 235) This follows from

$$F = \ker(p_k \times): M \to M$$

so it must be preserved under an isomorphism as well as $T = M/F$, so it is uniquely defined. The factor $R_1 = R/(g_1)$ is the kernel of
$$g_1 \times : T \to T$$
Define $T_j = T/R_{j-1}$ — this must be preserved under any isomorphism of the original M. The factor R_{j+1} is the kernel of
$$g_{j+1} \times : T_j \to T_j$$
so it is also preserved under isomorphisms of M.

Chapter 6, 6.3 Exercise 4 (p. 235) It is
$$\begin{bmatrix} 2 & 0 & 0 \\ 0 & 6 & 0 \\ 0 & 0 & 12 \end{bmatrix}$$
with
$$S = \begin{bmatrix} 0 & 0 & 1 \\ 0 & 1 & 0 \\ 1 & 0 & -5 \end{bmatrix}, T = \begin{bmatrix} 3 & -2 & 4 \\ -1 & 3 & -2 \\ 2 & -2 & 3 \end{bmatrix}$$

Chapter 6, 6.3 Exercise 6 (p. 240) This follows immediately from exercise 18 on page 201.

Chapter 6, 6.3 Exercise 7 (p. 241) This follows by a straightforward induction. It is clearly true for $n = 1$. If we assume the result for $n-1$ and multiply
$$\begin{bmatrix} \lambda^{n-1} & 0 & 0 & \cdots & 0 \\ (n-1)\lambda^{n-1} & \lambda^{n-1} & \ddots & & \vdots \\ 0 & (n-1)\lambda^{n-1} & \ddots & \ddots & 0 \\ \vdots & \cdots & \ddots & \lambda^{n-1} & 0 \\ 0 & \cdots & 0 & (n-1)\lambda^{n-1} & \lambda^{n-1} \end{bmatrix} \begin{bmatrix} \lambda & 0 & 0 & \cdots & 0 \\ 1 & \lambda & \ddots & \ddots & \vdots \\ 0 & 1 & \ddots & \ddots & 0 \\ \vdots & \cdots & \ddots & \lambda & 0 \\ 0 & \cdots & 0 & 1 & \lambda \end{bmatrix}$$
we get the expected result.

Chapter 6, 6.3 Exercise 8 (p. 241) We know that
$$P^{-1}AP = \begin{bmatrix} 1 & 0 & 0 & 0 & 0 \\ 0 & 2 & 0 & 0 & 0 \\ 0 & 1 & 2 & 0 & 0 \\ 0 & 0 & 0 & 3 & 0 \\ 0 & 0 & 0 & 1 & 3 \end{bmatrix} = J$$
with
$$P = \begin{bmatrix} 1 & 0 & 1 & 0 & 1 \\ -3 & 1 & -1 & 1 & -1 \\ 2 & -1 & -1 & -\frac{3}{4} & -\frac{3}{2} \\ -3 & 0 & -4 & -\frac{1}{2} & -4 \\ -1 & 0 & 0 & 0 & 0 \end{bmatrix}$$
and that
$$A^n = PJ^n P^{-1}$$
Exercise 7 on page 240 implies that
$$J^n = \begin{bmatrix} 1 & 0 & 0 & 0 & 0 \\ 0 & 2^n & 0 & 0 & 0 \\ 0 & n2^{n-1} & 2^n & 0 & 0 \\ 0 & 0 & 0 & 3^n & 0 \\ 0 & 0 & 0 & n3^{n-1} & 3^n \end{bmatrix}$$

We can easily calculate the result with SAGE. The columns of $B = A^n$ are, respectively

$$B_{*,1} = \begin{pmatrix} -8 \cdot 3^{n-1} n + 9 \cdot 2^{n-1} n - 8 \cdot 3^n + 9 \cdot 2^n \\ 8 \cdot 3^{n-1} n - 9 \cdot 2^{n-1} n \\ 12 \cdot 3^{n-1} n - 9 \cdot 2^{n-1} n + 18 \cdot 3^n - 18 \cdot 2^n \\ 32 \cdot 3^{n-1} n - 36 \cdot 2^{n-1} n + 36 \cdot 3^n - 36 \cdot 2^n \\ 0 \end{pmatrix}$$

$$B_{*,2} = \begin{pmatrix} 2^{n-1} n - 2 \cdot 3^n + 2 \cdot 2^n \\ -2^{n-1} n + 2 \cdot 3^n - 2^n \\ -2^{n-1} n + 3 \cdot 3^n - 3 \cdot 2^n \\ -4 \cdot 2^{n-1} n + 8 \cdot 3^n - 8 \cdot 2^n \\ 0 \end{pmatrix}$$

$$B_{*,3} = \begin{pmatrix} -2 \cdot 3^n + 2 \cdot 2^n \\ 2 \cdot 3^n - 2 \cdot 2^n \\ 3 \cdot 3^n - 2 \cdot 2^n \\ 8 \cdot 3^n - 8 \cdot 2^n \\ 0 \end{pmatrix}$$

$$B_{*,4} = \begin{pmatrix} -2 \cdot 3^{n-1} n + 2 \cdot 2^{n-1} n - 3^n + 2^n \\ 2 \cdot 3^{n-1} n - 2 \cdot 2^{n-1} n - 3^n + 2^n \\ 3 \cdot 3^{n-1} n - 2 \cdot 2^{n-1} n + 3 \cdot 3^n - 3 \cdot 2^n \\ 8 \cdot 3^{n-1} n - 8 \cdot 2^{n-1} n + 5 \cdot 3^n - 4 \cdot 2^n \\ 0 \end{pmatrix}$$

and

$$B_{*,5} = \begin{pmatrix} -2 \cdot 3^{n-1} n - 3 \cdot 3^n + 4 \cdot 2^n - 1 \\ 2 \cdot 3^{n-1} n + 3^n - 4 \cdot 2^n + 3 \\ 3 \cdot 3^{n-1} n + 6 \cdot 3^n - 4 \cdot 2^n - 2 \\ 8 \cdot 3^{n-1} n + 13 \cdot 3^n - 16 \cdot 2^n + 3 \\ 1 \end{pmatrix}$$

Chapter 6, 6.3 Exercise 9 (p. 241) Theorem 6.3.26 on page 237 states that there exists an invertible matrix, P, such that

$$J = P^{-1} M P = \begin{bmatrix} J_{\alpha_1}(\lambda_1) & 0 & \cdots & 0 \\ 0 & \ddots & & 0 \\ \vdots & & \ddots & \vdots \\ 0 & \cdots & 0 & J_{\alpha_m}(\lambda_m) \end{bmatrix}$$

is the Jordan Canonical Form. Furthermore

$$J^n = P^{-1} M^n P = I = \begin{bmatrix} J_{\alpha_1}(\lambda_1)^n & 0 & \cdots & 0 \\ 0 & \ddots & & 0 \\ \vdots & & \ddots & \vdots \\ 0 & \cdots & 0 & J_{\alpha_m}(\lambda_m)^n \end{bmatrix}$$

Exercise 7 on page 240 shows that a Jordan block $J_\alpha(\lambda)^n = I$ if and only if $\alpha = 1$. The statement about the eigenvalues is clear.

Chapter 6, 6.3 Exercise 10 (p. 247) This is basic linear algebra: U is the nullspace of the linear map $V \to W$ and the image is all of W.

Chapter 6, 6.3 Exercise 11 (p. 247) Suppose V is a vector-space over an infinite field, k, and
$$V = \bigcup_{i=1}^{n} V_i$$
where the V_i are proper subspaces. Without loss of generality, assume this decomposition is *minimal* (none of the V_i's are contained in a union of the others).

If $x \in V_1$ and $y \in V \setminus V_1$, then $x + r \cdot y \in V$ as $r \in k$ runs over an (infinite) number of nonzero values. The Pigeonhole Principle implies that there is a j such that
$$x + r \cdot y \in V_j$$
for an infinite number of values of r. This means that there exist $r_1 \neq r_2 \in k$ with $x + r_1 \cdot y, x + r_2 \cdot y \in V_j$ which implies that $(x + r_1 \cdot y) - (x + r_2 \cdot y) = (r_1 - r_2) \cdot y \in V_j$ so $y \in V_j$. We conclude that $x \in V_j$ as well. Since x was an *arbitrary* element of V_1, this means that
$$V_1 \subset \bigcup_{i=2}^{n} V_i$$
which contradicts the assumption that the original decomposition was *minimal*.

Chapter 6, 6.3 Exercise 12 (p. 247) A finite-dimensional vector-space over a finite field has a finite number of elements, hence is the (finite) union of the one-dimensional subspaces generated by these elements,

Chapter 6, 6.3 Exercise 13 (p. 247) Just apply proposition 6.3.16 on page 228 to the diagram

$$\begin{array}{ccc} & & P \\ & {}^g\nearrow & \| \\ M & \xrightarrow{f} & P \end{array}$$

Chapter 6, 6.3 Exercise 14 (p. 247) If $x \in R$ annihilates M_2, it annihilates any submodule and quotient so
$$\text{Ann}(M_2) \subset \text{Ann}(M_1) \cap \text{Ann}(M_3)$$
If $x \in \text{Ann}(M_1), y \in \text{Ann}(M_3)$, and $m \in M_2$, then $y \cdot m \in M_1$, since its image in M_3 is zero. Then $x \cdot (y \cdot m)) = 0$, so
$$\text{Ann}(M_1) \cdot \text{Ann}(M_3) \subset \text{Ann}(M_2)$$

Chapter 6, 6.3 Exercise 15 (p. 248) We claim that the map
$$(q, h): U \oplus W \to V$$
is an isomorphism. Suppose (u, v) maps to 0 in V, so $q(u) = -h(v)$. If we map this via p, we get $p \circ q(u) = -p \circ h(v)$. Since $p \circ q = 0$, we get $p \circ h(v) = 0$ which implies that $v = 0$ (since $p \circ h = 1$). Since q is injective, this also implies that $u = 0$. So the map, (q, h), is *injective*.

Suppose $v \in V$ is any element and let $z = v - h \circ p(v)$. Then $p(z) = p(v) - p \circ h \circ p(v) = p(v) - p(v) = 0$. This implies that $z = q(u)$ for some $u \in U$ and
$$v = (q, h)(z, h(v))$$
so the map is also *surjective*.

Chapter 6, 6.3 Exercise 16 (p. 248) This follows immediately from exercises 13 on page 247 and 15 on page 248.

Chapter 6, 6.4 Exercise 2 (p. 251) This follows immediately from the Ascending Chain Condition (in proposition 5.4.2 on page 122) and corollary 6.4.8 on page 250.

Chapter 6, 6.4 Exercise 3 (p. 251) The non-zero-divisors of $R \times S$ is the set (r,s) where r is a non-zero-divisor of R and s is a non-zero-divisor of S.

Chapter 6, 6.4 Exercise 4 (p. 251) Let $\mathrm{ann}(m) \subset R$ be the annihilator of m — an ideal. Then m goes to 0 in $M_{\mathfrak{a}}$ if and only if $\mathrm{ann}(m) \not\subset \mathfrak{a}$ (see definition 6.4.9 on page 250). But proposition 5.2.11 on page 115 shows that *every* ideal is contained in *some* maximal ideal.

Chapter 6, 6.4 Exercise 5 (p. 251) This follows immediately from exercise 4 on page 251.

Chapter 6, 6.4 Exercise 6 (p. 251) The general element of F is of the form
$$f = \frac{a_0 + a_1 X + \cdots}{b_0 + b_1 X + \cdots}$$
Suppose a_i is the lowest indexed coefficient in the numerator that is nonzero and b_j is the corresponding one in the denominator. Then
$$f = \frac{X^i(a_i + a_{i+1}X + \cdots)}{X^j(b_j + b_{j+1}X + \cdots)} = X^{i-j}\frac{a_i + a_{i+1}X + \cdots}{b_j + b_{j+1}X + \cdots}$$
where $b_j \neq 0$ so that the denominator is a unit in R (see proposition 5.1.9 on page 109). Set $\alpha = i - j$ and $r = (a_i + a_{i+1}X + \cdots)\left(b_j + b_{j+1}X + \cdots\right)^{-1}$.

Chapter 6, 6.5 Exercise 1 (p. 259) Suppose $s^{-1}x \in S^{-1}T$ is integral over $S^{-1}R$. Then it satisfies an equation
$$(s^{-1}x)^n + a_{n-1}(s^{-1}x)^{n-1} + \cdots + a_0 = 0$$
where the $a_i \in S^{-1}R$. Let $\bar{s} \in S$ be able to clear the denominators of all of the a_i. Multiplying this equation by $(s\bar{s})^n$ gives
$$(\bar{s}x)^n + a_{n-1}s\bar{s}(\bar{s}x)^{n-1} + \cdots + s^n \bar{s} a_0 = 0$$
so $\bar{s}x \in R$ is integral over R, and $(s\bar{s})^{-1}(\bar{s}x) = s^{-1}x$ is integral over $S^{-1}R$.

Chapter 6, 6.5 Exercise 2 (p. 259) Let
$$x^n + a_{n-1}x^{n-1} + \cdots + a_0 = 0$$
be the minimal polynomial (see definition 7.1.8 on page 263) of x with $a_i \in F$. If $s \in R$ can clear the denominators of the a_i, multiply this equation by s^n to get
$$\begin{aligned} s^n x^n + s^n a_{n-1} x^{n-1} + \cdots + s^n a_0 &= (sx)^n + a_{n-1}s(sx)^{n-1} + \cdots + s^n a_0 \\ &= 0 \end{aligned}$$
so sx is integral over R.

Chapter 6, 6.5 Exercise 3 (p. 259) Clearly, any element $x \in F$ that is integral over R is also integral over T. On the other hand, if x is integral over T, it is also integral over R because of statement 2 of proposition 6.5.5 on page 253 (the degree of the monic polynomial over R will usually be higher than that over T). It follows that the integral closures will be the same.

Chapter 7, 7.1 Exercise 1 (p. 265) Let $p(X) \in F[X]$ be the minimum polynomial of α. Its being of degree n implies that $F[X]/(p(X)) = F[\alpha] = F(\alpha)$ is a degree-n extension of F. The conclusion follows from proposition 7.1.6 on page 263.

Chapter 7, 7.1 Exercise 2 (p. 265) Whenever one gets X^3, it must be replaced by $-3 - 3X$, giving the result
$$ad - 3ce - 3bf + X(ae + bd - 3ce - 3cf - 3bf) + X^2(af + be + cd - 3cf)$$

Chapter 7, 7.2 Exercise 1 (p. 271) We claim that $F(X) = F(X^2) \cdot 1 \oplus F(X^2) \cdot X$ as a vector space. If
$$u = \frac{p(X)}{q(X)}$$
we can write
$$q(X) = a(X^2) + X \cdot b(X^2)$$
— just separate the terms with odd powers of X from the others. Now, we get
$$u = \frac{p(X)}{q(X)} \cdot \frac{a(X^2) - X \cdot b(X^2)}{a(X^2) - X \cdot b(X^2)} = \frac{p(X)(a(X^2) - X \cdot b(X^2))}{a(X^2)^2 - X^2 \cdot b(X^2)^2}$$
Now, write the numerator as
$$p(X) = c(X^2) + X \cdot d(X^2)$$
so we get
$$u = \frac{R(Y)}{S(Y)} + X \cdot \frac{T(Y)}{S(Y)}$$
where
$$\begin{aligned} S(Y) &= a(Y)^2 - Y \cdot b(Y)^2 \\ R(Y) &= c(Y)a(Y) - Y \cdot b(Y)d(Y) \\ T(Y) &= a(Y)d(Y) - c(Y)b(Y) \end{aligned}$$

Chapter 7, 7.2 Exercise 2 (p. 271) The number $2^{1/3}$ satisfies the equation
$$X^3 - 2 = 0$$
and Eisenstein's Criterion (theorem 5.6.8 on page 149) shows that this is irreducible. It follows that $X^3 - 2$ is the minimal polynomial of $2^{1/3}$.

Set $\mathbb{Q}(2^{1/3}) = \mathbb{Q}[X]/(X^3 - 2)$. We would like to find the multiplicative inverse of the polynomial
$$X^2 - X + 1$$
modulo $X^3 - 2$. We can use the extended Euclidean algorithm (algorithm 3.1.12 on page 17) for this. Dividing $X^3 - 2$ by $X^2 - X + 1$ gives a quotient of $q_1(X) = X + 1$ and a remainder of -3. We're done since
$$(X+1) \cdot (X^2 - X + 1) - 1 \cdot (X^3 - 2) = 3$$
or
$$\frac{1}{3}(X+1) \cdot (X^2 - X + 1) - \frac{1}{3} \cdot (X^3 - 2) = 1$$
so, modulo $X^3 - 2$, we get $\frac{1}{3}(X+1) \cdot (X^2 - X + 1) = 1$ which implies that
$$\frac{1}{3}(2^{1/3} + 1) = \frac{1}{2^{2/3} - 2^{1/3} + 1} \in \mathbb{Q}(2^{1/3}) = \mathbb{Q}[2^{1/3}]$$

Chapter 7, 7.2 Exercise 3 (p. 271) Just follow the proof of theorem 7.2.13 on page 270. The minimal polynomials of $\sqrt{2}$ and $\sqrt{3}$ are, respectively, $X^2 - 2$ and $X^2 - 3$. Their roots (i.e., the α_i and β_j in the proof) are
$$\pm\sqrt{2}, \pm\sqrt{3}$$
and the set of elements of $\mathbb{Q}$ we must avoid are
$$\frac{\sqrt{2}}{\sqrt{3}} \in \mathbb{Q}[\sqrt{2}, \sqrt{3}] \setminus \mathbb{Q}$$
Since this is not a rational number, it follows that we can pick *any* nonzero rational number for our c. We pick $c = 1$ and $\gamma = \sqrt{2} + \sqrt{3}$.

So
$$\mathbb{Q}[\sqrt{2}, \sqrt{3}] = \mathbb{Q}[\sqrt{2} + \sqrt{3}]$$
To find the minimal polynomial, we refer to example 7.3.1 on page 272.

Chapter 7, 7.2 Exercise 4 (p. 271) One obvious root of $X^3 - 2 \in \mathbb{Q}[X]$, is $X = 2^{1/3}$, so we try the field extension
$$\mathbb{Q}[2^{1/3}]$$
Since $2^{1/3}$ is a root of $X^3 - 2$, we get $(X - 2^{1/3})|(X^3 - 2)$ with a quotient of
$$X^2 + 2^{1/3} \cdot X + 2^{2/3}$$
and if we set $X = Y \cdot 2^{1/3}$, this becomes
$$2^{2/3} \cdot (Y^2 + Y + 1)$$
The roots of
$$Y^2 + Y + 1 = 0$$
are
$$\omega, \omega^2 = \frac{-1 \pm \sqrt{-3}}{2}$$
which are the cube-roots of 1 (other than 1 itself). So our splitting field is
$$\mathbb{Q}[2^{1/3}, \omega]$$
of degree 6 over $\mathbb{Q}$.

Chapter 7, 7.3 Exercise 1 (p. 273) We find a Gröbner basis for $\mathfrak{a} = (X^2 - 2, Y^3 - 2, A - X - Y)$ with lexicographic order with
$$X \succ Y \succ A$$
to get
$$\begin{aligned}\mathfrak{a} = (&-4 - 24A + 12A^2 - 6A^4 - 4A^3 + A^6, \\ &-364 + 152A - 156A^2 + 9A^4 \\ &- 160A^3 + 24A^5 + 310Y, \\ &364 - 462A + 156A^2 - 9A^4 \\ &+ 160A^3 - 24A^5 + 310X)\end{aligned}$$
so the minimal polynomial of α is
$$\alpha^6 - 6\alpha^4 - 4\alpha^3 + 12\alpha^2 - 24\alpha - 4 = 0$$

Chapter 7, 7.3 Exercise 2 (p. 276) The minimal polynomial is $X^3 - 2$ and we get a basis of $\{1, 2^{1/3}, 2^{2/3}\}$. If $\gamma = a + b2^{1/3} + c2^{2/3}$, then the effect of γ on the basis is given by
$$\begin{aligned}\gamma \cdot 1 &= a + b2^{1/3} + c2^{2/3} \\ \gamma \cdot 2^{1/3} &= 2c + a2^{1/3} + b2^{2/3} \\ \gamma \cdot 2^{2/3} &= 2b + 2c2^{1/3} + a2^{2/3}\end{aligned}$$
which gives a matrix
$$m_\gamma = \begin{bmatrix} a & 2c & 2b \\ b & a & 2c \\ c & b & a \end{bmatrix}$$
with a determinant
$$N_{H/F}(\gamma) = a^3 - 6acb + 2b^3 + 4c^3$$

and characteristic polynomial
$$\chi_\gamma(X) = X^3 - 3aX^2 - \left(6cb - 3a^2\right)X - a^3 + 6acb - 2b^3 - 4c^3$$

Chapter 7, 7.3 Exercise 3 (p. 276) In this case, our basis for H over $\mathbb{Q}$ is $\{1, \sqrt{2}, \sqrt{3}, \sqrt{6}\}$. If $\gamma = a + b\sqrt{2} + c\sqrt{3} + d\sqrt{6}$ is a general element, its effect on a basis is

$$\begin{aligned}
\gamma \cdot 1 &= a + b\sqrt{2} + c\sqrt{3} + d\sqrt{6} \\
\gamma \cdot \sqrt{2} &= 2b + a\sqrt{2} + 2d\sqrt{3} + c\sqrt{6} \\
\gamma \cdot \sqrt{3} &= 3c + 3d\sqrt{2} + a\sqrt{3} + b\sqrt{6} \\
\gamma \cdot \sqrt{6} &= 6d + 3c\sqrt{2} + 2b\sqrt{3} + a\sqrt{6}
\end{aligned}$$

which gives a matrix
$$m_\gamma = \begin{bmatrix} a & 2b & 3c & 6d \\ b & a & 3d & 3c \\ c & 2d & a & 2b \\ d & c & b & a \end{bmatrix}$$

with a determinant
$$N_{H/F}(\gamma) = a^4 - 4a^2b^2 + 48adbc - 12d^2a^2 - 6a^2c^2 \\ + 4b^4 - 24d^2b^2 - 12b^2c^2 + 9c^4 - 36c^2d^2 + 36d^4$$

and characteristic polynomial
$$\chi_{H/F}(\gamma) = X^4 - 4aX^3 + \left(-12d^2 - 6c^2 - 4b^2 + 6a^2\right)X^2 \\ + \left(-48dbc + 24d^2a + 12ac^2 + 8ab^2 - 4a^3\right)X \\ a^4 - 4a^2b^2 + 48adbc - 12d^2a^2 - 6a^2c^2 \\ + 4b^4 - 24d^2b^2 - 12b^2c^2 + 9c^4 - 36c^2d^2 + 36d^4$$

Chapter 7, 7.4 Exercise 1 (p. 279) If $\alpha \mid \beta$ in $\mathbb{Z}[\gamma]$, then $\beta = \alpha \cdot t$ where $t \in \mathbb{Z}[\gamma]$ and
$$t = \sum_{i=0}^{n-1} c_i \gamma^i$$
with $c_i \in \mathbb{Z}$. Since $\alpha \in \mathbb{Z}$, we have
$$\alpha \cdot t = \sum_{i=0}^{n-1} \alpha c_i \gamma^i$$
and this is in $\mathbb{Z} \subset \mathbb{Z}[\gamma]$ if and only if $c_i = 0$ for $i > 0$. This means that $\alpha \mid \beta$ in $\mathbb{Z}$.

Chapter 7, 7.4 Exercise 2 (p. 279) $\deg \Phi_n(X) = \phi(n)$ — see definition 3.3.1 on page 22.

Chapter 7, 7.5 Exercise 1 (p. 287) Suppose
$$f(X) = X^n + a_{n-1}X^{n-1} + \cdots + a_0$$
(after dividing by a_n if necessary) and embed F in its algebraic closure, $\bar{F}$. We get
$$f(X) = \prod_{j=1}^{n}(X - \alpha_j)$$
and the splitting field of $f(X)$ is just
$$F[\alpha_1, \ldots, \alpha_n] \subset \bar{F}$$
which is *unique* in $\bar{F}$. The conclusion follows from theorem 7.5.3 on page 283.

Chapter 7, 7.5 Exercise 2 (p. 287) The characteristic polynomial of γ was computed in the solution to 2 on page 276 (setting $c = 0$)

$$\chi_\gamma(X) = X^3 - 3aX^2 + 3a^2X - a^3 - 2b^3$$

and this is also the minimal polynomial of γ. One factor of this must be $X - \gamma$, so we take the quotient

$$\frac{X^3 - 3aX^2 + 3a^2X - a^3 - 2b^3}{X - \gamma}$$
$$= X^2 + X(\gamma - 3a) + \gamma(\gamma - 3a) + 3a^2$$
$$= X^2 + (-2a + b2^{1/3})X + a^2 - ab2^{1/3} + b^2 2^{2/3}$$

regarding it as a polynomial with coefficients in $\mathbb{C}$. The roots of this quadratic equation are

$$X = a + b\left(\frac{-1 \pm \sqrt{-3}}{2}\right) 2^{1/3} = a + b2^{1/3}\omega^j$$

where $\omega = e^{2\pi i/3}$ is a primitive cube root of 1 and $j = 1, 2$. Since $\omega \notin \mathbb{Q}[2^{1/3}]$, these conjugates *do not* lie in $\mathbb{Q}[2^{1/3}]$. This is why we need to go to an algebraic closure of the field.

Chapter 7, 7.5 Exercise 3 (p. 287) Yes! The algebraic numbers, $\bar{\mathbb{Q}}$, are an algebraic extension of $\mathbb{Q}$ but of infinite degree.

Chapter 7, 7.5 Exercise 4 (p. 287) Oddly, yes. If

$$\bar{\mathbb{Q}} = \mathbb{Q}(\alpha_1, \dots)$$

then

$$\bar{\mathbb{Q}} = \bigcup_{i=1}^{\infty} \mathbb{Q}(\alpha_1, \dots, \alpha_i) = \bigcup_{i=1}^{\infty} \mathbb{Q}[\alpha_1, \dots, \alpha_i] = \mathbb{Q}[\alpha_1, \dots]$$

Chapter 7, 7.6 Exercise 2 (p. 290) $R(x) = x^7 + x^6$, $S(x) = x^2 + x + 1$

Chapter 7, 7.7 Exercise 1 (p. 295) Suppose π is algebraic. Then so is πi and $e^{\pi i}$ should be transcendental, by the Lindemann–Weierstrass theorem. But $e^{\pi i} = -1$ (see Euler's Formula, theorem 2.1.2 on page 7), which is algebraic. This is a contradiction.

Chapter 7, 7.7 Exercise 2 (p. 295) The algebraic closure of $\mathbb{Q}$ is the algebraic numbers, $\bar{\mathbb{Q}}$, which is countable. If F is a countable field $F(X)$ is also a countable field and a simple induction shows that

$$F(X_1, \dots, X_n)$$

is also countable for any n. If $\mathbf{S} = \{X_1, \dots\}$ is a countable set of indeterminates, then

$$F(\mathbf{S}) = \bigcup_{i=1}^{\infty} F(X_1, \dots, X_i)$$

is also countable. It follows that an uncountable field like $\mathbb{C}$ must have an uncountable degree of transcendence over $\mathbb{Q}$.

Chapter 8, 8.3 Exercise 1 (p. 303) This follows immediately from the definition of $F(\alpha_1, \dots, \alpha_k)$: if $x \in F(\alpha_1, \dots, \alpha_k)$ then

$$x = \frac{\sum_{i=1}^{k} \sum_{j=0}^{n_i} c_{i,j} \alpha_i^j}{\sum_{i=1}^{k} \sum_{j=0}^{m_i} d_{i,j} \alpha_i^j}$$

where $c_{i,j}, d_{i,j} \in F$. All of the elements here are fixed by f.

Chapter 8, 8.6 Exercise 1 (p. 313) Suppose

(15.3.9) $$F \subset F_1 \subset \cdots \subset F_k = E$$

is a tower of radical extensions for E. We will *refine* it to one with the required properties — i.e. we will insert new terms between the old ones.

If $F_{i+1} = F_i(\alpha)$, where $\alpha^n = 1$ and $n = p \cdot m$, where p is a prime, set $E_1 = F_i(\beta)$ where $\beta = \alpha^m$, and consider the series

$$F_i \subset E_1 \subset F_{i+1}$$

Since $\beta^p = 1$, $F_i \subset E_1$ is pure of type p.

Now $F_{i+1} = E_1(\alpha)$, where $\alpha^m \in F_1$. If m is not a prime, continue this construction, inserting new subfields and stripping off a prime in each step.

We will eventually arrive at a sequence of subfields

$$F_i \subset E_1 \subset \cdots \subset E_k \subset F_{i+1}$$

where each extension is pure of prime type.

Apply this construction to all terms of equation 15.3.9 to get the required result.

Chapter 8, 8.6 Exercise 2 (p. 313) Since $F \subset E$ is finite, it is *algebraic* (see proposition 7.2.2 on page 266), so

$$E = F(\alpha_1, \ldots, \alpha_k)$$

If $f_i(X) \in F[X]$ is the minimal polynomial of α_i set

$$f(X) = \prod_{i=1}^{k} f_i(X)$$

and define G to be the splitting field of $f(X)$ (see corollary 7.2.4 on page 266).

Chapter 8, 8.6 Exercise 4 (p. 313) Since G is the splitting field of

$$f(X) = \prod_{i=1}^{k} f_i(X)$$

it contains the *conjugates* of each of the α_i — and these are the images of the α_i under the Galois group of G over F.

If $\mathrm{Gal}(G/F) = \{g_1 \ldots, g_n\}$ we define $E_i = g_i(E)$.

Chapter 8, 8.6 Exercise 5 (p. 314) Since

$$\begin{array}{c} E \\ | \\ F \end{array}$$

is a radical extension, we use exercise 1 on page 313 to conclude that it can be resolved into a sequence of pure extensions via primes, p_i. Each has the form $E_{i+1} = E_i(\alpha_i)$, where $\alpha_i^{p_i} \in E_i$. The conjugates of α_i are precisely $\alpha_i \cdot \omega_i^j$, where $\omega_i^j = 1$ — i.e. we must adjoin appropriate roots of unity to get a splitting field. These adjunctions are *also* pure of prime order, so the extension

$$\begin{array}{c} G \\ | \\ F \end{array}$$

must also be radical.

Chapter 8, 8.8 Exercise 1 (p. 321) We prove this by induction on $|G|$. If $|G| > 1$, then $Z(G) > 1$ by Burnside's Theorem 4.7.14 on page 79. If $Z(G) = G$, then G is abelian, so it is solvable. Otherwise, $G/Z(G)$ is a smaller p-group, which is solvable by the inductive hypothesis. Since $Z(G)$ is also solvable (because it's abelian), we use exercise 6 on page 87 to conclude that G is solvable.

Chapter 9, 9.1 Exercise 1 (p. 324) If $y = (a, b) \in A \oplus A$, then $y^* = (a^*, -b)$ and equation 9.1.1 on page 324 gives

$$yy^* = aa^* + bb^*$$
$$y^*y = a^*a + b^*b$$

so the right inverse is

$$y^{-1} = \frac{y^*}{yy^*}$$

and the left inverse is

$$\frac{y^*}{y^*y} = y^{-1}$$

Chapter 9, 9.1 Exercise 2 (p. 324) If $u = (a, b)$ and $v = (c, d)$, then

$$uv = (ac - d^*b, da + bc^*)$$

and

(15.3.10) $\qquad (uv)^* = (c^*a^* - b^*d, -da - bc^*)$

using equation 9.1.2 on page 324. Now $u^* = (a^*, -b)$ and $v^* = (c^*, -d)$, and direct computation, using equation 9.1.2 on page 324 shows that v^*u^* is equal to the right side of equation 15.3.10.

Chapter 9, 9.1 Exercise 3 (p. 324) Suppose a, b are nonzero and

$$ab = 0$$

If we have $a^{-1} \cdot a = 1$ and multiply the equation above by a^{-1}, we get

$$a^{-1}(ab) = 0 \neq (a^{-1}a)b = b$$

In other words, if the algebra fails to be *associative*, this can easily happen.

Chapter 9, 9.1 Exercise 4 (p. 324) If $a = x \cdot b$ and $u \cdot b = 0$ with $u \neq 0$, then $a = (x + u) \cdot b$ so that quotients are not unique.

Chapter 9, 9.1 Exercise 5 (p. 324) If $a \in A$ is nonzero, then $a \times : A \to A$ defines a linear transformation. Since A has no zero-divisors, the kernel of this linear transformation is zero. Corollary 6.2.33 on page 182 implies that $a \times$ is 1-1, so it maps some element, a^{-1}, to 1.

Chapter 9, 9.1 Exercise 6 (p. 325) If $c \in A$, then $m_x(c) = x \cdot c$, in general. The definition of matrix-product implies that

$$m_x m_y = x \cdot (y \cdot c))$$

— see section 6.2.2 on page 169. If A is associative, $x \cdot (y \cdot c)) = (x \cdot y) \cdot c$, so that $m_{x \cdot y} = m_x m_y$. Otherwise, we have $m_{x \cdot y} \neq m_x m_y$ in general, and there is *no* obvious relationship.

Chapter 9, 9.2 Exercise 1 (p. 332) No! If it was valid, it would imply that multiplication of exponentials is *commutative*, whereas equation 9.2.7 on page 329 implies that

$$e^{\alpha \cdot \mathbf{u}_1} \cdot e^{\beta \cdot \mathbf{u}_2} - e^{\beta \cdot \mathbf{u}_2} \cdot e^{\alpha \cdot \mathbf{u}_1} = 2 \sin \alpha \sin \beta \cdot \mathbf{u}_1 \times \mathbf{u}_2 \neq 0$$

in general.

Chapter 9, 9.2 Exercise 2 (p. 332) In several places, including the claim that $xy = yx \implies x^{-1}y = yx^{-1}$, and equation 9.2.8 on page 331.

Chapter 9, 9.2 Exercise 3 (p. 332) It suffices to check it on quaternion units with the correspondence:
$$1 \leftrightarrow (1,0)$$
$$i \leftrightarrow (i,0)$$
$$j \leftrightarrow (0,1)$$
$$k \leftrightarrow (0,i)$$

Chapter 9, 9.2 Exercise 4 (p. 332) The matrix is of the form
$$\begin{bmatrix} 0 & 0 & 0 & 0 \\ 0 & C_1 & C_2 & C_3 \end{bmatrix}$$
so the rotation is given by (see equation 6.2.7 on page 170)
$$M = \begin{bmatrix} C_1 & C_2 & C_3 \end{bmatrix}$$
where
$$C_1 = x \cdot i \cdot x^{-1} = 0 + i(a^2 + b^2 - c^2 - d^2) + j(2ad + 2bc) + k(2bd - 2ac)$$
$$= 0 + i(1 - 2c^2 - 2d^2) + j(2ad + 2bc) + k(2bd - 2ac)$$
$$C_2 = x \cdot j \cdot x^{-1}$$
$$C_3 = x \cdot k \cdot x^{-1}$$
or
$$M = \begin{bmatrix} 1 - 2(c^2 + d^2) & 2(bc - ad) & 2(bd + ac) \\ 2(ad + bc) & 1 - 2(b^2 + d^2) & 2(cd - ab) \\ 2(bd - ac) & 2(cd + ab) & 1 - 2(b^2 + c^2) \end{bmatrix}$$

Chapter 9, 9.3 Exercise 1 (p. 337) Just write $\alpha \mathbf{v} = \alpha |\mathbf{v}|(\mathbf{v}/|\mathbf{v}|)$ and apply equation 9.2.6 on page 328. We get $(\alpha \mathbf{v})^2 = -\alpha^2 |\mathbf{v}|^2$.

Chapter 9, 9.3 Exercise 2 (p. 337) We need to prove an analogue of equation 9.2.6 on page 328. If $\mathbf{u} \in \mathbb{O}$ is a purely imaginary unit octonion, then $\mathbf{u} = (\mathbf{a}, \mathbf{b})$ in the Cayley-Dickson construction, where $\mathbf{a}$ is a purely imaginary quaternion and $|\mathbf{a}|^2 + |\mathbf{b}|^2 = 1$. The Cayley-Dickson construction (definition 9.1.2 on page 324) implies that
$$(\mathbf{a}, \mathbf{b})^2 = \left(\mathbf{a}^2 - \mathbf{b}^* \mathbf{b}, \mathbf{b}(\mathbf{a} + \mathbf{a}^*) \right)$$
If $\mathbf{a}$ is a purely imaginary quaternion, $\mathbf{a} + \mathbf{a}^* = 0$, and we get
$$(\alpha(\mathbf{a}, \mathbf{b}))^2 = (-\alpha^2 |\mathbf{a}|^2 - \alpha^2 |\mathbf{b}|^2, 0) = -\alpha^2 \in \mathbb{R}$$
which is precisely what equation 9.2.6 on page 328 says. We answer the final question in the affirmative by a simple induction.

Chapter 10, 10.1 Exercise 1 (p. 344) This is literally a direct restatement of the definition of a product: every pair of morphisms $f \in \hom_{\mathscr{C}}(W, A)$ and $g \in \hom_{\mathscr{C}}(W, B)$ induces a unique morphism $f \times g \in \hom_{\mathscr{C}}(W, A \times B)$ that makes the diagram 10.1.2 on page 340 commute.

Chapter 10, 10.1 Exercise 3 (p. 344) Suppose $g_1, g_2 \in \hom_{\mathscr{C}}(C, A)$ map to the same element of $\hom_{\mathscr{C}}(C, B)$, then $f \circ g_1 = f \circ g_2 \colon C \to B$, which implies (by the definition of monomorphism) that $g_1 = g_2$.

Chapter 10, 10.1 Exercise 4 (p. 344) If
$$f \colon A \to B$$
has a kernel, the inclusion of distinct elements of A that differ by an element of the kernel are distinct morphisms whose composite with f are the same. The other conclusion follows by a similar argument.

Chapter 10, 10.2 Exercise 1 (p. 349) It is A^t, the transpose of A.

Chapter 10, 10.2 Exercise 2 (p. 349) It is an equivalence because of the natural isomorphism in example 10.2.5 on page 348. It is not an isomorphism of categories because the finite-dimensional vector-space V^{**} is not *identical* to V. If V is infinite-dimensional, it is not even isomorphic.

Chapter 10, 10.3 Exercise 1 (p. 351) Every pair of morphisms
$$\begin{aligned} y &\to x \\ z &\to x \end{aligned}$$
— i.e., every morphism $(y, z) \to \Delta x$ — corresponds to a *unique* morphism
$$y \coprod z \to x$$
so we get an equivalence
$$\hom_{\mathscr{C} \times \mathscr{C}}((y,z), \Delta x) = \hom_{\mathscr{C}}(y \coprod z, x)$$

Chapter 10, 10.4 Exercise 1 (p. 360) It is given in theorem 4.10.5 on page 88.

Chapter 10, 10.4 Exercise 2 (p. 360) Definition 10.4.10 on page 356 implies that there is a natural equivalence
$$\hom_{\mathscr{C}_\infty}(\Delta_\infty x, y) = \hom_{\mathscr{C}}(x, \varprojlim y)$$
for all $x \in \mathscr{C}$ and $y \in \mathscr{C}_\infty$.

Chapter 10, 10.4 Exercise 3 (p. 360) This follows immediately from the definition of the equivalence relation $\sim$ in definition 10.4.1 on page 351.

Chapter 10, 10.5 Exercise 2 (p. 364) If $a \in A$ is torsion-free then $a \mapsto x$ for any $x \neq 0$ defines a nonzero homomorphism. If a is of order n then $a \mapsto 1/n$ defines a nonzero map.

Chapter 10, 10.5 Exercise 3 (p. 365) The sub-object $\operatorname{im} A \subset B$ maps to A in a straightforward way. The injective property of A implies that this extends to all of B.

Chapter 10, 10.5 Exercise 4 (p. 365) We already know that
$$0 \to \hom_{\mathscr{A}}(D, A) \xrightarrow{\hom_{\mathscr{A}}(1,r)} \hom_{\mathscr{A}}(D, B)$$
is exact, by exercise 3 on page 344, so we must still show that
$$\hom_{\mathscr{A}}(D, A) \xrightarrow{\hom_{\mathscr{A}}(1,r)} \hom_{\mathscr{A}}(D, B) \xrightarrow{\hom_{\mathscr{A}}(1,s)} \hom_{\mathscr{A}}(D, C)$$
is exact. If $f \in \hom_{\mathscr{A}}(D, B)$ maps to 0 in $\hom_{\mathscr{A}}(D, C)$, then $s \circ f = 0$. Since $r = \ker s$, we have a *unique* morphism $D \to A$ that makes
$$\begin{array}{ccc} D & \xrightarrow{f} & B \\ & \searrow^{v} & \uparrow r \\ & & A \end{array}$$
commute. This is precisely the element of $\hom_{\mathscr{A}}(D, A)$ that maps to f.

Chapter 10, 10.6 Exercise 1 (p. 376) The hypotheses imply that the composites
$$R \hookrightarrow M \xrightarrow{p_1} M/S \to R$$
$$S \hookrightarrow M \xrightarrow{p_2} M/R \to S$$
are the identity maps, so we get a well-defined homomorphism
$$M \xrightarrow{(p_1, p_2)} R \oplus S$$
and this is easily verified to be an isomorphism.

Chapter 10, 10.6 Exercise 2 (p. 376) If $Q = M/N$, we get a short exact sequence
$$0 \to N \to M \to Q \to 0$$
and the conclusion follows from the fact that the sequence
$$0 \to A \otimes_R N \to A \otimes_R M \to A \otimes_R Q \to 0$$
is also exact (because A is flat).

Chapter 10, 10.6 Exercise 3 (p. 376) In the top formula, multilinear maps
$$M \times N \times T \to A$$
where A is an arbitrary R-module factor through $M \otimes_R (N \otimes_R T)$ and $(M \otimes_R N) \otimes_R T$ so the universal property of $\otimes$ implies that they are isomorphic.

To see the second equality, regard $M \otimes_R N$ and $N \otimes_R M$ as quotients of $\mathbb{Z}[M \times N]$ by the ideal generated by the identities in definition 10.6.3 on page 366 and noting that this ideal is symmetric with respect to factors.

Chapter 10, 10.6 Exercise 4 (p. 377) Corollary 10.6.8 on page 369 implies that
$$k^n \otimes_k k^m = k^{n \cdot m}$$
so the dimensions are as claimed.

If $\{e_i\}$ is a basis for V and $\{f_j\}$ is a basis for W then it is not hard to see that $\{e_i \otimes f_j\}, i = 1, \ldots, n, j = 1, \ldots, m$ spans $V \otimes_k W$ — just use the identities in definition 10.6.3 on page 366 to express any $v \otimes w$ in terms of them. The fact that $V \otimes_k W$ is $n \cdot m$-dimensional shows that these elements must be linearly independent too.

To prove the final statement, we must show that the set $\{e_i \otimes f_j\}$ is linearly independent even if there are an infinite number of basis elements. Suppose we have some linear combination

(15.3.11)
$$\sum_{t=1}^{n} a_t (e_{i_t} \otimes f_{k_t}) = 0$$

for $a_t \in k$. Since only a finite number of terms are involved, this equation really involves finite-dimensional subspaces of V and W, namely the span of the $\{e_{i_t}\}$ in V and the span of the $\{f_{j_t}\}$ in W. We have already seen that the $\{e_{i_t} \otimes f_{k_t}\}$ are linearly independent in this case, so all of the $a_t = 0$ in equation 15.3.11.

Chapter 10, 10.6 Exercise 5 (p. 377) This is an $ns \times mt$ matrix called the *Kronecker product* of A and B. If

$$A = \begin{bmatrix} a_{1,1} & \cdots & a_{1,n} \\ \vdots & \ddots & \vdots \\ a_{m,1} & \cdots & a_{m,n} \end{bmatrix}$$

Then

$$A \otimes B = \begin{bmatrix} a_{1,1}B & \cdots & a_{1,n}B \\ \vdots & \ddots & \vdots \\ a_{m,1}B & \cdots & a_{m,n}B \end{bmatrix}$$

Chapter 10, 10.6 Exercise 6 (p. 377) We construct a homomorphism
$$M^* \otimes N \to \hom_k(M, N)$$
Let M have basis $\{e_i\}$ with $i = 1, \ldots, m$, and let N have basis $\{f_j\}$ with $k = 1, \ldots, n$. Then M^* has the dual basis $\{e^i\}$ defined by
$$e^i(e_j) = \begin{cases} 0 & \text{if } i \neq j \\ 1 & \text{if } i = j \end{cases}$$

and $M^* \otimes N$ has a basis $\{e^i \otimes f_j\}$. Given an element, $m \in M$, we can evaluate each basis element
$$e^i(m) \cdot f_j$$
to define an element of $\hom_k(M, N)$. Since this is bilinear and $\dim M^* \otimes N = nm = \dim \hom_k(M, N)$, it follows that it defines an isomorphism.

Chapter 10, 10.6 Exercise 7 (p. 377) This follows immediately from proposition 10.6.11 on page 370:
$$\frac{R}{\mathfrak{a}} \otimes_R \frac{R}{\mathfrak{b}} = \left(\frac{R}{\mathfrak{a}}\right) / \mathfrak{b} \cdot \left(\frac{R}{\mathfrak{a}}\right) = \frac{R}{\mathfrak{a} + \mathfrak{b}}$$

Chapter 10, 10.6 Exercise 8 (p. 377) We always have a surjective natural map
$$\mathfrak{a} \otimes_R M \to \mathfrak{a} \cdot M$$
$$a \otimes m \mapsto a \cdot m$$
If M is flat, this map is also *injective* since it is
$$\mathfrak{a} \hookrightarrow R$$
$\otimes M$.

Chapter 10, 10.6 Exercise 9 (p. 377) For each i, take the natural maps
$$z_i \colon M_i \to \varinjlim M_j$$
$$Z_i \colon M_i \otimes_R N \to \varinjlim (M_j \otimes_R N)$$
and form the tensor product of M_i with N to get
$$z_i \otimes 1 \colon M_i \otimes_R N \to \left(\varinjlim M_j\right) \otimes_R N$$
The universal property of direct limits implies the existence of a *unique* map
$$v \colon \varinjlim (M_j \otimes_R N) \to \left(\varinjlim M_j\right) \otimes_R N$$
that makes the diagram
$$\begin{array}{ccc} \varinjlim (M_j \otimes_R N) & \xrightarrow{v} & \left(\varinjlim M_j\right) \otimes_R N \\ Z_i \uparrow & & \uparrow z_i \otimes 1 \\ M_i \otimes_R N & =\!=\!=\!= & M_i \otimes_R N \end{array}$$
commute. If $m \otimes n \neq 0 \in \left(\varinjlim M_j\right) \otimes_R N$, then m is the image of some $m_i \in M_i$ and $m \otimes n = v \circ Z_i(m_i \otimes n)$. It follows that v is surjective.

If $w \in \varinjlim (M_j \otimes_R N)$ is in the kernel of v, then $w = Z_i(m_i \otimes n)$ for some i. the commutativity of the diagram implies that $(z_i \otimes 1)(m_i \otimes n) = z_i(m_i) \otimes n = 0$, which implies that $Z_i(m_i \otimes n) = 0$ — so v is injective.

Chapter 10, 10.6 Exercise 10 (p. 377) The hypotheses imply that
$$\begin{array}{ccccccccc} 0 & \longrightarrow & \varinjlim A_i & \xrightarrow{\varinjlim f_i} & \varinjlim B_i & \xrightarrow{\varinjlim g_i} & \varinjlim C_i & \longrightarrow & 0 \\ & & \bar{a}_i \uparrow & & \uparrow \bar{b}_i & & \uparrow \bar{c}_i & & \\ 0 & \longrightarrow & A_i & \xrightarrow{f_i} & B_i & \xrightarrow{g_i} & C_i & \longrightarrow & 0 \end{array}$$
commutes for all i, where we don't know whether the top row is exact. If $x \in \varinjlim C_i$, then x is the image of some $x_i \in C_i$. The commutativity of this diagram implies that x is in the image of $\bar{b}_i(g_i^{-1}(x_i))$. It follows that $\varinjlim g_i$ is surjective. If $x \in \varinjlim B_i$ such that $x \in \ker \varinjlim g_i$, then $x = \bar{b}_i(x_i)$ for some i and $\bar{c}_i \circ g_i(x_i) = 0$. The definition

of direct limit implies that there exists $N > i$ such that $c_N \circ \cdots \circ c_i(x) = 0$, so $b_N \circ \cdots \circ b_i(x_i) \in \ker g_N$. The exactness of the original sequences implies that $b_N \circ \cdots \circ b_i(x_i) = f_N(y_N)$ and the commutativity of the diagram above implies that $\bar{a}_N(y_N) = x$. A similar argument implies that the left end of the upper row is also exact.

Chapter 10, 10.6 Exercise 11 (p. 377) Consider the map $g\colon (\alpha) \hookrightarrow R$. This is an inclusion and, since S is flat over R

$$(\alpha) \otimes_R S \xrightarrow{g \otimes 1} R \otimes_R S = S$$

is also an inclusion. Since α is a non-zero-divisor in R, $(\alpha) \cong R$ and the isomorphism $R \to R$ induced by the inclusion is multiplication by α. This implies that

$$S = R \otimes_R S \xrightarrow{(\times \alpha) \otimes 1 = \times f(\alpha)} R \otimes_R S = S$$

is also injective, which implies that $f(\alpha) \in S$ is a non-zero-divisor.

Chapter 10, 10.6 Exercise 12 (p. 377) Let

(15.3.12) $$0 \to U_1 \to U_2 \to U_3 \to 0$$

be a short exact sequence of modules over S. Since we can compose the action of S on these modules with the homomorphism, f, it is also a short exact sequence of modules over R. If we take the tensor product with $M \otimes_R S$, we get

$$\begin{array}{ccccccccc}
0 & \to & U_1 \otimes_S (M \otimes_R S) & \to & U_2 \otimes_S (M \otimes_R S) & \to & U_3 \otimes_S (M \otimes_R S) & \to & 0 \\
& & \| & & \| & & \| & & \\
& & U_1 \otimes_R M & & U_2 \otimes_R M & & U_3 \otimes_R M & &
\end{array}$$

which is exact since equation 15.3.12 is an exact sequence of R-modules.

Chapter 10, 10.6 Exercise 13 (p. 378)

Chapter 10, 10.6 Exercise 14 (p. 378) For any commutative ring U and any module A over U, $U \otimes_U A = A$. This (and the associativity of tensor products) implies that

$$\begin{aligned}
(S^{-1}R \otimes_R M) \otimes_{S^{-1}R} (S^{-1}R \otimes_R N) &= (S^{-1}R \otimes_R M) \otimes_{S^{-1}R} S^{-1}R \otimes_R N \\
&= (S^{-1}R \otimes_R M) \otimes_R N \\
&= S^{-1}R \otimes_R M \otimes_R N \\
&= S^{-1}R \otimes_R (M \otimes_R N)
\end{aligned}$$

Chapter 10, 10.6 Exercise 15 (p. 378) Since M is projective, it is a direct summand of a free module, F, so $F = M \oplus N$ for some other projective module, N. Then

$$F^* = M^* \oplus N^*$$

Chapter 10, 10.7 Exercise 1 (p. 386) First of all, note that the map $V \to W$ is split, i.e. there exists a left-inverse $t\colon W \to V$ so that $g \circ t = 1$. Any two such splitting maps differ by a map from V to U. Now define $1 - t \circ g\colon W \to \ker g = \operatorname{im} U$ or $f^{-1} \circ (1 - t \circ g)\colon W \to U$. We get an isomorphism

$$(f^{-1} \circ (1 - t \circ g), g)\colon W \cong U \oplus V$$

Given a commutative diagram like the one in the statement of the problem, we can lift a map $t_2\colon W_2 \to V_2$ to get a map $t_1\colon W_1 \to V_2$ so we get a natural isomorphism from W_i to $U_i \oplus V_i$. The conclusion follows from proposition 10.7.8 on page 381.

Chapter 10, 10.7 Exercise 2 (p. 386)

Chapter 10, 10.7 Exercise 3 (p. 386) Just compute

$$
\begin{aligned}
(e_2 + 2e_4) \wedge e_3 \wedge 2e_1 \wedge (e_2 - e_4) &= -e_2 \wedge e_3 \wedge 2e_1 \wedge e_4 \\
&\quad -2e_4 \wedge e_3 \wedge 2e_1 \wedge e_2 \\
&= -2e_1 \wedge e_2 \wedge e_3 \wedge e_4 \\
&\quad +4e_1 \wedge e_2 \wedge e_3 \wedge e_4 \\
&= 2e_1 \wedge e_2 \wedge e_3 \wedge e_4
\end{aligned}
$$

so the determinant is 2.

Chapter 11, 11.1 Exercise 2 (p. 392) If $x \neq 0 \in M$, then $R \cdot x \subset M$ is a submodule.

Chapter 11, 11.1 Exercise 4 (p. 392) Any sub-representation would be one dimensional, there is no invariant one-dimensional subspace.

Chapter 11, 11.1 Exercise 5 (p. 392) If $|G| = n$, then $g^n = 1$ (see definition 4.1.11 on page 38 and theorem 4.4.2 on page 43). It follows that $\rho(g)^n = I$ so the conclusion follows from exercise 9 on page 241.

Chapter 11, 11.1 Exercise 6 (p. 392) Exercise 6 on page 377 implies that there's a natural isomorphism

$$\operatorname{Hom}_k(M, N) \cong M^* \otimes_k N$$

"Natural" means it preserves group-actions, i.e. kG-module structures. So we must analyze the group-actions on M^* and N. The definition of the action on $M^* = \operatorname{Hom}(M, k)$ has $g \in G$ acting via

$$\operatorname{Hom}(g^{-1}, 1)\colon \operatorname{Hom}_k(M, k) \to \operatorname{Hom}_k(M, k)$$

and exercise 1 on page 349 implies that the matrix for the action of $\operatorname{Hom}(g^{-1}, 1)$ is the *transpose* of that for g^{-1}. The conclusion follows from equation 11.1.2 on page 391.

Chapter 11, 11.1 Exercise 7 (p. 392) If $f \in \operatorname{Hom}_k(M_1, M_2)$, the action of an element, $g \in G$, on it is given by

$$(g \cdot f)(x) = g \cdot f(g^{-1} \cdot x)$$

where $x \in M_1$. If $(g \cdot f)(x) = f(x)$ for all $g \in G$ and all $x \in M_1$, then $f(g^{-1} \cdot x) = g^{-1} f(x)$, so f preserves the group action and $f \in \operatorname{Hom}_{kG}(M_1, M_2)$.

Chapter 11, 11.1 Exercise 8 (p. 392) Recall that the trace is an invariant of the linear transformation defined by P, not the matrix used to compute it (see exercise 18 on page 201).

The statement that $P|W = 1\colon W \to W$ is equivalent to $P^2 = P$.

If $Q = I - P$, then $Q|W = 0$ and $P = I - Q$. If $Z = \operatorname{im} Q$ then $Q^2 = (I - P)^2 = I - 2P + P^2 = Q$. It follows that $Q|Z = 1\colon Z \to Z$. If we combine a basis of W with one for Q, we get a basis for V in which

$$P = \begin{bmatrix} 0 & 0 \\ 0 & I \end{bmatrix} \colon Q \oplus W \to Q \oplus W$$

so the trace of P is the dimension of W.

Chapter 11, 11.1 Exercise 9 (p. 392) The isomorphism $f\colon V \to V$ is an element of $GL(V)$. The reasoning in theorem 6.2.50 on page 194 implies the conclusion.

Chapter 11, 11.1 Exercise 10 (p. 392) Let $f\colon \rho_1(G) \to \rho_2(G)$ be the conjugation-isomorphism. Then $\varphi = \rho_2^{-1} \circ f \circ \rho_1$.

Chapter 11, 11.1 Exercise 11 (p. 393) Since they are irreducible $U_1 \cap U_2 = 0$. The conclusion follows from exercise 1 on page 376.

Chapter 11, 11.1 Exercise 12 (p. 393) If we have
$$M \cong N \oplus U$$
then N and U also have the property that every submodule of them has a complement, i.e. if $N' \subset N$, then N' has a complement, $V \subset M$ and and $N \cap V$ is a complement in N. If we write $M \cong N \oplus U$, and N is not simple (i.e., irreducible), we can split further $N = N_1 \oplus N_2$ — and this process will proceed until we have written M as a direct sum of simple modules.

On the other hand, if
$$M = S_1 \oplus \cdots \oplus S_n$$
where the S_i are simple, and $N \subset M$, then
$$N \cap S_i = \begin{cases} S_i & \text{or} \\ 0 \end{cases}$$
so we can define U to be the direct sum of the S_i *not* contained in N.

Chapter 11, 11.1 Exercise 13 (p. 393) This follows immediately from exercise 12 on page 393.

Chapter 11, 11.1 Exercise 14 (p. 393) If R is semisimple, then $R \oplus \cdots \oplus R$ will also be semisimple — in other words any *free* R-module will be semisimple. The conclusion follows from proposition 6.3.10 on page 225 and exercise 13 on page 393.

Chapter 11, 11.2 Exercise 1 (p. 398) Transpose defines an isomorphism
$$M_n^{\text{op}}(k) \xrightarrow{\text{tr}} M_n(k)$$

Chapter 11, 11.2 Exercise 2 (p. 398) Given a matrix algebra $M_n(D)$, we can form the opposite
$$M_n(D)^{\text{op}} = M_n^{\text{op}}(D^{\text{op}})$$
If D is a division ring, so is D^{op}. The conclusion follows from exercise 1 on page 398.

Chapter 11, 11.3 Exercise 1 (p. 408) If g has order n, then $\rho(g)^n = I$. The conclusion follows from exercise 9 on page 241.

Chapter 11, 11.3 Exercise 2 (p. 408) Since $\rho(g)$ is diagonalizable with eigenvalues $\lambda_1, \ldots, \lambda_k$, we get $\chi(g) = \lambda_1 + \cdots + \lambda_k$. Since $\rho(g)^n = I$, we conclude that $\lambda_i^n = 1$ for all i.

Chapter 11, 11.3 Exercise 3 (p. 408) As in exercise 2 on page 409, $\chi(g) = \lambda_1 + \cdots + \lambda_k$, and $\chi(1) = k$, the degree of the representation. The triangle inequality shows that
$$|\lambda_1 + \cdots + \lambda_k| \leq k$$
with equality only if $\lambda = \lambda_1 = \cdots = \lambda_k$. So $\rho(g) = \lambda \cdot I$.

Chapter 11, 11.3 Exercise 4 (p. 409) This follows immediately from exercise 2 on page 409 above. This reasoning was used in the proof of proposition 11.3.11 on page 401.

Chapter 11, 11.3 Exercise 5 (p. 409) The first statement follows immediately from exercise 4 on page 409 above, which implies that $\chi(g) = \overline{\chi(g)}$ for all $g \in G$. The second follows from the fact that a character-table is *square*: so a conjugacy class is *uniquely* determined by its characters.

Chapter 12, 12.1 Exercise 1 (p. 418) The closed sets of $\mathbb{A}^1$ are:
  (1) the empty set,
  (2) all of $\mathbb{A}^1$,
  (3) finite sets of point (roots of polynomials).

It follows that the closed sets in the product-topology on $\mathbb{A}^1 \times \mathbb{A}^1$ consist of

(1) all of $\mathbb{A}^1 \times \mathbb{A}^1$
(2) {finite set} $\times \mathbb{A}^1$
(3) $\mathbb{A}^1 \times$ {finite set}
(4) {finite set} $\times$ {finite set}

and the Zariski topology on $\mathbb{A}^2$ has *many* more closed sets, like the set of points that satisfy

$$x^2 + y^2 = 1$$

or even the diagonal line

$$y = x$$

Chapter 12, 12.1 Exercise 2 (p. 418) Both V and ℓ are closed sets of $\mathbb{A}^n$, so their intersection is also a closed set and a closed subset of $\ell = \mathbb{A}^1$. The only closed sets of ℓ (in the Zariski topology) are:

(1) $\emptyset$
(2) ℓ
(3) finite sets of points.

Since $p \in \ell$ and $p \notin V$, case 2 is ruled out.

Chapter 12, 12.1 Exercise 3 (p. 418) This follows from exercise 14 on page 247, which shows that

$$\text{Ann}(M_1) \cdot \text{Ann}(M_3) \subset \text{Ann}(M_2) \subset \text{Ann}(M_1) \cap \text{Ann}(M_3)$$

and proposition 12.1.2 on page 417.

Chapter 12, 12.1 Exercise 4 (p. 418) We can simplify the ideal $(X_1^2 + X_2^2 - 1, X_1 - 1)$ considerably. Since $X_1^2 - 1 = (X_1 + 1)(X_1 - 1)$, we subtract $X_1 + 1$ times the second generator from the first to get $(X_2^2, X_1 - 1)$. It follows that V consists of the single point $(0, 1)$ and $\mathcal{I}(V) = (X_1 - 1, X_2)$.

Chapter 12, 12.1 Exercise 5 (p. 418) In characteristic 2, $(X_1 + X_2 + X_3)^2 = X_1^2 + X_2^2 + X_3^2$, so V is the plane defined by

$$X_1 + X_2 + X_3 = 0$$

and $\mathcal{I}(V) = (X_1 + X_2 + X_3)$.

Chapter 12, 12.1 Exercise 6 (p. 418) This is XY, since $XY = 0$ implies $X = 0$ or $Y = 0$.

Chapter 12, 12.1 Exercise 7 (p. 418) If we use the results of the previous exercise, $XY = 0$ so $\mathcal{V}((XY))$ is the $P_{XZ} \cup P_{YZ}$ where P_{XZ} denotes the XZ-plane and P_{YZ} denotes the YZ-plane. Similarly, $\mathcal{V}((XZ)) = P_{XY} \cup P_{YZ}$ so that

$$\mathcal{V}((XY, XZ)) = (P_{XZ} \cup P_{YZ}) \cap (P_{XY} \cup P_{YZ}) = P_{YZ}$$

Since $\mathcal{V}((YZ)) = P_{XY} \cup P_{XZ}$, we get

$$\mathcal{V}((XY, XZ, YZ)) = (P_{XY} \cup P_{XZ}) \cap P_{YZ}$$
$$= (P_{XY} \cap P_{YZ}) \cup (P_{XZ} \cap P_{YZ})$$

and each of these terms are equal to the union of the axes.

Chapter 12, 12.1 Exercise 8 (p. 418) In $k[V] = k[X, Y]/(Y^2 - X^3)$ the identity $Y^2 = X^3$ holds, so every occurrence of Y^2 can be replaced by X^3.

Chapter 12, 12.2 Exercise 1 (p. 426) If $n = \max(n_1, \ldots, n_k)$ then $(p_1 \cdots p_k)^n \in (x)$ so that $(p_1 \cdots p_k) \in \sqrt{(x)}$.

Chapter 12, 12.2 Exercise 2 (p. 426) Suppose that $\mathfrak{p} \subset k[X_1,\ldots,X_n]$ is prime and suppose that $a^n \in \mathfrak{p}$. If we write $a^n = a \cdot a^{n-1}$, then the defining property of a prime ideal implies that either $a \in \mathfrak{p}$ or $a^{n-1} \in \mathfrak{p}$. In the first case, the claim is proved. In the second case, we do downward induction on n.

Chapter 12, 12.2 Exercise 3 (p. 426) Suppose $\mathfrak{a} \subset k[X_1,\ldots,X_n]$ is a proper ideal. The strong form of the Nullstellensatz says that $\mathcal{IV}(\mathfrak{a}) = \sqrt{\mathfrak{a}}$.

We claim that if $\mathfrak{a} \neq k[X_1,\ldots,X_n]$ then the same is true of $\sqrt{\mathfrak{a}}$. The statement that $1 \in \sqrt{\mathfrak{a}}$, is equivalent to saying that $1^n \in \mathfrak{a}$ for some n. But $1^n = 1$ so $1 \in \sqrt{\mathfrak{a}}$ implies that $1 \in \mathfrak{a}$.

Since $\sqrt{\mathfrak{a}} \neq k[X_1,\ldots,X_n]$, we conclude that $V(\mathfrak{a}) \neq \emptyset$.

Chapter 12, 12.2 Exercise 4 (p. 426) Set $n = m = 1$. In this case, the Zarski-closed sets are finite sets of points or the empty set or all of $\mathbb{A}^1$. The maps that swaps two points (like 1 and 2) but leaves all other points fixed is Zarski-continuous, but is clearly not regular.

Chapter 12, 12.2 Exercise 5 (p. 426) Since the determinant, z, can never vanish and since it is a polynomial over $X_1,\ldots,X_n$, Hilbert's Nullstellensatz implies that it must be a constant (any nonconstant polynomial has a zero somewhere).

Chapter 12, 12.2 Exercise 6 (p. 426) The second equation implies that

$$XZ - Z = Z(X-1) = 0$$

so $X = 1$ or $Z = 0$. Plugging each of these cases into the first equation gives:
Case 1:
If $X = 1$ then the first equation becomes $YZ = 1$ which generates a hyperbola.
Case 2:
If $Z = 0$ then the first equation becomes $X = 0$ and Y is unrestricted — i.e., we get the Y-axis. So the two components are the

(1) hyperbola $X = 1$, $YZ = 1$ and
(2) the Y-axis.

Chapter 12, 12.2 Exercise 7 (p. 426) Yes. If we compute Gröbner basis of $\mathfrak{a} + (1 - T(X+Y))$ (with respect to *any* ordering) we get (1). Using the `NormalForm` command in Maple gives

$$(X+Y)^2 \to_\mathfrak{a} 4Y^2$$
$$(X+Y)^3 \to_\mathfrak{a} 0$$

so $(X+Y)^3 \in \mathfrak{a}$.

Chapter 12, 12.3 Exercise 1 (p. 430) If we add the two equations, we get

$$2X^2 = -1$$

so

$$X = \pm \frac{i}{\sqrt{2}}$$

If we plug this into the second equation, we get

$$Y^2 + Z^2 = \frac{1}{2}$$

so V consists of two disjoint circles.

Chapter 12, 12.3 Exercise 2 (p. 430) We will prove that the negations of these statements are equivalent. If there exists a nonempty proper subset $S \subset X$ that is both open and closed then $X \setminus S$ is a closed set and

$$X = S \cup (X \setminus S)$$

is a decomposition with $S \cap (X \setminus S) = \emptyset$.

Conversely, if there exists a decomposition
$$X = X_1 \cup X_2$$
with X_1, X_2 nonempty closed sets and $X_1 \cap X_2 = \emptyset$, then $X \setminus X_1 = X_2$, so X_2 is open as well as closed.

Chapter 13, 13.1 Exercise 1 (p. 439) The cohomology groups of the third cochain complex will appear in the long exact sequence in proposition 13.1.9 on page 436 sandwiched between the zero-groups of the other two.

Chapter 13, 13.1 Exercise 2 (p. 439) In the long exact sequence
$$\cdots \to H^i(C) \xrightarrow{f^*} H^i(D) \xrightarrow{g^*} H^i(E) \xrightarrow{c} H^{i+1}(C) \to H^{i+1}(D) \to \cdots$$
we have $H^i(D) = H^{i+1}(D)$ so the exact sequence reduces to
$$\cdots \to H^i(C) \xrightarrow{f^*} 0 \xrightarrow{g^*} H^i(E) \xrightarrow{c} H^{i+1}(C) \to 0 \to \cdots$$

Chapter 13, 13.1 Exercise 3 (p. 439) If the chain-homotopy between f and g is Φ, simply use $F(\Phi)$ as the chain-homotopy between $F(f)$ and $F(g)$.

Chapter 13, 13.1 Exercise 4 (p. 440) If $b \in B$ maps to 0 under v, then its image under s must map to 0 under w. But w is an isomorphism so $s(b) = 0$. The exactness of the top row implies that $b = r(a)$ for some $a \in A$. If $a \neq 0$, then it maps to something nonzero under u (since u is an isomorphism) and therefore to something nonzero under r', which gives a contradiction. It follows that b must have been 0 to start with. So v is injective. Proof of surjectivity is left to the reader.

Chapter 13, 13.1 Exercise 5 (p. 446) If $B \otimes_R M \to C \otimes_R M$ is *not* surjective, it has a cokernel, K and there is a homomorphism
$$C \otimes_R M \to K$$
Make $D = K$ in the original exact sequence of hom's. Then this map goes to the 0-map in $\hom_R(B \otimes_R M, D)$, a contradiction.

The fact that $\beta \circ \alpha = 0$ implies that $\beta \otimes 1 \circ \alpha \otimes 1 = 0$, so $\operatorname{im} \alpha \otimes \subset \operatorname{im} \beta \otimes 1$ Now let $D = B \otimes_R M / \operatorname{im}(\alpha \otimes 1)$ with
$$p: B \otimes_R M \to D$$
the projection. Then $p \in \hom_R(B \otimes_R M, D)$ and $p \in \ker \hom(\alpha \otimes 1, 1)$ (by *construction*) which means
$$p \in \operatorname{im} \hom(\beta \otimes 1, 1)$$
so there exists a map $\phi: C \otimes_R M \to D$ such that $p = \beta \otimes 1 \circ \phi$, so
$$\operatorname{im} \beta \otimes 1 = \ker p = \ker(\beta \otimes 1 \circ \phi) = (\beta \otimes 1)^{-1}(\ker \phi)$$
which contains $\ker \beta \otimes 1$.

Chapter 13, 13.1 Exercise 6 (p. 446) First note that $\hom_{\mathscr{A}b}(\mathbb{Z}/n \cdot \mathbb{Z}, \mathbb{Q}/\mathbb{Z}) = \mathbb{Z}/n \cdot \mathbb{Z}$. The conclusion follows from the finite direct sum
$$A = \bigoplus \frac{\mathbb{Z}}{n_i \cdot \mathbb{Z}}$$
so
$$\hom_{\mathscr{A}b}(A, \mathbb{Q}/\mathbb{Z}) = \prod \frac{\mathbb{Z}}{n_i \cdot \mathbb{Z}} = A$$
since the product is *finite*.

The second statement follows from looking at the injective resolution of $\mathbb{Z}$ in exercise 16 on page 447.

Chapter 13, 13.1 Exercise 8 (p. 446) This follows from the corresponding property of hom.

Chapter 13, 13.1 Exercise 9 (p. 446) This follows from the corresponding property of hom.

Chapter 13, 13.1 Exercise 11 (p. 446) In this case N is its *own* injective resolution. And all others are chain-homotopy equivalent to it, so they have the same cohomology.

Chapter 13, 13.1 Exercise 12 (p. 446) We can identity $\hom_{R\text{-mod}}(R,M) = \hom_R(R,M) = M$. It follows that, applied to any resolution of M, we just recover the resolution.

Chapter 13, 13.1 Exercise 13 (p. 446) A projective module is a direct summand of a free module, so we get

$$\operatorname{Ext}^i_R(P \oplus Q, M) = \operatorname{Ext}^i_R(P, M) = \operatorname{Ext}^i_R(Q, M) = \operatorname{Ext}^i_R(F, M)$$

and $\operatorname{Ext}^i_R(F, M) = 0$, by exercise 12 on page 446.

Chapter 13, 13.1 Exercise 14 (p. 446) The proof is very much like that of corollary 13.1.23 on page 445 except that you "reverse the arrows."

Chapter 13, 13.1 Exercise 15 (p. 446) We get the induced long exact sequence (see proposition 13.1.9 on page 436)

$$0 = \hom(\mathbb{Z}_n, \mathbb{Z}) \to \hom_\mathbb{Z}(\mathbb{Z}_n, \mathbb{Z}_n) = \mathbb{Z}_n \xrightarrow{\cong} \operatorname{Ext}^1_\mathbb{Z}(\mathbb{Z}_n, \mathbb{Z}) \to \operatorname{Ext}^1_\mathbb{Z}(\mathbb{Z}_n, \mathbb{Z}) = 0$$

so $\operatorname{Ext}^1_\mathbb{Z}(\mathbb{Z}_n, \mathbb{Z}) = \mathbb{Z}_n$.

Chapter 13, 13.1 Exercise 16 (p. 446) This is just

$$\mathbb{Q} \to \mathbb{Q}/\mathbb{Z}$$

because both of these groups are injective objects (see propositions 10.5.6 on page 363 and 4.6.18 on page 69).

Chapter 13, 13.1 Exercise 17 (p. 447) This follows immediately from proposition 10.5.7 on page 363.

Chapter 13, 13.1 Exercise 18 (p. 447) Since each of the P_i are *projective*, we have $\operatorname{Ext}^j_R(P_i, *) = 0$ for $j > 0$. We prove this by induction on n. If M is any R-module, the long exact sequence in cohomology gives

$$\cdots \to \operatorname{Ext}^j_R(P_i, M) = 0 \to \operatorname{Ext}^j_R(\operatorname{im} P_{i+1}, M) \to \operatorname{Ext}^{j+1}_R(\operatorname{im} P_i, M) \to \operatorname{Ext}^{j+1}_R(P_i, M) = 0 \to \cdots$$

so

$$\operatorname{Ext}^j_R(\operatorname{im} P_{i+1}, M) = \operatorname{Ext}^{j+1}_R(\operatorname{im} P_i, M)$$

as long as $j > 0$ and $i \geq 1$. At the end of the induction step

$$\operatorname{Ext}^j_R(\operatorname{im} P_2, M) = \operatorname{Ext}^{j+1}_R(\operatorname{im} P_1, M) = \operatorname{Ext}^{j+1}_R(A, M)$$

Chapter 13, 13.2 Exercise 1 (p. 451) If M is projective, it is its own projective resolution so $\operatorname{Ext}^1_R(M, N) = 0$ for any N. On the other hand, suppose $\operatorname{Ext}^1_R(M, N) = 0$ for any N. If

$$W \to M$$

is a surjective map with kernel K the W is an extension of M by K. Since $\operatorname{Ext}^1_R(M, K) = 0$, this extension is *split*, i.e. $W = K \oplus M$ — which is the definition of a projective module (see 6.3.15 on page 227).

Chapter 13, 13.2 Exercise 2 (p. 451) The reasoning associating an extension to an element of $\operatorname{Ext}_R^1(A,B)$ is exactly the same as before.
If
$$\cdots \to P_2 \to P_1 \to P_0 \to A \to 0$$
is a projective resolution of A, and $x \in \operatorname{Ext}_R^1(A,B)$, then $x\colon P_1 \to B$ is a homomorphism that vanishes on the image of P_2 which is the kernel of $P_1 \to P_0$. It follows that it induces a map $\bar{x}\colon \operatorname{im}(P_1) \to B$ and we can form the *push-out*

$$\begin{array}{ccc} \operatorname{im}(P_1) & \longrightarrow & P_0 \\ \bar{x}\downarrow & & \downarrow \\ B & \cdots\cdots\rightarrow & E \end{array}$$

We clearly have a homomorphism
$$B \to E$$
which is injective since $\operatorname{im}(P_1) \hookrightarrow P_0$ is injective. We also have an isomorphism
$$\frac{E}{B} \cong \frac{P_0}{\operatorname{im}(P_1)} \cong A$$

Chapter 13, 13.2 Exercise 4 (p. 451) If $x \ne 0 \in \operatorname{Ext}_\mathbb{Z}^1(\mathbb{Z}_p, \mathbb{Z})$ the extension
$$0 \to \mathbb{Z} \xrightarrow{\iota} \mathbb{Z} \xrightarrow{\pi} \mathbb{Z}_p \to 0$$
corresponding to x is defined by:
(1) $\iota(1) = p$
(2) $\pi(1) = x$

Note that we are using the fact that p is a prime here (so every nonzero element of $\mathbb{Z}_p$ is a generator).

Chapter 13, 13.3 Exercise 3 (p. 461) This follows immediately from exercise 1 on page 93 and theorem 13.3.11 on page 461.

Chapter 13, 13.3 Exercise 4 (p. 461) This follows immediately from exercise 3 on page 461 and theorem 13.1.23 on page 445.

Glossary

$\mathscr{A}b$ The category of abelian groups.

diffeomorphic   Two topological spaces, X and Y, are *diffeomorphic* if there exist smooth maps $f\colon X \to Y$ and $g\colon Y \to X$ whose composites are the *identity maps* of the two spaces. Note the similarity to homeomorphism.

homeomorphic   Two topological spaces, X and Y, are *homeomorphic* if there exist continuous maps $f\colon X \to Y$ and $g\colon Y \to X$ whose composites are the *identity maps* of the two spaces.

SM The symmetric algebra generated by a module. See definition 10.7.4 on page 379.

$\mathrm{Assoc}(R)$ The set of associated primes of a ring R. See definition 6.3.27 on page 241.

$\mathbb{C}$ The field of complex numbers.

$\mathbb{H}$ The division algebra of quaternions.

$\mathbb{O}$ The division algebra of octonions.

$\mathscr{C}h$ The category of chain-complexes. See definition 13.1.1 on page 433.

$\mathscr{C}o$ The category of cochain-complexes. See definition 13.1.2 on page 434.

$\mathrm{coker}\, f$ The cokernel of a homomorphism. See definition 6.3.6 on page 224.

dense subset   A subset $S \subset X$ of a topological space is *dense* if, for any open set $U \subset X$, $S \cap U \neq \emptyset$.

$\mathrm{Ext}^i_R(A,B)$ The Ext-functor — see section 13.1.19 on page 443.

$\Lambda^i M$ The ith exterior power of a module, M. See definition 10.7.7 on page 380.

$\mathbb{F}_{p^n}$ The unique finite field of order p^n, where p is a prime number. See section 7.6 on page 287.

$x^{\underline{n}}$ The falling factorial or Pochhammer symbol, defined by $x^{\underline{n}} = x(x-1)\cdots(x-n+1)$ for $n \geq 1$.

$\overline{\mathbb{F}}_p$ The algebraic closure of the field $\mathbb{F}_p$. See theorem 7.6.7 on page 289.

$\mathcal{F}_p$ The Frobenius homomorphism of a finite field of characteristic p. See definition 7.6.5 on page 289.

inj-dim M The injective dimension of a module. See definition 13.1.12 on page 440.

$\mathfrak{J}(R)$ The Jacobson radical of a ring. See definition 5.7.1 on page 154.

$k[[X]]$ Power series ring, defined in definition 5.1.7 on page 109.

$k\{\{X\}\}$ Field of Puiseux series, defined in example 15.1.5 on page 468.

$\varinjlim A_n$ The direct limit of a sequence of objects and morphisms. See definition 10.4.1 on page 351.

$\varprojlim R_n$ The inverse limit of a sequence of objects and morphisms. See definition 10.4.10 on page 356.

LT(∗) Leading monomial of a polynomial with respect to some ordering. Used in computing Gröbner bases.

A(f) The algebraic mapping cone of a chain-map, f. See definition 13.1.11 on page 438.

\ A difference between *sets*, so $A \setminus B$ is the elements of A that are not contained in B.

R-mod The category of modules over a ring, R. See statement 7 on page 341.

$\hom_C(A, B)$ The set of morphisms between objects of a category C. See definition 10.1.2 on page 340.

$\mathfrak{N}(R)$ The nilradical of a ring. See definition 12.2.7 on page 421.

◁ A symbol indicating that a subgroup is normal — see definition 4.4.5 on page 44.

proj-dim(M) The projective dimension of a module. See definition 13.1.13 on page 441.

$\mathbb{Q}$ The field of rational numbers.

$\mathbb{R}$ The field of real numbers.

$R^i F$ Right derived functors of F. See definition 13.1.18 on page 443.

$\sqrt{*}$ Radical of an ideal. See definition 12.2.4 on page 421.

rank(A) Rank of a matrix, A.

$\Gamma(\mathfrak{a})$ The Rees algebra of a ring with respect to an ideal. See definition 15.3.4 on page 472.

Res(f, g) Resultant of two polynomials. See definition 6.2.41 on page 188.

$\text{Tor}^i_R(A, B)$ The Tor-functor — see definition 13.2.1 on page 447.

Tr(A) The trace of a square matrix (the sum of its diagonal elements).

$R^\times$ The group of units of a ring or field.

$\mathbb{Z}$ The ring of integers.

$\mathbb{Z}_{(p)}$ p-adic numbers. See example 10.4.14 on page 359.

Index

NIELS HENRIK ABEL, 314
Abel's Impossibility Theorem, 314
abelian category, 359
abelian group, 35
abelianization of a group, 49
absolute value of a complex number, 7
ACC, 122
action of a group on a set, 74
acyclic cochain complex, 433
additive functor, 359
adjoint functors, 347
adjoint of a matrix, 183
adjugate of a matrix, 183
adjunction, 347
affine k-algebra, 428
affine group, 201
affine regular mapping, 423
affine space, 413
algebra, 321
   alternative, 331
   power-associative, 331
algebra over a field, 260
algebraic closure
   construction, 282
algebraic closure of a field, 281
algebraic element of a field, 261
algebraic extension of fields, 263
algebraic independence, 289
algebraic mapping cone, 436
algebraic number, 120
   minimal polynomial, 120
algebraic numbers, 283
algebraic set, 413
algebraically closed field, 281
alternating group, 54
alternative algebra, 331
annihilator of an element in a module, 239
EMIL ARTIN, 158
Artin-Wedderburn theorem, 395
Artinian ring, 157
Artinian module, 224
ascending chain condition, 122

associated prime, 239
Aut(G), 48
automorphism, 40, 111
   inner, 48
   outer, 48
automorphism group, 48

Baer Sum, 449
Baer's Criterion for injectivity, 361
Baer-Specker group, 60
bar resolution, 454
basis
   ideal, 111
basis for a vector space, 164
ÉTIENNE BÉZOUT, 14
Bézout's Identity, 14
bijective function, 39
bilinear form, 254
bilinear map, 365
Bruno Buchberger, 127
WILLIAM BURNSIDE, 76
Burnside's Theorem, 409

GEROLAMO CARDANO, 297
category, 338
   concrete, 339
   discrete, 339
   equivalence, 346
AUGUSTIN-LOUIS CAUCHY, 76
Cauchy sequence, 469
ARTHUR CAYLEY, 49
Cayley Numbers., 330
Cayley-Dickson Construction, 322
Cayley-Hamilton Theorem, 197
center of a group, 46
center of a ring, 403
centralizer of an element of a group, 79
chain complex, 431
chain-homotopic chain-maps, 433
chain-homotopy equivalent, 434
chain-map, 431, 432
character, 302
character of a group-representation, 396

characteristic of a field, 259
characteristic polynomial in a finite
    extension, 271
characteristic polynomial of a matrix, 193
characteristic subgroup, 47
characters
    inner product, 399
Chinese Remainder Theorem, 116
class equation, 78
class equation of a group, 78
class function, 397
closure of a set, 415
cochain complex, 432
cofinal subsequence, 352
cohomology groups, 433
cokernel of a homomorphism, 113, 222
column-vectors of a matrix, 168
combinatorial group theory, 87
commutative ring, 107
commutator, 47
commutator subgroup, 47
compact topological space, 423
complement of a submodule, 392
complete metric space, 469
complex conjugate, 7
complex number
    conjugate, 7
composition factors, 85
composition series, 85
compositum of fields, 311
concrete category, 339
congruence modulo a number, 18
conjugacy problem, 90
conjugate
    complex number, 7
    quaternion, 324
conjugate closure of a subgroup, 46
conjugates of an algebraic element, 283
contravariant functor, 344
convergent sequence, 469
coordinate ring, 425
coproduct in a category, 340
Correspondence Theorem for groups, 48
covariant functor, 344
GABRIEL CRAMER, 176
Cramer's Rule, 175
cross product of vectors, 209
cross-product as wedge-product, 381
cycle, 50
cyclic group, 37
cyclotomic polynomial, 274

decomposable elements of an exterior
    algebra, 380
decomposable tensors, 365
JULIUS WILHELM RICHARD DEDEKIND,
    6
definition

group, 34
    symmetric group, 34
degenerate bilinear form, 254
degree of a field extension, 260
MAX DEHN, 91
dependent characters, 303
derived functors, 438
derived series, 84
determinant, 170
determinantal variety, 415
dihedral group
    geometry, 57
dihedral group, D_8, 37
dimension of a representation, 386
dimension of a vector space, 165
direct limit, 349
direct product, 59
direct sum, 363
direct sum of group-representations, 389
direct sum of groups, 38
directed set, 350
JOHANN LEJEUNE DIRICHLET, 279
Dirichlet's theorem, 279
discrete category, 339
discrete valuation, 465
discrete valuation ring, 466
discriminant
    cubic polynomial, 144
    quadratic polynomial, 144
discriminant of a polynomial, 144
disjoint cycles, 51
distinguished open sets, 415
divisible group, 69
division algebra, 321
division algorithm, 128
division ring, 107
domain of a function, 39
dot-product, 206
dual of a module, 374
dual of a representation, 389

echelon form of a matrix, 171
edges of a graph, 74
eigenspace, 196
eigenvalue, 193
eigenvector, 193
Eisenstein's irreducibility criterion, 149
elementary matrix
    type 1, 174
elementary row operation
    type 1, 174
elementary symmetric functions, 142
elliptic curve, 414
enough injectives, 360
enough projectives, 360
epimorphism
    category theoretic definition, 342
equivalence of categories, 346

equivalent extensions of groups, 456
Euclid Algorithm, 13
Euclid algorithm
　extended, 17
Euclidean ring, 118
　norm, 118
Euler
　ϕ-function, 22
LEONHARD EULER, 7
Euler formula
　quaternionic, 326
Euler's Criterion, 28
exact sequence, 222
exact sequence of cochain complexes, 434
$\text{Ext}_R^i(M, N)$-functor, 441
Extended Euclid algorithm, 17
extension
　degree, 260
n-fold extension, 450
extension of fields, 259
extension problem for groups, 456
extension ring, 110
exterior algebra, 378
　decomposable elements, 380

F -acyclic object, 443
faithful functor, 345
faithful representation, 385
Fermat number, 278
Fermat Prime, 278
Fermat Primes, 278
Fermat's Little Theorem, 22
field, 259
　algebraic closure, 281
　algebraic extension, 263
　algebraically closed, 281
　characteristic, 259
　compositum, 311
　extension, 259, 260
　of fractions of a ring, 260
　perfect, 268
　prime subfield, 259
　radical extension, 311
　rational function, 260
field extension
　Galois group, 298
field of fractions of a ring, 260
filtered colimit, 350
a-filtration, 471
finite fields, 285
First Sylow Theorem, 81
fixed field of a set of automorphisms, 304
flat module, 371
forgetful functors, 348
free abelian group, 59
free basis, 59
free group, 88
　defining property, 89

free module, 221
freely equal strings, 88
FERDINAND GEORG FROBENIUS, 328
Frobenius map, 287
Frobenius's Theorem
　associative division algebras, 328
full functor, 345
function, 39
　bijective, 39
　domain, 39
　image, 39
　injective, 39
　range, 39
　surjective, 39
functor, 343
　faithful, 345
　full, 345
　isomorphism, 344
fundamental theorem of algebra, 319

ÉVARISTE GALOIS, 298
Galois extension, 306
Galois group of an extension, 298
Galois Theory, 284
JOHANN CARL FRIEDRICH GAUSS, 146
Gauss's Lemma, 147
Gaussian Elimination, 176
Gaussian Integers, 250
gcd, 13
general linear group, 200
generating set of a module, 223
generators in a presentation, 89
generators of a group, 37
gimbal problem, 328
graded algebra, 469
graded ideal, 469
graded module, 470
graded reverse lexicographic ordering,
　128
graded ring, 377
Gram-Schmidt orthonormalization, 218
graph, 74
Grassmann algebras, 378
greatest common divisor, 13
Gröbner basis, 127
　leading term, 127
group
　p-primary component, 68
　abelian, 35
　abelianization, 49
　action, 74
　　orbit, 74
　affine, 201
　alternating, 54
　automorphism, 40
　Baer-Specker, 60
　center, 46
　centralizer of an element, 79

class equation, 78
commutator subgroup, 47
composition factors, 85
composition series, 85
correspondence theorem, 48
cyclic, 37
definition, 34
derived series, 84
dihedral, 37
direct sum, 38
divisible, 69
extension problem, 456
free
   defining property, 89
free abelian, 59
general linear, 200
generators, 37
homomorphism, 40
infinite cyclic, 41
isomorphism, 40
Jordan-Hölder Theorem, 85
Klein 4-group, 38
Lagrange's theorem, 43
left cosets, 43
normal subgroup, 44
order, 35
order of an element, 38
presentation, 89
   generators, 89
   relations, 89
   relators, 89
quotient, 45
representation, 385, 388
second isomorphism theorem, 48
semidirect product, 79
simple, 55
solvable, 87
special linear, 200
special orthogonal, 200
sub-representation, 386
subgroup, 37
   normalizer, 46
subnormal series, 84
symmetric, 34
third isomorphism theorem, 48
torsion free, 67
group of units, 108
group-action
   transitive, 74
group-algebra, 387
group-representation
   character, 396
   dimension, 386
   direct sum, 389
   homomorphism, 385
   isomorphism, 385
   tensor product, 389

group-ring, 109
grout
   orthogonal, 200

WILLIAM ROWAN HAMILTON, 198
KURT WILHELM SEBASTIAN HENSEL, 358
CHARLES HERMITE, 215
Hermitian matrix, 215
Hermitian transpose, 215
DAVID HILBERT, 417
Hilbert Basis Theorem, 123
Hilbert Nullstellensatz
   weak form, 417
Hilbert rings, 155
Hilbert space, 213
   L^2, 214
Hodge duality, 382
homogeneous ideal, 469
homology groups, 432
homomorphism
   kernel, 111
homomorphism of groups, 40
homomorphism of modules, 221
homomorphism of representations, 385
homomorphism of rings, 111
homomorphism of vector spaces, 167

ideal, 111
   generated by a set of elements, 111
   left, 111
   maximal, 111
   prime, 111
   principal, 111
   product, 111
   radical, 419
   right, 111
   two-sided, 111
ideal basis, 111
identity matrix, 168
image of a function, 39
imaginary part of a complex number, 7
imaginary part of a quaternion, 323
$\in$-minimal element (of a set), 462
independent group characters, 303
index of a subgroup, 43
inductive limit, 350
infinite cyclic group, 41
injective dimension, 438
injective function, 39
injective object, 360
injective resolution, 438
Inn(G), 48
inner automorphism, 48
inner product, 213
inner product of characters, 399
integers, 108
   unique factorization, 16

integral closure, 252
integral domain, 117
integral elements, 250
integral extension of rings, 251
integrally closed ring, 252
invariant set of a group-element, 75
inverse limit, 354
invertible matrix, 169
involution
    algebra, 321
irreducible element, 117
irreducible representation, 386
isomorphism, 111
isomorphism of algebraic sets, 428
isomorphism of graphs, 74
isomorphism of groups, 40
isomorphism of representations, 385
isomorphism problem, 90

Jacobi's Conjecture, 424
Jacobson radical, 154
Jacobson ring, 154, 428
MARIE ENNEMOND CAMILLE JORDAN, 233
Jordan Canonical Form, 236
Jordan-Hölder Theorem, 85

kernel of a homomorphism, 111
kernel of a homomorphism of groups, 40
Klein 4-group, 38
Kronecker product, 389, 508
Krull-Azumaya Theorem, 244

JOSEPH-LOUIS LAGRANGE, 43
Lagrange's formula
    triple cross product, 212
Laurent polynomials, 425
Law of Cosines, 206
lcm, 13
least common multiple, 13
Lebesgue measure
    outer, 189
left-adjoint, 347
left-cosets, 43
left-exact functor, 440
left-ideal, 111
Legendre symbol, 27
lexicographic ordering, 128
limit of an infinite sequence, 469
Lindemann–Weierstrass theorem, 289, 292
linear dependence, 163
linear independence, 163
linear transformation of vector spaces, 167
local ring, 113, 248
localization at a prime, 248
locally free modules, 373

lower-triangular matrix, 170

HEINRICH MASCHKE, 393
Maschke's Theorem, 391
matrix, 168
    adjoint, 183
    adjugate, 183
    characteristic polynomial, 193
    column-vectors, 168
    determinant, 170
    echelon form, 171
    elementary, 174
    Hermitian, 215
    identity, 168
    invertible, 169
    lower-triangular, 170
    minor, 182
    trace, 194
    transpose, 168
    unitary, 215
    upper-triangular, 170
maximal ideal, 111
metric space, 468
CLAUDE GASPARD BACHET DE MÉZIRIAC, 14
minimal polynomial, 120, 261
minimal polynomial of a matrix, 227
minor of a matrix, 182
module
    dual, 374
    socle, 389
module over a ring, 220
monomial
    graded reverse lexicographic ordering, 128
    lexicographic ordering, 128
monomorphism
    category theoretic definition, 342
multiplicative group of $\mathbb{Z}_n$, 20
multiplicative set in a ring, 246

Nakayama's Lemma, 243
natural isomorphism, 345
natural numbers, 5
natural transformation, 345
nilpotent element of a ring, 419
nilradical, 419
EMMY NOETHER, 121
Emmy Noether, 121
Noether Normalization Theorem, 417
noetherian module, 224
noetherian ring, 121
non-unique factorization, 120
norm
    Euclidean ring, 118
norm of a finite extension, 271
norm of a quaternion, 325
norm of a vector, 206

normal closure of a subgroup, 46
normal extension, 306
normal ring, 252
normal subgroup, 44
normalizer of a subgroup, 46
nullspace of a linear transformation, 167
Nullstellensatz
   strong form, 419
   weak form, 417

octonions, 330
orbit, 74
order of a group, 35
order of a group element, 38
ordering of monomials, 127
NICOLE ORESME, 2
orthogonal group, 200
orthonormal set of vectors, 216
outer automorphisms, 48
outer Lebesgue measure, 189

p-adic integers, 357
p-adic numbers, 357
p-adic valuation, 466
perfect field, 268
ϕ-function, 22
PID, 119
pivot element of a matrix, 170
polynomial
   discriminant, 144
   primitive, 147, 288
polynomial ring, 108
polynomially-closed, 156
power-associative algebra, 331
power-series ring, 109
powerset of a set, 11
presentation of a group, 89
prime element of a ring, 117
prime factors of a module, 240
prime field, 259
prime filtration, 240
prime ideal, 111
prime number, 15
primitive element, 268
primitive element theorem, 268
primitive polynomial, 147, 288
primitive root of unity, 301
principal ideal, 111
principal ideal domain, 119
product
   direct, 59
product of ideals, 111
projection to a quotient group, 45
projective dimension, 439
projective module, 225
projective object, 360
projective resolution, 439
Prüfer group, 38, 72

Puiseux series, 466
pull-back, 341
Puma 560 robot arm, 203
push-out, 340

Quadratic Reciprocity Theorem, 28
quadratic residues, 27
quaternion
   imaginary part, 323
   norm, 325
   scalar part, 323
   vector notation, 324
   vector part, 323
quaternion conjugate, 324
quaternion group, 106
quaternion units, 323
Quaternionic Euler Formula, 326
quaternions, 323
quotient group, 45
quotient ring, 112

Rabinowich Trick, 156
radical extension of a field, 311
radical of an ideal, 419
radical tower, 311
range of a function, 39
rank-variety, 414
rational function field, 260
real part of a complex number, 7
reduced form of a word, 88
reduced ring, 428
Rees algebra of an ideal, 470
KURT WERNER FRIEDRICH
   REIDEMEISTER, 99
Reidemeister rewriting process, 99
Reidemeister-Schreier rewriting process, 101
relators in a presentation, 89
representation
   character, 396
   dimension, 386
   group, 385, 388
   homomorphism, 385
   irreducible, 386
   isomorphism, 385
   simple, 386
   stable subspace, 388
representation of S_n, 387
representation space, 386
resultant, 186
rewriting process, 94
   Reidemeister, 99
right derived functors, 441
right resolution, 438
right-adjoint, 347
right-exact, 440
right-exact functor, 369
right-ideal, 111

ring, 107
    Artinian, 157
    center, 403
    commutative, 107
    discrete valuation, 466
    Euclidean, 118
    extension, 110
    field of fractions, 260
    homomorphism, 111
    ideal, 111
    integral domain, 117
    integrally closed, 252
    irreducible element, 117
    local, 113
    multiplicative set, 246
    noetherian, 121
    normal, 252
    PID, 119
    polynomial, 108
    prime element, 117
    principal ideal domain, 119
    quotient, 112
    reduced, 428
    subring, 107
    trivial, 107
    UFD, 119
    unit, 107
ring of fractions
    universal property, 353
root of unity
    primitive, 301
PAOLO RUFFINI, 314
BERTRAND ARTHUR WILLIAM RUSSELL, 461

S-polynomial, 131
scalar part of a quaternion, 323
OTTO SCHREIER, 101
Schreier coset representative function, 101
Schreier transversal, 103
ISSAI SCHUR, 388
second isomorphism theorem for groups, 48
Second Sylow Theorem, 81
sedenions, 334
semidirect product, 79
semisimple, 390
separable element of a field, 267
separable extension of fields, 268
separable polynomial, 267
set
    powerset, 11
short exact sequence, 222
sign-representation, 386
simple group, 55
simple representation, 386
Smith Normal Form, 233

socle of a group, 47
socle of a module, 389
solvability by radicals, 312
solvable group, 87
span of a set of vectors, 164
special linear group, 200, 414
special orthogonal group, 200
split short exact sequence, 246
splitting field, 264
stabilizer, 74
stable $\mathfrak{a}$-filtration, 471
stable subspace of a representation, 388
standard inner product, 213
standard representation of S_n, 387
sub-representation of a group, 386
subgroup, 37
    normalizer, 46
submodule, 221
subnormal series, 84
subring, 107
surjective function, 39
PETER LUDWIG MEJDELL SYLOW, 81
Sylow Theorem
    First, 81
    Second, 81
    Third, 81
JAMES JOSEPH SYLVESTER, 186
Sylvester Matrix, 186
symmetric algebra, 377
symmetric bilinear form, 254
symmetric group, 34
    cycle, 50
    standard representation, 387
    transposition, 50

NICCOLÒ FONTANA TARTAGLIA, 296
tensor algebra, 376
tensor product, 364
    universal property, 364
tensor product of group-representations, 389
third isomorphism theorem, 48
Third Sylow Theorem, 81
Tietze transformations, 91
torsion free group, 67
total quotient ring, 249
totient, 22
trace form, 255
trace of a finite field extension, 271
trace of a matrix, 194
transcendence basis, 292
transcendence degree, 291, 292
transcendental element of a field, 261
transcendental extension of fields, 263
transcendental number, 120
transitive group-action, 74
transpose of a matrix, 168
transposition, 50

transversal, 97
    Schreier, 103
trivial ring, 107
two-sided ideal, 111

UFD, 119
unique factorization domain, 119
unique factorization of integers, 16
unit, 107
unit vector, 206
unitary matrix, 215
units
    group, 108
upper-triangular matrix, 170

valuation, 465
Vandermonde matrix, 145
variety of a movement problem, 139
vector
    norm, 206
    unit, 206
vector notation for a quaternion, 324
vector space, 163
    basis, 164
    dimension, 165
    homomorphism, 167
vector-part of a quaternion, 323
vectors
    orthonormal, 216
vertices of a graph, 74

Weierstrass Division Theorem, 151
Weierstrass Preparation Theorem, 152
Wilson's Theorem, 27
word problem, 90

X_n-general power series, 151

Yoneda extensions, 450

$Z(G)$, 46
Oscar Zariski, 415
Zariski closure, 415, 420
Zariski topology, 415
    distinguished open sets, 415
zero-divisor, 108
Zorn's Lemma, 463

Bibliography

[1] Selman Akbulut. Algebraic Equations for a Class of P.L. Spaces. *Math. Ann.*, 231:19–31, 1977.

[2] Emil Artin. Zur Arithmetik hyperkomplexer Zahlen. *Abhandlungen aus dem Mathematischen Seminar der Universität Hamburg*, 5:261–289, 1928.

[3] Emil Artin. Zur Theorie der hyperkomplexer Zahlen. *Abhandlungen aus dem Mathematischen Seminar der Universität Hamburg*, 5:251–260, 1928.

[4] Michael Aschbacher. The Status of the Classification of the Finite Simple Groups. *Notices of the AMS*, 51(7):736–740, 2004.

[5] Michael Francis Atiyah. *K-theory*. Advanced Book Classics. Addison-Wesley, 2nd edition, 1989.

[6] Sergej I. Adian (Author), John Lennox (Translator), and James Wiegold (Translator). *The Burnside problem and identities in groups*, volume 95 of *Ergebnisse der Mathematik und ihrer Grenzgebiete*. Springer, 2011.

[7] Gorô Azumaya. On maximally central algebras. *Nagoya Math. J.*, 2:119–150, 1951. Available online from http://projecteuclid.org.

[8] R. Baer. Abelian groups that are direct summands of every containing abelian group. *Bull. Amer. Math. Soc.*, 46:800–806, 1940.

[9] Reinhold Baer. Abelian groups without elements of finite order. *Duke Mathematical Journal*, 3:68–122, 1937.

[10] John Baez. Higher-Dimensional Algebra II: 2-Hilbert Spaces. *Adv. Math.*, 127:125–189, 1997. available from arXiv:q-alg/9609018v2.

[11] John C. Baez and John Huerta. Superstrings, Geometry, Topology, and C*-algebras. In G. . Doran, Friedman, and J. Rosenberg, editors, *Proc. Symp. Pure Math.*, volume 81 of 65-80. AMS, 2010.

[12] Alan Baker. *Transcendental Number Theory*. Cambridge University Press, 1975.

[13] W. W. Rouse Ball. *A short account of the history of mathematics*. Project Gutenberg, http://www.gutenberg.org, e-book number 31246 edition, 2010.

[14] Étienne Bézout. Sur le degré des équations résultantes de l'évanouissement des inconnues et sur les moyens qu'il convient d'employer pour trouver ces équations. *Histoire de l'académie royale des sciences*, pages 288–338, 1764.

[15] William W. Boone. The word problem. *Proceedings of the National Academy of Sciences*, 44:1061 – 1065, 1958.

[16] R. Bott and J. Milnor. On the Parallelizability of the Spheres. *Bull. Amer. Math. Soc.*, 75:366–368, 1968.

[17] Kenneth S. Brown. *Cohomology of groups*, volume 87 of *Graduate Texts in Mathematics*. Springer-Verlag, 1982.

[18] R. Brown and T. Porter. On the Schreier theory of non-abelian extensions: generalisations and computations. *Proc. Roy. Irish Acad. Sect. A*, 96:213 – 227, 1996.

[19] R.L. Bryant, S.S. Chern, R.B. Gardner, H.L. Goldschmidt, and P.A. Griffiths. *Exterior differential systems*. Springer, 1991.

[20] Bruno Buchberger. *Ein Algorithmus zum Auffinden der Basiselemente des Restklassenrings nach einem nulldimensionalen Polynomideal*. PhD thesis, Johannes Kepler University of Linz (JKU), RISC Institute., 1965.

[21] Bruno Buchberger. Some properties of Gröbner bases for polynomial ideals. *ACM SIGSAM Bulletin*, 10:19–24, 1976.

[22] Bruno Buchberger. A criterion for detecting unnecessary reductions in the construction of Gröbner bases. In *Proceedings of the International Symposium on Symbolic and Algebraic Manipulation (EUROSAM '79)*, 1979.

[23] William Burnside. *Theory of Groups of Finite Order*. Cambridge University Press, 1897. Available online from Project Gutenberg.

[24] P. J. Cohen. *Set theory and the continuum hypothesis*. Addison-Wesley, 1966.

[25] Gabriel Cramer. *Introduction à l'analyse des lignes courbes algébraique*. Geneve: Freres Cramer & Cl. Philbert, 1750. Available for free from Google Books.

[26] John N. Crossley and Anthony W. C. Lun. *The Nine Chapters on the Mathematical Art: Companion and Commentary*. Oxford University Press, first edition, 2000.

[27] P. A. M. Dirac. *The Principles of Quantum Mechanics*. www.snowballpublishing.com, 2013.

[28] Peter Gustav Lejeune Dirichlet. Beweis des Satzes, dass jede unbegrenzte arithmetische Progression, deren erstes Glied und Differenz ganze Zahlen ohne gemeinschaftlichen Factor sind, unendlich viele Primzahlen enthält. *Abhandlungen der Königlichen Preußischen Akademie der Wissenschaften zu Berlin,*, 48:45–71, 1837.

[29] P.G.L. Dirichlet and R. Dedekind. *Lectures on Number Theory*. American Mathematical Society, 1999. translated by John Stillwell. Original scanned version available online at http://gdz.sub.uni-goettingen.de.

[30] B. Eckmann and A. Schopf. Über injective Modulen. *Archiv der Math.*, 3:75–78, 1953.

[31] Samuel Eilenberg and Saunders MacLane. General theory of natural equivalences. *Trans. of the Amer. Math. Soc*, 58(2):231–294, 1945. available from Google Scholar.

[32] Samuel Eilenberg and Saunders MacLane. On the groups $H(\Pi, n)$. I. *Ann. of Math. (2)*, 58:55–106, 1954.

[33] Samuel Eilenberg and Saunders MacLane. On the groups $H(\Pi, n)$. II. *Ann. of Math. (2)*, 60:49–139, 1954.

[34] David Eisenbud, Daniel Grayson, Michael Stillman, and Berndt Sturmfels, editors. *Computations in algebraic geometry with Macaulay 2*, volume 8 of *Algorithms and Computations in Mathematics*. Springer, 2001. Can be downloaded from http://www.math.uiuc.edu/Macaulay2/Book/.

[35] Arno van den Essen. *Polynomial Automorphisms: and the Jacobian Conjecture*. Progress in Mathematics. Birkhäuser Basel, 2000.

[36] Euclid. *The Thirteen Books of the Elements, Vol. 2*. Dover Publications, 1956.

[37] Euclid. *The Thirteen Books of the Elements, Vol. 3*. Dover Publications, 1956.

[38] Leonhard Euler. Observationes de theoremate quodam fermatiano aliisque af numeros primos spectantibus. *Comentarii academiae Petropolitanae*, 6:103–107, 1738.

[39] Edward FitzGerald. *The Rubaiyat of Omar Khayyam*. Number 246 in ebook. Project Gutenberg, http://www.gutenberg.org/ebooks/246, 1995.

[40] William Fulton and Joe Harris. *Representation theory. A first course*, volume 129 of *Graduate Texts in Mathematics*. Springer, 1991.

[41] Johann Carl Friedrich Gauss. *Disquisitiones Arithmeticae*. Yale University Press, 1965. Translated by Arthur A. Clarke. Original available online at http://resolver.sub.uni-goettingen.de/purl?PPN235993352.

[42] Walter Gautschi. Leonhard Euler: His Life, the Man, and His Works. *SIAM Review*, 2008. https://www.cs.purdue.edu/homes/wxg/Euler/EulerLect.pdf.

[43] A. O. Gelfond. *Transcendental and Algebraic Numbers*. Dover Phoenix Editions. Dover Publications, 2003.

[44] Sophie Germain. *Recherches sur la théorie des surfaces élastiques*. Courcier, 1821. Available online from http://books.google.com.

[45] Larry Joel Goldstein. *Abstract Algebra, A First Course*. Prentice Hall, 1971.

[46] Paul Gordan. Neuer Beweis des Hilbertschen Satzes über homogene Funktionen. *Nachrichten König. Ges. der Wiss. zu Gött.*, 3:240–242, 1899.

[47] Hermann Grassmann. *Die Lineale Ausdehnungslehre — Ein neuer Zweig der Mathematik*. Otto Wigand Verlag, 1844. Available from Google Books.

[48] David J. Griffiths. *Introduction to Quantum Mechanics*. Cambridge University Press, 2nd edition, 2016.

[49] R. Grone. Decomposable tensors as a quadratic variety. *Proc. Amer. Math. Soc.*, 64:227–230, 1977.
[50] Wolfgang Gröbner. Über die Eliminationstheorie. *Monatshefte für Mathematik*, 54:71–78, 1950.
[51] Allen Hatcher. *Algebraic Topology*. Cambridge University Press, 2002.
[52] Kurt Hensel. Über eine neue Begründung der Theorie der algebraischen Zahlen. *Jahresbericht der Deutschen Mathematiker-Vereinigung*, 6:83–88, 1897. Available online from http://www.digizeitschriften.de.
[53] David Hilbert. *The theory of algebraic number fields*. Springer-Verlag, 1998.
[54] H. Hopf. Über die Abbildungen von Sphären auf Sphären niedriger Dimension,. *Fund. Math.*, 25:427–440, 1935.
[55] Karel Hrbacek and Thomas Jech. *Introduction to Set Theory, Third Edition, Revised and Expanded*. Number 220 in CRC Pure and Applied Mathematics. Marcel Dekker, 1999.
[56] Lars Hörmander. *An Introduction to Complex Analysis in Several Variables*. North-Holland Publishing Company, 1973.
[57] Peter Høeg. *Smilla's Sense of Snow*. Delta; Reprint edition, 1995.
[58] Maurice Janet. Les systèmes d'équations aux dérivées partielles. *Journal de Mathématiques Pures et. Appliquées*, 3:65, 1920.
[59] Daniel Kan. Adjoint functors. *Trans. Amer. Math. Soc.*, 87:294–329, 1958.
[60] Israel Kleiner. *A History of Abstract Algebra*. Birkhäuser, 2007.
[61] Helmut Koch. *Number theory: algebraic numbers and functions*. American Mathematical Society, 2000.
[62] Wolfgang Krull. Allgemeine Bewertungstheorie. *J. Reine Angew. Math.*, 167:160–196, 1932.
[63] Wolfgang Krull. Beiträge zur Arithmetik kommutativer Integritätsbereiche. *Math. Z.*, 43:768–782, 1938.
[64] Wolfgang Krull. Jacobsonsche Ringe, Hilbertscher Nullstellensatz, Dimensionstheorie. *Math. Z.*, 54:354–387, 1951.
[65] Casimir Kuratowski. Une méthode d'élimination des nombres transfinis des raisonnements mathématiques. *Fundamenta Mathematicae*, 3:76–108, 1922.
[66] Y. S. Kwoh, J. Hou, E. A. Jonckheere, and S. Hayall. A robot with improved absolute positioning accuracy for CT guided stereotactic brain surgery. *IEEE Trans. Biomed. Engng.*, 35:153–161, 1988.
[67] Joseph Louis Lagrange. Demonstration d'un théorème nouveau concernant les nombres premiers. *Nouveaux Mémoires de l'Académie Royale des Sciences et Belles-Lettres*, 2:125–137, 1771.
[68] Joseph-Louis Lagrange. Suite des réflexions sur la résolution algébrique des équations. *Nouveaux Mémoires de l'Académie Royale des Sciences et Belles-Lettres de Berlin*, page 138–254, 1771. Available online at https://books.google.com/books.
[69] Serge Lang. *Algebra*. Graduate Texts in Mathematics. Springer-Verlag, 2002.
[70] S. MacLane. *Homology*. Springer-Verlag, 1995.
[71] Saunders MacLane. *Categories for the Working Mathematician*. Graduate Texts in Mathematics. Springer, second edition, 2010.
[72] Wilhelm Magnus, Abraham Karrass, and Donald Solitar. *Combinatorial Group Theory*. Interscience Publishers, 1966.
[73] Ernst W. Mayr. Some complexity results for polynomial ideals. *Journal of Complexity*, 13:301–384, 1997.
[74] John Milnor. *Introduction to Algebraic K-Theory*. Princeton University Press, 1972.
[75] Teo Mora. An algorithm to compute the equations of tangent cones. In *Proc. EUROCAM 82*, number 144 in Lecture Notes in Computer Science, pages 158–165. Springer, 1982.
[76] Teo Mora. An algorithmic approach to local rings. In *Proc. EUROCAL 85*, volume 204 of *Lecture Notes in Computer Science*, pages 518–525. Springer, 1985.
[77] David Mumford. *The Red Book of Varieties and Schemes*, volume 1358 of *Lecture Notes in Mathematics*. Springer-Verlag, second expanded edition, 1999.
[78] Tadashi Nakayama. A remark on finitely generated modules, II. *Nagoya Math. J.*, 9:21–23, 1955. Available online from http://projecteuclid.org.

[79] John Nash. Real algebraic manifolds. *Annals of Mathematics, Second Series,*, 56:405–421, 1952. Available online at http://www.jstor.org/pss/1969649.
[80] Jürgen Neukirch and Norbert Schappacher (Translator). *Algebraic Number Theory*, volume 322 of *Grundlehren der mathematischen Wissenschaften*. Springer, 1991.
[81] Isaac Newton. *The correspondence of Isaac Newton. II.* Cambridge University Press, 1960. Letter to Oldenburg dated Oct 24, 1676.
[82] Emmy Noether. Idealtheorie in Ringbereichen. *Mathematische Annalen*, 83:24–66, 1921.
[83] Emmy Noether. Der Endlichkeitsatz der Invarianten endlicher linearer Gruppen der Charakteristik p. *Nachrichten von der Gesellschaft der Wissenschaften zu Göttingen*, 1926:28–35, 1926. Available online from http://gdz.sub.uni-goettingen.de.
[84] Emmy Noether. Abstrakter Aufbau der Idealtheorie in algebraischen Zahl und Funktionenkörpern. *Math. Ann.*, 92:26–61, 1927.
[85] Pyotr S Novikov. On the algorithmic unsolvability of the word problem in group theory. *Proceedings of the Steklov Institute of Mathematics (in Russian)*, 44:1 – 143, 1955.
[86] Krzysztof Jan Nowak. A simple proof of Puiseux's theorem. *Univ. Iagel. Acta. Math.*, 32:199–200, 1995.
[87] Christof Paa and Jan Pelzl. *Understanding Cryptography: A Textbook for Students and Practitioners*. Springer, 2010.
[88] R. S. Palais. The Classification of Real Division Algebras. *The American Mathematical Monthly*, 75(4):366 – 368, 1968.
[89] Michael E. Peskin and Dan V. Schroeder. *An Introduction To Quantum Field Theory*. Frontiers in Physics. Westview Press, 1995.
[90] Benjamin C. Pierce. *Basic Category Theory for Computer Scientists*. The MIT Press, 1991.
[91] Victor Alexandre Puiseux. Recherches sur les fonctions algébriques. *J. Math. Pures Appl.*, 15:365–480, 1850.
[92] S. Rabinowich. Zum Hilbertschen Nullstellensatz. *Math. Ann.*, 102:520, 1929.
[93] K. Reidemeister. Knoten und Gruppen. *Abhandlungen Hamburg*, 5:7–23, 1926.
[94] Edward Rosen. *Kepler's Conversation with Galileo's Sidereal messenger. First Complete Translation, with an Introduction and notes*. Johnson Reprint Corp., 1965.
[95] George Rousseau. On the quadratic reciprocity law. *J. Austral. Math. Soc. Ser. A*, (3):423–425, 1991.
[96] Pierre Samuel. On unique factorization domains. *Illinois J. of Math.*, 5:1–17, 1961.
[97] O. Schreier. Die Untergruppen der freien Gruppen. *Abhandlungen Hamburg*, 5:161–183, 1927.
[98] Jean-Pierre Serre. Géométrie algébrique et géométrie analytique. *Université de Grenoble. Annales de l'Institut Fourier*, 6:1–42, 1956.
[99] Justin R. Smith. *Introduction to Algebra Geometry*. Five Dimensions Press, 2014.
[100] Ernst Specker. Additive Gruppen von Folgen ganzer Zahlen. *Portugaliae Math.*, 9:131–140, 1950.
[101] Michael Spivak. *A Comprehensive Introduction to Differential Geometry*, volume 1. Publish or Perish, 3rd edition, 1999.
[102] Ernst Steinitz. Algebraische Theorie der Körper. *Journal für die reine und angewandte Mathematik*, 137:167–309, 1910. Available online at http://gdz.sub.uni-goettingen.de/.
[103] Jean-Pierre Tignol. *Galois' Theory of Algebraic Equations*. World Scientific, Singapore, 2001.
[104] A. Tognoli. Su una congettura di Nash. *Ann. Scuola Norm. Sup. Pisa*, 27:167–185, 1973.
[105] Edward Waring. *Meditationes Algebraicae*. Cambridge, England:, 1770.
[106] Charles Weibel. *The K-book: an introduction to algebraic K-theory*, volume 145 of *Graduate Studies in Math*. American Mathematical Society, 2013.
[107] Karl Weierstrass. *Mathematische Werke. II. Abhandlungen 2*. Mayer & Müller, 1895.
[108] Alfred North Whitehead and Bertrand Russell. *Principia Mathematica (3 vols.)*. Cambridge University Press, 1910-1913. Available online from https://archive.org/.
[109] Charles Wieibel. *An introduction to homological algebra*. Cambridge University Press, 1994.
[110] Oscar Zariski. *Algebraic surfaces*. Classics in mathematics. Springer-Verlag, second edition, 2004. originally published in 1935.
[111] Max Zorn. A remark on method in transfinite algebra. *Bulletin of the American Mathematical Society*, 41:667–670, 1935.

www.ingramcontent.com/pod-product-compliance
Lightning Source LLC
Chambersburg PA
CBHW081423220526
45466CB00008B/2248